Donated in memory of
Dr. Aydin Ungan
1 /12/ 01

INTRODUCTION TO HEAT TRANSFER

VEDAT S. ARPACI

SHU-HSIN KAO

AHMET SELAMET

PRENTICE HALL
Upper Saddle River, NJ 07458

Library of Congress Cataloging-in-Publication Data

Arpacı, Vedat S., 1928.
    Introduction to Heat Transfer/
    Vedat S. Arpacı, Shu-Hsin Kao, and Ahmet Selamet/
        p.   cm.
    Includes bibliographical references and index.
    ISBN: 0–13–391061–X
    1. Heat Transmission,   I. Kao, Shu-Hsin,   II. Selamet, Ahmet   III. Title.
    TJ260.A742000
    621.4022–de21   99-053323 CIP

Editor-in-chief: **Marcia Horton**
Acquisitions editor: **Laura Curless**
Production editor: **Irwin Zucker**
Executive managing editor: **Vince O'Brien**
Managing editor: **David A. George**
Manufacturing buyer: **Pat Brown**
Copy editor: **Robert Lentz**
Director of production and manufacturing: **David W. Riccardi**
Cover director: **Jayne Conte**
Editorial assistant: **Lauri Friedman**
Composition: **PreT_EX, Inc.**

© 1999 by Prentice-Hall, Inc.
Upper Saddle River, NJ 07458

All rights reserved. No part of this book may be reproduced, in any form or by any means, without permission in writing from the publisher.

The author and publisher of this book have used their best efforts in preparing this book. These efforts include the development, research, and testing of the theories and programs to determine their effectiveness. The author and publisher make no warranty of any kind, expressed or implied, with regard to these programs or the documentation contained in this book. The author and publisher shall not be liable in any event for incidental or consequential damages in connection with, or arising out of, the furnishing, performance, or use of these programs.

Printed in the United States of America

10 9 8 7 6 5 4 3 2 1

ISBN 0-13-391061-X

Prentice-Hall International (UK) Limited, *London*
Prentice-Hall of Australia Pty. Limited, *Sydney*
Prentice-Hall Canada Inc., *Toronto*
Prentice-Hall Hispanoamericana, S.A., *Mexico*
Prentice-Hall of India Private Limited, *New Delhi*
Prentice-Hall of Japan, Inc., *Tokyo*
Pearson Education Asia Pte. Ltd., *Singapore*
Editora Prentice-Hall do Brasil, Ltda., *Rio de Janeiro*

CONTENTS

Preface — viii

▶ 1 FOUNDATIONS OF HEAT TRANSFER — 1

1.1 Place of Heat Transfer in Engineering — 1
1.2 Formulation of Heat Transfer — 3
1.3 First Law of Thermodynamics — 4
1.4 Control Surface — 12
1.5 Origin of Heat Transfer. Particular Laws — 13
    1.5.1 Original Problem of Conduction   14
    1.5.2 Fourier's Law of Conduction   15
    1.5.3 Thermal Conductivity   17
    1.5.4 Newton's Definition of Convection   19
    1.5.5 Stefan-Boltzmann's Law of Radiation   24
1.6 Heat Transfer Modes Combined — 27
1.7 Methods of Formulation — 32
1.8 Five-Step Inductive Formulation — 32
    References — 34
    Computer Program Appendix — 35
    Exercises — 37

▶ 2 STEADY CONDUCTION — 40

2.1 Variable Conductivity and Variable Area — 40

		2.1.1 Variable Conductivity 42	
		2.1.2 Variable Area 43	
	2.2	Composite Structures. Critical Radius	45
		2.2.1 Composite Slabs 45	
		2.2.2 Composite Cylinders 49	
		2.2.3 Critical Thickness for Cylindric Insulation 52	
		2.2.4 Composite Spheres. Critical Thickness for Spherical Insulation 57	
	2.3	Energy Generation (Heat Source)	58
		2.3.1 Flat Plate (Key Problem) 58	
		2.3.2 Cylinder and Sphere (Key Problem) 70	
	2.4	Extended Surfaces (Fins, Pins)	74
		2.4.1 Thermal Length 80	
		2.4.2 Performance 89	
	2.5	Two Key Problems of Convection	90
		2.5.1 First Key Problem 90	
		2.5.2 Second Key Problem 95	
	2.6	Solar Collector ⊕	99
	2.7	Reactor Core ⊕	102
		References	109
		Computer Program Appendix	110
		Exercises	113

▶ 3 UNSTEADY/STEADY, MULTIDIMENSIONAL CONDUCTION — 125

3.1	Lumped Problems ($Bi \leq 0.1$)	126
3.2	Periodic Problems ⊕	141
3.3	Distributed Problems ($Bi > 0.1$). Differential Formulation	144
3.4	Steady Periodic Solution ⊕	149
3.5	Integral Formulation. Approximate Solution ⊕	152
3.6	Charted Exact Solutions	156
	3.6.1 Flat Plate (Key Problem) 156	
	3.6.2 Solid Cylinder (Key Problem) 161	
	3.6.3 Solid Sphere (Key Problem) 163	
	3.6.4 Semi-infinite Plate 165	
3.7	Mixed (Differential-Difference) Formulation. Analog Solution	168
	3.7.1 Active Circuit Elements. High-Gain DC Amplifiers 168	
	References	178
	Computer Program Appendix	179
	Exercises	180

▶ 4 COMPUTATIONAL CONDUCTION — 184

4.1	Discrete Formulation	184
	4.1.1 Exact Discrete Formulation 185	
	4.1.2 Finite-Difference/Finite-Volume Formulation 186	
4.2	Multidimensional Formulation	194

	4.2.1	Nonuniform Grid Spacing ⊕ 202	
	4.2.2	Effect of Enthalpy Flow ⊕ 207	
4.3	Truncation Error		209
4.4	Unsteady Conduction		212
	4.4.1	Explicit Finite-Difference Formulation 212	
	4.4.2	Stability of Explicit Scheme 213	
	4.4.3	Truncation Error of Explicit Scheme 218	
	4.4.4	Implicit Scheme 220	
	4.4.5	Crank-Nicolson Method 222	
4.5	Euler's Method		224
4.6	Concluding Remarks		226
	References		227
	Computer Program Appendix		228
	Exercises		236

▶ 5 FOUNDATIONS OF CONVECTION — 240

5.1	Boundary-Layer Concept. Laminar Forced Convection ○	244
5.2	Laminar Natural Convection ○	258
5.3	Dimensional Analysis ○	266
5.4	A Forced Flow ○	270
5.5	A Free Fall ○	273
5.6	Forced Convection	275
5.7	Natural Convection	278
	References	282
	Exercises	283

▶ 6 CORRELATIONS FOR CONVECTION — 288

6.1	Friction Factor, Drag Coefficient ○		289
6.2	Forced Convection		295
	6.2.1	Internal Flow 295	
	6.2.2	Computation of the Heat Transfer Coefficient of Internal Flow 297	
	6.2.3	External Flow 301	
	6.2.4	Computation of the Heat Transfer Coefficient for External Flow 305	
6.3	Natural Convection		312
	6.3.1	Computation of the Heat Transfer Coefficient for Given T_w 314	
	6.3.2	Computation of the Heat Transfer Coefficient for Given q_w 320	
	References		330
	Computer Program Appendix		332
	Exercises		341

▶ 7 HEAT EXCHANGERS — 346

7.1	Thermal Design. LMTD Method	349
7.2	Correction Factor	359

7.3	Condenser. Evaporator (Boiler) ○	365
7.4	Performance. NTU Method	370
7.5	Fouling Factor. Variable Coefficient of Heat Transfer. Closure ○	385
	References	388
	Computer Program Appendix	388
	Exercises	391

▶ 8 FOUNDATIONS OF RADIATION ○ — 396

8.1	Origin of Radiation. Electromagnetic Waves ○	396
8.2	Approximation of Radiation. Optical Rays ○	400
8.3	Monochromatic Radiation. Quantum Mechanics ○	405
8.4	Properties of Radiation ○	413
	References	425
	Computer Program Appendix	425
	Exercises	428

▶ 9 ENCLOSURE RADIATION — 430

9.1	View Factor	434
9.2	Electrical Analogy	443
9.3	Net Radiation	472
9.4	Combined Heat Transfer	475
	References	484
	Computer Program Appendix	484
	Exercises	492

▶ 10 GAS RADIATION ⊕ — 506

10.1	Balance of Radiation Energy	507
10.2	Radiation Properties of Gases	509
10.3	Distributed Gas Radiation	517
	10.3.1 Thin Gas 518	
	10.3.2 Thick Gas 519	
	10.3.3 Effect of Boundaries 521	
	References	533
	Exercises	534

▶ 11 PHASE CHANGE ⊕ — 535

	An Illustrative Example	535
11.1	Laminar Two-Phase	537
11.2	A Dimensionless Number	544

		11.2.1 A Dimensional Approach 546	
11.3	Regimes of Boiling		548
	References		554
	Exercises		554

▶ A CORRELATIONS — 555

▶ B THERMOPHYSICAL PROPERTIES — 578

▶ C SI UNITS — 598

▶ D HEISLER CHARTS — 601

Index — 609

PREFACE

This text is an introduction to engineering heat transfer. The philosophy of the text is based on the development of an **inductive** approach, earlier introduced by the author (*Conduction Heat Transfer*, 1966), to the **formulation** and **solution** of applied problems. Since the greatest difficulty a student faces is how to formulate rather than how to solve a problem, the formulation of problems is stressed from the beginning and throughout the entire text. This is done by first noting that heat transfer rests on but goes beyond thermodynamics, and taking as a basis the well-known form of the first law of thermodynamics for a system,

$$E_2 - E_1 = AQ - AW ,$$

developing the **rate** of the first law for a control volume,

$$\boxed{\underbrace{\frac{dE_{CV}}{dt}}_{\text{Energy rate}} = \underbrace{\sum_{i=1}^{N} \dot{m}_i h_i^\circ}_{\text{Enthalpy flow}} + \underbrace{\dot{Q}_{CV}}_{\text{Heat flux}} - \underbrace{\dot{W}_{CV}}_{\text{Work rate}}} ,$$

in Chapter 1. Then, the discussion of every problem in the text begins with the interpretation of this law in terms of an appropriate control volume (or a system). After stressing the fact that thermodynamics provides no information about $\dot{Q}_{CV}$, the three (conduction, convection and radiation) laws of heat transfer are introduced by relating $\dot{Q}_{CV}$ to temperature. This philosophy is different from that of most existing textbooks.

The aim constantly is, not to obtain a speedy general formulation, but to teach mastery of a few basic and simple tools by which each problem can be individually formulated.

Although the present text is much less voluminous, though well within the range of other textbooks on the subject, still the material is more than can be covered in an introductory one-semester course. This was done purposely **(1)** to trigger the curiosity of students who are interested in furthering themselves beyond the minimum requirements, **(2)** to leave some flexibility to instructors in the selection of the material, and **(3)** to speculate and incorporate now some of the future material. For a one-semester introductory course, we suggest

1. exclusion of a chapter or a section of a chapter marked with ⊕,
2. partial coverage of sections marked with ○ in a chapter to fit personal taste.

With addition of this material, the text may be considered for an intermediate course.

The text has a number of novel parts:

The concept behind the two key problems of convection (cooling with a film coefficient of fluids flowing in a pipe and heating with an applied heat flux of fluids flowing in a pipe) is demonstrated in Chapter 2. Thermocouple selection for the measuring of unsteady temperatures, depending on the time constant of the problem under consideration, and the concept of analog solution based on active electric-circuit elements, are discussed in Chapter 3. In an introductory text, numerical methods are usually explored in terms of the finite-difference method, which is relatively less involved and easier to learn than others. Among these, for example, a finite-element method based on variational calculus is beyond the scope of the text, but the finite-element method based on an integral formulation is quite straightforward and is introduced in Chapter 4. This informs the student about the availability of other numerical methods for future considerations. With boundary-layer (penetration-depth) concepts introduced to conduction, the transition to convection boundary layers is facilitated in Chapter 5. Also, complexities involving higher-order velocity and temperature profiles in boundary-layer analyses are eliminated by using first-order profiles which often lead to reasonably accurate solutions. The difficulties encountered in the application of the Π-Theorem to the dimensional analysis of heat transfer problems are avoided by considering successive rather than the simultaneous elimination of the fundamental units. As an alternative method, the application of physical similitude to dimensional analysis of heat transfer problems is explored. Among a few casual possibilities, the most physically significant nondimensionalization of natural convection leads, in terms of Rayleigh and Prandtl numbers, to

$$Nu = f(Ra, Pr).$$

A fundamental dimensionless number based on a combination of Ra and Pr,

$$\Pi_N \sim \frac{Ra}{1 + Pr^{-1}},$$

is introduced in Chapter 6 and is used to correlate the data on natural convection. A five-step approach to the use of correlations is demonstrated in terms of two for forced convection and two for natural convection, leaving the rest of the literature to an appendix. A rapid mastery to be gained with a minimum number of correlations can easily be extended to other correlations. Original relations for the heat transfer area and for the ratio of mass flows in heat exchangers involving two-phase flows are introduced in Chapter 7. The solid angle relation between thermomechanics and optics,

$$\text{Thermomechanics} = \int_\Omega (\text{Optics}) \, d\Omega,$$

usually overlooked among other relations resulting from extensive manipulations, is emphasized in Chapter 8. A five-step approach to the solution of enclosure radiation problems is utilized in Chapter 9. In view of the extensive contemporary research on gas radiation, an introduction to this timely subject is provided in Chapter 10. The use of the dimensionless number for natural convection, Π_N, is extended to film boiling in Chapter 11.

A great majority of the examples worked in the text and the problems left to the students, in particular those clearing and extending a fundamental point, are our own invention. In general, homework problems are designed to supplement and extend the text. Repetitive problems are avoided. One of the difficulties of our educational system is the selection of the system of units. There is no bigger obstacle to learning than a text which suggests the use of more than one system of units. The system of units used throughout the world, and in most of our educational institutions, is the metric International System, which is also used in this text. Furthermore, a contemporary danger to the learning process is the temptation to rely on commercial software programs before mastering a subject. A proficiency developed via the repetitive use of a program, resting on a shaky background, leads to a rapid obsoleteness. Yet, a judicious use of these programs is essential to contemporary learning. Individual FORTRAN programs are developed for some of the illustrative examples in the text. The interested readers may parametrically study these examples by changing various values of the given data.

The text is a result of about four decades of teaching an introductory heat transfer course at the University of Michigan. Our goal was to produce an undergraduate "textbook" rather than a voluminous "handbook." During the past decade, Dr. Ahmet Selamet was instrumental in the earlier improvement of my original class notes. The manuscript could not have been completed, however, without Dr. Shu-Hsin Kao, who, with unusual dedication, helped me prepare the final form of the text. We are grateful to Dr. Laila Guessous for her numerous suggestions, which led to significant improvements, and to the reviewers for their useful comments on the several parts of the final manuscript.

VEDAT S. ARPACI
Ann Arbor, Michigan

CHAPTER 1

FOUNDATIONS OF HEAT TRANSFER

The foundations of an engineering discipline may be best understood by considering the place of that discipline in relation to other engineering disciplines. Therefore, our first concern in this chapter will be to determine the place of heat transfer among engineering disciplines. Next, we shall proceed to a review of the general principles needed for heat transfer. Finally, we shall discuss the three modes of heat transfer—conduction, convection, and radiation—and introduce a five-step methodology for an inductive formulation.

1.1 PLACE OF HEAT TRANSFER IN ENGINEERING

Let us first review a well-known problem taken from mechanics. For this problem let us consider two formulations, based on different assumptions. Our concern will be with the nature of the physical laws employed in these formulations. (At this stage our discussion will be somewhat conventional; the philosophy of the text will be set forth at the end of this chapter).

Example: Free fall of a body. Consider a body of mass m in a vacuum falling freely under the effect of the gravitational field g. We wish to determine the instantaneous location of this body.

Formulation of the problem. Newton's second law of motion,

$$F = ma, \tag{1.1}$$

F being the sum of external forces and a the acceleration vector, gives in terms of vertical distance x

$$mg = m\frac{d^2x}{dt^2} \tag{1.2}$$

subject to appropriate initial conditions.

In our second formulation of the problem, let us include the resistance to the motion of the body from the surroundings. With this consideration, we have

$$mg - R = m\frac{d^2x}{dt^2}, \tag{1.3}$$

which is not complete without further information about the resistance force R. If, for example, this force is assumed to be proportional to the square of velocity of the body,— that is, if

$$\frac{R}{m} = k\left(\frac{dx}{dt}\right)^2 \tag{1.4}$$

—then Eq. (1.3) gives

$$g - k\left(\frac{dx}{dt}\right)^2 = \frac{d^2x}{dt^2}, \tag{1.5}$$

where k is a constant.

As demonstrated by the foregoing two formulations, some problems taken from mechanics can be formulated by using only Newton's laws of motion; these are called mechanically determined problems. The dynamics of rigid bodies in the absence of friction, statically determined problems of rigid bodies, and mechanics of ideal fluids provide examples of this class. Some other mechanics problems, however, require knowledge beyond Newton's laws of motion. These are called mechanically undetermined problems. The dynamics of rigid bodies with friction and the mechanics of deformable bodies provide examples of this class.

Thermal problems may be similarly divided into two classes. Some of these can be solved by employing only the laws of thermodynamics; they are called thermodynamically determined problems. Some others, however, require knowledge beyond these laws; these are called thermodynamically undetermined problems. Gas dynamics and heat transfer are two major thermodynamically undetermined disciplines. In addition to the general laws of thermodynamics and fluid mechanics, gas dynamics depends on equation of state while heat transfer requires knowledge on **conduction, convection,** and **radiation phenomena,** which we shall now introduce. Each of these phenomena relates heat to temperature, the same way that stress must be related to strain in mechanics.

Phenomenologically speaking, conduction is the transfer of thermal energy from a point of higher temperature to an **adjacent** point of lower temperature in matter. At the microscopic level, the mechanism of conduction is visualized as an exchange of energy between **adjacent** matter particles. Consequently, conduction is local and, being directional, is irreversible, and it can only happen through matter. Burning a hand on a hot plate is a result of conduction.

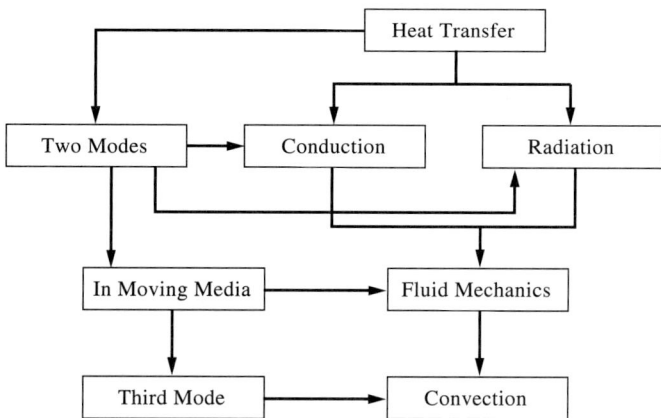

Figure 1.1 Three modes of heat transfer.

Again, phenomenologically speaking, radiation is the transfer of thermal energy by electromagnetic waves in a particular wavelength range from a point of higher temperature to a **distant** point of lower temperature in matter. At the microscopic level, the mechanism of radiation is visualized as the transport of energy by radiation particles (photons) traveling with the speed of light. Acting at a distance, radiation is global, and is reversible through vacuum. Feeling warm before an open fire is a result of radiation. From a conceptual viewpoint, **convection** is not a basic mode of heat transfer, but, rather, is conduction and/or radiation in moving media. Blowing on food to cool it is a process of convection. Therefore, fluid mechanics plays an important role in convection. For only customary reasons, we shall hereafter refer to conduction of heat in moving (or stationary) rigid media as conduction and to conduction in moving deformable media as convection (Fig. 1.1).

Having gained some appreciation of the three modes of heat transfer we proceed now to the methodology adopted in this text. We shall return to the three modes of heat transfer in Section 1.5, elaborate on conduction, and make further remarks on convection and radiation.

1.2 FORMULATION OF HEAT TRANSFER

In the preceding section we established the place of heat transfer among the engineering disciplines and distinguished the modes of heat transfer—conduction, convection, and radiation. We proceed now to the formulation of heat transfer.

The formulation of an engineering discipline such as heat transfer is based on **definitions of concepts** and **statements of natural laws** in terms of these concepts. The natural laws of heat transfer, like those of other disciplines, can be neither proved nor disproved but are arrived at inductively, on the basis of evidence collected from a wide variety of experiments. As we continue to increase our understanding of the universe, the present statements of natural laws will be refined and generalized. For the time being, however, we shall refer to these statements as the available approximate descriptions of nature and employ them for the solution of current problems of engineering.

The natural laws may be classified as (1) general laws, and (2) particular laws. **A general law is independent of the nature of the medium**. Examples are the law of conservation of mass, Newton's laws of motion, the first and second laws of thermodynamics, Lorentz's force law, Ampere's circuit law, and Faraday's induction law. The problems of nature which can be formulated completely by using only general laws are called mechanically, thermodynamically, or electromagnetically **determined** problems. On the other hand, the problems which can not be formulated completely by means of general laws alone are called mechanically, thermodynamically, or electromagnetically **undetermined** problems. Each problem of the latter category requires, in addition to the general laws, one or more conditions stated in the form of particular laws. **A particular law depends on the nature of a medium**. Examples are Hooke's law of elasticity, Newton's law of viscosity, the ideal gas law, Fourier's law of conduction, Stefan-Boltzmann's law of radiation, and Ohm's law of electricity.

In this text we shall employ two general laws,

(a) the conservation of mass,

(b) the first law of thermodynamics,

and three particular laws,

(c) Fourier's law of conduction,

(d) Newton's definition of convection,

(e) Stefan-Boltzmann's law of radiation,

each with a different degree of importance. Since all thermal problems (thermodynamically determined or undetermined) begin with the general laws of thermodynamics, and since the first law of thermodynamics is vitally important for heat transfer, the next section is devoted primarily to a review of this law. The conservation of mass, because of its lesser significance, will be mentioned briefly. We shall assume that the **definition of concepts** such as system, control volume, property, state, process, cycle, work, heat, temperature and others are known to the student (see, for example, Van Wylen, Sonntag and Borgnakke[1]).

1.3 FIRST LAW OF THERMODYNAMICS

The first step in the statement of the first law (or any general law) **is the selection of a system or control volume**. Without this step it is meaningless to speak of such concepts as heat, work, internal energy, and others, which are the terms used in statements of the first law. Although the well-known, simple form of the first law is always written for a system, the use of this form of the law becomes inconvenient when dealing with continua in motion, because it is often difficult to identify the boundaries of a moving system for any appreciable length of time. The control-volume approach is therefore generally preferred for continua in motion.

---

[1] Reference 5.

Consider a thermal machine consisting of an insulated piston-cylinder assembly attached to a container as shown in Fig. 1.2. Initially, the matter in the cylinder is separated from that of the container by a partition. The partition is ruptured and, following an infinitesimal process, the mass Δm_i within the cylinder is slowly pushed by the piston into the container. Assume the container to be a **control volume**. During this process, the heat received and the shaft work done by the control volume, respectively, are ΔQ_{cv} and ΔW_{cv}, subscript cv denoting the control volume. We wish to find the rate of the first law of thermodynamics for this control volume.

Since the well-known familiar forms of general principles have been deduced and always written for a system, consider first a system coinciding with the control volume at the final state while including in the initial state the piston-cylinder assembly as well as the control volume. Let E_1, E_2 and E'_{cv}, E''_{cv} denote the initial and final values of the total energy of the system and the control volume, respectively. The first law of thermodynamics for the system undergoing a differential process is[2]

$$E_2 - E_1 = \Delta Q - \Delta W, \tag{1.6}$$

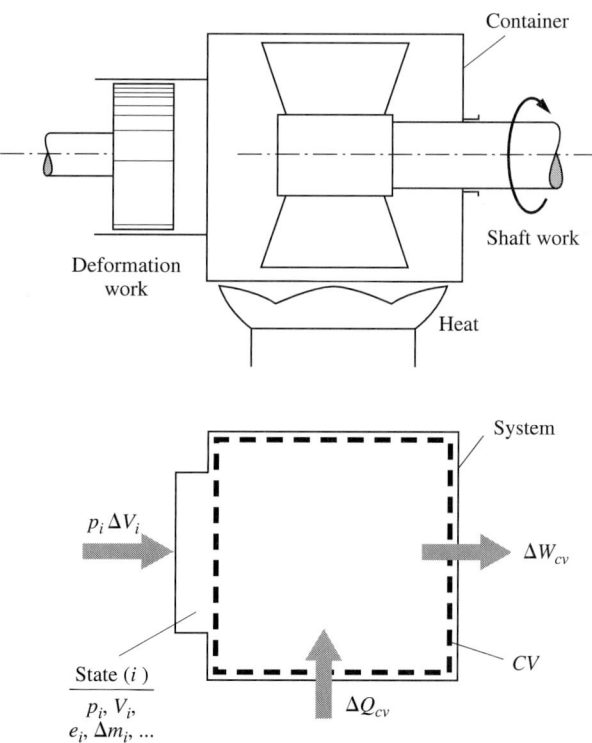

Figure 1.2 The first law for a control volume.

---

[2] More explicitly, $E_2 - E_1 = \Delta Q - (\pm \Delta W)$, where the minus sign in parenthesis is for work done **on** the system, and the plus sign for work done **by** the system.

where ΔQ and ΔW denote respectively the heat received and the work done by the system. Here, after neglecting the heat loss from the piston-cylinder assembly,

$$\Delta Q = \Delta Q_{cv}, \quad \Delta W = \Delta W_{cv} - p_i \Delta V_i, \tag{1.7}$$

$$E_2 = E''_{cv}, \quad E_1 = E'_{cv} + \Delta m_i e_i, \tag{1.8}$$

subscript i denoting a property (or a quantity[3]) associated with mass Δm_i in state i, $p_i \Delta V_i$ denoting the deformation work done by the piston-cylinder assembly. Now, rearranging Eq. (1.6) in terms of Eq. (1.7), and rearranging $p_i \Delta V_i$ with $\Delta V_i = v_i \Delta m_i$, gives

$$E''_{cv} - E'_{cv} = \Delta m_i (e_i + p_i v_i) + \Delta Q_{cv} - \Delta W_{cv}. \tag{1.9}$$

Finally, introducing the definition of stagnation enthalpy

$$h^o = e + pv,$$

assuming N differential masses enter or leave the control volume, letting $E''_{cv} - E'_{cv} = \Delta E_{cv}$, dividing each term of Eq. (1.6) by Δt, and letting $\Delta t \to 0$, we get the **rate of the first law of thermodynamics for a control volume**,

$$\boxed{\frac{dE_{cv}}{dt} = \sum_{i=1}^{N} \dot{m}_i h_i^o + \dot{Q}_{cv} - \dot{W}_{cv}}, \tag{1.10}$$

where enthalpy flow **into** the control volume is assumed to be **positive** and enthalpy flow **out** of the control volume is to be **negative**, $\dot{m}_i$ is the mass flow rate, $\dot{Q}_{cv}$ is the rate of net heat received by the control volume, and $\dot{W}_{cv}$ is the power (rate of net work) done by the control volume. Explicitly,

$$\underbrace{\frac{dE_{cv}}{dt}}_{\text{Rate of change of energy in CV}} = \underbrace{\sum_{\text{in}} \dot{m}_{\text{in}} h_{\text{in}}^o}_{\text{Enthalpy flow into CV}} - \underbrace{\sum_{\text{out}} \dot{m}_{\text{out}} h_{\text{out}}^o}_{\text{Enthalpy flow out of CV}}$$

$$+ \underbrace{(\dot{Q}_{cv})_{\text{in}}}_{\substack{\text{Heat received} \\ \text{by CV}}} - \underbrace{(\dot{Q}_{cv})_{\text{out}}}_{\substack{\text{Heat rejected} \\ \text{by CV}}} + \underbrace{(\dot{W}_{cv})_{\text{in}}}_{\substack{\text{Work done} \\ \text{on CV}}} - \underbrace{(\dot{W}_{cv})_{\text{out}}}_{\substack{\text{Work done} \\ \text{by CV}}}. \tag{1.11}$$

Recall from thermodynamics that the total energy E includes internal, kinetic, potential, chemical and nuclear energy,

$$E = U + \frac{1}{2} m V^2 + mgz + U_{\text{chem}} + U_{\text{nucl}}, \tag{1.12}$$

the stagnation enthalpy h^o is

$$h^o = h + \frac{1}{2} V^2 + gz, \tag{1.13}$$

---

[3] quantity $\equiv$ nonproperty.

Sec. 1.3 First Law of Thermodynamics   7

h being the enthalpy, and the power $\dot{W}$ is composed of displacement, shaft, and electrical power,

$$\dot{W} = \dot{W}_d + \dot{W}_s + \dot{W}_e. \tag{1.14}$$

Note that p/ρ, $V^2/2$, and gz have units of m²/s²=J/kg=energy/mass, while p, $\rho V^2/2$, and ρgz have N/m²=J/m³=energy/volume, $\rho = 1/v$ being the density. Since the electric energy U_e generated in a system is thermodynamically equivalent to work done on the system,

$$\dot{W}_e = -U_e. \tag{1.15}$$

For the **rate of the first law of thermodynamics for a system**, $\dot{m}_i = 0$ and Eq. (1.10) is reduced to

$$\boxed{\frac{dE}{dt} = \dot{Q} - \dot{W}}, \tag{1.16}$$

An integration of Eq. (1.16) over a time interval converts this equation back to Eq. (1.6).

The conservation of mass, the balance of momentum, and the second law of thermodynamics also may play, although to a reasonably lesser degree of importance, a role in heat transfer. In terms of Fig. 1.2, the **conservation of mass for a control volume** is

$$\boxed{\frac{dm_{cv}}{dt} = \sum_{i=1}^{N} \dot{m}_i}, \tag{1.17}$$

or, explicitly,

$$\underbrace{\frac{dm_{cv}}{dt}}_{\text{Rate of change of mass in CV}} = \underbrace{\sum_{\text{in}} \dot{m}_{\text{in}}}_{\text{Mass flow into CV}} - \underbrace{\sum_{\text{out}} \dot{m}_{\text{out}}}_{\text{Mass flow out of CV}}$$

We proceed now to a couple of examples illustrating the application of the conservation of mass and/or the first law.

EXAMPLE 1.1 Steady one-dimensional flow

Consider the steady one-dimensional flow of a frictionless incompressible fluid through a pipe of constant cross section and a diffuser of the same length (Fig. 1.3). The pipe and diffuser are subjected to the same uniform heat flux q'' (W/m²). The inlet diameter and inlet velocity of the diffuser are identical to those of the pipe.

1. We wish to determine whether the exit temperature of the diffuser is higher or lower than that of the pipe on the basis of physical reasoning rather than mathematics.
2. We wish to support our conclusion with a simple analysis.

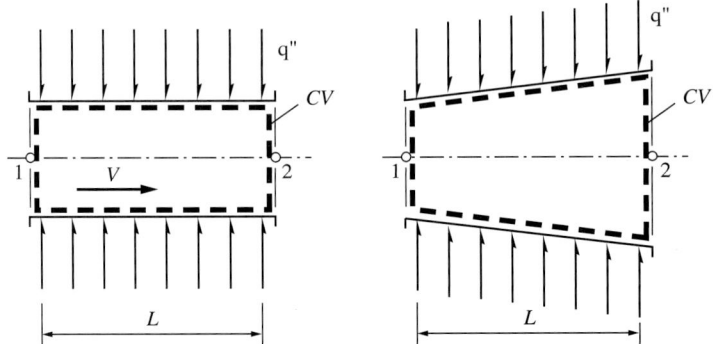

Figure 1.3 Control-volume configuration.

1. The exit temperature of the diffuser is higher because of the higher heat input resulting from the larger peripheral area of the diffuser. The slowdown of the diffuser velocity has a minor effect which will be discussed later in the next part.

2. Let us follow, for the time being somewhat informally, a couple of basic steps in the formulation of the problem.[4] As a **first step**, consider the control volumes shown in Fig. 1.3. As a **second step**, consider the conservation of mass and the first law for these control volumes. The conservation of mass (Eq. 1.17) gives, for steady flow of an incompressible fluid,

$$\dot{m} = \rho A V = \text{Const.}, \tag{1.18}$$

which reduces for the pipe to

$$V_1 = V_2 \tag{1.19}$$

and for the diffuser to

$$A_1 V_1 = A_2 V_2. \tag{1.20}$$

Under steady conditions, the first law given by Eq. (1.10), noting also the absence of any power terms, yields

$$0 = \dot{m}(h_1^o - h_2^o) + \dot{Q}_{cv.} \tag{1.21}$$

(Since $\dot{m}$ = Const. for both cases, the slowdown of flow in the diffuser does have a minor effect on temperature only through the kinetic energy in the stagnation enthalpy.)

Recalling from Eq. (1.13)

$$h^o = h + V^2/2 \tag{1.22}$$

and, for an incompressible flow through a short pipe or diffuser, neglecting the effect of pressure drop,

$$dh \cong c \, dT$$

and with

$$A_p = \int P(x) dx, \quad \dot{Q}_{cv} = q'' A_p, \tag{1.23}$$

---

[4] At the end of this chapter, these steps together with three more steps will be formalized as an inductive method based on a **five-step formulation**.

where A_p is the peripheral area, Eq. (1.21) may be rearranged for the diffuser as

$$0 = \dot{m}c(T_1 - T_2) + \frac{1}{2}\dot{m}(V_1^2 - V_2^2) + q''A_{p,\text{diff}},$$

which readily gives

$$T_1 - T_2 = \frac{1}{\dot{m}c}\left[\frac{1}{2}\dot{m}(V_1^2 - V_2^2) + q''A_{p,\text{diff}}\right]. \tag{1.24}$$

For the pipe, Eq. (1.19) and the fact that $A_{p,\text{diff}} \to A_{p,\text{pipe}}$ simplifies Eq. (1.24) into

$$T_2 - T_1 = \frac{1}{\dot{m}c}(q''A_{p,\text{pipe}}). \tag{1.25}$$

For the diffuser, since $A_2 > A_1$, Eq. (1.20) yields $V_2 < V_1$. Then

$$\frac{1}{2}\dot{m}(V_1^2 - V_2^2) > 0 \quad \text{and also} \quad A_{p,\text{diff}} > A_{p,\text{pipe}},$$

and, it follows from the comparison of Eqs. (1.24) and (1.25),

$$(T_2 - T_1)_{\text{diff}} > (T_2 - T_1)_{\text{pipe}}.$$

As a practical application,[5] let

$$D_1 = 5 \text{ cm}, \quad D_2 = 25 \text{ cm}, \quad L = 10 \text{ m},$$
$$T_1 = 300 \text{ K}, \quad V_1 = 2 \text{ m/s}, \quad q'' = 10 \text{ kW/m}^2,$$
$$\rho = 1{,}000 \text{ kg/m}^3, \quad c = 4{,}000 \text{ J/kg·K}.$$

Introduce

$$\Delta T_t = \frac{1}{\dot{m}c}(q''A_p),$$

$$\Delta T_m = \frac{1}{2c}(V_1^2 - V_2^2),$$

where subscripts t and m denote thermal and mechanical contributions, respectively. To determine ΔT_m in the diffuser, we need V_2, which follows from Eq. (1.20)

$$V_2 = V_1\left(\frac{D_1}{D_2}\right)^2 = \frac{V_1}{25} = 0.08 \text{ m/s}.$$

Then

$$\Delta T_m = \frac{1}{2 \times 4{,}000 \text{ kg·m}^2/\text{s}^2/\text{kg·K}}\left[2^2 - (0.08)^2\right] \text{m}^2/\text{s}^2 \cong 5 \times 10^{-4} \text{K}.$$

To evaluate ΔT_t in the pipe and diffuser, we need $\dot{m}$, $A_{p,\text{pipe}}$, and $A_{p,\text{diff}}$. Recalling Eq. (1.18),

$$\dot{m} = \rho A_1 V_1 (= \rho A_2 V_2),$$

---

[5] The FORTRAN program EX1–1.F is listed in the appendix of this chapter.

$$\dot{m} = 1{,}000 \text{ kg/m}^3 \times \frac{\pi(0.05\text{m})^2}{4} \times 2\text{m/s} \cong 3.927 \text{ kg/s}.$$

Peripheral areas,
$$A_{p,\text{pipe}} = \pi D_1 L,$$

$$A_{p,\text{diff}} = \int_0^L \pi D(x)\,dx = \int_0^L \pi \left[D_1 + (D_2 - D_1)\frac{x}{L}\right] dx$$

$$= \pi D_1 L + \frac{1}{2}\pi(D_2 - D_1)L = \frac{1}{2}\pi(D_1 + D_2)L.$$

Accordingly,
$$A_{p,\text{pipe}} = \pi 0.05 \text{ m} \times 10 \text{ m} = 1.57 \text{ m}^2,$$
$$A_{p,\text{diff}} = 0.5\pi(0.05 + 0.25) \text{ m} \times 10\text{m} = 4.71 \text{ m}^2,$$

which, for this case, implies $\Delta T_{t,\text{diff}} \cong 3\Delta T_{t,\text{pipe}}$. Then

$$\Delta T_{t,\text{pipe}} = \frac{1}{\dot{m}c}(q'' A_{p,\text{pipe}}) = \frac{10{,}000\text{W/m}^2 \times 1.57\text{m}^2}{3.927\text{kg/s} \times 4{,}000\text{W}\cdot\text{s/kg}\cdot\text{K}}, \cong 1\text{K}.$$

$$\Delta T_{t,\text{diff}} = 3 \times 1 = 3\text{K}.$$

Note that in the diffuser
$$\frac{\Delta T_t}{\Delta T_m} = \frac{3}{5 \times 10^{-4}} \cong 6 \times 10^3,$$

demonstrating that the effect of kinetic energy on temperature compared to that of imposed heat flux is negligible. However, this may not always be the case. ◆

EXAMPLE 1.2 An unsteady problem

A thermally insulated electric wire of diameter D, length ℓ, density ρ, specific heat c and electrical resistance R is initially at ambient temperature T_∞. Let electric potential V^* be suddenly applied to this wire. We wish to find the time required for the wire to reach its melting temperature, T_m.

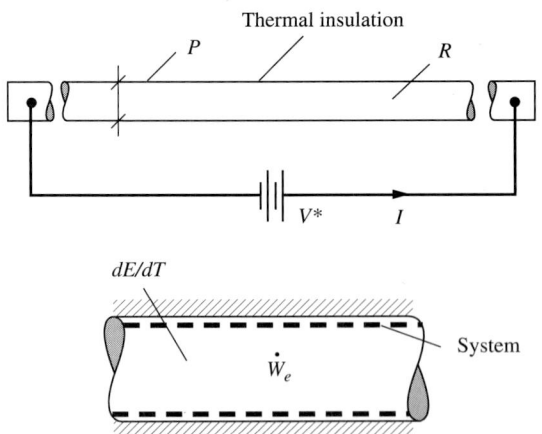

Figure 1.4 Electrically heated wire.

Step 1: Assume the entire wire be the system, as shown in Fig. 1.4.
Step 2: In the absence of heat loss to the ambient, the first law given by Eq. (1.16) is reduced to

$$\frac{dE}{dt} = -\dot{W}. \tag{1.26}$$

Furthermore, for a solid, assume p and ρ are uniform, and

$$dE = mc\,dT, \tag{1.27}$$

where $m = \rho V$, and m and V are the mass and the volume of the wire, respectively. Also, recall that electric energy generated in the wire is thermodynamically identical to power applied to the system, and

$$\dot{W} = -\dot{W}_e, \tag{1.28}$$

where

$$\dot{W}_e = V^* I = I^2 R = V^{*2}/R. \tag{1.29}$$

Then Eq. (1.26) becomes

$$\rho c V \frac{dT}{dt} = V^{*2}/R. \tag{1.30}$$

Contrary to the algebraic nature of the preceding example, here we end up with a differential equation. Apparently, when the two steps of formulation lead to a differential equation, we need another step to determine the integration constant of the equation (we shall elaborate this step later). The wire is initially at temperature

$$T(0) = T_\infty. \tag{1.31}$$

Equation (1.30) together with Eq. (1.31) completes the formulation of our problem.

The integration of Eq. (1.30) with Eq. (1.31) gives the solution

$$T - T_\infty = \left(\frac{V^{*2}/R}{\rho V c}\right) t, \tag{1.32}$$

where $V = (\pi D^2/4)\ell$.

As a practical application,[6] let us consider a copper wire with the following specifications:

$\ell = 1\text{m},$ $\qquad D = 1\text{mm},$
$T_\infty = 270\text{ K},$ $\qquad T_m \cong 1{,}400\text{ K},$
$V^* = 100\text{ volt},$ $\qquad R^* = R/\ell \cong 0.1\text{ ohm/m},$
$\rho \cong 9{,}000\text{ kg/m}^3,$ $\qquad c \cong 400\text{ J/kg·K}$

Solving for t in Eq. (1.32) and substituting R and V explicitly,

$$t = (T - T_\infty)/\left(\frac{V^{*2}/R^*\ell}{\rho c \dfrac{\pi D^2}{4}\ell}\right) = \rho c \frac{\pi D^2}{4}\ell^2 \left(\frac{T_m - T_\infty}{V^{*2}/R^*}\right),$$

---

[6] The FORTRAN program EX1–2.F is listed in the appendix of this chapter.

$$t = 9{,}000 \text{ kg/m}^3 \times 400 \text{ J/kg·K} \times \frac{\pi(1 \times 10^{-3})^2}{4}\text{m}^2 \times (1)^2\text{m}^2 \times \frac{1{,}400 - 270\text{K}}{100^2/0.1 \text{ W·m}},$$

$$t \cong 3.2 \times 10^{-2} \text{ s}.$$

◆

1.4 CONTROL SURFACE

In the preceding section we developed the first law of thermodynamics for a control volume. In this section we wish to develop the same law for a **control surface**.

Define a control surface as a control volume with zero volume surrounding a moving interface or a stationary boundary (Fig. 1.5). The first law of thermodynamics for this control surface can be readily obtained by eliminating the volumetric term, dE_{cv}/dt, replacing the stagnation enthalpy with enthalpy in Eq. (1.10), and interpreting the remaining terms with Fig. 1.5. Thus, in the absence of $\dot{W}_{cv}$,

$$0 = \dot{m}(h_1 - h_2) + \dot{Q}_1 - \dot{Q}_2. \tag{1.33}$$

The application of Eq. (1.33) to a stationary boundary readily gives

$$0 = \dot{Q}_1 - \dot{Q}_2. \tag{1.34}$$

$\dot{Q}_1$ and $\dot{Q}_2$ will be later related to temperature. The application of Eq. (1.33) to a moving interface is illustrated below.

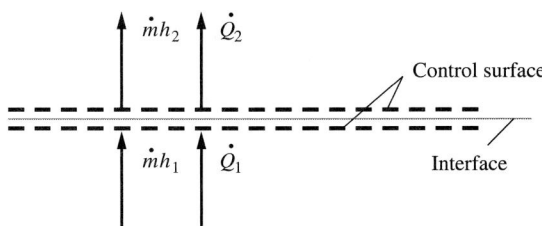

Figure 1.5 First law for a moving interface.

EXAMPLE 1.3 Unsteady one-dimensional flow

Consider an interface separating a saturated liquid layer from its vapor, as shown in Fig. 1.6. The bottom of the fluid layer is insulated, while its top absorbs a specified heat flux $\dot{Q}$ acting at a distance. The initial thickness of the layer is X_0. The liquid and its vapor are at the saturation temperature T_s. We wish to determine the unsteady thickness $X(t)$ of the evaporating liquid.

For an observer fixed to the interface, Eq. (1.33) yields

$$0 = \dot{m}(h_f - h_g) + \dot{Q}, \tag{1.35}$$

where subscripts f and g stand for liquid and vapor, respectively. Also, the conservation of mass at the interface gives

$$\rho_f A V_f = \rho_g A V_g = \dot{m} = \text{Const.}, \tag{1.36}$$

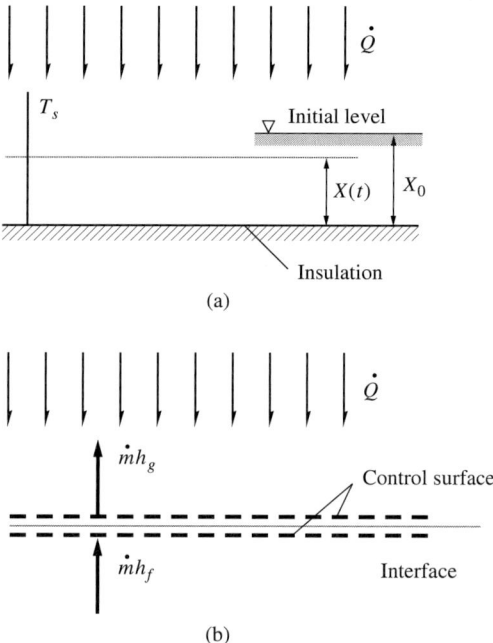

Figure 1.6 (a) Evaporating liquid layer, (b) first law for a control surface.

where

$$V_f = -\frac{dX}{dt}. \tag{1.37}$$

In terms of the latent heat of evaporation $h_{fg} = h_g - h_f$ and $q'' = \dot{Q}/A$, Eqs. (1.35), (1.36), and (1.37) lead to

$$0 = \rho_f h_{fg} \frac{dX}{dt} + q''. \tag{1.38}$$

The initial value of the liquid layer thickness is

$$X(0) = X_0. \tag{1.39}$$

The integration of Eq. (1.38) readily gives, after the consideration of Eq. (1.39),

$$X(t) = X_0 - \frac{q''}{\rho_f h_{fg}} t, \tag{1.40}$$

which shows the linear decrease of the thickness of the fluid layer with time. ◆

1.5 ORIGIN OF HEAT TRANSFER. PARTICULAR LAWS

In Section 1.1 we classified the problems of mechanics, extended this classification to thermal problems, and distinguished between thermodynamically determined and undetermined problems. Then we stated the need for particular laws of heat transfer for thermodynamically undetermined problems. Here, in terms of the original problem of conduction, we shall demonstrate why for a thermodynamically undetermined thermal problem some knowledge **beyond** thermodynamics is needed.

1.5.1 Original Problem of Conduction

Consider a flat plate of thickness ℓ whose surfaces are kept at temperatures T_1 and T_2. We wish to find the steady heat transfer through this plate.

Under steady conditions, and in the absence of any mass flow and power input, the first law stated by Eq. (1.16) reduces to

$$\dot{Q}(\text{Net}) = 0. \tag{1.41}$$

Interpretation of this result for the **differential system** shown in Fig. 1.7, gives

$$\dot{Q} - \left(\dot{Q} + \frac{d\dot{Q}}{dx}dx\right) = 0,$$

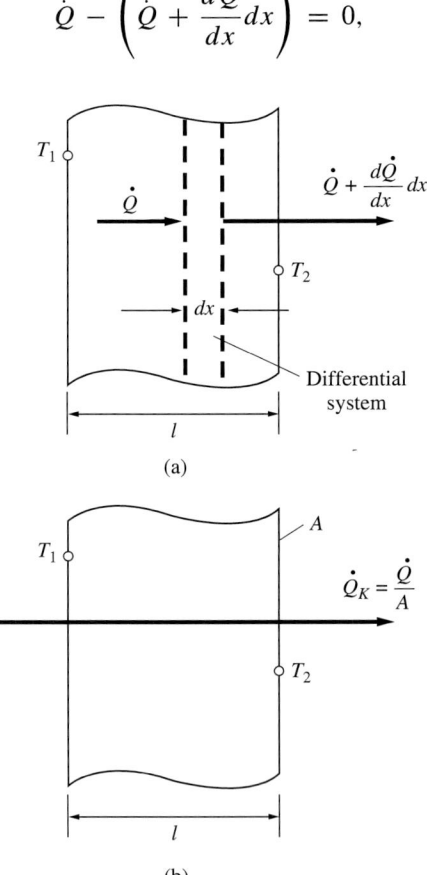

Figure 1.7 (a) Origin of conduction, (b) Fourier's law of conduction.

or
$$-\frac{d\dot{Q}}{dx}dx = 0 \qquad (1.42)$$

which integrates to
$$\dot{Q} = \text{Const.} \qquad (1.43)$$

That is, the heat flux is constant at any cross section of the plate. However, for the size of a heat transfer device, say for a heater providing this flux through the walls of a room to be heated, we need the specific value of this constant. **Thermodynamics is silent to this need**. The attempt to find an answer for this need is the origin of (conduction) heat transfer. Since the statement of our example specifies the temperatures of two surfaces, we need **a relation between heat flow and temperature**, $\dot{Q} = f(T)$, which is **phenomenologically provided by heat transfer**. Observations show that **any relation of this nature** is dependent on the medium it applies to and, consequently, **is a particular law**. The remainder of this section is devoted to particular laws of heat transfer. We begin with the particular law associated with our illustrative example.

1.5.2 Fourier's Law of Conduction

Experimental observations on different solids lead us to the temperature dependence of Eq. (1.43) as

$$\boxed{q_K = \frac{\dot{Q}_K}{A} = k\frac{T_1 - T_2}{\ell}}, \qquad (1.44)$$

which is **Fourier's law for homogeneous media** [Fig. 1.7(b)]. The proportionality constant k is called the **thermal conductivity** of the plate material and has units of W/m·K. Equation (1.44) continues to be valid for a fluid (liquid or gas) placed between two plates separated a distance ℓ apart, provided suitable precautions are taken to eliminate convection and radiation.

EXAMPLE 1.4 Conduction heat loss

Consider a human being with a total feet area $A = 2 \times 10 \times 30$ cm^2 standing on the ground. The thickness of the shoe leather soles is $\ell = 0.5$ cm. Assuming the temperature of the feet to be $T_1 = 37$ °C and the temperature of the ground to be $T_2 = 0$ °C, we wish to determine the heat loss to ground. Assume $k_{\text{leather}} \cong 0.0135$ W/m·K.

The first **two steps** of formulation are identical to those of the original conduction problem, which leads to
$$\dot{Q} = \text{Const.} \qquad (1.45)$$

Here, we need a **third step** for the evaluation of the particular value of this constant in terms of temperature. Fourier's law,

$$q_K = \frac{\dot{Q}_K}{A} = k\frac{T_1 - T_2}{\ell} = \text{Const.} \qquad (1.46)$$

provides this step.[7]

---

[7] Actually, there is a hidden **fourth step** in Eq. (1.44). This step will be clarified in Section 1.8.

In terms of the total area A.

$$A = 2 \times 10 \times 30 \text{ cm}^2 \times 10^{-4} \text{ m}^2/\text{cm}^2 = 0.06 \text{ m}^2,$$

the total heat transfer

$$\dot{Q}_K = q_K A = 0.0135 \text{ W/m·K} \times \frac{(37 - 0)\text{K}}{0.5 \times 10^{-2}\text{m}} \times 0.06 \text{ m}^2.$$

$$\dot{Q}_K = 6 \text{ W}.$$

Note that the total heat transfer from a human body is approximately 200 W. ◆

Thermal conductivity may be helpful for thermal classification of media. A medium is said to be **thermally homogeneous** if its conductivity does not vary from point to point within the medium, and **thermally heterogeneous** if there is such a variation. A medium is said to be **thermally isotropic** if its conductivity is the same in all directions and **thermally anisotropic** if there exists directional variation. It becomes clear after the foregoing classifications that solids used in experiments which suggest Fourier's law stated by Eq. (1.44) must necessarily be homogeneous. Also, a homogeneous material must necessarily be isotropic, but an isotropic material may be homogeneous or heterogeneous.

Let us see now what happens to Eq. (1.44) for heterogeneous and isotropic media. Assuming **a globally heterogeneous material to be locally homogeneous**, Eq. (1.44) can be used for a plate thickness of Δx as $\Delta x \to 0$. In terms of Fig. 1.8, letting $T_1 = T$ and $T_2 = T + \Delta T$ in Eq. (1.44), we get

$$q_x = -k \lim_{\Delta x \to 0} \frac{\Delta T}{\Delta x} = -k \frac{\partial T}{\partial x}, \quad (1.47)$$

which is **Fourier's law for heterogeneous isotropic media**. Note that Eq. (1.47) holds regardless of the actual temperature distribution. For example, in Fig. 1.9(a) we have $q_x > 0$ for $\partial T/\partial x < 0$, in Fig. 1.9(b) we have $q_x < 0$ for $\partial T/\partial x > 0$, and both cases agree with the second law of thermodynamics, which states that the heat is transferred in a direction from higher to lower temperatures. **Fourier's law for anisotropic media**

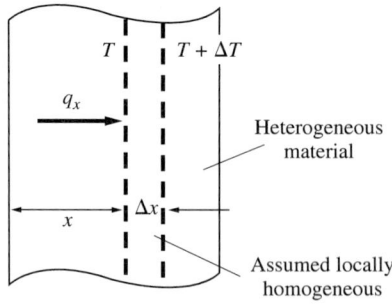

Figure 1.8 Fourier's law for heterogeneous material.

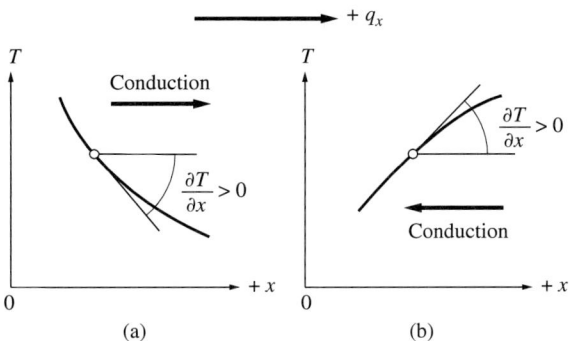

Figure 1.9 Sign of Fourier's law.

requires nine components for thermal conductivity and goes beyond the scope of the text.

1.5.3 Thermal Conductivity

It is appropriate here to make some remarks on the physical foundations of thermal conductivity. The dependence of thermal conductivity on temperature has been experimentally recognized. However, there is no universal theory explaining this dependence. Gases, liquids, conducting and insulating solids can each be explained with somewhat different microscopic considerations. Although the text is on the continuum aspects of heat transfer, the following remarks are made for some appreciation of the microscopic aspects of thermal conductivity.

For dilute gases, molecules are assumed to be independent from each other, and thermal conductivity is explained by means of kinetic theory, which analytically leads to $k \sim T^{1/2}$. Experimental results, however, indicate that for real gases

$$k \sim T^n, \tag{1.48}$$

n being greater than 1/2 and depending on the nature of the gas.

In solids the interaction between particles is strong and the system of particles become arranged in a lattice of definite crystalline structure. The collective motions involving many particles are then interpreted as **sound waves** propagating through solids. The quantized sound waves act like weakly interacting quasi-particles called **phonons**. Electron scattering by phonons becomes predominant at higher temperatures. At even higher temperatures the effect of scattered electrons continues to make a major contribution to conductivity, while a secondary effect appears from phonon scattering by the lattice. Thus

$$k = C_3 + \frac{C_4}{T}, \tag{1.49}$$

where the constant term of the righthand side (known as the Wiedeman-Franz law) gives the electron contribution and the second one the phonon contribution. Figure 1.10 shows a sketch of Eq. (1.49) as well as other forms of dependence of thermal conductivity on temperature.

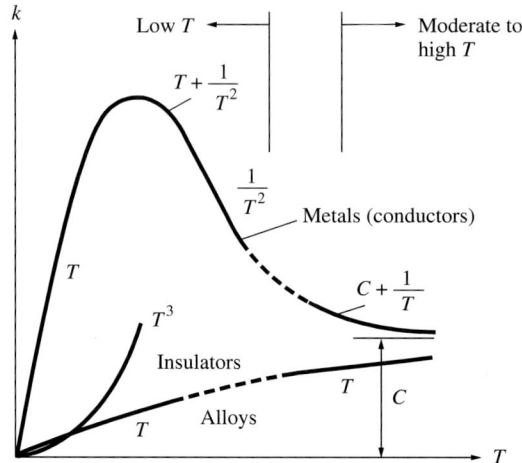

Figure 1.10 Thermal conductivity of solids.

Temperature dependence of thermal conductivity for liquids, metal alloys, and nonconducting solids is more complicated than those mentioned above. Because of these complexities, the temperature dependence of thermal conductivity for a number of materials, as illustrated in Fig. 1.11, does not show a uniform trend. Typical ranges for the thermal conductivity of these materials are given in Table 1.1. We now proceed to a discussion on the foundations of convective and radiative heat transfer.

Table 1.1 Typical values of thermal conductivity.

Material	$k \left(\dfrac{W}{m \cdot K} \right)$,
Gases	
at atmospheric pressure	0.007–0.2
Insulation material	0.03–0.2
Nonmetallic liquids	0.08–0.7
Nonmetalic solids	
brick, stone, cement	0.03–3
Liquid metals	9–50
Alloys	14–20
Pure metals	50–400

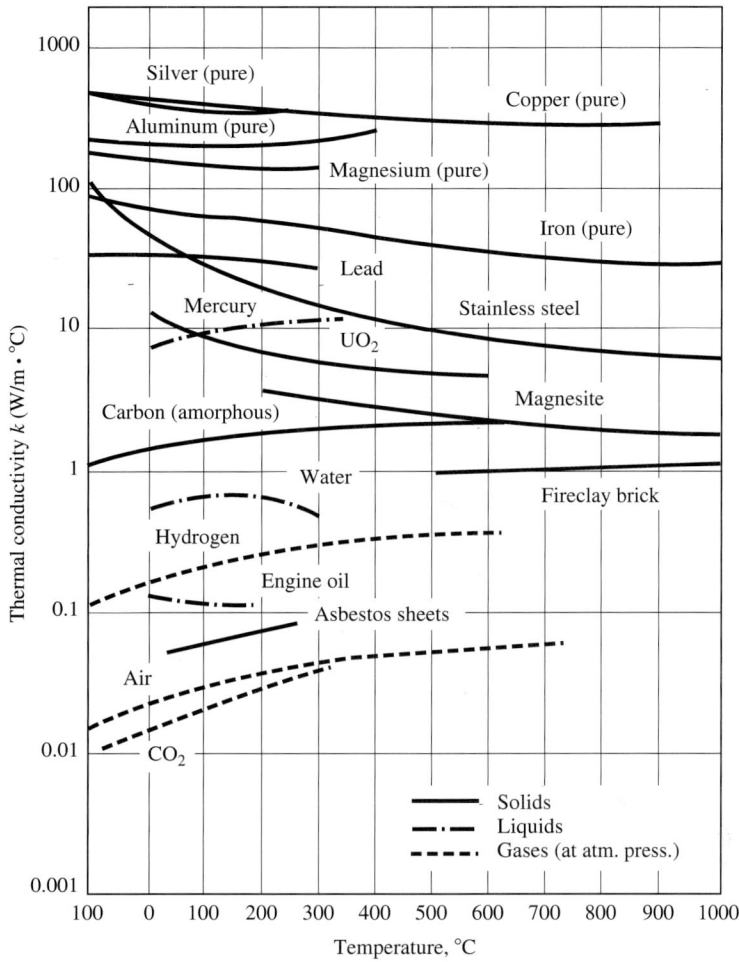

Figure 1.11 Thermal conductivity of some materials.

1.5.4 Newton's Definition of Convection

Consider two solid bodies each with a flat surface, kept at temperatures T_1 and T_2. Let the flat surfaces be separated by a distance ℓ. If the space between the flat surfaces were (ideally) fitted with a solid or filled with a stagnant fluid, the heat from one body to the other, as we already learned, would be transferred by conduction (Fig. 1.12), as stated by Eq. (1.44).

Let the medium between the flat surfaces of two bodies (now a fluid because of practical reasons) flow with a mean velocity V (Fig. 1.13). This flow results from either an imposed pressure drop or an induced buoyancy, respectively called **forced** and **natural** convection. Let the inlet temperature of the fluid be T_2. (Note that the fluid temperature need not be T_2. Selection of T_2 for this temperature eliminates temperature gradient near plate 2 and simplifies the following development.) The **convection heat transfer** from plate 1 is defined as the conduction in the fluid next to plate 1 (in view of the fact

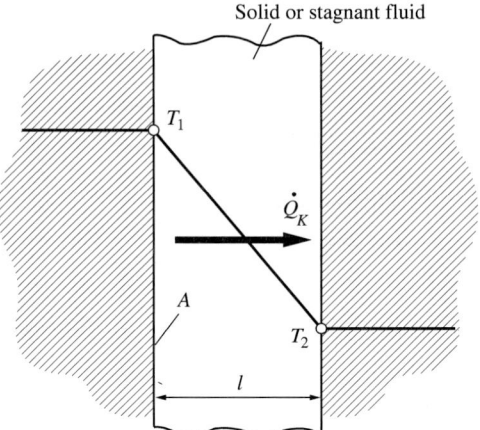

Figure 1.12 Conduction in homogeneous medium.

that the fluid next to the plates remains stagnant because of friction),

$$q_C = (q_K)_w = -k \left(\frac{\partial T}{\partial y} \right)_w, \qquad (1.50)$$

where y denotes the coordinate normal to the walls, and subscript w refers to location $y = 0$. It would be convenient to describe the same convection in terms of a heat

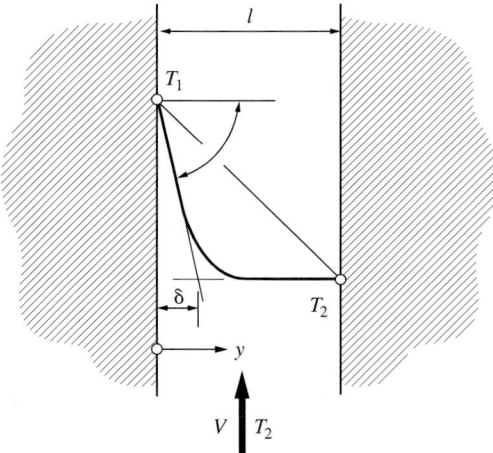

Figure 1.13 Nusselt number in terms of wall gradient of fluid temperature or in terms of thermal boundary-layer thickness.

transfer coefficient h and the temperatures T_1 and T_2 as

$$q_C = \frac{\dot{Q}_C}{A} = h(T_1 - T_2), \tag{1.51}$$

which is known as **Newton's law of cooling**. Equating Eqs. (1.50) and (1.51), we get

$$\underbrace{q_C}_{\text{Convection}} = \underbrace{-k\left(\frac{\partial T}{\partial y}\right)_w}_{\text{Foundation}} = \underbrace{h(T_1 - T_2)}_{\text{Definition}} \tag{1.52}$$

which can be rearranged with the help of Fig. 1.13 as

$$h = \frac{(q_K)_w}{T_1 - T_2} = k\frac{\partial}{\partial y}\left(\frac{T_1 - T}{T_1 - T_2}\right)_w, \tag{1.53}$$

or, in terms of a characteristic length ℓ (say the distance between the plates), as

$$\frac{h\ell}{k} = \frac{\partial}{\partial y^*}\left(\frac{T_1 - T}{T_1 - T_2}\right)_w = \frac{(q_K)_w}{k(T_1 - T_2)/\ell}, \quad y^* = \frac{y}{\ell}. \tag{1.54}$$

In terms of Eq. (1.44), this result may be interpreted as

$$Nu = \frac{h\ell}{k} = \frac{(q_K)_w}{q_K} = \frac{q_C}{q_K}. \tag{1.55}$$

Here q_K denotes the conduction in stagnant fluid, $(q_K)_w$ the wall value of conduction in moving fluid, and their ratio introduces the definition of the **Nusselt number**, which is dimensionless.

Clearly, in each convection problem, the Nusselt number is the **wall gradient of the dimensionless fluid temperature**, which needs to be analytically or computationally evaluated or to be experimentally determined from

$$Nu = \frac{(q_K)_w}{k(T_1 - T_2)/\ell} \tag{1.56}$$

with the measured values of $(q_K)_w$, T_1, and T_2.

In reality the variation of the fluid temperature under the influence of motion is confined to a thin **thermal boundary layer** δ. Then,

$$\frac{\partial}{\partial y}(T_1 - T) \cong \frac{T_1 - T_2}{\delta}, \tag{1.57}$$

and Eqs. (1.53) and (1.55) are reduced to

$$h \cong \frac{k}{\delta}, \tag{1.58}$$

and

$$Nu \cong \frac{\ell}{\delta}, \tag{1.59}$$

where

$$\delta = f(\text{flow}), \tag{1.60}$$

and, consequently,

$$h = f(\text{flow}). \tag{1.61}$$

Thus, unlike k (which is a thermal property), h is merely a definition and depends on flow (conditions). That is, unlike thermal conductivity, the heat transfer coefficient cannot be tabulated and needs to be determined for each flow condition. Accordingly, Chapters 5 and 6 are devoted to elaboration of Eqs. (1.60) and (1.61) and the solution of convection problems in terms of a heat transfer coefficient. Here, for some appreciation, an order-of-magnitude range of each heat transfer coefficient corresponding to natural or forced convection in different fluids is given in Table 1.2. The order-of-magnitude difference between the h values for natural convection and forced convection resulting from flow of the same fluid should be noted.

Table 1.2 Typical values of heat transfer coefficient.

Condition		$h \left(\frac{W}{m^2 \cdot K} \right)$
Natural Convection	Gases	5–12
	Oils	10–120
	Water	100–1,200
	Liquid metals	1,000–7,000
Forced Convection	Gases	10–300
	Oils	50–1,200
	Water	300–12,000
	Liquid metals	5,000–120,000
Phase Change	Boiling	3,000–50,000
	Condensation	5,000–120,000

EXAMPLE 1.5 Convection heat loss

An engine delivers 100 hp to a transmission. The efficiency of this transmission is 95% and its outer surface area is $A = 0.6 \text{ m}^2$. The ambient air temperature is $T_\infty = 25\,°C$ and the coefficient of heat transfer is $h = 150 \text{ W/m}^2 \cdot \text{K}$. We wish to determine the steady surface temperature of the transmission.

Step 1: Let the entire transmission box be the **system** as illustrated in Fig. 1.14.

Step 2: Under steady operation, the **first law**, Eq. (1.16), applied to this system reduces to

$$0 = \dot{Q} - \dot{W},$$

where, for the transmission box above,

$$\dot{Q} = -\dot{Q}_C$$

and

$$\dot{W} = -\dot{W}_i + \dot{W}_0,$$

$\dot{W}_i$ and $\dot{W}_0$ respectively being the rates of incoming and outgoing shaft work. The relationship between $\dot{W}_i$ and $\dot{W}_0$ is described in terms of the efficiency η as

$$\dot{W}_0 = \eta \dot{W}_i = 0.95 \dot{W}_i.$$

Substitution into first law readily gives

$$0 = (1 - \eta)\dot{W}_i - \dot{Q}_C.$$

Step 3: Employ **Newton's law of cooling** as the particular law (Eq. 1.47),

$$\dot{Q}_C = hA(T_w - T_\infty),$$

Step 4: **Combine** the preceding two steps to obtain the governing equation

$$hA(T_w - T_\infty) = (1 - \eta)\dot{W}_i.$$

In terms of the given data, this relation yields

$$150 \text{W/m}^2\text{K} \times 0.6 \text{m}^2 \times (T_w - 25\,°C) = 0.05 \times 100 \text{hp} \times 745.7 \text{W/hp},$$

$$T_w \cong 66\,°C. \qquad \blacklozenge$$

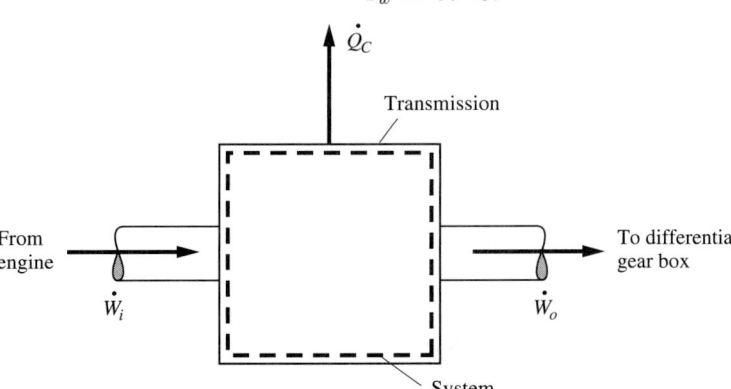

Figure 1.14 System configuration.

1.5.5 Stefan-Boltzmann's Law of Radiation

So far, we have talked about conduction in solids and stagnant fluids and convection in moving fluids. These modes of heat transfer depend on matter, and they disappear in the absence of matter. A third and final mode of heat transfer is **thermal radiation**. This mode of heat transfer, unlike conduction and convection, is hindered by matter and is at its best in a vacuum. Some aspects of thermal radiation, being a manifestation of the wide spectrum of natural phenomena including AM-FM radio waves, UHF-VHF television waves, optics, X-rays, γ-rays, cosmic rays, etc., can be explained in terms of electromagnetic waves. Other aspects of this radiation, being a manifestation of the many forms of particle interaction, can be explained in terms of radiation quanta (photons). As a theoretical (electromagnetic or quantum) and/or experimental fact, assume an ideal surface at absolute temperature T emitting thermal radiation energy E_b according to the **Stefan-Boltzmann law** (Fig. 1.15a),

$$E_b = \sigma T^4 \tag{1.62}$$

which defines the surface of a **black body**. Here E_b is called the black-body emissive power, and

$$\sigma = 5.67 \times 10^{-8} \text{W/m}^2 \cdot \text{K}^4 (0.171 \times 10^{-8} \text{Btu/ft}^2 \cdot \text{hr} \cdot {}^\circ R^4) \tag{1.63}$$

is the Stefan-Boltzmann constant. Black-body emissive power represents the maximum amount of heat that can be radiated by a body with a temperature T. A justification of this fact requires background on isotropic versus anisotropic radiation which is beyond the scope of this text. Surfaces other than that of a black body emit a fraction of this energy.

So far we have introduced three parameters, k, h, and σ, in the statement of the three particular laws of heat transfer. Note the fundamental difference among these parameters: k is a thermophysical property, h is a definition depending on flow, and σ is a universal constant.

Now reconsider two bodies each with a flat **black** surface at absolute temperatures T_1 and T_2 separated ℓ distance apart in a vacuum, as shown in Fig. 1.15b. Surface 1 has emissive power E_{b1}, surface 2 has emissive power E_{b2}, and the radiation heat transfer between these bodies is

$$q_R = \frac{\dot{Q}_R}{A} = E_{b1} - E_{b2}, \tag{1.64}$$

or, in terms of the Stefan-Boltzmann law,

$$\boxed{q_R = \frac{\dot{Q}_R}{A} = \sigma(T_1^4 - T_2^4)}. \tag{1.65}$$

In general,

$$q_R = F_{12}\sigma(T_1^4 - T_2^4), \tag{1.66}$$

where the **correction factor** F_{12} characterizes surfaces other than those of a black body and their relative position.

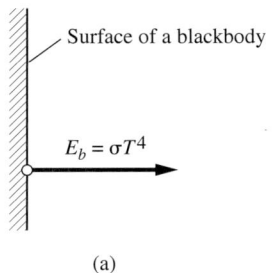

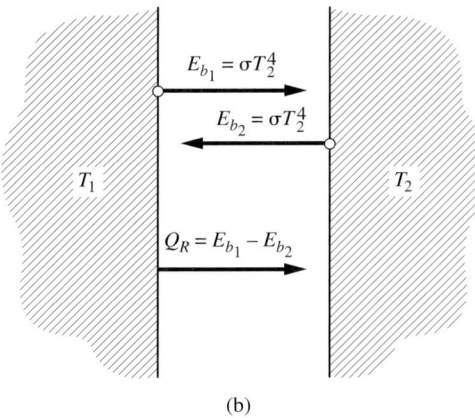

Figure 1.15 (a) The Stefan-Boltzmann law, (b) radiation heat transfer between two parallel black surfaces.

When there is a transparent medium between two bodies, Eq. (1.66) continues to apply, but the total heat transfer q_T now includes the effect of q_K (or q_C) as well as q_R. The total Nusselt number in the latter case is

$$Nu_T = Nu(1 + q_R/q_K), \tag{1.67}$$

Nu being the usual definition of the Nusselt number. When there is an absorbing medium, q_R includes the radiation effect of the medium as well as that of the surrounding surfaces. These facts will be elaborated in Chapters 8 and 9 on radiation.

For small temperature differences, expanding both T_1^4 and T_2^4 into a Taylor series about a characteristic temperature T_0 and subtracting,

$$T_1^4 - T_2^4 \cong 4T_0^3(T_1 - T_2), \tag{1.68}$$

Eq. (1.66) may be approximated as

$$q_R = 4\sigma T_0^3 F_{12}(T_1 - T_2), \tag{1.69}$$

or, introducing a **radiative heat transfer coefficient**

$$h_R = 4\sigma T_0^3 F_{12}, \tag{1.70}$$

as
$$q_R \cong h_R(T_1 - T_2). \tag{1.71}$$

Although not valid for large temperature differences, this linearized form of the radiative heat flux is frequently used because of its convenience, especially in problems dealing with a combination of all three modes of heat transfer.

EXAMPLE 1.6 Radiation heat loss

A satellite in space is required to dissipate 5,000 W/m² at a steady rate. Determine the steady temperature of the satellite, assuming that the satellite behaves as a black body.

Step 1: **System** consisting of the satellite as shown in Fig. 1.16.

Step 2: **First law** from Eq. (1.16), noting that the process is steady and no motion is involved

$$0 = \dot{Q} - \dot{W}.$$

Step 3: Space may be considered as a vacuum at a temperature of 0 K, hence the only possible mode of heat transfer is radiation. Then take the **particular law** from Eq. (1.69),

$$\dot{Q}_R = \sigma A(T_w^4 - 0),$$

which is related to $\dot{Q}$ simply as

$$\dot{Q} = -\dot{Q}_R.$$

Step 4: **Combination** of the two preceding steps readily gives

$$\dot{Q}_R = -\dot{W},$$

$$\sigma A T_w^4 = -\left(-5{,}000 \text{W/m}^2\right) \times A(\text{m}^2).$$

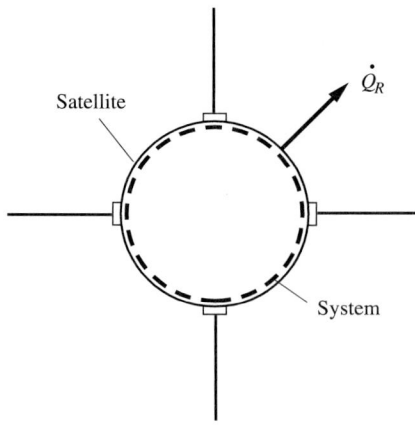

Figure 1.16 System configuration.

Eliminating A's from both sides and inserting $\sigma = 5.67 \times 10^{-8} \text{W/m}^2 \cdot \text{K}^4$ from Eq. (1.63)

$$T_w^4 = \frac{5{,}000 \text{W/m}^2}{5.67 \times 10^{-8} \text{W/m}^2 \cdot \text{K}^4} \cong 8.82 \times 10^{10} \text{K}^4.$$

Solving for T,

$$T \cong 545 \text{K} (\cong 272\,°\text{C}).$$

◆

1.6 HEAT TRANSFER MODES COMBINED

In the preceding section we learned about the three modes of heat transfer. In practical situations, as in the cases of fossil fuel and nuclear power plants, internal combustion engines, jet engines, and rocket motors (Fig. 1.17), heat is transferred by an appropriate combination of these modes. As to be expected, each part of a power plant, engine, or motor is more involved than the schematic representation shown in the figure. For example, a boiler includes also a superheater, an economizer, and an air heater (Fig. 1.18). In the boiler of a conventional power plant, hot gases resulting from combustion evaporate the water to its saturation temperature. Since the temperature of the hot gases is much higher than this saturation temperature, steam is superheated by passing these gases through a heat exchanger called a superheater. Also, with the remaining thermal energy in hot gases, the feed water from the condenser and the air intake are heated

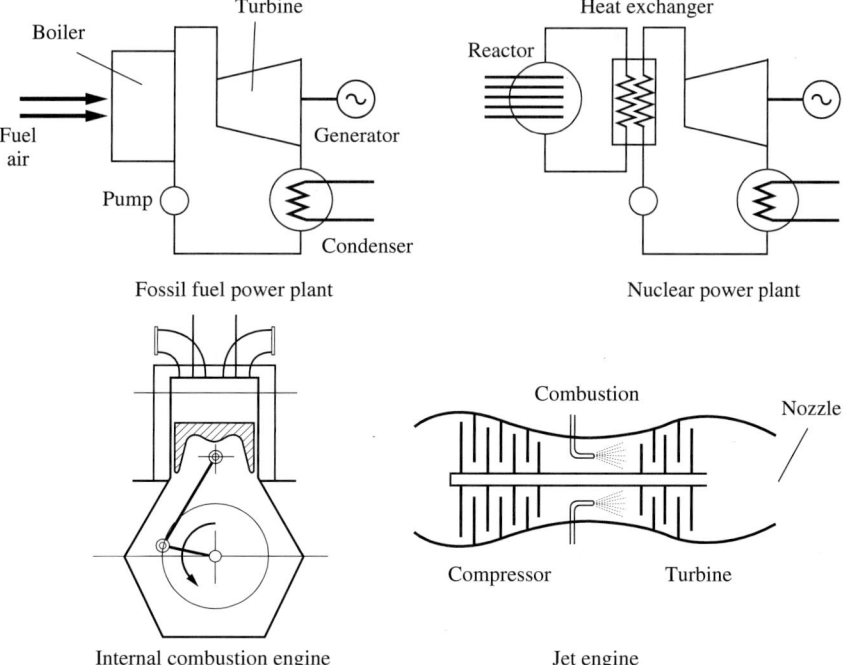

Figure 1.17 Examples involving combined modes of heat transfer.

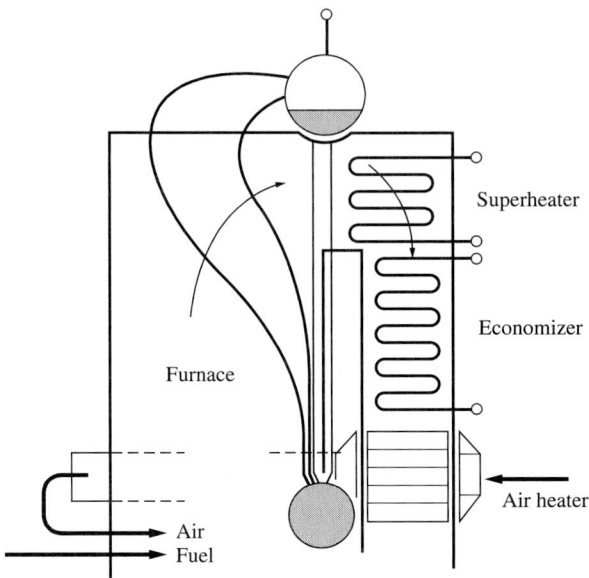

Figure 1.18 Components of a boiler.

through heat exchanger respectively called economizer and air heater. Figure 1.19 shows the design details of a contemporary boiler. In all of these cases, a chemical or nuclear fuel is converted into heat, which is transferred through a wall (or clad) to a carrier fluid or coolant (Fig. 1.20a). Products of the chemical reaction may involve gases such as CO, CO_2, and/or H_2O vapor, which appreciably emit and absorb radiation, or temperatures may reach levels at which radiation compared to convection becomes important-maybe even the dominant mode of heat transfer.

EXAMPLE 1.7 Combined-mode heat transfer

Let a flat wall of thickness ℓ separate a hot (combusting) ambient at temperature T_i from a coolant at temperature T_0. In addition to convection, include the effect of radiation on both sides. We wish to determine the heat transfer to the coolant.

Step 1: Consider the two **systems** shown in Fig. 1.20(b).

Step 2: After neglecting unsteady and power terms, apply the **first law of thermodynamics**, Eq. (1.16), to these systems. The result is

$$(\dot{Q}_C + \dot{Q}_R)_i = \dot{Q}_K = (\dot{Q}_C + \dot{Q}_R)_0 = \dot{Q}, \tag{1.72}$$

$\dot{Q}$ denoting the constant value of heat at each cross section (which cannot be determined by thermodynamical considerations).

Step 3: Recall the **particular laws** of heat transfer, Eqs. (1.44), (1.51), and (1.71),

$$(\dot{Q}_C + \dot{Q}_R)_i = (h + h_R)_i A(T_i - T_1), \tag{1.73}$$

$$\dot{Q}_K = kA(T_1 - T_2)/\ell, \tag{1.74}$$

$$(\dot{Q}_C + \dot{Q}_R)_0 = (h + h_R)_0 A(T_2 - T_0). \tag{1.75}$$

Sec. 1.6 Heat Transfer Modes Combined   29

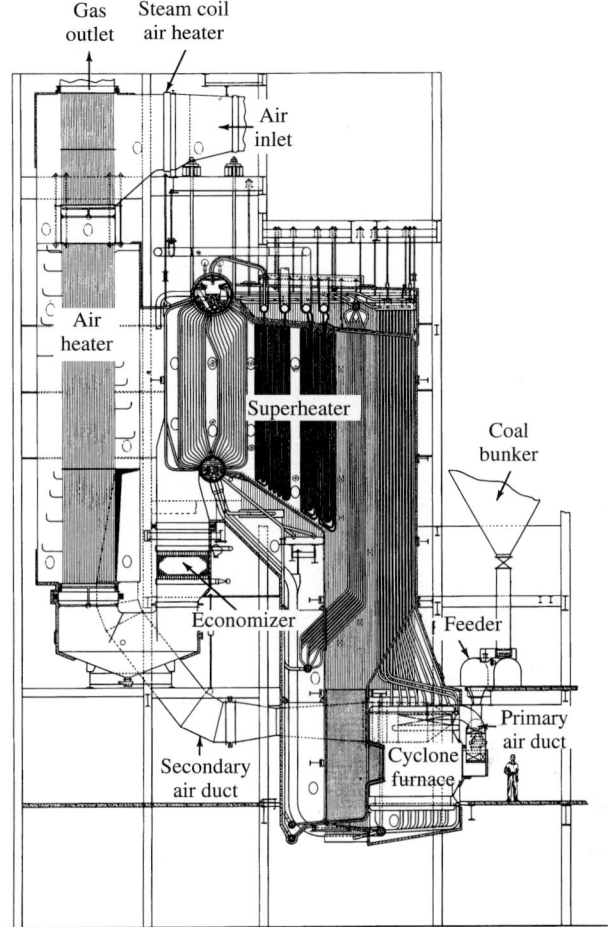

Figure 1.19 Details of a contemporary boiler (from Babcock & Wilcox [12]).

Step 4: Combine steps 2 and 3 to get

$$\left. \begin{array}{r} (h + h_R)_i A(T_i - T_1) \\ \dfrac{kA}{\ell}(T_1 - T_2) \\ (h + h_R)_0 A(T_2 - T_0) \end{array} \right\} = \dot{Q} \qquad (1.76)$$

which completes the formulation. Since T_1 and T_2 are unknown, we wish to solve for $\dot{Q}$ in terms of given T_i and T_0.

For the solution, rearrange each relation in terms of its temperature difference as

$$(T_i - T_1) = \frac{\dot{Q}}{(h + h_R)_i A},$$

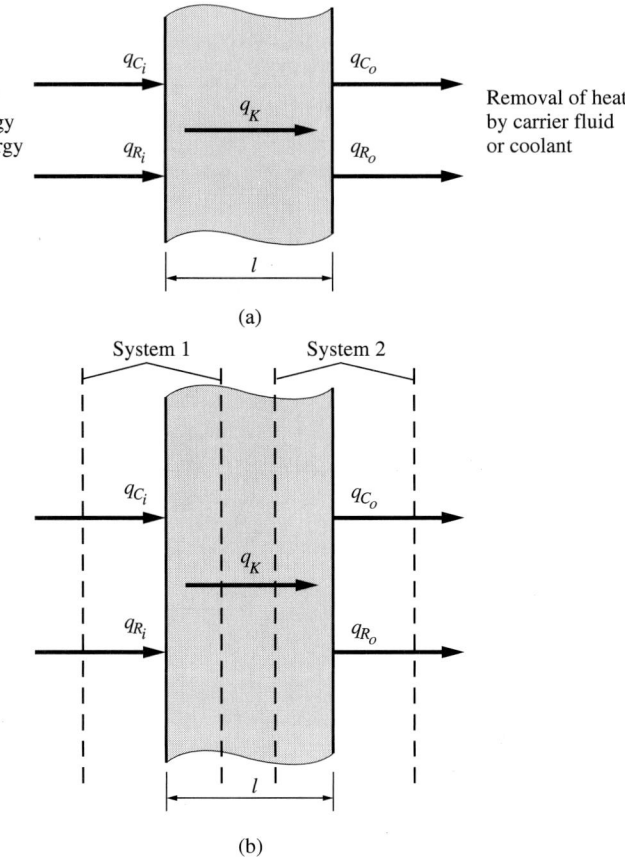

Figure 1.20 (a) A model for combined modes of heat transfer, (b) two systems for the model.

$$(T_1 - T_2) = \frac{\dot{Q}}{\frac{kA}{\ell}}, \tag{1.77}$$

$$(T_2 - T_0) = \frac{\dot{Q}}{(h + h_R)_0 A}.$$

Add both sides of Eq. (1.77) to get

$$T_i - T_0 = \left[\frac{1}{(h + h_R)_i A} + \frac{\ell}{kA} + \frac{1}{(h + h_R)_0 A}\right] \dot{Q},$$

or,

$$\dot{Q} = \frac{T_i - T_0}{\frac{1}{(h + h_R)_i A} + \frac{\ell}{kA} + \frac{1}{(h + h_R)_0 A}}, \tag{1.78}$$

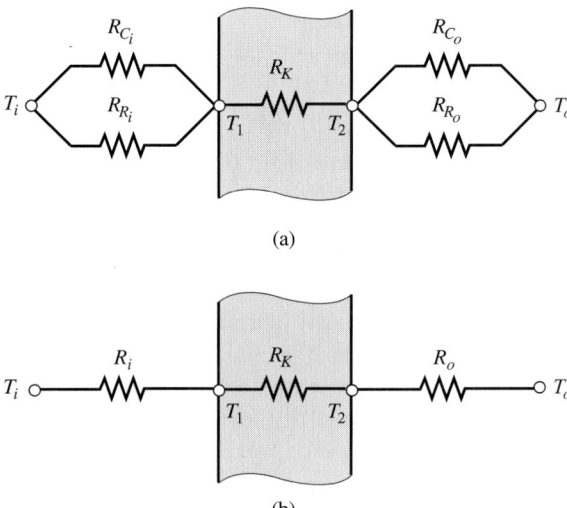

Figure 1.21 (a) Actual circuit, (b) equivalent circuit.

which can be rewritten as,

$$\dot{Q} = \frac{T_i - T_0}{R_i + R_K + R_0}, \tag{1.79}$$

where

$$R_i = \frac{1}{(h + h_R)_i A}, \quad \text{total (convective + radiative) inside resistance,}$$

$$R_K = \frac{\ell}{kA}, \quad \text{conductive resistance,}$$

$$R_0 = \frac{1}{(h + h_R)_0 A}, \quad \text{total (convective + radiative) outside resistance.}$$

The relative magnitude of these resistances varies over a large range. For example, $h_i \ll h_{Ri}$ and $h_0 \gg h_{Ro}$ in the case of radial heat transfer from the vertical tubes of a boiler furnace and from the walls of a rocket engine.

Clearly, Eq. (1.79) readily admits an interpretation in terms of the electric-circuit theory as shown in Fig. 1.21, where temperatures are analogous to potentials (voltages) and heat flux is analogous to electric current. In Section 2.2 of Chapter 2 dealing with composite structures we shall again utilize this analogy. ◆

So far we have determined the place of heat transfer as a thermodynamically undetermined subject among thermal disciplines, thus learning the need of knowledge beyond thermodynamics. Having established the three modes of heat transfer for this need, we are ready for an individual study of these modes. Here, let us set forth the method we shall follow in this text.

1.7 METHODS OF FORMULATION

In the formulation of a specific problem, we have the choice of following a method of **reduction** or of **induction**. The method of reduction is based on a simplification of the general formulation in accordance with the particular nature of the problem. Although desirable from an advanced point of view, the formalism involved with this method makes it inconvenient for use in an introductory text. By contrast, the method of induction treats each problem individually from the beginning with special emphasis on physics at each step of the formulation, and it is suitable for our objective. For example, in the thermodynamically determined problems discussed in Section 1.3, we recognized the important fact that a general law (such as the first law of thermodynamics) can be correctly applied only when it is considered for a clearly defined system or control volume. Then in that section, we carefully defined an appropriate system and two control volumes as a **first step** in the formulation of the problems in Exs. 1.1, 1.2, and 1.3. Then as a **second step**, we expressed the first law in terms of these system and control volumes.

For thermodynamically undetermined problems, however, we need to know a $\dot{Q} = f(T)$ relation to relate $\dot{Q}$ of the first law to temperature. So far we have learned three $\dot{Q} = f(T)$ relations as the three modes of heat transfer (or as particular laws) which need to be considered as a **third step** in the formulation of these problems. In Exs. 1.4 through 1.7, inserting the particular laws into the first law (a general law), we obtained as a **fourth step** the **governing equation** of the formulation. If this step leads to an algebraic equation (as in the case of Exs. 1.4, 1.5, and 1.6), or to a set of algebraic equations (as in the case of Ex. 1.7), the formulation of the problem is completed with this step, to be followed by the **solution** of algebraic equations. However, in heat transfer problems, the fourth step most frequently leads to a differential equation (as in the cases of Exs. 1.2 and 1.3) or to a set of differential equations. The formulation of these problems requires as a **fifth step** the specification of some initial and boundary conditions for the evaluation of the integration constants resulting from the temporal and spatial integrations of the differential equations. For convenience and later reference, the method of formulation by induction, elaborated in terms of the foregoing five steps, is summarized below:

1. **Define a system or control volume.**
2. **State general law(s) for (1).**
3. **State particular law(s) for (2).**
4. **Obtain governing equation by inserting (3) into (2).**
5. **Specify initial and/or boundary conditions pertinent to (4).**

As an illustration for the use of these steps, let us reconsider the original problem of conduction. This time, however, we wish to know the temperature distribution in, as well as the heat transfer through, the plate.

1.8 FIVE-STEP INDUCTIVE FORMULATION

Step 1: Consider a differential **system** because of the distributed nature of the problem. Since distributed problems are always in terms of a coordinate system, select a cartesian coordinate normal to the plate (Fig. 1.22).

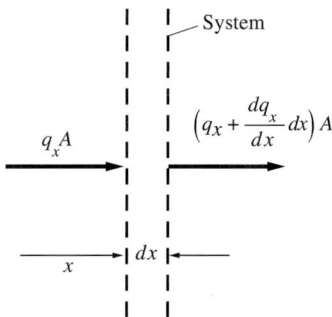

Figure 1.22 First law for a differential system.

Step 2: The **general law**, Eq. (1.16), is reduced, in terms of Fig. 1.22, to

$$0 = +q_x A - \left(q_x + \frac{dq_x}{dx}dx\right) A, \tag{1.80}$$

or

$$-\frac{dq_x}{dx} = 0. \tag{1.81}$$

Since this result does not lead to a temperature distribution for the plate, the problem is thermodynamically undetermined and needs a particular law.

Step 3: The **particular law**, Eq. (1.47),

$$q_x = -k\frac{dT}{dx} \tag{1.47}$$

relates q_x to T.

Step 4: Insert Eq. (1.47) into Eq. (1.81) to obtain the **equation governing** the temperature distribution,

$$\frac{d}{dx}\left(k\frac{dT}{dx}\right) = 0, \tag{1.82}$$

or, for $k = \text{Const.}$,

$$\frac{d^2 T}{dx^2} = 0. \tag{1.83}$$

Step 5: Up to this step there is no need to decide on the origin of the coordinate system. Let now this origin be at the surface with temperature T_1. Then

$$T(0) = T_1, \quad T(\ell) = T_2 \tag{1.84}$$

are **boundary conditions**.

Equation (1.82) or (1.83) subject to Eq. (1.84) completes the formulation of our problem. Next we consider its solution.

SOLUTION

Integrate twice the homogeneous case given by Eq. (1.83). The result,

$$T = C_1 x + C_2, \qquad (1.85)$$

combined with Eq. (1.84) yields

$$T_1 = C_2,$$
$$T_2 = C_1 \ell + T_1. \qquad (1.86)$$

In terms of Eq. (1.86), Eq. (1.85) becomes

$$\frac{T - T_1}{T_2 - T_1} = \frac{x}{\ell}. \qquad (1.87)$$

which shows that the temperature varies linearly between T_1 and T_2 across the plate. Also, inserting Eq. (1.87) into Eq. (1.47), we get the heat flux,

$$q_x = k \frac{T_1 - T_2}{\ell},$$

which recovers the experimental result stated by Eq. (1.44).

So far, we have learned the foundations of heat transfer. We are now ready to proceed to individual problems controlled by conduction, which is the simplest of the three modes.

■ REFERENCES

1.1 V.S. Arpacı, *Conduction Heat Transfer*. Addison-Wesley, Reading, MA., 1966.

1.2 V.S. Arpacı and P.S. Larsen, *Convection Heat Transfer*. Prentice-Hall, Englewood Cliffs, NJ, 1984.

1.3 A.H. Shapiro, *The Dynamics and Thermodynamics of Compressible Flow*. Robert E. Krieger, Florida, 1985.

1.4 G.H. Hatsopulos and J.H. Keenan, *Principles of General Thermodynamics*. Wiley, New York,1965.

1.5 G.J. Van Wylen, R.E. Sonntag, and C. Borgnakke, *Fundamentals of Classical Thermodynamics*. 4th ed., Wiley, New York, 1994.

1.6 M.J. Moran and H.N. Shapiro, *Fundamentals of Engineering Thermodynamics*, 3d ed., Wiley, New York, 1996.

1.7 Y.A. Cengel and M.A. Boles, *Thermodynamics-An Engineering Approach*, 3d ed., WCB/McGraw-Hill, New York, 1998.

1.8 C.Y. Ho, R.W. Powell, and P.E. Liley, *Thermal Conductivity of Elements*, Vol. 1, First Supplement to Journal of Physical and Chemical Reference Data (1972), American Chemical Society, Washington, D.C.

1.9 Y.S. Touloukian, R.W. Powell, C.Y. Ho, and P.G. Klemens, *Thermophysical Properties of Matter*, Vol.1: *Thermal Conductivity, Metallic Elements and Alloys*. IFI/Plenum, New York, 1970.

1.10 F. Reif, *Fundamentals of Statistical and Thermal Physics*. McGraw-Hill, New York, 1965.

1.11 N. Ozısık, *Heat Transfer: A Basic Approach*. McGraw-Hill, New York, 1985.

1.12 *Steam: Its Generation and Use*. Babcock & Wilcox, New York, 1978.

1.9 APPENDIX

```
C----------------------------------
C     EX1-1.F (START)
C----------------------------------
      PROGRAM MAIN
      IMPLICIT REAL*8 (A-H,K-Z)
      PI=4*ATAN(1.)
      WRITE(*,*) 'EXAMPLE 1.1....'
C--------------------------------------------------
C     INPUT DATA
C--------------------------------------------------
      WRITE(*,*) 'INPUT THE FOLLOWING DATA...'
      WRITE(*,*) 'D_1: cm'
      READ(*,*) D1
      WRITE(*,*) 'D_2: cm'
      READ(*,*) D2
      WRITE(*,*) 'L: m'
      READ(*,*) L
      WRITE(*,*) 'T_1: K'
      READ(*,*) T1
      WRITE(*,*) 'V_1: m/m'
      READ(*,*) V1
      WRITE(*,*) 'Q: kW/m^2'
      READ(*,*) Q
      WRITE(*,*) 'RHO: kg/m^3'
      READ(*,*) RHO
         WRITE(*,*) 'C: J/kg.K'
      READ(*,*) C
C--------------------------------------------------
C     UNIT CONVERSION
C--------------------------------------------------
      D1=0.01*D1
      D2=0.01*D2
      Q=1000*Q
C--------------------------------------------------
C     CALCULATION
C--------------------------------------------------
      V2=V1*(D1/D2)**2
      DTM=(V1**2-V2**2)/(2*C)
      A1=PI*D1**2/4
      DOTM=RHO*A1*V1
      APPIPE=PI*D1*L
      APDIFF=PI*(D1+D2)*L/2
      DTTPIPE=Q*APPIPE/(DOTM*C)
      DTTDIFF=APDIFF/APPIPE*DTTPIPE
C--------------------------------------------------
C     ANSWER
```

```
C-----------------------------------------------------
      WRITE(*,*) 'TEMPERATURE RISE DUE TO THE MECHANICAL CONTRIBUTION'
      WRITE(*,*) 'OF THE DIFFUSER IS'
      WRITE(*,*) DTM,' K'
      WRITE(*,*) 'TEMPERATURE RISE DUE TO THE THERMAL CONTRIBUTION'
      WRITE(*,*) 'OF THE DIFFUSER IS'
      WRITE(*,*) DTTDIFF,' K'
      WRITE(*,*) 'TEMPERATURE RISE DUE TO THE THERMAL CONTRIBUTION'
      WRITE(*,*) 'OF THE PIPE IS'
      WRITE(*,*) DTTPIPE,' K'
      STOP
      END
C-----------------------------------
C     EX1-1.F (END)
C-----------------------------------
C-----------------------------------
C     EX1-2.F (START)
C-----------------------------------
      PROGRAM MAIN
      IMPLICIT REAL*8 (A-H,K-Z)
      PI=4*ATAN(1.)
      WRITE(*,*) 'EXAMPLE 1.2....'
C-----------------------------------------------------
C     INPUT DATA
C-----------------------------------------------------
      WRITE(*,*) 'INPUT THE FOLLOWING DATA...'
      WRITE(*,*) 'L: m'
      READ(*,*) L
      WRITE(*,*) 'D: mm'
      READ(*,*) D
      WRITE(*,*) 'T_INFTY: K'
      READ(*,*) TINFTY
      WRITE(*,*) 'T_M: K'
      READ(*,*) TM
      WRITE(*,*) 'V^*: volt'
      READ(*,*) VS
      WRITE(*,*) 'R^*: ohm/m'
      READ(*,*) RS
      WRITE(*,*) 'RHO: kg/m^3'
      READ(*,*) RHO
         WRITE(*,*) 'C: J/kg.K'
      READ(*,*) C
C-----------------------------------------------------
C     UNIT CONVERSION
C-----------------------------------------------------
      D=0.001*D
C-----------------------------------------------------
C     CALCULATION
C-----------------------------------------------------
      T=RHO*C*PI*D**2*L**2*(TM-TINFTY)/(4*VS**2/RS)
C-----------------------------------------------------
C     ANSWER
C-----------------------------------------------------
      WRITE(*,*) 'TIME REQUIRED TO REACH MELTING TEMPERATURE IS'
      WRITE(*,*) T,' s'
      STOP
      END
C-----------------------------------
C     EX1-2.F (END)
C-----------------------------------
```

EXERCISES

1.1 The heat flux through a wall 10 cm thick is 20 W/m^2. The thermal conductivity is 0.1 W/m·K and the outer (cold) surface temperature of the wall is 0 °C. (a) Calculate the inner surface temperature of the wall. (b) Repeat the calculation for a different wall material with a thermal conductivity of 0.04 W/m·K.

1.2 The most common method of determining the conductivity of insulating and building materials that are available in the form of large plates is the so-called twin-plate method. It consists of placing a flat electrical heater between two identical flat plates made of the material to be tested. The plates and the heater are then placed between two flat coolers (Fig. 1P–1). In an experiment performed with two 3-cm-thick square (30 cm x 30 cm) flat plates, the temperature at the interface between the heater and the plates is 80 °C and the temperature at the interface between the plates and the coolers is 20 °C. These temperatures are maintained with 140 W supplied to the electric heater. Determine the thermal conductivity and, by the help of Table 1.1, the material type of these plates.

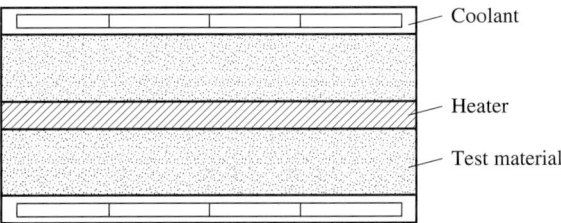

Figure 1P–1

1.3 A commercially available instrument, used to measure the heat flux directly, is known as a plug-type heat flux meter. The incident heat is conducted through the plug of area A and conductivity k to the cooling water on the other surface. Two thermocouples are embedded a distance ℓ apart. What is the total heat incident on the plug face?

1.4 Estimate the temperature of the bottom surface of a thin metal cup in which water under atmospheric conditions is boiled with a heat flux of 300,000 W/m^2. Assume a value of 30,000 W/m^2·K for the coefficient of heat transfer.

1.5 A 10 kW/m^2 electric heater is surrounded by an ambient at 20 °C.

Condition	Fluid	h [W/m^2·K]
	Air	10
Stagnant	Oil	100
	Water	1,000
	Air	250
Flowing	Oil	1,000
	Water	10,000

(a) Estimate the surface temperature of the heater for the following conditions:

(b) Compare the foregoing values of the heat transfer coefficient with those given in Table 1.2.

1.6 A particular kind of quantum detector (or photodetector) used in laboratories and industry is known as the thermopile detector. In an experiment, it is desired to measure the radiation from a black flat plate extending to infinity. A detector has been placed parallel to the plate. The catalog of the detector indicates that the maximum allowable incident radiation is 200 mW/cm^2. Determine the highest tolerable plate temperature which will not damage the thermopile detector. Neglect the radiation emitted by the detector itself.

1.7 A thin horizontal flat plate receives 1,200 W/m^2 of radiant heat from the sun. The upward and downward heat transfer coefficients are 10 and 2.5 W/m$^2\cdot$K. Determine the steady temperature of the plate if placed in ambient air at a temperature of 25 °C.

1.8 Draw the electric circuit for Prob. 1.7. Show the effect of the plate conductivity on the circuit.

1.9 A wall 10 cm thick separates an ambient at 150 °C from another ambient at 20 °C. The thermal conductivity of the wall is 0.6 W/m·K and the heat transfer coefficient on both sides of the wall is 10 W/m$^2\cdot$K. Evaluate the heat loss from the warm ambient to the cold ambient.

1.10 Air at 25 °C is blown past a thick wall whose surface is at 85 °C. The coefficient of heat transfer between the air and the surface is 100 W/m$^2\cdot$K. The thermal conductivity of the wall for three different materials is 0.3, 30, 300 W/m·K. Evaluate the temperature of the wall 2 cm from the surface for each material. Neglect the effect of radiation.

1.11 Consider a 20 cm thick brick wall separating a room at $T_i = 25$ °C from the outside air at $T_0 = -5$ °C. The heat transfer coefficient on both sides of the wall is 10 W/m$^2\cdot$K. The thermal conductivity of the brick is 0.7 W/m·K. Neglect the effect of radiation.

(a) Evaluate the inner surface temperature of the wall, T_i.

(b) Now, let the outer coefficient of heat transfer be increased to 100 W/m$^2\cdot$K due to a change in the condition of air outside from stagnant to windy. Reevaluate the inner surface temperature.

(c) Assume that a human standing in the room can be approximated as a cylinder 1.8 m tall, 30 cm/in diameter. Neglecting radiation, evaluate the steady heat loss from the human being by means of convection both for (a) and (b).

(d) Reevaluate (c) by including the effect of radiation as well as that of convection.

1.12 Steam is condensing inside a pipe at 2 atm (abs). The heat transfer coefficient on the steam side is $h_i = 6{,}000$ W/m$^2\cdot$K and on the outer surface is $h_0 = 10$ W/m$^2\cdot$K. The thermal conductivity and the wall thickness of the pipe are $k = 20$ W/m·K and $\delta = 2$ mm, respectively. The pipe is suspended in a room at 25 °C. (a) Show the relative importance of the radial temperature drop inside, outside and through the pipe. Neglect the effect of curvature. (b) Evaluate the heat loss per meter length of pipe. (c) Find the change of steam quality per 10 m length of a 20 mm OD and 16 mm ID pipe for a steam velocity of 3 m/s.

1.13 Reconsider Prob. 1.9. The heat loss will now be eliminated by attaching a flat-plate heater to the cold surface of the wall. The heat transfer coefficient between the heater and the cold ambient is also 10 W/m$^2\cdot$K. Evaluate the power need in W/m^2. Sketch the temperature distribution.

1.14 Power P per unit area is generated in a horizontal flat plate. The heat loss from the lower surface of the plate is reduced to a large extent by an insulator of thickness ℓ. Find the upward and downward heat transfer. Evaluate the temperature of the plate and the lower surface temperature of the insulator. Data: $P = 100$ W/m^2, $\ell = 4$ cm, $k = 0.4$ W/m·K, $h_1 = 10$ W/m$^2\cdot$K, $h_2 = 2.5$ W/m$^2\cdot$K, $T_\infty = 25$ °C.

1.15 Two vertical black plates are separated by a vacuum space. One of the plates is in contact with water at 90 °C while the other is exposed to ambient air at 0 °C. Estimate the heat loss from the water to the ambient air. Justify your assumption on the convection effects.

1.16 A thermocouple is used to measure the temperature of the gas flow in a combustor. The thermocouple reading is 1000 °C. The walls of the combustor are at 200 °C. The heat transfer coefficient between the thermocouple and the hot gases is 200 W/m²·K. Evaluate the actual temperature of the hot gases.

1.17 Evaluate the heat loss by natural convection, forced convection, and radiation from a flat plate at a uniform temperature T_w to ambient air or water at a temperature T_∞. The temperature difference between the wall and ambient is 100 K. The heat transfer coefficients for natural and forced convection in air are 10 and 200 W/m²·K, and in water are 500 and 10,000 W/m²·K, respectively. Plot the various heat losses from the plate as a function of $T_\infty/(T_w - T_\infty)$ for $T_\infty = 0, 400, 800,$ and 1200 K. Note the effect of convection relative to radiation as a function of temperature.

1.18 In a laboratory experiment, the horizontal surface of a plastic material with an area of $0.3 \times 0.3 \text{m}^2$ is deliberately set on fire. For combustion to take place, the solid plastic has to be gasified first. Following the initial external firing, this gasification energy is supplied to the surface by the hot flame region through convection q_C and radiation q_R. Due to a relatively high temperature, the surface at 400 °C also reradiates q_{RR} back to the ambient at 30 °C. The heat of gasification of this plastic is known to be 1.60 kJ/kg and q_R is measured to be 20 kW/m². Assuming the hot gases above the burning plastic pool to be at 1,200 K, and the heat transfer coefficient to be 10 W/m²·K, and neglecting the conduction to the solid plastic, determine (a) the burning rate $\dot{m}$ of plastic in g/m². (b) the change in height of the stationary plastic block in 2 hours.

CHAPTER 2

STEADY CONDUCTION

In this chapter we shall consider steady conduction in one-dimensional geometry. Although the main objective is conduction, convection described in terms of an assumed heat transfer coefficient will be included whenever it is pertinent. This may be the case when the heat transfer is desired in terms of ambient temperatures (Section 2.2) or when heat loss normal to the direction of conduction is essential, as in the case of extended surfaces (Section 2.4). Here we continue to employ the five-step formulation but somewhat less explicitly than the way we used it in Chapter 1. Each reader should tailor the degree of elaboration of this formulation to his or her particular needs.

2.1 VARIABLE CONDUCTIVITY AND VARIABLE AREA

In the development of the one-dimensional temperature distribution in a flat plate (Section 1.8), we assumed that the thermal conductivity, k, and the cross sectional area, A, were constant. However, as mentioned in Section 1.6, conductivity usually depends on the temperature. Also, except for cartesian geometry, the area of a geometry varies in the direction of heat transfer. We wish to examine now the steady, one-dimensional conduction, including the effects of variable conductivity and variable heat transfer area.

Consider part of a shell of constant wall thickness as shown in Fig. 2.1. Let the inner and outer surface temperatures be T_1 and T_2, respectively, and the thickness of the shell be $\ell = x_2 - x_1$. Following the **five-step formulation**, we assume first a differential system (Step 1). The first law of thermodynamics for this system (Step 2) yields

$$-d\dot{Q}_K = 0 \quad \text{or} \quad \dot{Q}_K = \text{Const}. \tag{2.1}$$

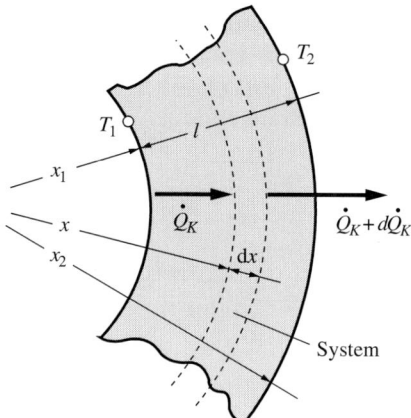

Figure 2.1 Variable area.

Fourier's law of conduction (Step 3) in light of Eq. (2.1) gives the governing equation (Step 4) as

$$\dot{Q}_K = -k(T)A(x)\frac{dT}{dx} = \text{Const.}, \tag{2.2}$$

subject to boundary conditions (Step 5),

$$T(x_1) = T_1 \quad \text{and} \quad T(x_2) = T_2. \tag{2.3}$$

After separating variables, integration of Eq. (2.2) yields, in terms of Eq. (2.3),

$$\dot{Q}_K \int_{x_1}^{x_2} \frac{dx}{A(x)} = -\int_{T_1}^{T_2} k(T)dT. \tag{2.4}$$

Introducing an **average heat transfer area**

$$\frac{1}{\bar{A}} = \frac{1}{\ell}\int_{x_1}^{x_2} \frac{dx}{A(x)}, \tag{2.5}$$

and an **average thermal conductivity**

$$\bar{k} = -\frac{1}{\Delta T}\int_{T_1}^{T_2} k(T)dT, \tag{2.6}$$

ΔT denoting $T_1 - T_2$, we may rearrange Eq. (2.4) as

$$\dot{Q}_K = \bar{k}\bar{A}\frac{\Delta T}{\ell} = \frac{\Delta T}{R_K}, \tag{2.7}$$

where

$$R_K = \frac{\ell}{\bar{k}\bar{A}} = -\left(\frac{\int_{x_1}^{x_2} dx/A(x)}{\int_{T_1}^{T_2} k(T)dT}\right)\Delta T, \tag{2.8}$$

which shows that the conductive resistance depends on both temperature and geometry. Note that, in terms of an average thermal conductivity and average heat transfer area, Eq. (2.7) is identical to Eq. (1.44) introduced for a flat plate with constant thermal conductivity. For a constant conductivity, Eq. (2.8) reduces to

$$R_K = \frac{1}{k} \int_{x_1}^{x_2} \frac{dx}{A(x)}, \quad (2.9)$$

and, for a constant heat transfer area (cartesian geometry) reduces to

$$R_K = -\frac{\ell}{A} \frac{\Delta T}{\int_{T_1}^{T_2} k(T)dT}. \quad (2.10)$$

We now proceed to a successive study of the effects of variable conductivity and heat transfer area.

2.1.1 Variable Conductivity

Let the temperature variation in a problem be large enough so that the assumption of a uniform conductivity is no longer valid. A first approximation may be

$$k = k_0(1 + \beta T), \quad (2.11)$$

where k_0 denotes the value of thermal conductivity at $T = 0$ and β is the temperature coefficient of conductivity. Inserting Eq. (2.11) into Eq. (2.6),

$$\bar{k} = \frac{-k_0}{T_1 - T_2} \int_{T_1}^{T_2} (1 + \beta T)dT,$$

which yields

$$\bar{k} = \frac{k_0}{T_2 - T_1} \left[(T_2 - T_1) + \frac{1}{2}\beta \left(T_2^2 - T_1^2 \right) \right]$$

or

$$\bar{k} = k_0(1 + \beta \bar{T}), \quad (2.12)$$

where $\bar{T} = \frac{1}{2}(T_1 + T_2)$ is the arithmetic mean of T_1 and T_2. Equation (2.11) is a good approximation for the conductivity of alloys at sufficiently low and high temperatures (recall Fig. 1.11). For small intervals of temperature, it may also approximate the conductivity of some other solids.

A more involved approximation,

$$k = k_0 \left[1 + \gamma \left(\frac{T_0}{T} \right) \right], \quad (2.13)$$

adequately describes the conductivity of metals at sufficiently high temperatures (again recall Fig. 1.11). Then, in terms of Eq. (2.6),

$$\bar{k} = \frac{-k_0}{(T_1 - T_2)} \int_{T_1}^{T_2} \left[1 + \gamma \left(\frac{T_0}{T}\right)\right] dT,$$

which yields

$$\bar{k} = \frac{k_0}{(T_2 - T_1)} \left[(T_2 - T_1) + \gamma T_0 \ln\left(\frac{T_2}{T_1}\right)\right],$$

or

$$\bar{k} = k_0 \left[1 + \gamma \left(\frac{T_0}{\bar{T}}\right)\right], \qquad (2.14)$$

where $\bar{T} = (T_1 - T_2)/\ln(T_1/T_2)$ is the logarithmic average of T_1 and T_2.

2.1.2 Variable Area

Radial transfer of heat through a hollow cylinder and a hollow sphere are two important applications of variable heat transfer area and are considered here.

For a hollow **cylinder** of length L, rearranging Eq. (2.5) according to Fig. 2.2, we have

$$\frac{1}{\bar{A}} = \frac{1}{r_2 - r_1} \int_{r_1}^{r_2} \frac{dr}{2\pi r L}, \qquad (2.15)$$

where $\ell = r_2 - r_1$ is the thickness and L is the axial length of the cylinder. Integration of Eq. (2.15) yields

$$\bar{A} = \frac{2\pi(r_2 - r_1)L}{\ln(r_2/r_1)}$$

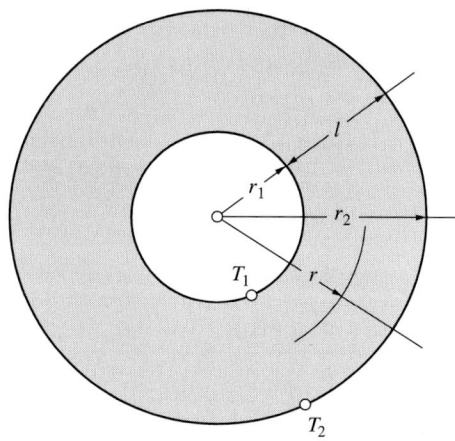

Figure 2.2 A hollow cylinder or a hollow sphere.

which may be rearranged, after multiplying the numerator and denominator of the logarithmic term by $2\pi L$, and noting $A_1 = 2\pi r_1 L$ and $A_2 = 2\pi r_2 L$, as

$$\bar{A} = \frac{A_2 - A_1}{\ln(A_2/A_1)}, \quad \text{for a \textbf{cylinder}}. \tag{2.16}$$

Equation (2.16) denotes the **logarithmic-mean average** of A_1 and A_2. For constant thermal conductivity, noting $\ell = r_2 - r_1 = (A_2 - A_1)/2\pi L$, we have from $R_K = \ell/k\bar{A}$ for a hollow cylinder

$$R_K = \frac{\ln(A_2/A_1)}{2\pi kL} = \frac{\ln(r_2/r_1)}{2\pi kL}. \tag{2.17}$$

For a hollow **sphere**, rearranging Eq. (2.5) according to Fig. 2.2, we have

$$\frac{1}{\bar{A}} = \frac{1}{r_2 - r_1} \int_{r_1}^{r_2} \frac{dr}{4\pi r^2}, \tag{2.18}$$

where $\ell = r_2 - r_1$ is the thickness of the sphere. Integration of Eq. (2.18) gives

$$\bar{A} = 4\pi r_1 r_2,$$

which may be rearranged, in terms of $A_1 = 4\pi r_1^2$ and $A_2 = 4\pi r_2^2$, as

$$\bar{A} = (A_1 A_2)^{1/2}, \quad \text{for a \textbf{sphere}}. \tag{2.19}$$

Equation (2.19) denotes the **geometric-mean average** of A_1 and A_2. For constant thermal conductivity, $R_K = \ell/k\bar{A}$ for a hollow sphere becomes

$$R_K = \frac{\ell}{k(A_1 A_2)^{1/2}} = \frac{r_2 - r_1}{4\pi k r_1 r_2}. \tag{2.20}$$

For convenience, the average heat transfer area and the conductive resistance for the foregoing three configurations are summarized in Table 2.1.

Table 2.1 Average heat transfer area and conductive resistance.

$\dot{Q}_K = \frac{\Delta T}{R_K},$	$R_K = \frac{\ell}{k\bar{A}} \left[\frac{K}{W}\right]$	
Geometry	$\bar{A}$	R_K
Flat Plate	A	$\frac{\ell}{kA}$
Cylinder	$\frac{A_2 - A_1}{\ln\frac{A_2}{A_1}}$	$\frac{\ln(r_2/r_1)}{2\pi kL}$
Sphere	$(A_1 A_2)^{1/2}$	$\frac{r_2 - r_1}{4\pi k r_1 r_2}$

It can be shown by appropriate series expansions in terms of powers of $\Delta A/A_1 = (A_2 - A_1)/A_1$ that Eqs. (2.16) and (2.19) both approach the **arithmetic-mean average**,

$$\bar{A} = \frac{1}{2}(A_1 + A_2), \qquad (2.21)$$

as $A_2 \to A_1$ (and $\Delta A/A_1 \to 0$). When $A_2/A_1 \leq 2$, Eq. (2.21) can be used to approximate Eq. (2.16) for a cylinder within an error of $\leq 4\%$ and Eq. (2.19) for a sphere within an error of $\leq 6\%$. We now proceed to one-dimensional conduction through composite structures and to the concept of critical radius.

2.2 COMPOSITE STRUCTURES. CRITICAL RADIUS

In many practical problems heat transfer takes place through a medium composed of several parallel layers, each with a different thickness and a different conductivity. For example, a typical brick house wall is usually composed of a layer of dry wall (or plaster), a layer of insulation, and a layer of bricks. Each one of these parallel layers has a different thickness and a different thermal conductivity. Heat is transferred by conduction through each layer of material and by convection to/from the inside/outside ambient air. If one wishes to determine the heat transfer from the inside of the house to the outside through the walls, typical information that might be provided are the dimensions and properties of each layer, the inside and outside air temperatures, and the convective heat transfer coefficients. Usually, the air temperatures are given because they are more meaningful and easier to measure than surface temperatures. This type of situation is not unique to brick walls. In fact, in many practical problems, heat transfer takes place through a medium of several parallel layers, each with a different thickness and a different thermal conductivity. These are all composite-structure problems, and the specific examples of heat transfer through layers of slabs, cylinders, and spheres are the subject matter of this section.

2.2.1 Composite Slabs

Consider a composite wall made of three parallel slabs of cross-sectional area A (Fig. 2.3). The thickness and thermal conductivity of these slabs are ℓ_1, ℓ_2, ℓ_3 and k_1, k_2, k_3, respectively. Heat is transferred from a hot fluid at temperature T_i through this composite wall to a cold fluid at temperature T_0. Coefficients of heat transfer on the hot and cold sides are h_i and h_0, respectively.

Electric analogy or the five-steps formulation elaborated in Ex. 1.7 readily lead to

$$\dot{Q} = \frac{T_i - T_0}{R_i + \sum_{K=1}^{3} R_K + R_0}, \qquad (2.22)$$

where $R_i = 1/h_i A$, $R_0 = 1/h_0 A$, $R_1 = \ell_1/k_1 A$, $R_2 = \ell_2/k_2 A$, and $R_3 = \ell_3/k_3 A$ are the convective and conductive resistances.

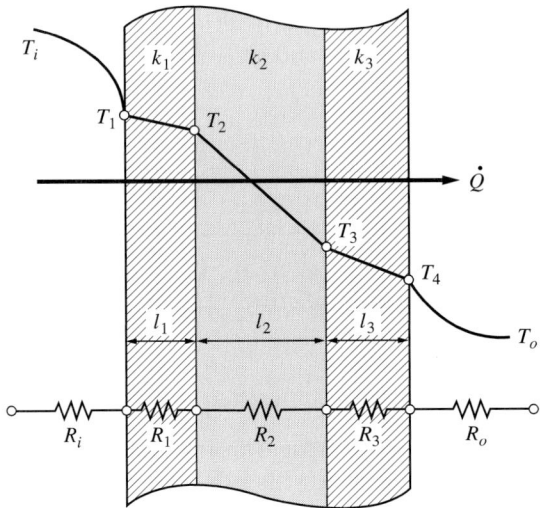

Figure 2.3 Composite wall.

Actually, there exists a thermal resistance, the so-called **contact resistance**, through each interface of a composite structure. This resistance depends on the contact pressure, the roughness of two surfaces, and the fluid in which these surfaces are pressed together. For more information on the subject the reader is referred to Reference 6 (and the references cited therein).

EXAMPLE 2.1

A glass plate is to be bonded to a plastic plate by melting a layer of powdered glue of negligible thickness between the plates. Assume the glass to be transparent and the plastic to be opaque to the radiant energy source q''. The melting temperature of the glue is T_m. The thickness and thermal conductivity of the glass and plastic are ℓ_1, ℓ_2 and k_1, k_2, respectively. The upward and downward heat transfer coefficients are h_1 and h_2. We wish to determine the radiant heat flux needed for this process.

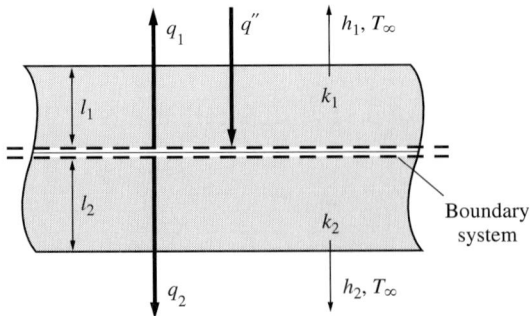

Figure 2.4 First law for interface.

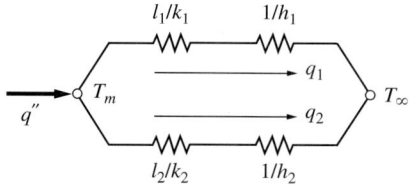

Figure 2.5 Analog for Ex. 2–1.

For a boundary system (Step 1) at the interface of two plates (Fig. 2.4), the first law of thermodynamics (Step 2) yields

$$0 = q'' - q_1 - q_2. \tag{2.23}$$

With Fourier's conduction and Newton's cooling laws (Step 3), expressing q_1 and q_2 now in terms of total (conductive plus convective) resistances over the temperature drop $T_m - T_\infty$ on both sides, we have

$$q_1 = \frac{T_m - T_\infty}{\ell_1/k_1 + 1/h_1}, \quad q_2 = \frac{T_m - T_\infty}{\ell_2/k_2 + 1/h_2}. \tag{2.24}$$

Insertion of Eq. (2.24) into Eq. (2.23) gives the governing equation (Step 4):

$$q'' = \left(\frac{1}{\ell_1/k_1 + 1/h_1} + \frac{1}{\ell_2/k_2 + 1/h_2} \right)(T_m - T_\infty), \tag{2.25}$$

which is the power need (per unit area) for the bonding process. The electric circuit analogous to the thermal problem is shown in Fig. 2.5.

In terms of following data,[1]

$$\begin{aligned}
&T_\infty = 25\,°\mathrm{C}, & &T_m = 175\,°\mathrm{C}, \\
&\ell_1 = 0.5\ \mathrm{cm}, & &\ell_2 = 1\ \mathrm{cm}, \\
&k_1 = 1.4\ \mathrm{W/m \cdot K}, & &k_2 = 0.4\ \mathrm{W/m \cdot K}, \\
&h_1 = 10\ \mathrm{W/m^2 \cdot K}, & &h_2 = 4\ \mathrm{W/m^2 \cdot K},
\end{aligned}$$

we obtain

$$\frac{\ell_1}{k_1} + \frac{1}{h_1} = \frac{0.005\ \mathrm{m}}{1.4\ \mathrm{W/m \cdot K}} + \frac{1}{10\ \mathrm{W/m^2 \cdot K}} = 0.104\ \mathrm{m^2 \cdot K/W}$$

$$\frac{\ell_2}{k_2} + \frac{1}{h_2} = \frac{0.01\ \mathrm{m}}{0.4\ \mathrm{W/m \cdot K}} + \frac{1}{4\ \mathrm{W/m^2 \cdot K}} = 0.275\ \mathrm{m^2 \cdot K/W}$$

and, inserting these total resistances into Eq. (2.3),

$$q'' = \left(\frac{1}{0.104} + \frac{1}{0.275} \right) \mathrm{W/\,m^2 \cdot K} \times (175 - 25)\ \mathrm{K},$$

---

[1] The FORTRAN program EX2–1.F is listed in the Appendix.

or,

$$q'' = (9.66 + 3.64) \times 150 \text{ W/m}^2,$$

or,

$$q'' \cong 2 \text{ kW/m}^2.$$

◆

In practice, a series-parallel combination of conduction paths is also encountered quite frequently. A typical example is a wall constructed from concrete blocks (Fig. 2.6). Note that the problem is actually two-dimensional but will be approximated here as one-dimensional. This is a simplified model useful for approximate heat-loss calculations in the building industry.

In terms of the electric analogy, the total (convective + conductive) resistance is

$$R = R_i + \cfrac{1}{\cfrac{1}{R_1} + \cfrac{1}{R_2}} + R_3 + R_0, \qquad (2.26)$$

where $R_i = 1/h_i A$, $R_1 = \ell_1/k_1 A_1$, $R_2 = \ell_1/k_2 A_2$, $R_3 = \ell_2/k_3(A_2 + A_3)$, $R_0 = 1/h_0 A$, $A_1 = b_1 L$, $A_2 = b_2 L$, and L is the thickness.

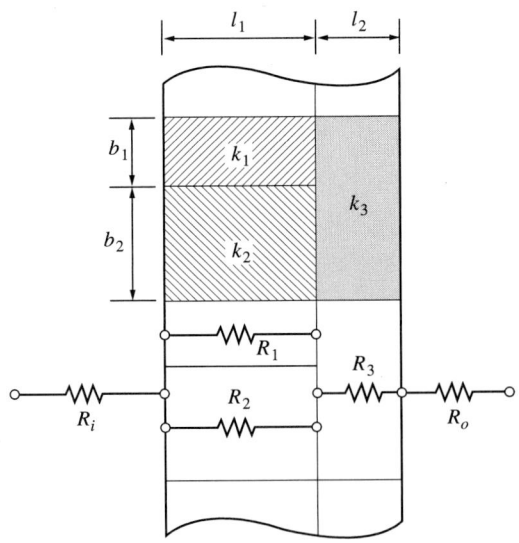

Figure 2.6 Composite (series and parallel) wall.

2.2.2 Composite Cylinders

Consider radial heat transfer through two concentric cylinders of different thickness and different conductivity (Fig. 2.7). A typical example is the heat loss from an insulated pipe. Let the inside and outside fluid temperatures and the inside and outside heat transfer coefficients be T_i, T_0 and h_i, h_0, respectively.

The electric analogy gives, in terms of resistances,

$$\dot{Q} = \frac{T_i - T_0}{R_i + R_1 + R_2 + R_0}, \tag{2.27}$$

or

$$\dot{Q} = \frac{T_i - T_0}{\dfrac{1}{h_i A_1} + \dfrac{\ell_{12}}{k_1 \bar{A}_{12}} + \dfrac{\ell_{23}}{k_2 \bar{A}_{23}} + \dfrac{1}{h_0 A_3}}, \tag{2.28}$$

where

$$R_i = \frac{1}{h_i A_1} = \frac{1}{2\pi r_1 L h_i}, \quad R_1 = \frac{\ell_{12}}{k_1 \bar{A}_{12}} = \frac{\ln(r_2/r_1)}{2\pi L k_1},$$

$$R_2 = \frac{\ell_{23}}{k_2 \bar{A}_{23}} = \frac{\ln(r_3/r_2)}{2\pi L k_2}, \quad R_0 = \frac{1}{h_0 A_3} = \frac{1}{2\pi r_3 L h_0}.$$

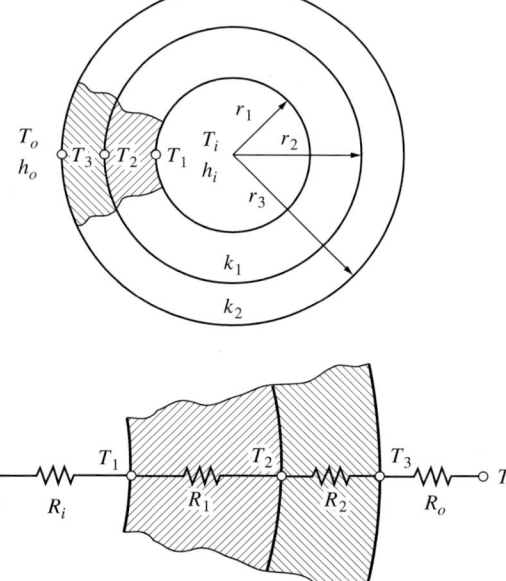

Figure 2.7 Composite cylinder or sphere.

Then, explicitly,

$$\dot{Q} = \frac{T_i - T_0}{\frac{1}{2\pi r_1 L h_i} + \frac{\ln(r_2/r_1)}{2\pi L k_1} + \frac{\ln(r_3/r_2)}{2\pi L k_2} + \frac{1}{2\pi r_3 L h_o}}, \quad (2.29)$$

and, in terms of an **overall heat transfer coefficient** based on the outer heat transfer area,

$$\dot{Q} = U A_3 (T_i - T_0), \quad A_3 = 2\pi r_3 L, \quad (2.30)$$

where

$$\frac{1}{U} = \frac{r_3/r_1}{h_i} + \frac{r_3 \ln(r_2/r_1)}{k_1} + \frac{r_3 \ln(r_3/r_2)}{k_2} + \frac{1}{h_o}. \quad (2.31)$$

EXAMPLE 2.2[2]

Steam at 2 atm (saturation temperature $T_s \cong 120\,°C$) is condensing in a 1-inch ($r_1 = 1.33$ cm and $r_2 = 1.67$ cm) stainless steel pipe. The inside and outside heat transfer coefficients are $h_i = 100{,}000$ W/m^2·K and $h_o = 10$ W/m^2·K, respectively. The thermal conductivity of the pipe is $k = 15$ W/m·K. The pipe is suspended in a room at 20 °C. **(a)** We wish to evaluate the radial temperature drop inside, outside, and across the thickness of the pipe walls. **(b)** The pipe is insulated with a fiberglass ($k = 0.04$ W/m·K) layer of thickness 15 mm. Evaluate the reduction in the heat loss per unit length of pipe.

(a) For the bare pipe, in terms of convective and conductive resistances,

$$\dot{Q} = \frac{T_s - T_\infty}{R_i + R_1 + R_0},$$

where

$$R_i = \frac{1}{h_i 2\pi r_1 L} = \frac{1}{100{,}000 \text{W/m}^2\cdot\text{K} \times 2\pi \times 1.33 \times 10^{-2}\text{m} \times L} = \frac{1.2 \times 10^{-4}}{L}\text{m·K/W}$$

and, in view of

$$\frac{A_0}{A_i} = \frac{r_2}{r_1} \cong 1.25 < 2,$$

$$R_1 \cong \frac{\ell}{k\bar{A}} = \frac{r_2 - r_1}{k\pi(r_2 + r_1)L} = \frac{(1.67 - 1.33) \times 10^{-2}\text{m}}{15\text{W/m·K} \times \pi(1.67 + 1.33) \times 10^{-2}\text{m} \times L}$$

$$= \frac{2.4 \times 10^{-3}}{L}\text{m·K/W}$$

and

$$R_0 = \frac{1}{h_0 2\pi r_2 L} = \frac{1}{10\text{W/m}^2\cdot\text{K} \times 2\pi \times 1.67 \times 10^{-2}\text{m} \times L} = \frac{0.953}{L}\text{m·K/W}.$$

---

[2] The FORTRAN program EX2–2.F is listed in the appendix of this chapter.

The heat loss per unit length of pipe is then

$$\frac{\dot{Q}}{L} = \frac{(120 - 20)\text{K}}{(1.2 \times 10^{-4} + 2.4 \times 10^{-3} + 0.953)\text{m}\cdot\text{K/W}} \cong 105\text{W/m}.$$

Note that the conductive and inside convective resistances are negligibly small compared with the outside convective resistance. Thus the temperature drop in the steam and that across the pipe walls,

$$T_s - T_{wi} = \dot{Q}R_i = 105\text{W/m} \times 1.2 \times 10^{-4}\text{m}\cdot\text{K/W} \cong 0.01\text{K},$$

$$T_{wi} - T_{wo} = \dot{Q}R_1 = 105\text{W/m} \times 2.4 \times 10^{-3}\text{m}\cdot\text{K/W} \cong 0.3\text{K},$$

are negligibly small, and the entire temperature drop between the steam and outside is confined to that between the pipe and outside.

(b) For the insulated pipe, neglecting R_i and R_1 (of pipe walls),

$$\dot{Q} = \frac{T_s - T_\infty}{R_2 + R_0},$$

R_2 being the conductive resistance of insulation. The radius of insulation is

$$r_3 = r_2 + 1.5\text{cm} = 1.67 + 1.50 = 3.17\text{cm}.$$

In view of

$$\frac{A_3}{A_2} = \frac{r_3}{r_2} = \frac{3.17\text{cm}}{1.67\text{cm}} = 1.9 < 2,$$

$$R_2 \cong \frac{\ell}{k\bar{A}_{23}} = \frac{r_3 - r_2}{k\pi(r_3 + r_2)L} = \frac{(3.17 - 1.67) \times 10^{-2}\text{m}}{0.04\text{W/ m}\cdot\text{K} \times \pi \times (3.17 + 1.67) \times 10^{-2}\text{m} \times L}$$

$$= \frac{2.47}{L}\text{m}\cdot\text{K/W}.$$

Also,

$$R_0 = \frac{1}{h_0 2\pi r_3 L} = \frac{1}{10\text{W/m}^2\cdot\text{K} \times 2\pi \times 3.17 \times 10^{-2}\text{m} \times L} \cong \frac{0.50}{L}\text{m}\cdot\text{K/W}.$$

Then

$$\frac{\dot{Q}}{L} = \frac{(120 - 20)\text{K}}{(2.47 + 0.50)\text{m}\cdot\text{K/W}} \cong 33.7\text{W/m}.$$

The insulation turns out to be reducing the heat loss by 68% approximately.

Inspection of Eq. (2.29) reveals that the conductive resistance of insulation increases logarithmically and the outside convective resistance decreases hyperbolically as the thickness of the insulation increases. This suggests the possibility of an extremum for the sum of these resistances, which we explore next. ◆

2.2.3 Critical Thickness for Cylindric Insulation

Consider now the heat loss from an insulated pipe as a function of the insulation thickness. Note from the preceding example that the conductive resistivity of the pipe walls, compared with the conductive resistivity of the insulation, and the inner convective resistivity, compared with the outer convective resistivity, can be neglected. Consequently, the inner surface of the insulation assumes approximately the temperature of the inner fluid, T_i.

For notational convenience let the thermal conductivity and inner and outer radii of the insulation be k, r_i, r, respectively, and the outer heat transfer coefficient be h. Equation (2.29) may be then rearranged as

$$\frac{\dot{Q}}{2\pi k L (T_i - T_0)} = \frac{1}{\ln(r/r_i) + \left(\dfrac{k}{hr_i}\right) / \left(\dfrac{r}{r_i}\right)}. \tag{2.32}$$

For an extremum of Eq. (2.32),

$$\frac{d\dot{Q}}{dr} = 0,$$

which gives

$$\frac{-\left[1 - \dfrac{1}{(r/r_i)\,Bi}\right]}{(r/r_i)\left[\ln(r/r_i) + \dfrac{1}{(r/r_i)\,Bi}\right]^2} = 0$$

leading, because of the positive denominator, to

$$\boxed{\left(\frac{r}{r_i}\right)_C = \frac{1}{Bi}} \quad \text{or} \quad \boxed{r_C = \frac{k}{h}}, \tag{2.33}$$

where

$$\boxed{Bi = \frac{hr_i}{k}} \tag{2.34}$$

introduces the definition of **Biot number** in cylindrical geometry. The physical significance of this number will be elaborated in Section 2.3.1.

The heat loss corresponding to the **critical radius**, inserting Eq. (2.33) into Eq. (2.32), is found to be

$$\frac{\dot{Q}}{2\pi k L (T_i - T_0)} = \frac{1}{1 + \ln\left(\dfrac{k}{hr_i}\right)}. \tag{2.35}$$

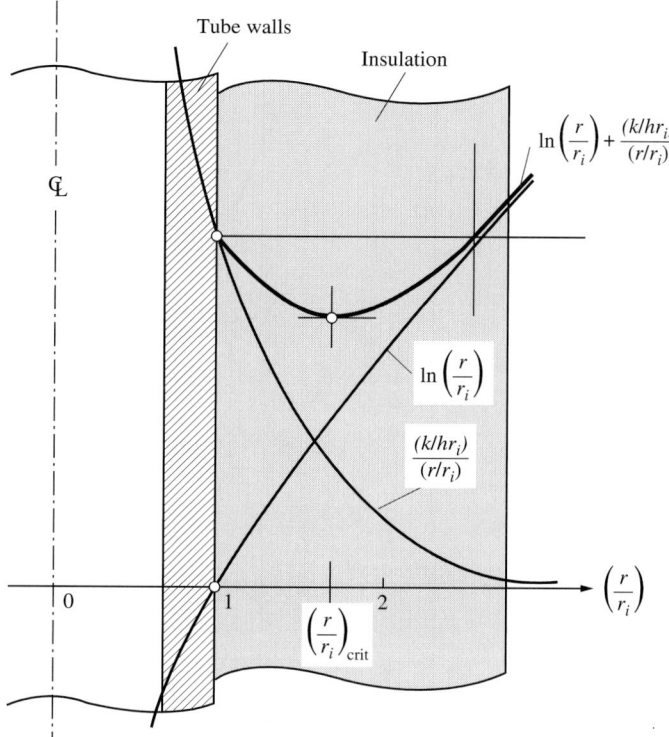

Figure 2.8 Critical radius.

It can also be shown that

$$\left.\frac{d^2 \dot{Q}}{dr^2}\right|_{r_C = k/h} < 0,$$

indicating that the heat loss given by Eq. (2.35) is a maximum. That is, the heat loss from a pipe may be increased by insulating the pipe! This fact may be clarified further by a sketch (Fig. 2.8) which shows the denominator of Eq. (2.32). The sum of the two terms comprising this denominator assumes a minimum value at the critical radius. Figure (2.8) also shows that for $(r/r_i)_s$, the heat transfer with insulation is the same as that without insulation. Thus, if the goal of insulating a pipe is to reduce the amount of heat transfer, an insulation-layer thickness of $(r/r_i) > (r/r_i)_s$ must be used. In Fig. 2.9 the heat loss from an insulated pipe is plotted against the insulation thickness for various values of $1/Bi = k/hr_i$. The critical radius finds also an electrical application. A thickness of insulation in the neighborhood of the critical radius provides more cooling for an insulated wire than for a bare wire. The lower the temperature of a wire, the lower its electrical resistance[3] and, in turn, the lower the Joulean heat loss of electrical energy from it.

---

[3] Note an approximate relation such as $R = R_0(1 + \beta T)$.

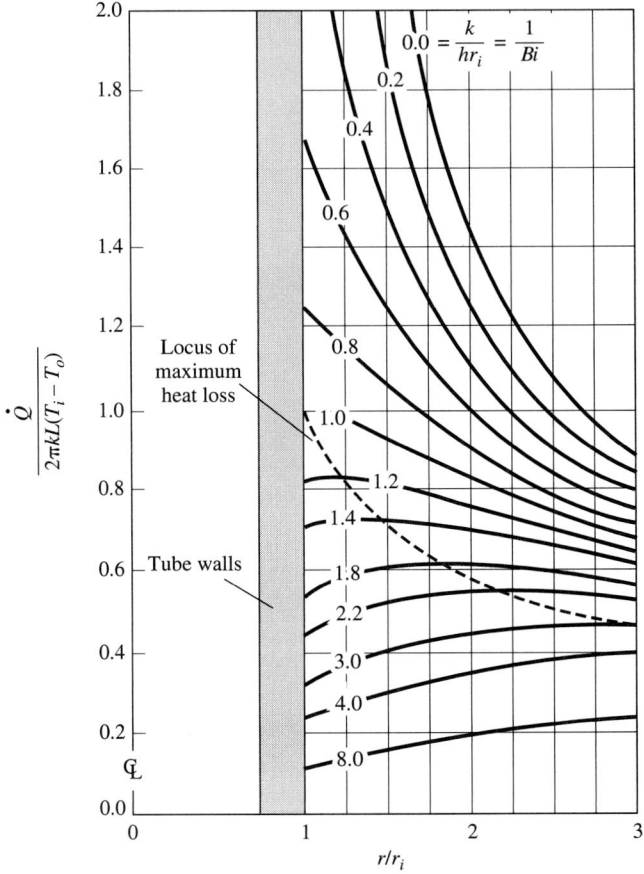

Figure 2.9 Heat loss against insulation thickness.

Example 2.3

Reconsider Ex. 2.2 for **(1)** fiberglass ($k = 0.04$ W/m·K), **(2)** plaster (gypsum) ($k \cong 0.20$ W/m·K). For each material we wish to determine the thickness of insulation for which the heat loss from the pipe is **(a)** maximum, **(b)** equal to that from the bare-pipe analysis.

(a) We learned with Eq. (2.33) that $r_C = k/h$ yields the maximum heat loss. For fiberglass insulation,

$$r_C = \frac{0.04 \text{W/m·K}}{10 \text{W/m}^2\text{·K}} = 0.004 \text{ m} = 0.4 \text{ cm}$$

and

$$\frac{r_C}{r_i} = \frac{0.4 \text{ cm}}{1.67 \text{ cm}} = 0.24 < 1.$$

The minimum of total resistance is at a radius less than the outside radius of pipe! So there is no physically realizable critical radius. The total resistance increases with increasing insulation and reduces the heat loss.

On the other hand, for plaster (gypsum) insulation,

$$r_C = \frac{k}{h} = \frac{0.20 \text{ W/m·K}}{10 \text{ W/m}^2\text{·K}} = 0.02 \text{ m} = 2 \text{ cm},$$

and, because

$$\frac{r_C}{r_i} = \frac{2 \text{ cm}}{1.67 \text{ cm}} \cong 1.2 = \frac{1}{Bi},$$

$$r_C - r_i = 2 - 1.67 = 0.33 \text{ cm}.$$

Thus, a 0.33-cm-thick insulation results in the maximum heat loss from the pipe.

(b) For the insulation thickness r yielding the same heat loss as the bare pipe, equating the bare and insulated pipe resistances,

$$\frac{1}{h 2\pi r_i L} = \frac{1}{h 2\pi r L} + \frac{\ln(r/r_i)}{k 2\pi L},$$

or

$$\frac{1}{Bi} = \frac{\frac{1}{Bi}}{r/r_i} + \ln(r/r_i),$$

or, in terms of $Bi = 1/1.2$,

$$1.2 = \frac{1.2}{(r/r_i)} + \ln(r/r_i),$$

After some trial and error,

$$r/r_i \cong 1.45,$$

which, in view of $r_i = 1.67$ cm, gives

$$r = 2.42 \text{cm}$$

or

$$r - r_i = 2.42 - 1.67 = 0.75 \text{cm},$$

as the thickness of plaster insulation leading to a total resistance equal to the convective resistance of the bare pipe. ◆

Example 2.4

A solid shaft rotates steadily with angular velocity ω in a sleeve (Fig. 2.10a). The pressure and the coefficient of dry friction between the shaft and sleeve are p and μ, respectively. We wish to determine the steady temperature of the interface and find the specific value of r when this temperature has the lowest value.

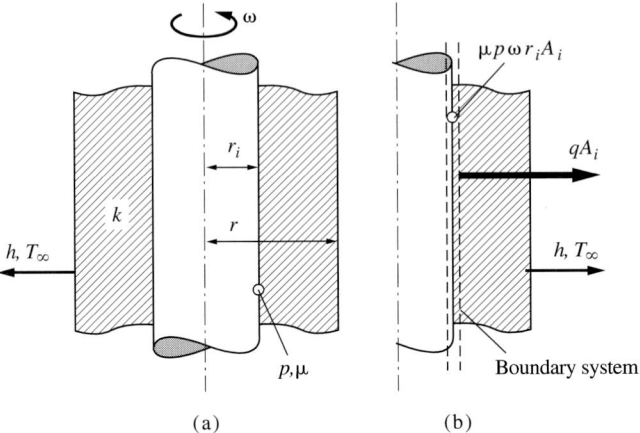

Figure 2.10 (a) Rotating shaft (b) the first law.

Here, some elaboration on friction power may be useful. First, recall the definitions of work and power:

$$\text{work} = \text{force} \times \text{displacement}$$

and its rate,

$$\text{power} = \text{force} \times \text{velocity}.$$

A normal force P acting on an interface between two solids (Fig. 2.11) creates a tangential friction force μP, μ being the (dynamic) coefficient of friction. Then, the friction power is

$$\text{Power} = \mu P V,$$

V being the velocity. If the motion is rotational rather than being rectilinear,

$$V = \omega r,$$

ω being the angular velocity and r the radius of motion,

$$\text{Power} = \mu P \omega r.$$

More specifically, in terms of interface pressure $p = P/A$, A being the peripheral area of the cylindrical surface of longitudinal length L,

$$\text{Power} = \mu p \omega r A,$$

where $A = 2\pi r L$. This power dissipates into heat within the interface.

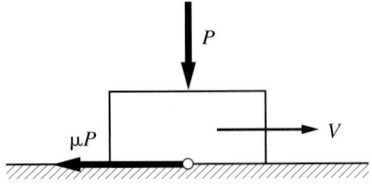

Figure 2.11 Friction power.

Consider a boundary system (Step 1) at the interface between the shaft and the sleeve (Fig. 2.10b). Under steady conditions, the shaft temperature is uniform (why?) and all energy generated at the interface is transferred to the ambient. The first law of thermodynamics (Step 2) then yields

$$0 = -qA_i + \mu p\omega r_i A_i, \qquad (2.36)$$

$A_i = 2\pi r_i L$ being the interface area, and the second term denoting the friction power dissipated into heat. With Fourier's conduction and Newton's cooling laws (Step 3), expressing the outward q in terms of total (conductive plus convective) resistances over the temperature drop $T_i - T_\infty$,

$$qA_i = \frac{T_i - T_\infty}{\ell/k\bar{A} + 1/hA}, \qquad (2.37)$$

where $\ell = r - r_i$, $\bar{A} = 2\pi(r - r_i)L/\ln(r/r_i)$ is the mean area per unit length, and $A = 2\pi r L$ is the outer surface area of sleeve per unit length. Insertion of Eq. (2.37) into Eq. (2.36) gives the governing equation (Step 4),

$$2\pi \mu p \omega r_i^2 L = \frac{T_i - T_\infty}{\dfrac{\ln(r/r_i)}{2\pi kL} + \dfrac{1}{2\pi hrL}}, \qquad (2.38)$$

or the interface temperature relative to ambient,

$$T_i - T_\infty = \mu p \omega r_i^2 \left[\frac{\ln(r/r_i)}{k} + \frac{1}{hr} \right]. \qquad (2.39)$$

The total resistance in brackets assumes a minimum value for $r_C = k/h$. This radius, coupled with Eq. (2.39), gives the lowest interface temperature

$$T_i - T_\infty = \frac{\mu p \omega r_i^2}{k}(1 - \ln Bi), \qquad (2.40)$$

where $Bi = hr_i/k$. ◆

2.2.4 Composite Spheres. Critical Thickness for Spherical Insulation

Stress considerations make the spherical geometry most convenient for the construction of calorimetric bombs, chemical containers, etc. For an insulated spherical shell, the electric analogy gives, in terms of Fig. 2.7 and Eq. (2.30),

$$\dot{Q} = UA_3(T_i - T_0),$$

or, explicitly,

$$\dot{Q} = \frac{T_i - T_0}{\dfrac{1}{4\pi r_1^2 h_i} + \dfrac{r_2 - r_1}{4\pi r_1 r_2 k_1} + \dfrac{r_3 - r_2}{4\pi r_2 r_3 k_2} + \dfrac{1}{4\pi r_3^2 h_0}}, \qquad (2.41)$$

58 Chap. 2 Steady Conduction

where $A_3 = 4\pi r_3^2$ and

$$\frac{1}{U} = \frac{(r_3/r_1)^2}{h_i} + \frac{r_3^2}{k_1}\left(\frac{1}{r_1} - \frac{1}{r_2}\right) + \frac{r_3^2}{k_2}\left(\frac{1}{r_2} - \frac{1}{r_3}\right) + \frac{1}{h_0}. \tag{2.42}$$

Also, the critical radius for spherical insulation is

$$\boxed{(r/r_i)_C = 2\left(\frac{k}{hr_i}\right) = \frac{2}{Bi}} \quad \text{or} \quad \boxed{r_C = 2\left(\frac{k}{h}\right)}. \tag{2.43}$$

2.3 ENERGY GENERATION (HEAT SOURCE)

Generation of internal energy finds many important applications in engineering. Examples are electric heaters, nuclear reactors, exothermic chemical reactions, etc. In this section we shall consider variable as well as uniform energy generation in flat plates, cylinders, and spheres. Also, the conductivity will be assumed variable in some of these considerations.

2.3.1 Flat Plate (Key Problem)

Let the rate of energy per unit volume $u'''(x)$ be generated in a flat plate and let the thickness and the thermal conductivity of the plate be 2ℓ and $k(T)$, respectively. Under steady conditions, the total energy generated in the plate is transferred, with a heat transfer coefficient h, to an ambient at temperature T_∞. This plate could be one of the fuel plates of a nuclear reactor core or one of the elements of an electric heater.[4]

Following the **five steps of formulation**, first we consider the differential system (Step 1) shown in Fig. 2.12(a). The first law of thermodynamics (Step 2), Eq. (1.16) interpreted in terms of Fig. 2.12(b), yields

$$0 = +q_x A - \left(q_x + \frac{dq_x}{dx}dx\right)A + u''' A dx,$$

or,

$$0 = -\frac{dq_x}{dx} + u'''(x), \tag{2.44}$$

where $u''' A dx = -\dot{W}_e$ can be identified as a power resulting from the electric **work done on the system.** An alternative form of Eq. (2.44) is

$$-u'''(x) = -\frac{dq_x}{dx}, \tag{2.45}$$

---

[4] For electrical applications, $I^2 R \equiv I^2 R_0(1 + \beta T)$ and the energy generation is more conveniently represented by $u'''(T)$.

Sec. 2.3 Energy Generation (Heat Source) 59

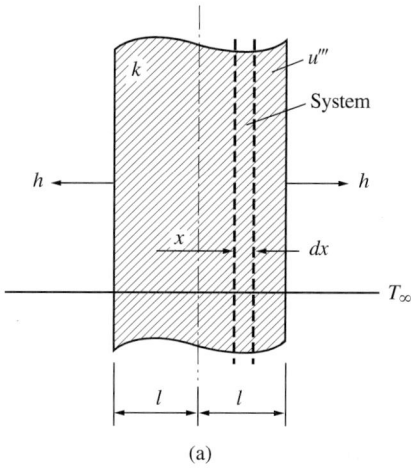

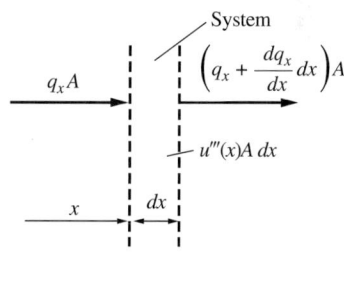

Figure 2.12 (a) Flat plate with energy generation, (b) first law for the system of flat plate.

where $-u''' A dx = dU_{nucl}/dt$ or dU_{chem}/dt can be identified as a **decrease of nuclear or chemical internal energy within the system**. Fourier's law of conduction (Step 3),

$$q_x = -k \frac{dT}{dx}, \qquad (2.46)$$

introduced into Eqs. (2.44)-(2.45) gives the governing equation (Step 4),

$$\frac{d}{dx}\left(k\frac{dT}{dx}\right) + u''' = 0, \qquad (2.47)$$

which, for a constant k, may be rearranged as

$$\frac{d^2 T}{dx^2} + \frac{u'''}{k} = 0. \qquad (2.48)$$

In the last step of our formulation, we need the origin of the coordinate axis. Because of the geometric as well as the thermal symmetry of the problem, this origin is assumed to be on the midplane. Then, the first boundary condition is (Step 5)

$$\frac{dT(0)}{dx} = 0, \qquad (2.49)$$

which is a result of the foregoing symmetries. A plane of symmetry is equivalent to an insulated surface. The second boundary condition is the result of the first four steps of formulation applied to a **boundary system** shown in Fig. 2.13 [recall Fig. 1.1 and Eq. (1.33)] as follows. For the **boundary system** of Fig. 2.13 (Step 1), consider the first law (Step 2),

$$+q_{x=\ell} - q_C = 0.$$

Insert Fourier's and Newton's laws (Step 3),

$$q_x = -k\frac{dT}{dx}, \quad q_C = h(T - T_\infty),$$

into Step 2 to obtain (Step 4) the second boundary condition as

$$-k\frac{dT(\ell)}{dx} = h[T(\ell) - T_\infty]. \tag{2.50}$$

Equation (2.47) or (2.48) together with Eqs. (2.49) and (2.50) complete the formulation of our problem. Next, we proceed to its solution.

For a distributed energy generation and a variable thermal conductivity, multiplying Eq. (2.47) by dx and integrating the result gives

$$k\frac{dT}{dx} = -\int u''' dx + C_1^*, \tag{2.51}$$

and repeating the integration process once more,

$$\int k(T) dT = -\int \left(\int u'''(x) dx\right) dx + C_1^* x + C_2^*. \tag{2.52}$$

Thus, the solution of a problem with $k(T)$ and $u'''(x)$ is reduced to a straightforward integration process.

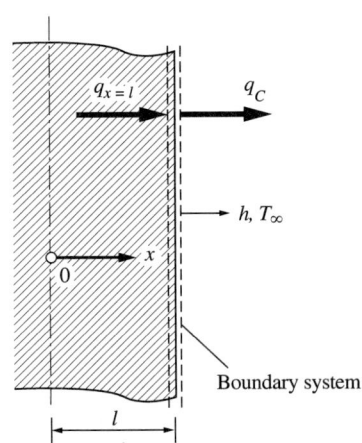

Figure 2.13 Boundary condition.

For a uniform energy generation and a constant thermal conductivity, we get[5] from Eq. (2.52), or from twice integration of Eq. (2.48),

$$T = -\frac{u''' x^2}{2k} + C_1 x + C_2. \qquad (2.53)$$

Here on we continue the solution process in terms of Eq. (2.53). The problem is symmetric with respect to the midplane, and the temperature distribution should be made of even functions. This fact makes $C_1 = 0$. The same result follows also from insertion of Eq. (2.53) into Eq. (2.49). The total power generated within the plate is transferred from its surfaces to the ambient,

$$u''' A 2\ell = 2hA [T(\ell) - T_\infty]. \qquad (2.54)$$

Insertion of Eq. (2.53) into Eq. (2.54) gives, after letting $C_1 = 0$,

$$u''' \ell = h\left(-\frac{u''' \ell^2}{2k} + C_2 - T_\infty\right),$$

or,

$$C_2 = \frac{u''' \ell}{h} + \frac{u''' \ell^2}{2k} + T_\infty.$$

The same result follows also from the insertion of Eq. (2.53) into Eq. (2.50). In terms of C_1 and C_2, Eq. (2.53) becomes

$$T(x) - T_\infty = \frac{u'''}{2k}(\ell^2 - x^2) + \frac{u''' \ell}{h}. \qquad (2.55)$$

Letting $x = 0$ in Eq. (2.55), we get the midplane temperature of the flat plate relative to the ambient temperature,

$$T(0) - T_\infty = \frac{u''' \ell^2}{2k} + \frac{u''' \ell}{h}, \qquad (2.56)$$

where $u''' \ell / h$ denotes the difference between the plate surface and ambient temperatures [recall Eq. (2.54)],

$$T(\ell) - T_\infty = \frac{u''' \ell}{h}, \qquad (2.57)$$

and $u''' \ell^2 / 2k$ denotes the difference between the midplane temperature and the surface temperature,

$$T(0) - T(\ell) = \frac{u''' \ell^2}{2k} = \frac{u''' \ell}{2(k/\ell)}, \qquad (2.58)$$

---

[5] Note that $C_1 = C_1^*/k$ and $C_2 = C_2^*/k$.

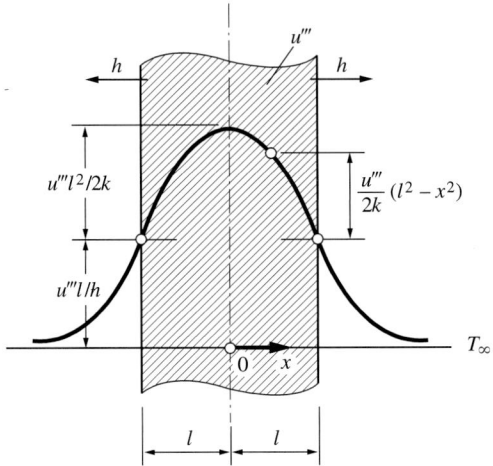

Figure 2.14 Temperature distribution in a flat plate with energy generation.

as shown in Fig. 2.14. Finally, Eq. (2.55) may be nondimensionalized as

$$\frac{T(\xi) - T_\infty}{u''' \ell^2 / k} = \frac{1}{2}(1 - \xi^2) + \frac{1}{Bi}, \qquad (2.59)$$

where $\xi = x/\ell$ is the dimensionless distance, and $Bi = h\ell/k$ is the **Biot number**.

Here we wish to examine the temperature distribution depending on geometry, thermal conductivity, and heat transfer coefficient. For a fixed energy generation, holding k/ℓ constant while letting $h \to \infty$ in Eq. (2.57) eliminates the temperature difference between the plate surface and the ambient; holding h constant while letting $k/\ell \to \infty$ in Eq. (2.58) eliminates the temperature difference between the midplane and the plate surface, thus leading to a uniform temperature on the plate; letting both $k/\ell \to \infty$ and $h \to \infty$ eliminates the entire temperature distribution, and the plate temperature becomes T_∞. Actually, the effects of k/ℓ and h are combined together in the definition of the **Biot number**. Nondimensionalization of Eq. (2.50) in terms of $\xi = x/\ell$ readily yields

$$-\frac{dT(1)}{d\xi} = Bi \left[T(1) - T_\infty \right]. \qquad (2.60)$$

The foregoing three special cases, and the general case, corresponding to moderate values of k and h, are sketched and interpreted with the Biot number in Fig. 2.15. In summary, **a large Bi** (say 10) **simplifies a boundary condition** by reducing Eq. (2.50) to $T(\ell) = T_\infty$; **a small Bi** (say 10^{-1}) **simplifies a governing equation** by reducing Eq. (2.56) to a uniform T. We shall refer to these facts on many later occasions.

Sec. 2.3 Energy Generation (Heat Source)

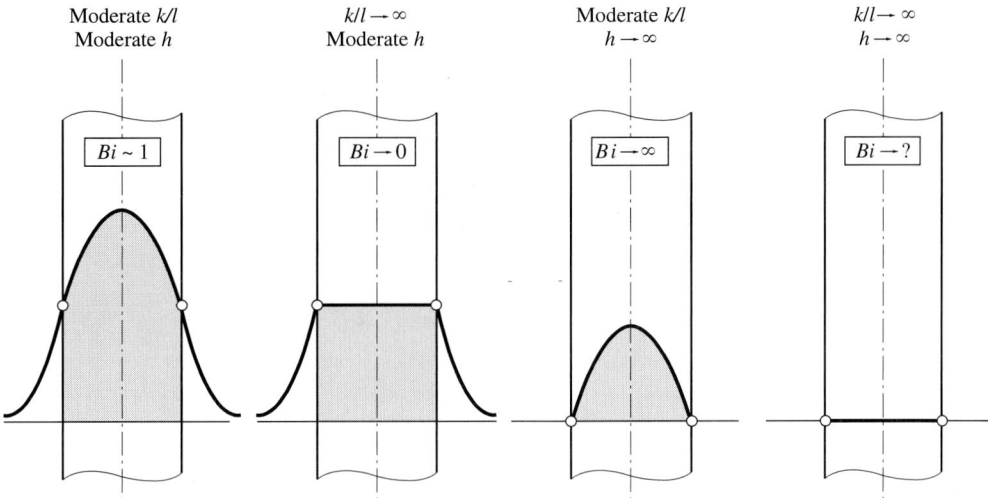

Figure 2.15 Effect of Biot number.

The apparent similarity of and the conceptual difference between the Nusselt and Biot numbers should be noted. Both numbers result from dimensional interpretation of a relation between conduction and convection,

$$\dot{Q}_K \sim \dot{Q}_C.$$

However, $\dot{Q}_K$ is evaluated from the solid side for Bi and from the fluid side for Nu. Thus, **Bi is a measure for the magnitude of temperature change in the solid side relative to the fluid side of a boundary** while **Nu is a measure for heat transfer from a boundary to a fluid in motion**. These important concepts will be further clarified in the following chapters. We proceed now to three examples illustrating the effect of energy generation on flat plates.

EXAMPLE 2.5 ⊕

A flat plate of thickness ℓ separates two ambients at temperatures T_i and $T_0 (< T_i)$. The heat transfer coefficient on both sides is h. We wish to eliminate the heat loss from the warm ambient.

We learned in Section 2.2 that the total resistance between two ambients can be increased by adding insulation to one or both surfaces of the plate. Regardless of the thickness and material of the insulation, however, the heat loss from the warm ambient cannot be completely eliminated by such an insulation. On the other hand, a proper amount of uniform internal energy generated electrically in the plate[6] may reduce the heat loss to zero. The problem is then reduced to finding the appropriate value of u'''.

---

[6] If the plate is not an electrical conductor or if electric current through the plate is undesirable, another electrically heated plate (guard heater), electrically insulated from but in good thermal contact with the original plate, can be utilized.

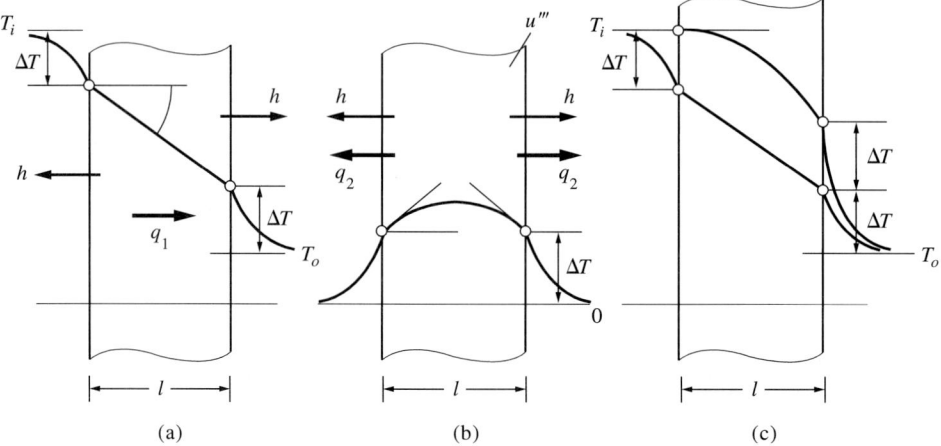

Figure 2.16 The principle of superposition. (a) first problem, (b) second problem, (c) two problems superimposed.

For a solution, consider the superposition[7] of two problems, one involving only the temperature difference $T_i - T_0$, the other involving only the internal energy generation [Figs. 2.16(a),(b)]. As we have seen in Section 1.6, the heat transfer at any cross section of the first problem is constant, and according to Eq. (1.78) with no radiation

$$q_1 = \frac{T_i - T_0}{2/h + \ell/k}. \tag{2.61}$$

The second corresponds to our key problem for the flat plate (of thickness ℓ). Because of the symmetry with respect to the middle plane, each surface of the plate transfers one-half of the energy generated within the plate, that is,

$$q_2 = \frac{u''' \ell}{2}. \tag{2.62}$$

To eliminate the heat loss from the left surface of the plate,

$$q_1 - q_2 = 0,$$

which yields, in terms of Eqs. (2.61) and (2.62)

$$\frac{T_i - T_0}{2/h + \ell/k} = \frac{u''' \ell}{2},$$

or, the particular value of the energy generation,

$$u''' = 2\frac{k(T_i - T_0)/\ell}{\ell[1 + 2(k/h\ell)]}. \tag{2.63}$$

---

[7] The concept of superposition finds important application in linear problems. However, rather than its casual use for some appreciation as done in this example, it requires elaboration beyond the scope of this text.

This energy generation eliminates the heat transfer from the left surface but doubles the heat transfer from the right surface of the plate (note that the objective was the complete elimination of heat transfer in one direction without other considerations). Also, for the first problem, expressing q_1 of Eq. (2.61) in terms of the temperature of the plate surface (on the T_i side) relative to the ambient, say ΔT,

$$q_1 = \frac{T_i - T_0}{2/h + \ell/k} = \frac{\Delta T}{1/h}, \qquad (2.64)$$

which may be rearranged as

$$\Delta T = \frac{T_i - T_0}{2 + h\ell/k}. \qquad (2.65)$$

For the second problem, from Eq. (2.57), for an ℓ-thick plate, we have

$$\Delta T = \frac{u''' \ell}{2h}, \qquad (2.66)$$

which, in view of Eq. (2.63), is identical to Eq. (2.65). In other words, the energy generation eliminates ΔT of the left side but doubles ΔT of the right side [Fig. 2.16(c)]. ◆

Example 2.6

Our objective in Ex. 1.3 was the demonstration of a moving (evaporating or condensing) boundary in terms of a simple problem. Accordingly, we assumed the entire solar energy to be absorbed at the top surface of a liquid layer. In the more realistic case associated with solar ponds, this energy is absorbed over a thickness which may even extend beyond the depth of the pond. The present example deals with a simple problem for this case based on the assumption of a constant depth.

Let the solar flux decay as $q_s'' = q_0'' e^{-\gamma x}$ in a solar pond (or a liquid layer) of thickness ℓ [Fig. 2.17(a)]. The lower surface of the pond is conductively insulated but radiatively transparent, while its upper surface loses heat to the ambient with a heat transfer coefficient h. The entire system is at temperature T_∞ in the absence of solar flux. **(a)** We wish to determine the steady temperature of the pond. **(b)** What would happen to this temperature if the lower surface were radiatively opaque? **(c)** Find the difference between the bottom surface temperature corresponding to parts (a) and (b).

(a) For a differential system (Step 1) as shown in Fig. 2.17(b), the first law of thermodynamics (Step 2) yields

$$0 = -\frac{dq_x}{dx} - \frac{dq_s''}{dx}. \qquad (2.67)$$

In terms of Fourier's law and the radiation law (Step 3),

$$q_x = -k\frac{dT}{dx}, \quad q_s'' = q_0'' e^{-\gamma x}, \qquad (2.68)$$

Equation (2.67) results in the governing equation (Step 4)

$$\frac{d^2 T}{dx^2} + \frac{\gamma q_0''}{k} e^{-\gamma x} = 0, \qquad (2.69)$$

66 Chap. 2 Steady Conduction

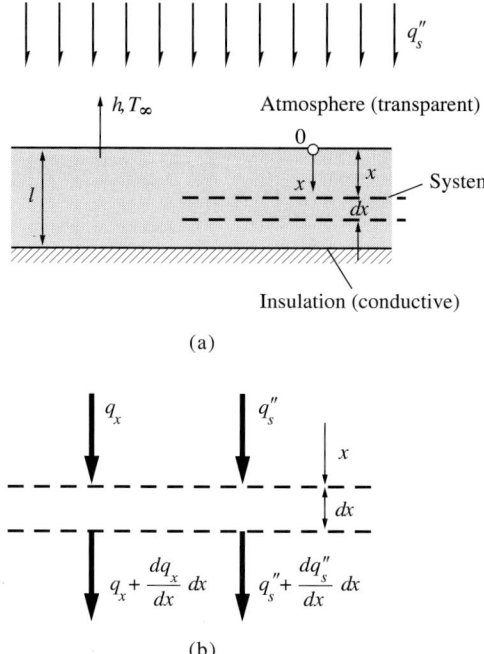

Figure 2.17 (a) Solar pond (b) first law for solar pond.

where

$$\frac{\gamma q_0''}{k} e^{-\gamma x} \equiv u'''(x)$$

effectively being a volumetric energy generation. Now, we proceed to boundary conditions (Step 5) in terms of the x coordinate fixed to the top surface of the solar pond. For the top surface (Fig. 2.18),

$$+k\frac{dT(0)}{dx} = h[T(0) - T_\infty]. \qquad (2.70)$$

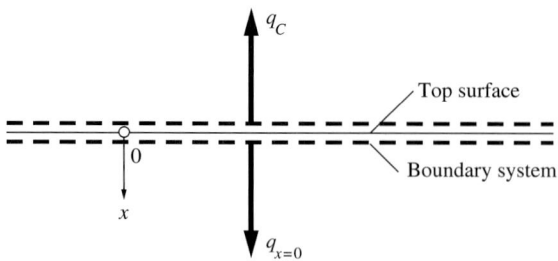

Figure 2.18 First law for top surface.

For the radiatively transparent but conductively insulated bottom surface,

$$\frac{dT(\ell)}{dx} = 0, \tag{2.71}$$

which completes (Step 5) and the formulation of the first case.
From the first integration of Eq. (2.69),

$$\frac{dT}{dx} = \frac{q_0''}{k}e^{-\gamma x} + C_1, \tag{2.72}$$

and from the second integration,

$$T = -\frac{q_0''}{\gamma k}e^{-\gamma x} + C_1 x + C_2. \tag{2.73}$$

Equation (2.72) subject to Eq. (2.71) gives

$$C_1 = -\frac{q_0''}{k}e^{-\gamma \ell}. \tag{2.74}$$

Now, rearrange Eqs. (2.72) and (2.73) with Eq. (2.74) and insert the resulting T and dT/dx into Eq. (2.70) to get

$$C_2 = \frac{q_0''}{h}(1 - e^{-\gamma \ell}) + \frac{q_0''}{\gamma k} + T_\infty. \tag{2.75}$$

For the boundary system (Step 1), consider the first law (Step 2),

$$-q_{x=0} - q_C = 0.$$

Insert the particular law (Step 3),

$$q_x = -k\frac{dT}{dx}, \quad \text{or} \quad q_C = h[T(0) - T_\infty],$$

into the first law to get the governing equation (Step 4) for the upper surface,

$$+k\frac{dT(0)}{dx} - h[T(0) - T_\infty] = 0,$$

or Eq. (2.70), which is a boundary condition (Step 5) for the solar pond.

Then, Eq. (2.73) gives, in terms of Eqs. (2.74) and (2.75), the first temperature distribution in the solar pond,

$$\frac{T_1(x) - T_\infty}{q_0''/\gamma k} = 1 - e^{-\gamma x} - \gamma x e^{-\gamma \ell} + \frac{\gamma k}{h}(1 - e^{-\gamma \ell}), \tag{2.76}$$

where $h/\gamma k$ is a Biot number based on the characteristic length $1/\gamma$.

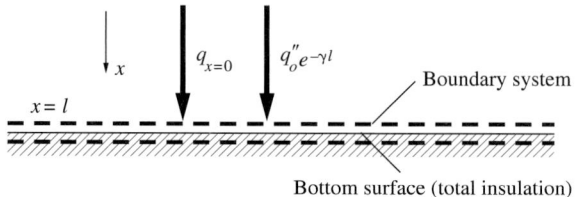

Figure 2.19 First law for bottom surface.

(b) For the radiatively opaque and conductively insulated bottom surface (Fig. 2.19),

$$+k\frac{dT(\ell)}{dx} = q_0'' e^{-\gamma \ell}, \tag{2.77}$$

which, together with Eq. (2.70) completes (Step 5) and the formulation of second case. Equation (2.72) subject to Eq. (2.77) gives

$$C_1 = 0. \tag{2.78}$$

Now, simplify Eqs. (2.72) and (2.73) with Eq. (2.78) and insert the results into Eq. (2.70) to get

$$C_2 = \frac{q_0''}{\gamma k}\left(1 + \frac{\gamma k}{h}\right) + T_\infty. \tag{2.79}$$

For the boundary system (Step 1), consider the first law (Step 2),

$$+q_{x=\ell} + q_R = 0.$$

Insert the particular law (Step 3) into the first law to get the governing equation (Step 4) for the bottom surface

$$-k\frac{dT(\ell)}{dx} + q_0'' e^{-\gamma \ell} = 0,$$

or Eq. (2.77), which is a boundary condition (Step 5) for the solar pond.

Then, Eq. (2.73) gives, in terms of Eqs. (2.78) and (2.79), the second temperature distribution in the solar pond,

$$\frac{T_2(x) - T_\infty}{q_0''/\gamma k} = 1 - e^{-\gamma x} + \frac{\gamma k}{h}. \tag{2.80}$$

Actually, this temperature distribution is a special case of the first temperature distribution and it can be directly obtained from Eq. (2.78) by eliminating the terms involving $e^{-\gamma \ell}$ [note the difference between Eqs. (2.71) and (2.77)].

(c) Because of the radiative, as well as conductive, insulation of the bottom surface, temperature levels are higher in the second case than the first case. The difference is

$$\frac{T_2(\ell) - T_1(\ell)}{q_0''/\gamma k} = \frac{\gamma k}{h}\left(1 + \frac{h\ell}{k}\right)e^{-\gamma \ell}. \tag{2.81}$$

(Why is this difference linear?) ◆

Sec. 2.3 Energy Generation (Heat Source) 69

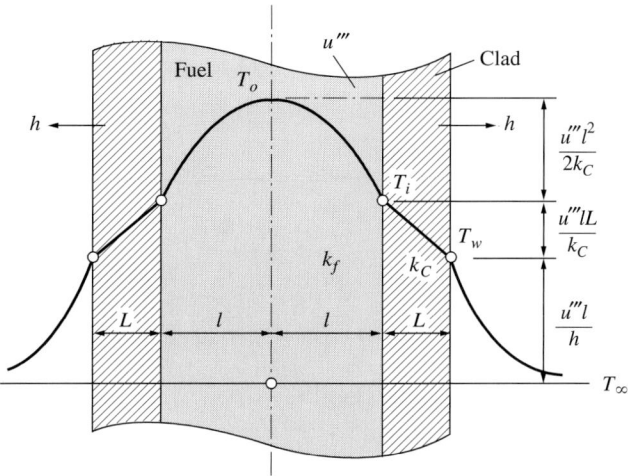

Figure 2.20 A flat-plate fuel element

EXAMPLE 2.7

The core of a pool reactor is composed of flat fuel plates of thickness 2ℓ. Both sides of each plate are covered with flat clads, each of thickness L. Assume that the gap between the fuel plates and the clads is negligible. Nuclear energy u''' is generated only in the fuel plates (Fig. 2.20). Under steady conditions, this energy is transferred, with a heat transfer coefficient h, from the clads to an ambient at temperature T_∞. We wish to know the maximum temperature in the fuel plates.

Under steady conditions, the total energy $u''' A 2\ell$ generated in the fuel plate is transferred to the ambient through the outside surface A of each clad [note the use of the same argument for Eq. (2.54) in the case of a single flat plate]. Let T_w be the outside surface temperature of the clads. This gives

$$u''' A 2\ell = h 2 A (T_w - T_\infty),$$

or the outside surface temperature of clads relative to the ambient temperature,

$$T_w - T_\infty = \frac{u''' \ell}{h}. \tag{2.82}$$

This is the same result as the one found for the flat-plate key problem, Eq. (2.57)! Adding the clads has no effect on the outside surface temperature for flat-plate fuel elements (what about cylindrical elements?). Let us now look at the inside temperatures. Since the same total energy balances the heat flux at any cross section of clads,

$$u''' A 2\ell = 2 k_C A \frac{T_i - T_w}{L},$$

and the temperature of the interface relative to the outside surface temperature of the clads is

$$T_i - T_w = \frac{u''' \ell L}{k_C}. \tag{2.83}$$

Also, Eq. (2.58) of the flat plate key problem gives the temperature of the midplane relative to the temperature of the interface,

$$T_0 - T_i = \frac{u''' \ell^2}{2k_f}, \qquad (2.84)$$

which is the location of the maximum temperature. The sum of Eqs. (2.82)-(2.84) yields the midplane temperature of the fuel plate relative to the ambient temperature,

$$T_0 - T_\infty = \frac{u''' \ell}{h}(1 + Bi_C + \frac{1}{2}Bi_f), \qquad (2.85)$$

where $Bi_C = hL/k_C$ and $Bi_f = h\ell/k_f$ are the Biot numbers related to conductivity and thickness of the clad and fuel, respectively. Thus, adding clads of thickness L to the flat plate increases the midplane temperature by $(u'''\ell/h)Bi_C$.

Having learned the effect of energy generation on flat plates, we proceed now to the effect of energy generation on cylinders and spheres. ◆

2.3.2 Cylinder and Sphere (Key Problem)

Let the rate of energy per unit volume generated in a solid cylinder or a solid sphere be $u'''(r) \equiv u'''_r$, the radius and the thermal conductivity of the cylinder or the sphere be R and $k(T) \equiv k_T$ (alternate notations u'''_r and k_T are used for convenience in the following formulation). Under steady conditions, the total energy generated in the cylinder or sphere is transferred, with a heat transfer coefficient h, to an ambient at temperature T_∞. This cylinder could be one of the fuel rods of a reactor core, or one of the elements of an electric heater, and the cylinder or sphere could be a bare, homogeneous reactor core. We wish to determine the radial temperature distribution.

Following the **five steps of formulation**, first we consider the differential system (Step 1) shown in Fig. 2.21(a). The first law of thermodynamics (Step 2), Eq. (1.16) interpreted in terms of Fig. 2.21(b), yields

$$0 = +q_r A_r - \left[q_r A_r + \frac{d}{dr}(q_r A_r)dr\right] + u'''_r A_r dr,$$

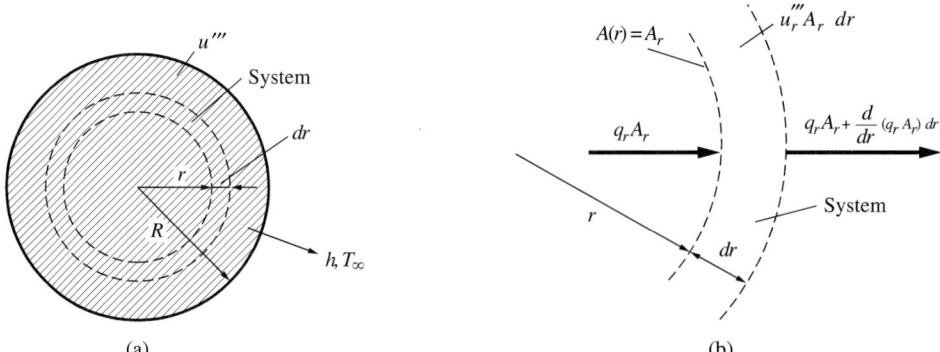

Figure 2.21 (a) Cylindrical or spherical differential system, (b) the first law.

or
$$0 = -\frac{d}{dr}(q_r A_r) + u_r''' A_r. \tag{2.86}$$

Note variable A_r is kept inside the derivative. Fourier's law of conduction (Step 3),
$$q_r = -k_T \frac{dT}{dr}, \tag{2.87}$$

introduced into Eq. (2.86) gives the governing equation (Step 4),
$$\frac{d}{dr}\left(k_T A_r \frac{dT}{dr}\right) + u_r''' A_r = 0, \tag{2.88}$$

which, for a constant k, may be rearranged as
$$\frac{d}{dr}\left(A_r \frac{dT}{dr}\right) + \frac{u_r'''}{k} A_r = 0, \tag{2.89}$$

where $A_r = 2\pi r L$ for a cylinder and $A_r = 4\pi r^2$ for a sphere, L being the axial length of the cylinder. Equations (2.88) and (2.89) require two boundary conditions (Step 5)
$$\frac{dT(0)}{dr} = 0 \quad \text{or} \quad T(0) = \text{Finite}, \tag{2.90}$$

$$-k\frac{dT(R)}{dr} = h\left[T(R) - T_\infty\right]. \tag{2.91}$$

The alternate condition in Eq. (2.90), $T(0) = $ Finite, turns out to be more convenient to use than $dT(0)/dr = 0$ in cases involving curvature. Note that mathematical solutions that would lead to an infinite temperature at the center are not physically meaningful.

For a distributed energy generation and a variable thermal conductivity, integrating Eq. (2.88) twice results in
$$\int k_T dT = -\int \frac{1}{A_r}\left(\int u_r''' A_r dr\right) dr + C_1^* \int \frac{dr}{A_r} + C_2^*, \tag{2.92}$$

which, depending on the particular explicit forms of k_T, u_r''', $A_r = 2\pi r L$ or $A_r = 4\pi r^2$, is reduced to a straightforward integration process.

For a uniform energy generation and constant thermal conductivity, from Eq. (2.92) with $A_r = 2\pi r L$ or from twice integration of Eq. (2.89) with $A_r = 2\pi r L$,
$$\frac{d}{dr}\left(r \frac{dT}{dr}\right) + \frac{u'''}{k} r = 0, \tag{2.93}$$

we get, **for a cylinder**,[8]
$$T = -\frac{u''' r^2}{4k} + C_1 \ln r + C_2. \tag{2.94}$$

---
[8] Note that $C_1 = C_1^*/2\pi L k$ and $C_2 = C_2^*/k$.

Also from Eq. (2.92) with $A_r = 4\pi r^2$, or from twice integration of Eq. (2.89) with $A_r = 4\pi r^2$,

$$\frac{d}{dr}\left(r^2 \frac{dT}{dr}\right) + \frac{u'''}{k} r^2 = 0, \tag{2.95}$$

we get, **for a sphere**,

$$T = -\frac{u''' r^2}{6k} + \frac{C_1}{r} + C_2. \tag{2.96}$$

For a solid cylinder and for a solid sphere, the condition of a finite center temperature readily gives $C_1 = 0$ (note that using $dT(0)/dr = 0$ leads to the same result).

For a solid cylinder, the second boundary condition, Eq. (2.91), or the fact that the total energy generated in the cylinder is transferred to the ambient, gives

$$u''' \pi R^2 L = h 2\pi R L [T(R) - T_\infty], \tag{2.97}$$

which, in terms of Eq. (2.94) with $C_1 = 0$, results in

$$C_2 = \frac{u''' R}{2h} + \frac{u''' R^2}{4k} + T_\infty,$$

and the temperature of the **cylinder**,

$$\boxed{T - T_\infty = \frac{u'''}{4k}(R^2 - r^2) + \frac{u''' R}{2h}}. \tag{2.98}$$

The same way, for a sphere,

$$u''' \frac{4}{3}\pi R^3 = h 4\pi R^2 [T(R) - T_\infty], \tag{2.99}$$

which, in terms of Eq. (2.96) with $C_1 = 0$, yields

$$C_2 = \frac{u''' R}{3h} + \frac{u''' R^2}{6k} + T_\infty,$$

and the temperature of the **sphere**,

$$\boxed{T - T_\infty = \frac{u'''}{6k}(R^2 - r^2) + \frac{u''' R}{3h}}. \tag{2.100}$$

The temperature distribution in a flat plate, solid cylinder, and solid sphere with internal energy generation given respectively by Eqs. (2.55), (2.98), and (2.100) are sketched in Fig. 2.22 for $\ell = R$. For the same u''', the temperature levels in the cylinder and sphere respectively are 1/2 and 1/3 of those in the flat plate. Increasing the effect of curvature increases the heat loss, as expected.

Having learned the effect of energy generation on the solid cylinder and solid sphere, we proceed now to an example on hollow cylinders in which $C_1 \neq 0$.

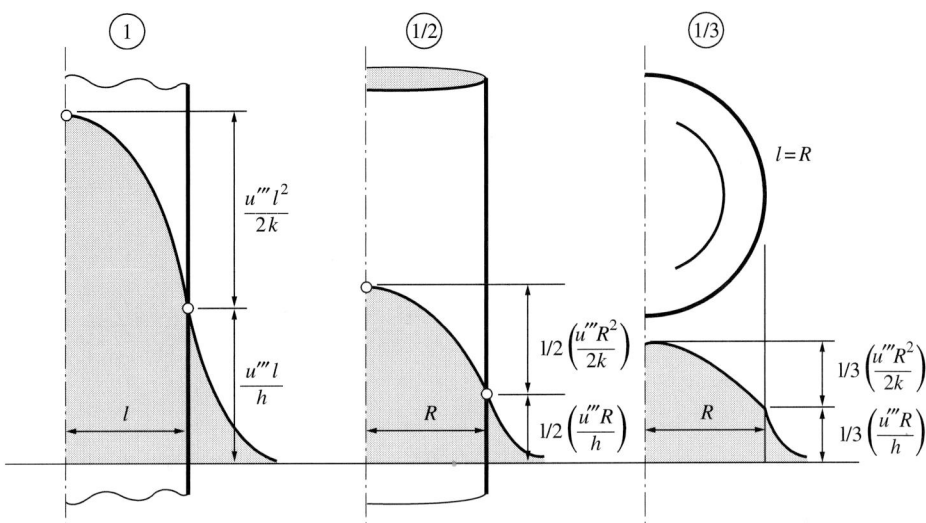

Figure 2.22 Comparison of temperatures.

EXAMPLE 2.8

The local heat transfer coefficient for forced convection inside tubes is determined from experiments conducted with electrically heated tubes. This coefficient requires that the inner temperature of the tube walls T_i be known. However, T_i turns out to be more difficult to measure than the outer temperature of the tube walls, T_0. The usual practice is to measure T_0 and relate it to T_i by an analytical expression. We wish to obtain this expression in terms of a tube with an inner radius R_1, an outer radius R_2, and internal energy generation u'''.

Let the internal energy u''' be uniformly generated within the tube walls, and the tube be insulated (Fig. 2.23). The boundary conditions are

$$\frac{dT(R_2)}{dr} = 0 \quad \text{and} \quad T(R_2) = T_o(\text{measured}). \tag{2.101}$$

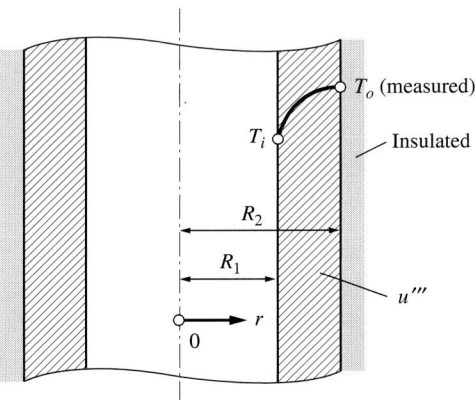

Figure 2.23 Electrically heated tube.

Applying the first law to a radially differential system within the pipe walls leads to the same governing equation as for the solid cylinder, with different boundary conditions. The solution given by Eq. (2.94), inserted into the first condition, gives

$$C_1 = \frac{u''' R_2^2}{2k},$$

and into the second condition yields, after the use of C_1,

$$C_2 = T_0 + \frac{u''' R_2^2}{4k} - \frac{u''' R_2^2}{2k} \ln R_2.$$

In terms of C_1 and C_2, Eq. (2.94) results in

$$T_0 - T = \frac{u''' R_2^2}{2k} \ln \frac{R_2}{r} - \frac{u''' R_2^2}{4k}\left(1 - \frac{r^2}{R_2^2}\right). \tag{2.102}$$

The temperature difference across the thickness of the tube walls is then

$$\Delta T = \frac{u''' R_2^2}{2k}\left[\ln \frac{R_2}{R_1} - \frac{1}{2}\left(1 - \frac{R_1^2}{R_2^2}\right)\right], \tag{2.103}$$

where $\Delta T = T_0 - T_i$ and $T_i = T(R_1)$. ◆

2.4 EXTENDED SURFACES (FINS, PINS)

The purpose of an extended surface such as a fin or a pin is to increase heat transfer from a surface to an ambient. This way we control the size and cost of a heat transfer device. For example, the design of a heat exchanger may be based on achieving the smallest possible heat transfer area (for lightness and compactness of moving vehicles) or it may be based on demanding the largest possible amount of heat transfer from a given size heat exchanger (for cost and compactness of stationary power plants).

Let us now search for means of increasing heat transfer from a surface. For this purpose, consider a surface at temperature T_w transferring heat by convection to an ambient at temperature T_∞ (Fig. 2.24). As we learned in Chapter 1, the rate of heat transfer from this surface may be evaluated in terms of Newton's cooling law [Eq. (1.53)],

$$\dot{Q}_C = hA(T_w - T_\infty). \tag{2.104}$$

Clearly, $\dot{Q}_C$ may be increased by increasing the temperature difference ΔT, by increasing the heat transfer coefficient, or by increasing the heat transfer area. The temperature difference is usually dictated by the nature of practical problems and cannot be altered; the control of the heat transfer coefficient by using different fluids and/or increased flow of these fluids is the subject of the convection heat transfer; the increase of the heat transfer area is the concern of this section.

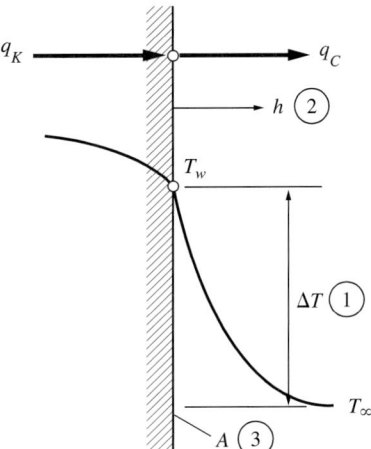

Figure 2.24 Convection control.

Extended surfaces are usually constructed in two ways: they are either extensions of the base material, obtained by a casting or extruding process [Fig. 2.25(a)], or they are attached to the base by pressing, soldering, or welding; in the latter case, they may or may not be made from the base material [Fig. 2.25(b)]. Generally weight and/or cost dictate the way of construction. The most common types of extended surfaces are straight fins, annular fins and pin fins [Fig. 2.26(a),(b),(c)]. Well-known applications are heating or cooling radiators, resistance heaters, liquid-to-gas or gas-to-gas heat exchangers, boilers, and air-cooled engines (Fig. 2.27).

Having learned the purpose, construction, types, and application of extended surfaces, we proceed now to the main objective of this section, the study of heat transfer from extended surfaces. Because of transversal convection, the temperature in extended surfaces varies longitudinally, and Eq. (2.104) cannot be used directly. We must first evaluate the temperature distribution, and then the heat transfer in terms of this temperature distribution.

Consider an extended surface with variable cross section [Fig. 2.28(a)]. Assume

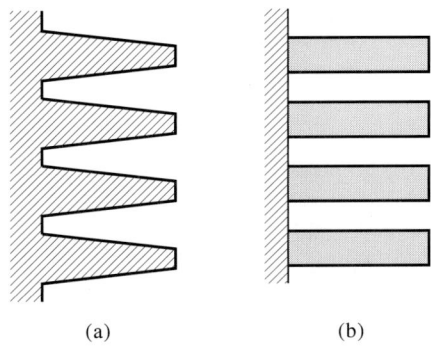

Figure 2.25 Manufacture of extended surfaces. (a) cast or extended, (b) welded, soldered, or pressed.

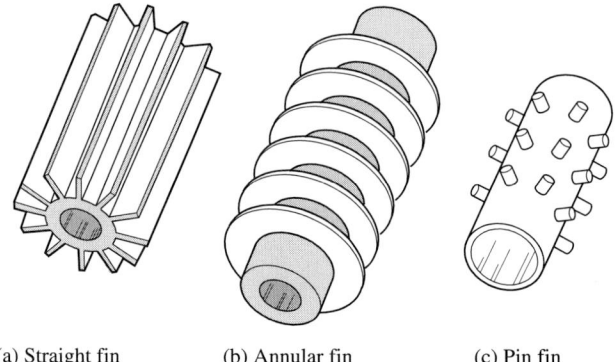

(a) Straight fin   (b) Annular fin   (c) Pin fin

Figure 2.26 Types of extended surfaces. (a) Straight fin, (b) annular fin, (c) pin fin.

that the Biot number based on a transversal characteristic length[9] ℓ is small,

$$Bi = \frac{h\ell}{k} \ll 1, \qquad (2.105)$$

such that the transversal temperature distribution is negligible [recall Fig. 2.15(b)]. Fins are typically thin, thus making this assumption quite realistic. Accordingly, assume a transversally lumped and longitudinally differential system (Step 1). Under steady conditions, the first law of thermodynamics (Step 2), interpreted for this system

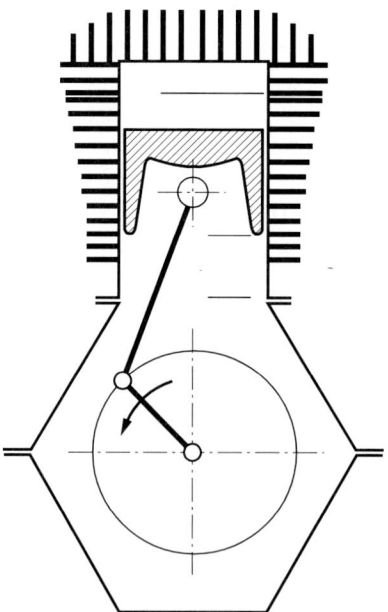

Figure 2.27 Fins of an air-cooled engine.

---

[9] This length will be clarified with Eq. (2.113).

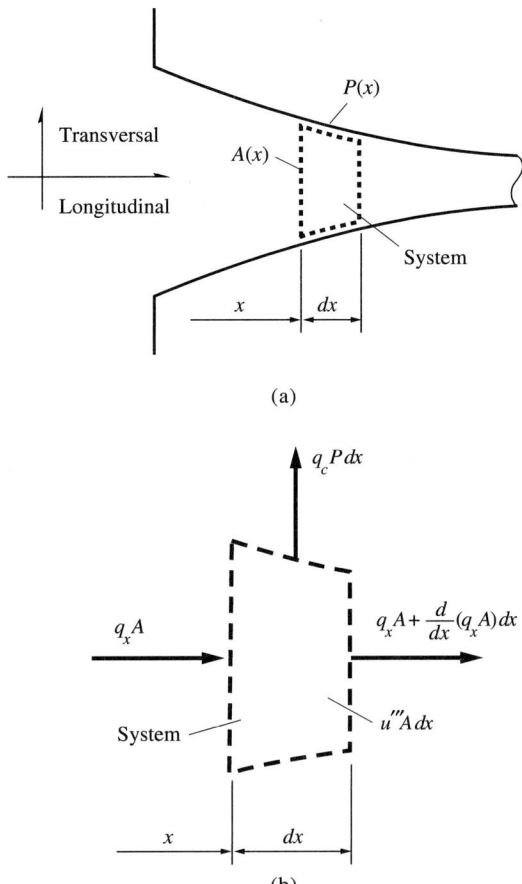

Figure 2.28 (a) Mixed (transversally lumped + axially differential) system, (b) the first law.

[Fig. 2.28(b)], results in

$$0 = q_x A - \left[q_x A + \frac{d}{dx}(q_x A)dx\right] - q_C P dx + u''' A dx,$$

or,

$$0 = -\frac{d}{dx}(q_x A) - q_C P + u''' A, \qquad (2.106)$$

where P is the local perimeter and u''' is the imposed internal energy generation. Fourier's conduction law and Newton's cooling law (Step 3),

$$q_x = -k\frac{dT}{dx}, \quad q_C = h(T - T_\infty), \qquad (2.107)$$

inserted into Eq. (2.106) give the governing equation (Step 4),

$$\frac{d}{dx}\left(kA\frac{dT}{dx}\right) - hP(T - T_\infty) + u'''A = 0, \tag{2.108}$$

where the generation term applies to problems involving electric, nuclear, or chemical energy. Equation (2.108) is a second-order differential equation with variable coefficients. Almost all forms of this equation lead to solutions in terms of Bessel functions. Since these functions are beyond the scope of this text, and since the variation of cross section does not affect our understanding of extended surfaces, hereafter we will consider only extended surfaces with constant cross section.

For a constant A (and a constant k), Eq. (2.108) reduces to

$$\frac{d^2T}{dx^2} - \frac{hP}{kA}(T - T_\infty) + \frac{u'''}{k} = 0, \tag{2.109}$$

which may be rearranged, in terms of $\theta = T - T_\infty$ and $hP/kA = m^2$, as

$$\frac{d^2\theta}{dx^2} - m^2\theta + \frac{u'''}{k} = 0, \tag{2.110}$$

or, for a constant u''', in terms of

$$\Theta = \theta - \frac{u''' A}{hP} \tag{2.111}$$

as

$$\frac{d^2\Theta}{dx^2} - m^2\Theta = 0, \tag{2.112}$$

subject to appropriate boundary conditions (Step 5), one related to the base and the other to the tip of the extended surface (note that the introduction of θ and Θ is done solely to simplify the mathematics).

On dimensional grounds, Eq. (2.112) yields

$$\frac{\Theta}{\ell^2} \sim m^2\Theta,$$

or

$$\ell \sim \frac{1}{m} = \sqrt{\frac{kA}{hP}}, \tag{2.113}$$

which is a **characteristic length** for extended surfaces. In terms of this length, Eq. (2.105) becomes

$$Bi = \frac{h\ell}{k} = \sqrt{\frac{hA}{kP}}, \tag{2.114}$$

the Biot number appropriate for extended surfaces. In order for a transversally lumped assumption to hold, $Bi \ll 1$.

The general solution[10] of Eq. (2.112) can be expressed as

$$\Theta(x) = C_1 e^{mx} + C_2 e^{-mx}, \tag{2.115}$$

or, equivalently,

$$\Theta(x) = C_3 \sinh mx + C_4 \cosh mx, \tag{2.116}$$

where $\sinh mx = (e^{mx} - e^{-mx})/2$ and $\cosh mx = (e^{mx} + e^{-mx})/2$. Clearly, $\Theta(x) \equiv \theta(x)$, when $u''' = 0$. Although these solutions equally apply to all cases, Eq. (2.115) turns out to be convenient for problems with infinite geometry, and Eq. (2.116) for problems with finite geometry. For a nonuniform u''', depending on the explicit form of $u'''(x)$, particular solutions need to be constructed from the theory of nonhomogeneous differential equations. After the foregoing general considerations, we proceed now to a number of illustrative examples.

EXAMPLE 2.9

Consider an infinitely long fin with a specified base temperature T_0 (Fig. 2.29). We wish to find the temperature distribution in and the heat transfer from the fin.

Assuming the base of the fin to be the origin of the x coordinate, and noting that the temperature of the fin approaches the temperature of the ambient as $x \to \infty$, the boundary conditions may be written as

$$\theta(0) = \theta_0, \tag{2.117}$$

$$\lim_{x \to \infty} \theta(x) \to 0, \tag{2.118}$$

where $\theta_0 = T_0 - T_\infty$. Equation (2.115) satisfies Eq. (2.118) only when $C_1 = 0$. The result, combined with Eq. (2.117), gives the temperature distribution in the fin,

$$\frac{\theta(x)}{\theta_0} = e^{-mx}. \tag{2.119}$$

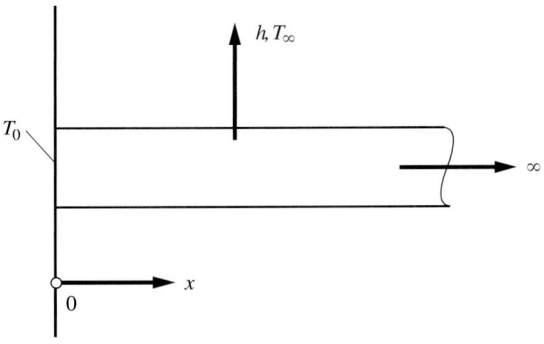

Figure 2.29 Infinitely long fin.

---

[10] Recall that linear differential equations with constant coefficients accept **exponential** solutions.

The heat transfer from the fin may be now evaluated in terms of Eq. (2.119) by simply integrating the local convection along the fin:

$$\dot{Q}_C = \int_0^\infty hP\theta\,dx = hP\theta_0 \int_0^\infty e^{-mx}\,dx = \theta_0(hPkA)^{1/2}. \tag{2.120}$$

Since the total heat transferred by convection from the fin is supplied by conduction through the base of the fin, the same result may be obtained from

$$\dot{Q}_K = -kA\left(\frac{d\theta}{dx}\right)_{x=0} = -kA\theta_0\left(-me^{-mx}\right)_{x=0} = \theta_0(hPkA)^{1/2}. \tag{2.121}$$

The second approach, requiring a simple differentiation, will be preferred hereafter for heat loss calculations from extended surfaces. ◆

2.4.1 Thermal Length

Reconsider the infinite fin of the preceding example. Since only a finite amount of energy is supplied by conduction through the base of this fin, the transversal heat loss by convection continuously decreases with increasing x, and for all practical purposes, it diminishes beyond a certain length of the fin. We wish to determine this length, say the **thermal length** of a fin, δ (this concept is identical to that of **thermal boundary-layer** or **penetration depth** introduced in Section 1.5.4 on convection). Using a fin of length greater than δ does not effectively increase the amount of heat transfer (compared to a fin of length δ) and is thus a waste of material.

Consider a transversally lumped and axially integral system (Step 1) as shown in Fig. 2.30. For this system, the first law of thermodynamics (Step 2) gives

$$-Aq_{x=\delta} - \int_0^\delta q_C P\,dx = 0, \tag{2.122}$$

where x is measured from the location a distance δ away from the base[11]. There is no appreciable conduction beyond δ. Employing the particular (Fourier's conduction and Newton's cooling) laws (Step 3),

$$q_x = -k\,dT/dx, \qquad q_C = h(T - T_\infty),$$

rearrange Eq. (2.126) to get the governing equation (Step 4):

$$kA\left.\frac{dT}{dx}\right|_{x=\delta} = hP\int_0^\delta (T - T_\infty)\,dx,$$

or,

$$\left.\frac{d\theta}{dx}\right|_{x=\delta} = m^2 \int_0^\delta \theta\,dx. \tag{2.123}$$

Note that Eq. (2.123) includes the effect of boundaries (Step 5). That is, a governing equation resulting from an integral formulation always combines Steps 4 and 5. In the

---

[11] This coordinate leads to the simplest form for the approximate temperature profile to be developed below.

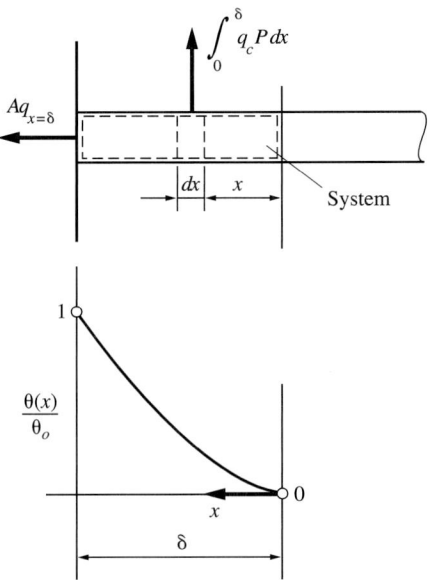

Figure 2.30 The thermal length.

absence of any energy generation, integration of Eq. (2.110) over the interval $(0, \delta)$ yields the same result, as expected.

Consider now an approximate temperature profile satisfying the apparent physics of the problem—that is, satisfying $\theta(0) = 0$, $d\theta(0)/dx = 0$, and $\theta(\delta) = \theta_0$ (see Fig. 2.30). For example, a parabola,

$$\theta(x) = Ax^2 + Bx + C,$$

can be rearranged to satisfy these conditions. The first of these conditions gives $C = 0$, the second, $B = 0$; and the third, $\theta_0 = A\delta^2$. Thus,

$$\theta(x)/\theta_0 = (x/\delta)^2. \tag{2.124}$$

Clearly, the origin for x selected in Fig. 2.30 leads to the simplest possible form of this parabola. Inserting Eq. (2.124) into Eq. (2.123) results in

$$\frac{2}{\delta} = m^2 \frac{\delta}{3},$$

or, the thermal length

$$\boxed{\delta = \frac{\sqrt{6}}{m}}. \tag{2.125}$$

This distance denotes approximately the longitudinal **penetration depth** (boundary layer) of heat by conduction through the fin. Also, on dimensional grounds, replacing ℓ with δ in Eq. (2.113), we have the same distance within a numerical constant,

$$\delta \sim \frac{1}{m} = \sqrt{\frac{kA}{hP}}.$$

For a thickness ℓ and width L, $A = \ell L$ and $P = 2L + 2\ell \cong 2L$, hence $m = \sqrt{hP/kA} = \sqrt{2h/k\ell}$ and $\delta = \sqrt{3k\ell/h}$. Assume the engine fins of a motorbike have $\ell = 0.2$ cm, $k = 180$ W/m·K (an aluminum alloy), and $h = 300$ W/m²·K (forced convection to air). These give $\delta \cong 6$ cm. However, the fin is usually cut somewhat shorter than δ; consequently some material and weight are saved without appreciably affecting the heat transfer from the fin. (In Fig. 2.30 note the slope of the parabola, which is a measure of the heat loss near the origin.)

EXAMPLE 2.10

Consider a fin of finite length $\ell < \delta$. The base temperature T_0 of the fin is specified and the tip of the fin is insulated (Fig. 2.31). We wish to find the temperature distribution in and the heat transfer from the fin.

Here the tip of the fin, being a point of symmetry, is more convenient for the origin of x. The boundary conditions are then

$$\frac{d\theta(0)}{dx} = 0, \qquad (2.126)$$

$$\theta(\ell) = \theta_0, \qquad (2.127)$$

where, as before, $\theta_0 = T_0 - T_\infty$. Since the fin is finite in length, we refer to the general solution given by Eq. (2.116). The use of Eq. (2.126), or, equivalently, the fact that the temperature distribution is symmetric with respect to x, hence is composed of even functions only, yields $C_3 = 0$. Next, the consideration of Eq. (2.127) gives $C_4 = \theta_0/\cosh m\ell$. Thus,

$$\frac{\theta(x)}{\theta_0} = \frac{\cosh mx}{\cosh m\ell}. \qquad (2.128)$$

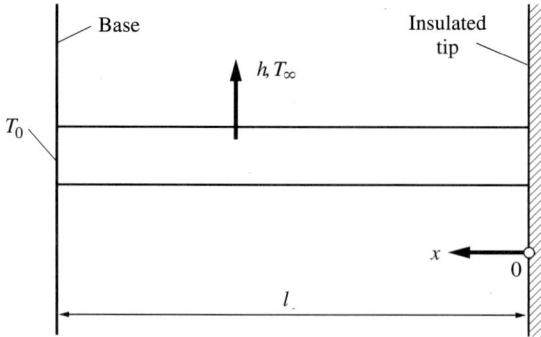

Figure 2.31 Finite fin with insulated tip.

The total heat loss from the fin, evaluated from conduction at the base of the fin, is

$$-\dot{Q}_K = kA \left(\frac{d\theta}{dx}\right)_{x=\ell} = \frac{kA\theta_0}{\cosh m\ell}(m \sinh mx)_{x=\ell} = \theta_0(hPkA)^{1/2} \tanh m\ell. \quad (2.129)$$

Since $\tanh x \to 1$ as $x \to \infty$, Eq. (2.129) approaches[12] Eq. (2.120) as $m\ell \to \infty$. That is, the heat transfer from a finite fin approaches that from an infinite fin as the length of finite fin increases indefinitely. This statement is independent of the boundary condition employed at the tip of the fin, since the effect of the tip diminishes as $\ell \to \infty$.

The heat loss from the tip of a fin is usually negligible. The ratio of this loss relative to peripheral loss is

$$\frac{\dot{Q}_t}{\dot{Q}_p} \sim \frac{h_t A_t \Delta T_t}{h_p A_p \Delta T_p},$$

subscripts t and p respectively indicating tip and periphery, ΔT_p being an average temperature difference. In cooling fins,

$$\Delta T_t \leq \Delta T_p.$$

Also, as we shall learn in the chapters on convection, $h_t \sim h_p$. Consequently,

$$\frac{\dot{Q}_t}{\dot{Q}_p} \leq \frac{A_t}{A_p} \ll 1,$$

and the heat from the tip of a fin can be safely neglected. ◆

EXAMPLE 2.11

Let a spoon in a cup of tea be approximated by a thin rod of constant cross section (Fig. 2.32). The thermal conductivity, length, periphery, and cross-sectional area of the spoon are k, 2ℓ, P, and A, respectively. The heat transfer coefficients are h_1 and h_2. Assume one-half of the spoon to be in the tea, the temperature of the tea to remain constant at T_0, and the ends of the spoon to be insulated. We wish to determine the steady temperature distribution in the spoon and to discuss the results in terms of the following data: $2\ell = 10$ cm, $A \cong 0.2$ cm^2, $P \cong 2$ cm, $k = 15$ W/m·K (stainless steel) or $k = 400$ W/m·K (silver), $h_1 = 1{,}000$ W/m^2·K, $h_2 = 10$ W/m^2·K, $T_\infty = 20\,°$C, $T_0 = 80\,°$C.

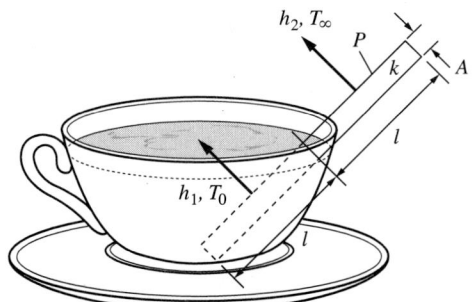

Figure 2.32 Cup of tea.

---

[12] The condition $m\ell \to \infty$ may also be interpreted as $m \to \infty$ for a given ℓ. This case, as readily seen from the definition of $m = (hP/kA)^{1/2}$, corresponds to $h \to \infty$ or $k \to 0$.

The Biot numbers [recall (Eq. 2.114)] for the half of the steel or silver spoon in the tea are, respectively,

$$Bi_1 = \left(\frac{h_1 A}{kP}\right)^{1/2} = \left(\frac{10\,\text{W/m}^2\cdot\text{K} \times 0.2 \times 10^{-4}\,\text{m}^2}{(15;\,400)\,\text{W/m}\cdot\text{K} \times 2 \times 10^{-2}\,\text{m}}\right)^{1/2} = 0.026;\,0.005$$

and for the other halves in air are

$$Bi_2 = \left(\frac{h_2 A}{kP}\right)^{1/2} = \left(\frac{1{,}000\,\text{W/m}^2\cdot\text{K} \times 0.2 \times 10^{-4}\,\text{m}^2}{(15;\,400)\,\text{W/m}\cdot\text{K} \times 2 \times 10^{-2}\,\text{m}}\right)^{1/2} = 0.26;\,0.05,$$

and the spoon temperatures can be **transversally lumped,** each with a different degree of approximation. Because of $h_1 \neq h_2$ and $T_0 \neq T_\infty$, the problem needs a two-domain formulation. Before this formulation, however, let us look into some physical facts.

Under steady conditions, neglecting the heat loss from the ends, the heat transfer from the tea to the spoon is balanced with that from the spoon to the air. That is

$$q = h_1 A \Delta T_1 = h_2 A \Delta T_2,$$

$A = P\ell$ being one-half of the peripheral area of the spoon, $\Delta T_1 = T_0 - T_1$ and $\Delta T_2 = T_2 - T_\infty$ respectively denoting the mean temperature difference between the tea and the spoon, and the spoon and the air (Fig. 2.33). Since $h_1 \gg h_2$,

$$\Delta T_1 \ll \Delta T_2,$$

and the temperature difference between the tea and the spoon can be neglected. Over the penetration depths δ_1 and δ_2 the spoon temperature drops continuously from T_1 to T_2. For the same geometry and material (same A, P, and k),

$$\frac{\delta_1}{\delta_2} = \sqrt{\frac{h_2}{h_1}},$$

or, in view of the fact that $h_1 \gg h_2$,

$$\frac{\delta_1}{\delta_2} \ll 1.$$

Also, the equality of the interface heat fluxes yields, for the same material,

$$\frac{T_1 - T_i}{\delta_1} \sim \frac{T_i - T_2}{\delta_2},$$

or

$$\frac{T_1 - T_i}{T_i - T_2} \sim \frac{\delta_1}{\delta_2} \ll 1,$$

that is, T_i is much closer to T_1 than T_2 (Fig. 2.34). The foregoing considerations suggest a spoon temperature in the tea approximately equal to the temperature of the tea. The spoon temperature in the air may then be evaluated with Eq. (2.128) in terms of a coordinate (say ξ) measured from the upper tip downward,

$$\frac{T_2(\xi) - T_\infty}{T_i - T_\infty} = \frac{\cosh m_2 \xi}{\cosh m_2 \ell}, \qquad (2.130)$$

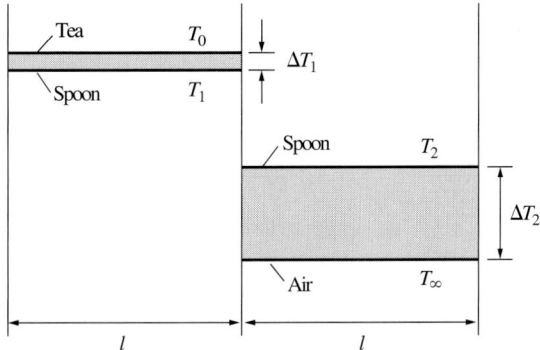

Figure 2.33 Averaged spoon temperature relative to tea and air.

where $m_2 = \sqrt{h_2 P/kA}$. Actually, ignoring only the heat loss from the lower tip of the spoon, we may obtain a temperature distribution for the part of the spoon in the tea. For a coordinate (say x) measured from the lower tip,

$$\frac{T_1(x) - T_0}{T_i - T_0} = \frac{\cosh m_1 x}{\cosh m_1 \ell}. \tag{2.131}$$

Equality of temperatures and heat fluxes of $T_1(x)$ and $T_2(\xi)$ at $(x = \ell, \xi = \ell)$ yields

$$T_i = \frac{T_0 m_1 \tanh m_1 \ell + T_\infty m_2 \tanh m_2 \ell}{m_1 \tanh m_1 \ell + m_2 \tanh m_2 \ell}.$$

In terms of the given data, we have for stainless steel,

$$m_1 = 258.2 \text{ m}^{-1}, \quad m_2 = 25.8 \text{ m}^{-1}, \quad T_i = 75.3°\text{C},$$

and for silver,

$$m_1 = 50 \text{ m}^{-1}, \quad m_2 = 5 \text{ m}^{-1}, \quad T_i = 78.5°\text{C},$$

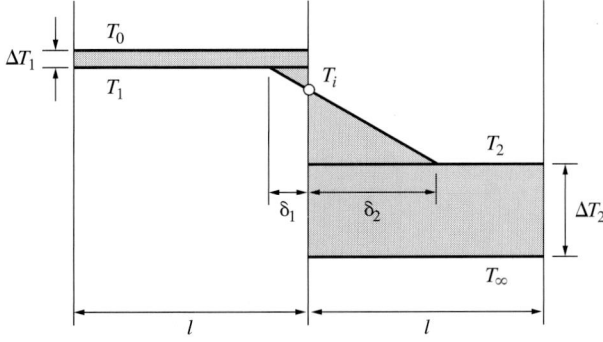

Figure 2.34 Penetration depths.

which proves our earlier assumption of a spoon temperature in tea approximately equal to the temperature of the tea.

Equipped with the insight gained on this problem, we now proceed to a problem with energy generation. ◆

EXAMPLE 2.12 ⊕

Uniform internal energy u''' is generated electrically in a vertical rod—say, an element of a heater [Fig. 2.35(a)]. The upper part, of length L, of this rod is in a stagnant ambient (say air) at temperature T_∞, the lower part, of length ℓ, is in another stagnant ambient (say water) at the same temperature. Let the heat transfer coefficient for the lower and upper parts be h_1 and h_2, respectively. The cross section, periphery, and diameter of the rod are A, P, and D. We wish to determine the steady temperature distribution in the rod and to discuss the results in terms of the following data: $\ell = 15$ cm, $L = 5$ cm, $D = 1$ cm, $k = 15$ W/m·K (stainless steel) or $k = 400$ W/m·K (copper), $h_1 = 1{,}000$ W/m²·K, $h_2 = 10$ W/m²·K, $T_\infty = 20\,°\text{C}$.

Assume the radial Biot numbers are small enough to allow a **radially lumped** analysis. Neglecting the heat loss from the ends and ignoring, for the time being, conduction across the interface, the energy generated at each part is balanced by the convective heat loss to the water and air. Thus,

$$u''' A \ell = h_1 P \ell \Delta T_1, \quad u''' A \ell = h_2 P \ell \Delta T_2$$

or

$$\Delta T_1 = u''' A / h_1 P \quad \text{and} \quad \Delta T_2 = u''' A / h_2 P, \tag{2.132}$$

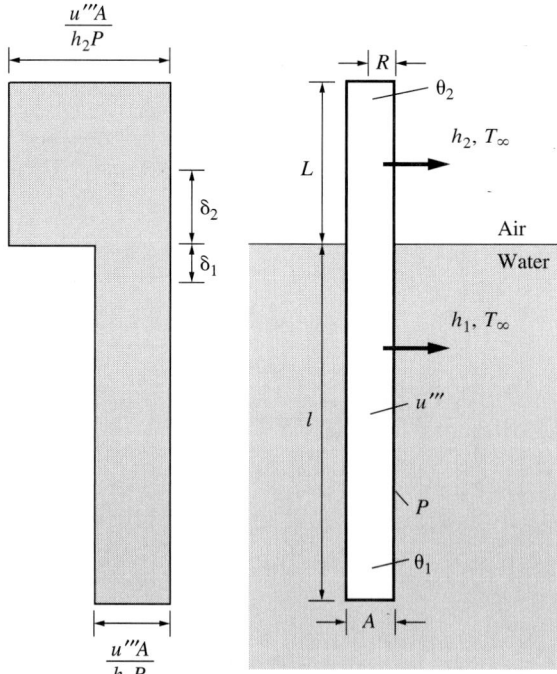

Figure 2.35 Energy generation in a rod.

which shows the average temperature difference between the rod and water or air (Fig. 2.35). In the actual case, the rod temperature drops continuously over the penetration depths δ_1 and δ_2. Since $\delta_1 \ll \delta_2$, the rod temperature in water may be assumed uniform (recall the preceding example). A better approximation, however, is to assume some temperature distribution for the part of the rod in water. Because of the lesser importance of this distribution relative to the temperature of the rod in air, we may further assume the rod length in water to be infinite. A model based on this approximation is given below.

A model. The temperature of the rod in air satisfying the insulated upper-tip boundary condition may be written, in terms of the coordinate axis measured from this tip downward, as

$$\theta_2(\xi) = \frac{u''' A}{h_2 P} + C_2 \cosh m_2 \xi. \tag{2.133}$$

Also, the temperature of the rod in the water, satisfying the condition of finite temperature at the lower tip (assumed extending to infinity), may be written, in terms of the coordinate axis measured from the interface downward, as

$$\theta_1(x) = \frac{u''' A}{h_1 P} + C_1 e^{-m_1 x}. \tag{2.134}$$

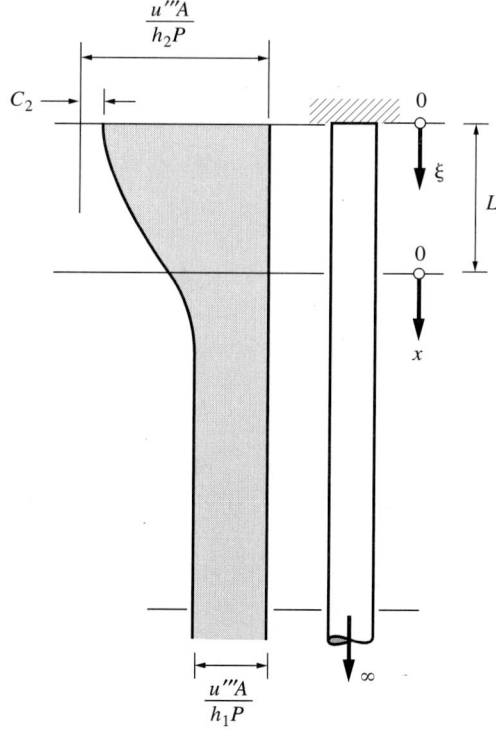

An approximate model

Figure 2.36 An approximate model for rod with energy generation.

The equality of interface heat fluxes,

$$\frac{d\theta_1(0)}{dx} = \frac{d\theta_2(L)}{d\xi},$$

gives

$$-m_1 C_1 = m_2 C_2 \sinh m_2 L$$

or

$$C_2 = -\left(\frac{m_1}{m_2}\right)\frac{C_1}{\sinh m_2 L}. \tag{2.135}$$

The equality of interface temperatures,

$$\theta_1(0) = \theta_2(L),$$

yields

$$\frac{u''' A}{h_1 P} + C_1 = \frac{u''' A}{h_2 P} + C_2 \cosh m_2 L,$$

or, in terms of Eq. (2.135),

$$\frac{u''' A}{h_1 P} + C_1 = \frac{u''' A}{h_2 P} - C_1 \frac{(m_1/m_2)}{\sinh m_2 L} \cosh m_2 L,$$

or

$$C_1 = \frac{\dfrac{u''' A}{h_2 P} - \dfrac{u''' A}{h_1 P}}{1 + \dfrac{m_1/m_2}{\tanh m_2 L}}, \tag{2.136}$$

which in turn gives, in terms of Eq. (2.135),

$$C_2 = -\frac{(m_1/m_2)}{\sinh m_2 L} \frac{\dfrac{u''' A}{h_2 P} - \dfrac{u''' A}{h_1 P}}{1 + \dfrac{m_1/m_2}{\tanh m_2 L}},$$

or

$$C_2 = -\frac{\left(\dfrac{u''' A}{h_2 P} - \dfrac{u''' A}{h_1 P}\right)/\cosh m_2 L}{1 + \dfrac{\tanh m_2 L}{m_1/m_2}}. \tag{2.137}$$

In terms of Eqs. (2.136) and (2.137), we obtain the entire rod temperature from Eqs. (2.133) and (2.134),

$$\frac{\theta_1(x) - \dfrac{u''' A}{h_1 P}}{\dfrac{u''' A}{h_2 P} - \dfrac{u''' A}{h_1 P}} = \frac{e^{-m_1 x}}{1 + \dfrac{m_1/m_2}{\tanh m_2 L}}, \qquad (2.138)$$

$$\frac{\dfrac{u''' A}{h_2 P} - \theta_2(\xi)}{\dfrac{u''' A}{h_2 P} - \dfrac{u''' A}{h_1 P}} = \frac{\cosh m_2 \xi / \cosh m_2 L}{1 + \dfrac{\tanh m_2 L}{m_1/m_2}}. \qquad (2.139)$$

◆

2.4.2 Performance

Here a basis may be introduced for the evaluation and comparison of extended surfaces. Such a basis is usually given in terms of one of the two following customary definitions: (1) **extended surface effectiveness** described as

$$\eta^* = \frac{\text{Actual heat transfer from extended surface}}{\text{Heat transfer from wall without fin}}, \qquad (2.140)$$

or, (2) **extended surface efficiency** described as

$$\eta = \frac{\text{Actual heat transfer from extended surface}}{\text{Heat transfer from extended surface at base}\,T}. \qquad (2.141)$$

The denominator of Eq. (2.140) denotes the heat transfer from an area of the wall equivalent to the base area of the extended surface; the heat transfer to be evaluated in the numerator and denominator is based on the same temperature difference, $T_0 - T_\infty$. Since the temperature of a wall and the heat transfer coefficient between the wall and the ambient are somewhat changed when an extended surface is attached to the wall, the efficiency defined by Eq. (2.140) is quite approximate. The error involved in this approximation depends on the length of and the space between the extended surfaces. Since the changes in the wall temperature and the heat transfer coefficient affect both the numerator and the denominator of Eq. (2.141), the efficiency defined by this equation is more realistic and is often preferred in practice. Furthermore, rather than using it only for one type of extended surface, this efficiency may be better utilized in the comparison of different extended surfaces. The particular value of the latter efficiency for Ex. 2–10 is

$$\eta = \frac{\theta_0 (h P k A)^{1/2} \tanh m\ell}{\theta_0 h P \ell} = \frac{\tanh m\ell}{m\ell}. \qquad (2.142)$$

In Fig. 2.37 this efficiency is compared with the efficiency of four different fins with variable cross section. Note that cost and manufacturing convenience may be more important factors than a 5–10% more efficient fin. For this reason, the efficiency of extended surfaces will not be further elaborated (for a detailed treatment see Ref. 7).

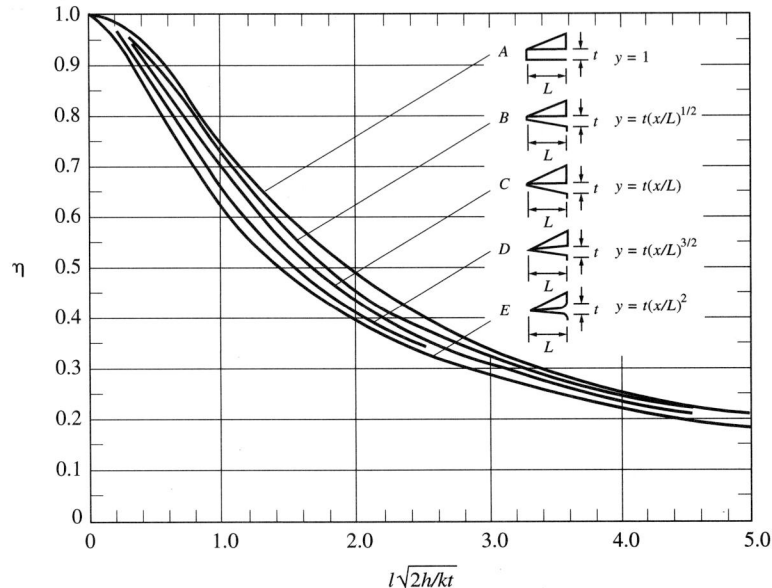

Figure 2.37 Efficiency of fins with variable cross section (from Gardner [7]).

In terms of the knowledge gained so far, we are ready now for a study of **two key problems** which will be utilized frequently in the chapters on convection.

2.5 TWO KEY PROBLEMS OF CONVECTION

A constant property fluid having velocity V and upstream temperature T_0 flows steadily through an infinitely long tube of cross sectional area A and periphery P. The upstream half of the tube is insulated, while the downstream half either transfers heat with a coefficient h to an ambient at temperature T_∞ [Fig. 2.38(a)] or is subjected to a peripheral heat flux q'' [Fig. 2.38(b)]. The wall thickness of the tube is negligible. Based on a radially lumped analysis, we wish to know the axial temperature distribution in the fluid.

2.5.1 First Key Problem

Consider the problem involving peripheral heat transfer to an ambient. Assume a radially lumped and axially differential **control volume** (Step 1) shown in Fig. 2.38(a). The first law[13] for this control volume (Step 2), interpreted in terms of Fig. 2.39 and with the conservation of mass,

$$\rho A V = \text{Const.}, \tag{2.143}$$

---

[13] The first law conserves the total energy, which, in the present case, involves the thermomechanical energy. The mechanical part of this energy leads, under certain conditions, to the Bernoulli equation, which can directly be obtained from Newton's law.

Sec. 2.5 Two Key Problems of Convection   91

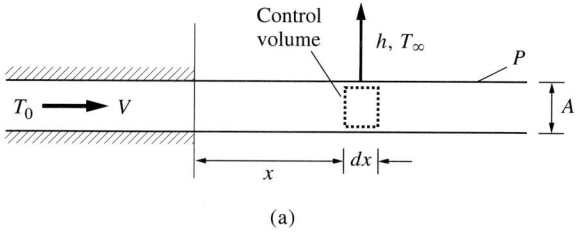

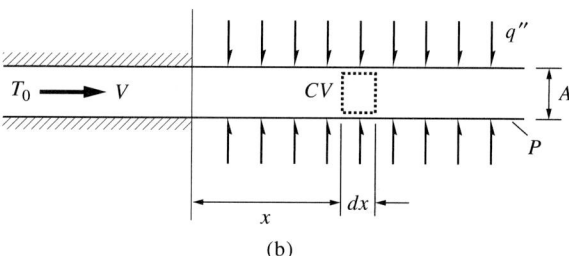

Figure 2.38 Two key problems of convection. (a) Heat transfer to ambient, (b) prescribed heat flux.

gives

$$0 = +\rho A V h^o - \rho A V \left(h^o + \frac{dh^o}{dx} dx \right) + q_x A - \left(q_x + \frac{dq_x}{dx} dx \right) A - q_C P \, dx,$$

or

$$0 = -\rho A V \frac{dh^o}{dx} - A \frac{dq_x}{dx} - q_C P. \tag{2.144}$$

Separately, the definition of the stagnation enthalpy,

$$h^o = h + \frac{1}{2} V^2 + gz, \tag{2.145}$$

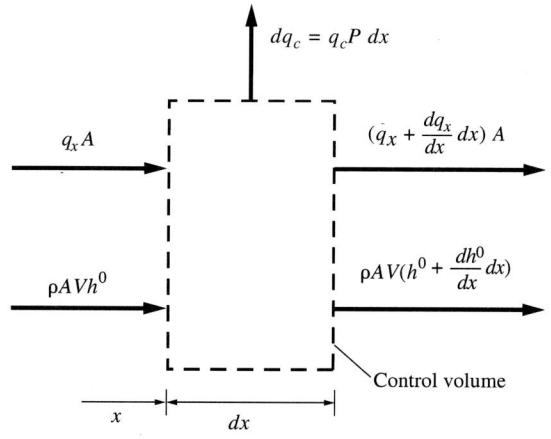

Figure 2.39 The first law for the first key problem of convection.

yields

$$\frac{dh^o}{dx} = \frac{dh}{dx} + \frac{d}{dx}\left(\frac{1}{2}V^2 + gz\right). \tag{2.146}$$

For an incompressible fluid, the conservation of mass for flow of this fluid through a tube of constant cross section implies $V = $ Constant. Furthermore, if the tube is horizontal, then $z = $ Constant. Thus

$$\frac{dh^o}{dx} = \frac{dh}{dx}. \tag{2.147}$$

Also, for this fluid, neglecting the small and usually ignored difference between the specific heat at constant pressure, c_p, and at constant volume, c_v,

$$c_p = c_v = c$$

and

$$dh = cdT. \tag{2.148}$$

Now Eq. (2.144) becomes, in terms of Eqs. (2.147) and (2.148),

$$0 = -\rho c A V \frac{dT}{dx} - A \frac{dq_x}{dx} - q_C P. \tag{2.149}$$

Inserting the particular laws (Step 3), $q_x = -kdT/dx$ and $q_C = U(T - T_\infty)$, into Eq. (2.149), we get the governing equation (Step 4)

$$kA\frac{d^2T}{dx^2} - \rho c A V \frac{dT}{dx} - UP(T - T_\infty) = 0, \tag{2.150}$$

where

$$U = \frac{1}{R_i + R_K + R_0},$$

U being the total heat transfer coefficient, R_i the inside convective resistance, R_K the conductive resistance of pipe walls, and R_0 the outside convective resistance. This result may be rearranged in terms of the characteristic length $1/m = \sqrt{kA/UP}$ for fins [recall Eq. (2.113)] introducing the **thermal diffusivity**, and $\theta = T - T_\infty$ as

$$\alpha = \frac{k}{\rho c}, \tag{2.151}$$

$$\frac{d^2\theta}{d\xi^2} - \frac{V(1/m)}{\alpha}\frac{d\theta}{d\xi} - \theta = 0, \tag{2.152}$$

where

$$\frac{V(1/m)}{\alpha} = Pe \tag{2.153}$$

is a dimensionless parameter called the **Peclet number** and $\xi = x/(1/m)$. This parameter is a measure of the importance of axial enthalpy flow relative to axial conduction,

$$\frac{\text{Enthalpy Flow}}{\text{Axial Conduction}} \sim \frac{\rho c A V \theta}{kA \dfrac{\theta}{(1/m)}} = \frac{V(1/m)}{\alpha}. \tag{2.154}$$

As $Pe \to \infty$, the contribution of axial conduction relative to axial enthalpy flow becomes negligible in Eqs. (2.150) and (2.152). For all practical problems

$$Pe \gg 1, \tag{2.155}$$

and the axial conduction can be neglected in the formulation of these problems. For example, for two fluids at temperature $T = 95\,°C$ flowing with velocity $V = 1$ m/s in a pipe of diameter $D = 2$ cm, Table 2.2 gives typical values for the Peclet number. This table shows that, except for slow-moving liquid metals, axial conduction does not play any appreciable role in convection problems. Then, Eq. (2.150) may be approximated by

$$\rho c A V \frac{d\theta}{dx} + U P \theta = 0, \tag{2.156}$$

subject to (Step 5)

$$\theta(0) = \theta_0 = T_0 - T_\infty. \tag{2.157}$$

Table 2.2 Range of Peclet number

Fluid	h [W/m²·K]	k [W/m·K]	m [m^{-1}]	α [m²/s]	Pe [–]
Water	10^4	0.678	1,718	1.68×10^{-7}	3,500
Sodium	10^5	86.2	482	671×10^{-7}	30

The solution of Eq. (2.156) may be readily obtained by the **separation of variables**,

$$\frac{d\theta}{\theta} = -\left(\frac{UP}{\rho c A V}\right) dx,$$

whose integration gives

$$\ln \theta = -\left(\frac{UP}{\rho c A V}\right) x + C_1,$$

or,

$$\theta = C \exp\left(-\frac{UPx}{\rho c A V}\right). \tag{2.158}$$

For the inlet condition [stated by Eq. (2.157)], Eq. (2.158) is reduced to

$$\theta_0 = C. \tag{2.159}$$

Then, the solution for the first key problem in terms of T is

$$\frac{T - T_\infty}{T_0 - T_\infty} = \exp\left(-\frac{UPx}{\rho c A V}\right). \tag{2.160}$$

EXAMPLE 2.13[14]

Water pressurized to 2 atm (abs) flows steadily with inlet velocity $V = 1$ m/s and inlet temperature $T_i = 97$ °C in an industrial heater made of 1-inch stainless steel pipes ($D_1 \cong 2.66$ cm and $D_2 = 3.34$ cm), elbowed and connected as shown in Fig. 2.40. The total pipe length is $L = 100$ m. Several of these heaters are attached to the side walls of an industrial plant to be kept at temperature $T_\infty = 27$ °C. The inside and outside heat transfer coefficients are $h_i = 10,000$ W/m²·K and $h_0 = 10$ W/m²·K, respectively. The thermal conductivity of the pipe is $k = 15$ W/m·K. We wish to determine the exit temperature of the water.

Noting that $h_i \gg h_0$ and

$$R_i \ll R_0,$$

R_i can be neglected relative to R_0. Also,

$$\frac{R_K}{R_0} = \frac{\ell_{12}/k\bar{A}_{12}}{1/hA_2} = \frac{h\ell_{12}}{k}\left(\frac{A_2}{\bar{A}_{12}}\right),$$

which, in view of $\ell_{12} = r_2 - r_1$, and $\bar{A}_{12} = (A_1 + A_2)/2$ because of

$$A_2/A_1 = D_2/D_1 = 3.34/2.66 = 1.26 < 2,$$

gives

$$\frac{R_K}{R_0} = \frac{2hr_2}{k}\frac{(r_2 - r_1)}{(r_2 + r_1)} = \frac{h_0 D_2}{k}\left(\frac{D_2 - D_1}{D_2 + D_1}\right) = \frac{1}{396},$$

and R_K can be neglected relative to R_0. Thus, for the present problem,

$$U \cong h_0.$$

To proceed further, we need ρ and c of water and P and A of pipe. Since only the inlet temperature and velocity are given, for $T_i = 97 + 273 = 370$ K, from Table B.3 for water,

$$v = 1.041 \times 10^{-3} \text{m}^3/\text{kg} \quad \text{or} \quad \rho \cong 961 \text{kg/m}^3,$$

$$c_p = 4.214 \times 10^3 \text{J/kg·K}.$$

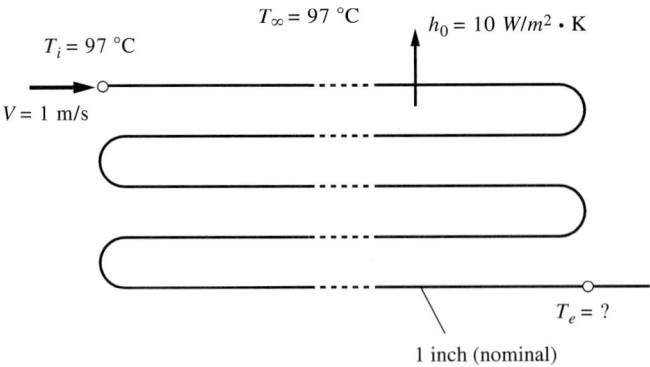

Figure 2.40 An industrial heater.

---

[14] The FORTRAN program EX2–13.F is listed in the appendix of this chapter.

Note that $\rho AV = $ Constant along the pipe because of the conservation of mass, and this constant can be evaluated at any location in the pipe. Also, from the pipe geometry,

$$P = \pi D_2 = \pi \times 3.34 \times 10^{-2} \text{m} = 0.105 \text{m},$$

$$A = \pi D_1^2/4 = \pi \times (2.66 \times 10^{-2})^2/4 = 0.556 \times 10^{-3} \text{ m}^2.$$

Then,

$$\frac{h_0 PL}{\rho c_p AV} = \frac{10 \frac{\text{W}}{\text{m}^2 \cdot \text{K}} \times 0.105 \text{m} \times 100 \text{m}}{961 \text{kg/m}^3 \times 0.556 \times 10^{-3} \text{m}^2 \times 1 \text{m/s} \times 4.214 \times 10^3 \text{J/kg} \cdot \text{K}} = 0.0466$$

and

$$\frac{T_e - 27}{97 - 27} = e^{-0.0466} = 0.954,$$

which gives

$$T_e \cong 94\,°\text{C}.$$

A temperature drop of only 3 °C along a pipe 100 m long may be a surprising result, especially in view of a slow water velocity of 1 m/s! However, recognition of the fact that natural convection in a gas leads to very low heat transfer coefficients eliminates the initial surprise. For all practical purposes, **a stagnant gas acts like an insulator**. By the computer program provided, the interested reader may parametrically study the exit temperature for various values of h_i and h_o taken from Table 1.2.

We proceed now to the second key problem. In view of the foregoing discussion, the axial conduction will be excluded from the beginning. ◆

2.5.2 Second Key Problem

Reconsider the control volume used for the first key problem. Since the axial conduction is neglected and the peripheral flux is specified, there is no need for any particular law. We now have a thermodynamically determined problem. The first law applied to the control volume shown in Fig. 2.41 directly gives the governing equation subject to the inlet boundary condition. The formulation is then

$$-\rho c A V \frac{dT}{dx} + q'' P = 0, \quad T(0) = T_0. \tag{2.161}$$

The separation of variables followed by a simple integration readily yields the solution of Eq. (2.161) as

$$\frac{T(x) - T_0}{q''/\rho c V} = \left(\frac{P}{A}\right) x = 4\left(\frac{x}{D}\right). \tag{2.162}$$

Figure 2.42 shows this (linear) temperature distribution.

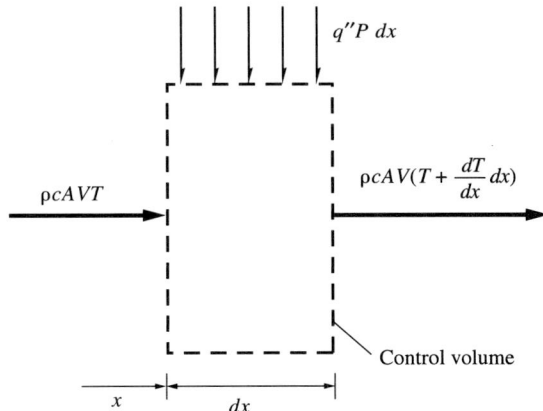

Figure 2.41 The first law for the second key problem of convection.

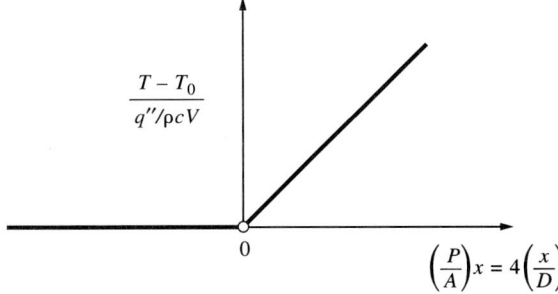

Figure 2.42 The second key problem of convection (excluding axial conduction).

EXAMPLE 2.14[15]

Water pressurized to 2 atm (abs) flows steadily with inlet velocity $V = 0.05$ m/s and inlet temperature $T_i = 20\,°C$ in an industrial evaporator made of a 1 inch (nominal) diameter and 20 m long insulated pipe (Fig. 2.43). The pipe is heated electrically. We wish to determine the power need for which the water evaporates completely at the exit of the pipe and the fraction of pipe length at which the water begins to evaporate.

Assume axial conduction to be negligible. Then, all electrical energy generated in the pipe walls is transferred to the fluid. For a control volume involving the entire fluid (Fig. 2.44), the first law of thermodynamics yields, after recognizing the fact that there is no change in kinetic and potential energies along the pipe,

$$0 = \dot{m}_i h_i - \dot{m}_e h_e + \dot{W}_e, \tag{2.163}$$

$\dot{W}_e$ being the electrical power input. Clearly, this power is related to the peripheral heat flux q'' and the rate of energy generation per unit volume u''' as

$$\dot{W}_e = q'' P \ell = u''' A_p \ell, \tag{2.164}$$

where P is the inside perimeter and A_p the cross-sectional area of the pipe.

---

[15] The FORTRAN program EX2–14.F is listed in the appendix of this chapter.

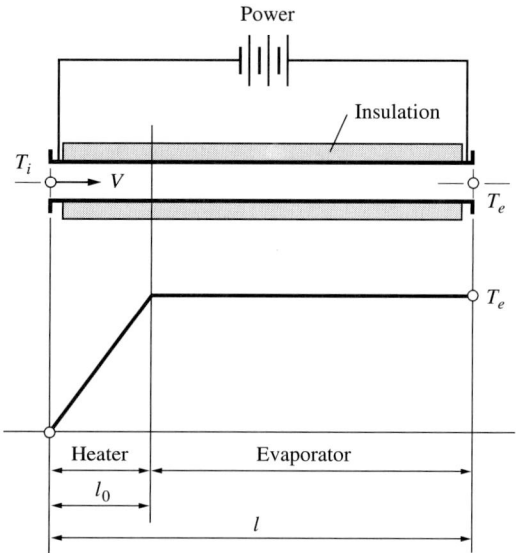

Figure 2.43 An industrial evaporator.

Also, the conservation of mass applied to the same control volume gives

$$\rho_i A V_i = \rho_e A V_e,$$

or,

$$\dot{m}_i = \dot{m}_e = \dot{m} = \text{Const}. \qquad (2.165)$$

Then, from the combination of Eqs. (2.163) and (2.165),

$$\dot{W}_e = \dot{m}(h_e - h_i),$$

or, explicitly,

$$\dot{W}_e = \dot{m}\left[(c_{pe}T_e - c_{pi}T_i) + (h_{ev} - h_{ew})\right], \qquad (2.166)$$

h_{ev} and h_{ew} denoting the enthalpy of vapor and liquid water at the saturation temperature T_e. Assuming

$$c_{pe} \cong c_{pi} \cong c_p = \text{Const},$$

letting

$$T_e - T_i = \Delta T,$$

and in terms of the usual notation of steam tables,

$$h_{ev} - h_{ew} = h_g - h_f = h_{fg},$$

Eq. (2.166) may be rearranged as

$$\dot{W}_e = \dot{m}\left(c_p \Delta T + h_{fg}\right), \qquad (2.167)$$

where the first and second terms on the right side respectively denote the sensible and latent parts of the enthalpy change. Also, the same two terms respectively correspond to the heater and evaporator parts of the pipe (see Fig. 2.43).

98 Chap. 2 Steady Conduction

For the heater part of the pipe alone

$$\dot{W}_{eo} = \dot{m}(c_p \Delta T), \qquad (2.168)$$

$\dot{W}_{eo}$ being the fraction of electrical power input for heating. The ratio of Eqs. (2.167) and (2.168) gives, in terms of Eq. (2.164),

$$\frac{\dot{W}_{eo}}{\dot{W}_e} = \frac{q'' P \ell_0}{q'' P \ell} = \frac{\dot{m}(c_p \Delta T)}{\dot{m}(c_p \Delta T + h_{fg})},$$

or, the fraction of pipe length corresponding to the heater part of the pipe,

$$\frac{\dot{W}_{eo}}{\dot{W}_e} = \frac{\ell_0}{\ell} = \frac{1}{1 + h_{fg}/c_p \Delta T}. \qquad (2.169)$$

Beyond length ℓ_0, the water begins to evaporate. For 2 atm (abs), the steam tables give the saturation temperature, $T_e \cong 120\,°C$, and the latent heat for this temperature, $h_{fg} = 2{,}201$ kJ/kg. Appendix B respectively gives for 20 °C and 120 °C water,

$$\rho_i = 998 \text{ kg/m}^3,$$

$$c_{pi} = 4.182 \text{ kJ/kg·K} \quad \text{and} \quad c_{pe} = 4.244 \text{ kJ/kg·K},$$

or, approximately,

$$c_p \cong 4.2 \text{ kJ/kg·K}.$$

Also, from the geometry of the pipe ($D_i = 2.66$ cm),

$$A_i = \pi D_i^2/4 = \pi \times (2.66 \times 10^{-2})^2/4 = 0.556 \times 10^{-3} \text{m}^2.$$

In terms of the foregoing results, the mass flow is

$$\dot{m} = \rho_i A_i V = 998 \text{ kg/m}^3 \times 0.556 \times 10^{-3} \text{m}^2 \times 0.05 \text{ m/s} \cong 2.77 \times 10^{-2} \text{kg/s}.$$

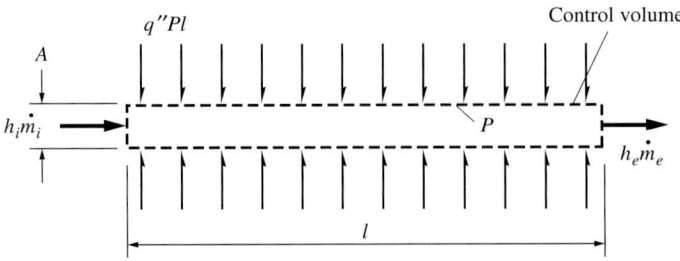

Figure 2.44 First Law for pipe control volume.

Also, noting

$$\Delta T = T_e - T_i = 120 - 20 = 100\,°C,$$

the need of electrical power is found from Eq. (2.167)

$$\dot{W}_e = (4.2 \times 100 + 2{,}201)\text{kJ/kg} \times 2.77 \times 10^{-2}\text{kg/s} \cong 73\,\text{kW},$$

and, the fraction of this power and of the pipe length for the heater section is obtained from Eq. (2.169),

$$\frac{\dot{W}_{eo}}{\dot{W}_e} = \frac{\ell_0}{\ell} = \frac{1}{1 + 2{,}201/4.2 \times 100} = 0.16$$

which gives

$$\dot{W}_{eo} = 0.16 \times 73 \cong 11.7\,\text{kW}$$

and

$$\ell_0 = 0.16 \times 20 = 3.2\,\text{m}.$$

We clearly know from Eq. (2.162) that the temperature increases linearly in the heater part of the pipe. Once it begins to evaporate, the temperature remains constant and equal to the saturation temperature.

The knowledge accumulated so far on steady, one-dimensional conduction, especially that of Sections 2.3 (energy generation) and 2.4 (extended surfaces), finds important applications in solar collectors and in (nuclear) reactor cores, which we consider next. ◆

2.6 SOLAR COLLECTOR ⊕

First consider the solar collector because of its relative simplicity. A solar collector may be demonstrated in terms of a pipe or channel flow heated by a distant energy source (solar radiation). Depending on direct or concentrated use of the energy source, we distinguish between **flat-plate** and **concentrating** collectors. In a flat-plate collector the area absorbing solar radiation is the same as the area intercepting this radiation. In a concentrating collector the (pipe or channel) area absorbing solar radiation is much smaller than the (reflector) area intercepting and focusing this radiation.

Here, we consider a concentrating collector as shown in Fig. 2.45(a). With proper interpretation, however, the formulation given below may be applied to flat-plate as well as concentrating collectors.

Neglecting the effect of axial conduction, considering a radially lumped analysis, and assuming a peripherally uniform q'', this formulation [obtained by an appropri-

100 Chap. 2 Steady Conduction

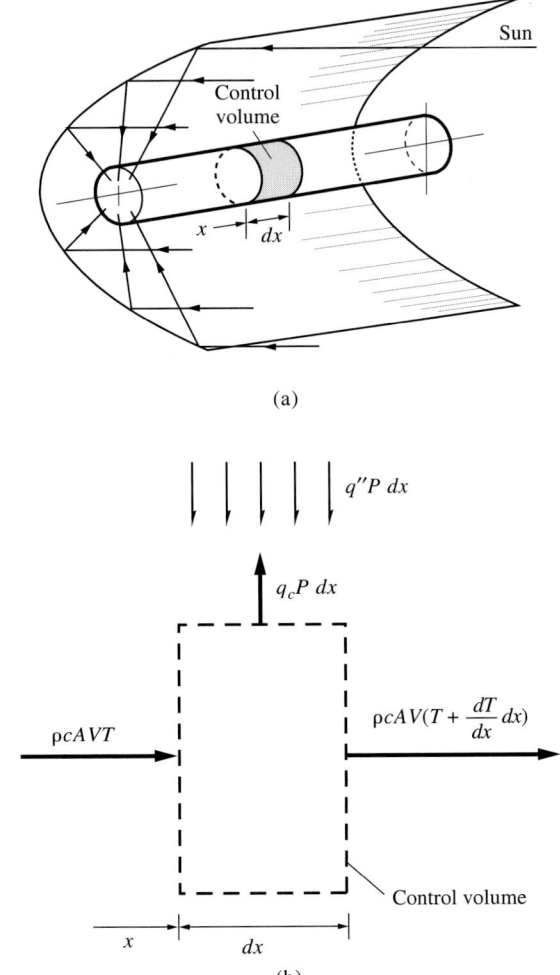

Figure 2.45 (a) A concentrating collector, (b) the first law for a collector.

ate combination of the first and second key problems of convection or by the help of Fig. 2.45(b)] may be written as

$$\rho c A V \frac{dT}{dx} - q'' P + hP(T - T_\infty) = 0, \qquad (2.170)$$

$$T(0) = T_0 \qquad (2.171)$$

[a direct superposition of the two key problems does *not* lead to Eq. (2.170)]. For a solution, rearrange Eq. (2.170) as

$$\rho c A V \frac{dT}{dx} + hP\left(T - T_\infty - \frac{q''}{h}\right) = 0, \qquad (2.172)$$

or, in terms of

$$\Theta(x) = T - T_\infty - \frac{q''}{h},$$

as

$$\frac{d\Theta}{dx} + \left(\frac{hP}{\rho c A V}\right)\Theta = 0 \qquad (2.173)$$

subject to

$$\Theta(0) = T_0 - T_\infty - \frac{q''}{h}. \qquad (2.174)$$

We have already learned the solution of Eq. (2.173) in connection with the first key problem. From the development leading to Eq. (2.158), we have

$$\Theta = T - T_\infty - \frac{q''}{h} = C \exp\left(-\frac{hPx}{\rho c A V}\right), \qquad (2.175)$$

which gives, for Eq. (2.174),

$$\Theta(0) = T_0 - T_\infty - \frac{q''}{h} = C. \qquad (2.176)$$

The combination of Eqs. (2.175) and (2.176) yields the temperature distribution in the solar collector,

$$\frac{T - T_\infty - \frac{q''}{h}}{T_0 - T_\infty - \frac{q''}{h}} = \exp\left(-\frac{hPx}{\rho c A V}\right), \qquad (2.177)$$

which can be rearranged as follows:

$$T(x) - T_\infty = (T_0 - T_\infty)\exp\left(-\frac{hPx}{\rho c A V}\right) + \frac{q''}{h}\left[1 - \exp\left(-\frac{hPx}{\rho c A V}\right)\right], \qquad (2.178)$$

where the first righthand term gives the temperature distribution resulting from the difference between the fluid inlet and the ambient temperatures, $T_0 - T_\infty$, and the second righthand term gives the temperature distribution resulting from the distant heat flux, q''. Note that the first term is the solution of the first key problem but the second term is *not* the solution of the second key problem. The reason for this difference is that q'' of the second key problem acts on the pipe and disallows any convection, while q'' of the present problem acts at a distance and allows convection.

2.7 REACTOR CORE ⊕

A nuclear power reactor produces heat for the thermal cycle of a steam power plant that generates electrical energy. A schematic diagram of such plant is shown in Fig. 2.46. The thermal analysis involving the temperature distribution and heat transfer in a core (like the one shown in Fig. 2.47) plays an important role in reactor design. The core design must provide the desired thermal power without exceeding temperature limitations on core components that might lead to fuel failure and the release of radioactive material into the coolant. More specifically, the capacity of a cooling system limits the core power density, which determines the core size necessary to meet the desired reactor output of thermal power.

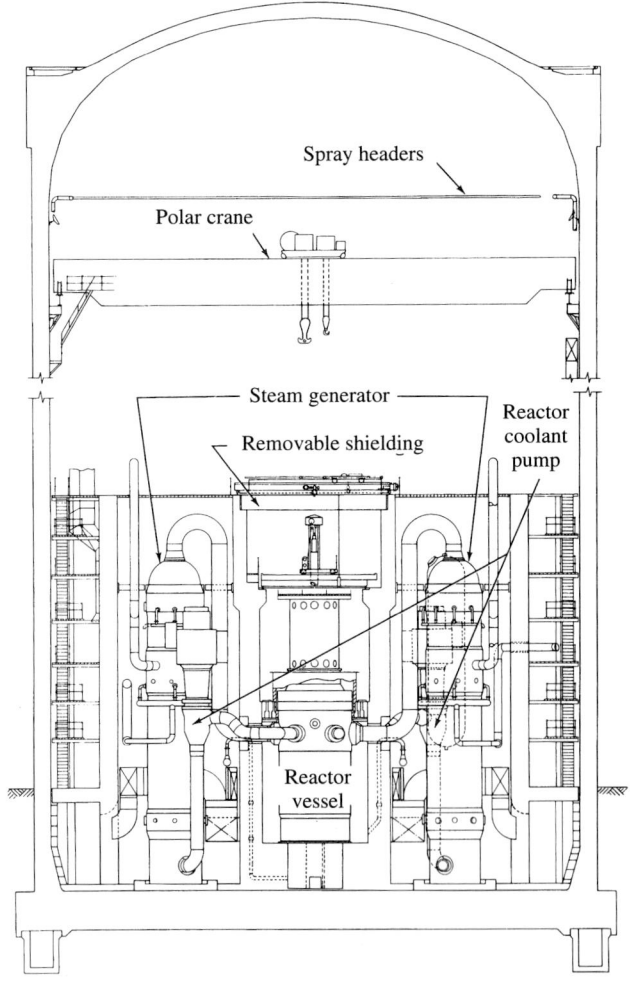

Figure 2.46 A schematic diagram of a nuclear power plant (from Babcock & Wilcox [8]).

The energy released by nuclear fission reactions appears primarily as kinetic energy of the various fission products. Most of this fission energy is converted rapidly to internal energy in the neighboring fuel material. This energy is then transferred by conduction through the fuel element, through the gap separating the fuel from the clad, and then through the clad to the clad surface. It is then transferred by convection from the clad surface to a coolant (Fig. 2.48). The moving coolant then carries the thermal energy up and out of the reactor core, either as sensible heat (increased temperature) or latent heat (boiling).

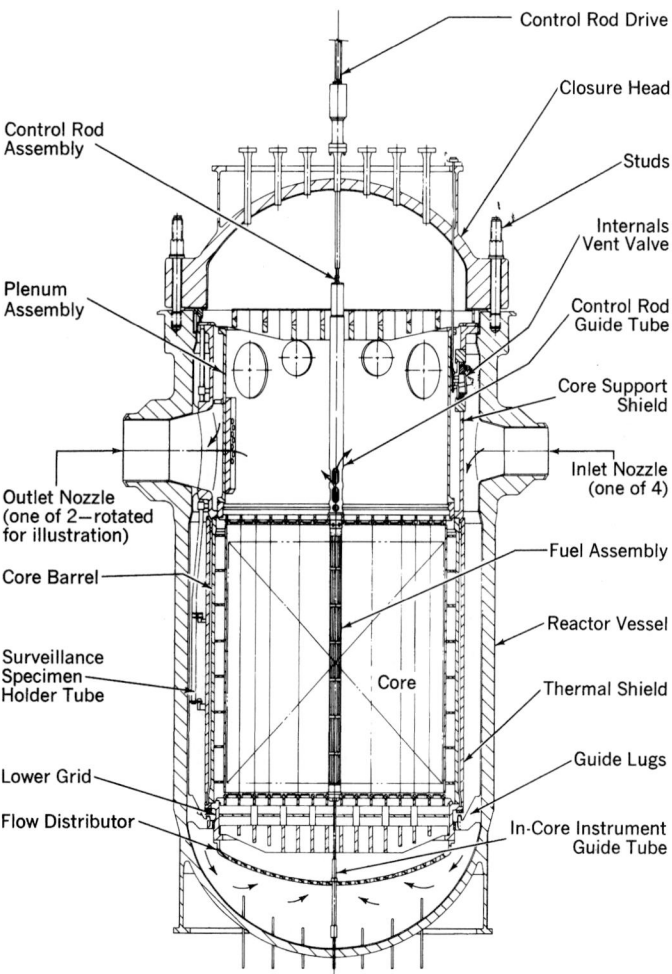

Figure 2.47 Core of a nuclear reactor (from Babcock & Wilcox [8]).

The simplest model for fission energy generation corresponds to a bare, homogeneous core. The geometry of most practical importance is the cylindrical core, for which the distribution of (radial and axial) energy generation is given by

$$u''' = u_0''' J_0\left(2.405\frac{r}{r_E}\right) \sin \pi \left(\frac{x}{H_E}\right), \qquad (2.179)$$

where J_0 is the Bessel function of the first kind of order zero, and r_E and H_E are the effective core dimensions. Figure 2.49 shows a sketch of the fission energy distribution. How this energy distribution is related to the neutron flux is the concern of the nuclear engineer. Here, we are interested in the temperature distribution in a core resulting from energy generation.

The assumptions to be made in the formulation are:

1. The temperature of the coolant is lumped radially.

2. The cross section of a fuel element is small compared with that of the core and the energy generation affecting the fuel element is uniform radially.

3. Axial conduction in the coolant is negligible compared to axial enthalpy flow.

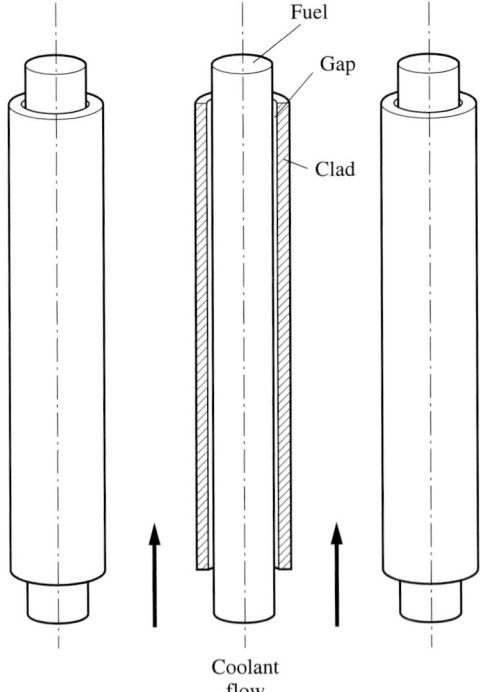

Figure 2.48 Fuel elements of a nuclear core.

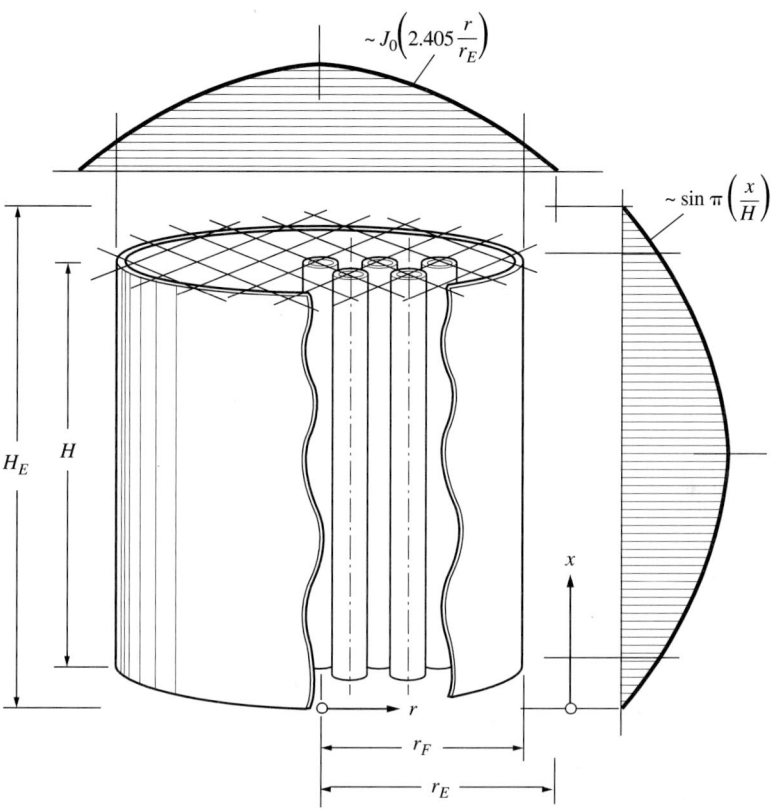

Figure 2.49 Fission energy distribution in a nuclear core.

4. The axial temperature gradient is an order of magnitude smaller than the radial temperature gradient, hence the axial conduction in the fuel, gap, and clad is negligible.

5. Radial and axial extrapolation of energy generation is negligible, hence $r_E \cong r_F$ and $H_E \cong H$.

6. The gap thickness and the clad thickness are small compared to the radius of the fuel.

Under steady conditions and assumption (4), the first law applied to lumped systems for the fuel, gap, and clad states that all the fission energy u''' generated in the fuel is transferred from the fuel, through the gap and through the clad, to the coolant. Hence,

$$u''' A = q_F'' P_F = q_G'' P_G = q_C'' P_C = \text{Const.} \tag{2.180}$$

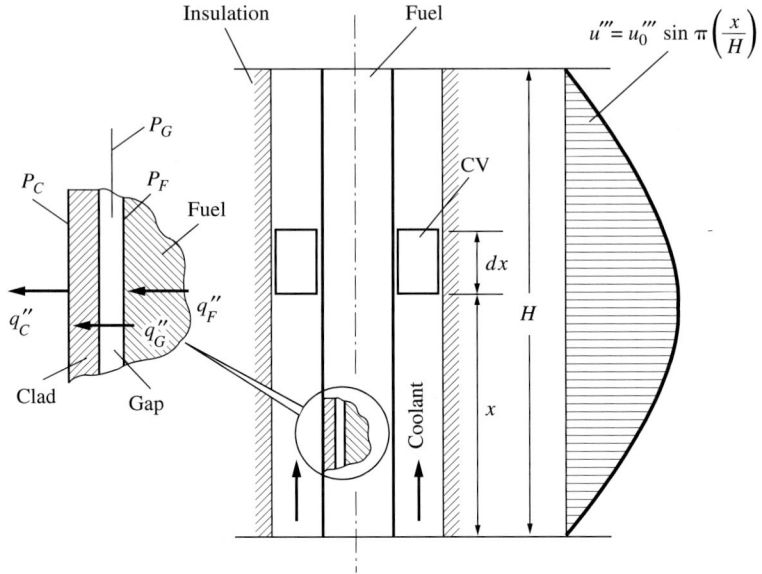

Figure 2.50 Control volume for coolant.

The first law applied to the control volume for the coolant shown in Fig. 2.50 yields, in terms of Fig. 2.51 (the second key problem of convection),

$$\dot{m} c \frac{dT_C}{dx} = q''_C P_C = u''' A,$$

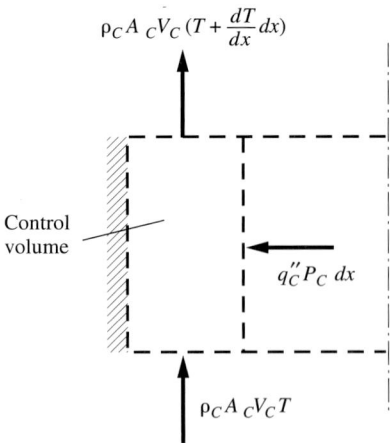

Figure 2.51 The first law for coolant.

which may be rearranged, after ignoring the radial distribution of energy generation, as

$$\dot{m}c\frac{dT_C}{dx} = u_0''' A \sin\pi\left(\frac{x}{H}\right), \quad T_C(0) = T_{Ci}, \tag{2.181}$$

where $\dot{m} = \rho_C A_C V$ is the mass flow rate of the coolant.

Separation of variables followed by an integration readily gives the solution of Eq. (2.181) as

$$\dot{m}c(T_C - T_{Ci}) = u_0''' A \left(\frac{H}{\pi}\right)\left(1 - \cos\pi\frac{x}{H}\right),$$

which, in terms of the outlet temperature of the coolant, obtained by letting $x = H$ and also noting $\cos\pi = -1$,

$$\dot{m}c(T_{Co} - T_{Ci}) = 2u_0''' A \left(\frac{H}{\pi}\right), \tag{2.182}$$

gives the dimensionless coolant temperature

$$\frac{T_C - T_{Ci}}{T_{Co} - T_{Ci}} = \frac{1}{2}\left(1 - \cos\pi\frac{x}{H}\right). \tag{2.183}$$

Under assumption (6), Eq. (2.180) first expressed in terms of Newton's law, $h\Delta T_{\text{Coolant}}$, yields the temperature drop in the coolant,

$$\Delta T_{\text{Coolant}} = \frac{u''' A}{h P_F}; \tag{2.184}$$

next expressed in terms of Fourier's law for the clad, $k_C \Delta T_{\text{Clad}}/t_C$, k_C and t_C being the thermal conductivity and thickness of the clad, yields the temperature drop across the clad,

$$\Delta T_{\text{Clad}} = \frac{u''' A}{P_F}\left(\frac{t_C}{k_C}\right); \tag{2.185}$$

finally expressed in terms of the total heat transfer coefficient U_G for the gap yields the temperature drop across the gap,

$$\Delta T_{\text{Gap}} = \frac{u''' A}{U_G P_F}. \tag{2.186}$$

Also, the temperature drop in the fuel (recall Fig. 2.22) is

$$\Delta T_{\text{Fuel}} = \frac{u'''r_F^2}{4k_F} = \frac{u'''r_F A}{2k_F P_F}, \qquad (2.187)$$

k_F being the thermal conductivity of the fuel and $A/P_F = r_F/2$.

The sum of Eqs. (2.184)-(2.187) results in the temperature drop between the fuel center and the coolant:

$$\Delta T_{\text{Fuel center - Coolant}} = \Delta T_{\text{Fuel}} + \Delta T_{\text{Gap}} + \Delta T_{\text{Clad}} + \Delta T_{\text{Coolant}}$$

$$= \frac{u''' A}{P_F} \left(\frac{r_F}{2k_F} + \frac{1}{U_G} + \frac{t_C}{k_C} + \frac{1}{h} \right),$$

which, after nondimensionalizing in terms of Eq. (2.182), using $u'''/u_0 = \sin \pi(x/H)$, and rearranging, gives

$$\frac{\Delta T_{\text{Fuel center - Coolant}}}{T_{Co} - T_{Ci}}$$

$$= \frac{\dot{m}c}{8Hk_F} \left(1 + 2\frac{k_F}{U_G r_F} + 2\frac{t_C}{r_F}\frac{k_F}{k_C} + 2\frac{k_F}{hr_F} \right) \sin \pi \frac{x}{H}. \qquad (2.188)$$

The axial temperature rise in the coolant, Eq. (2.183), the radial temperature drop and the axial temperature distribution in the fuel, the gap, the clad, and the coolant, Eq. (2.188), are sketched in Fig. 2.52. Some typical values encountered in practice for the radial temperature drop are: $\Delta T_{\text{Fuel}} \sim 1500\,°C$, $\Delta T_{\text{Gap}} \sim 150-300\,°C$, $\Delta T_{\text{Clad}} \sim 50\,°C$, and $\Delta T_{\text{Coolant}} \sim 5\,°C$ (for water). Also, some values for the geometry, thermal conductivity and heat transfer coefficient are:

$t_G \sim 0.005$ cm, $\quad t_C \sim 0.05$ cm,

$k_F \sim 4$ W/m·K, $\quad k_C \sim 15$ W/m·K,

$U_G \sim 6{,}000 - 12{,}000$ W/m²·K, $\quad h \sim 30{,}000 - 60{,}000$ W/m²·K.

The reader should keep in mind the rather approximate nature of the foregoing consideration. Variable properties, fuel swelling, nonuniform fission energy generation resulting from loading or age, radial depression of the energy generation in the fuel resulting from self-shielding, coolant phase change, etc., may play a significant role in actual cases.

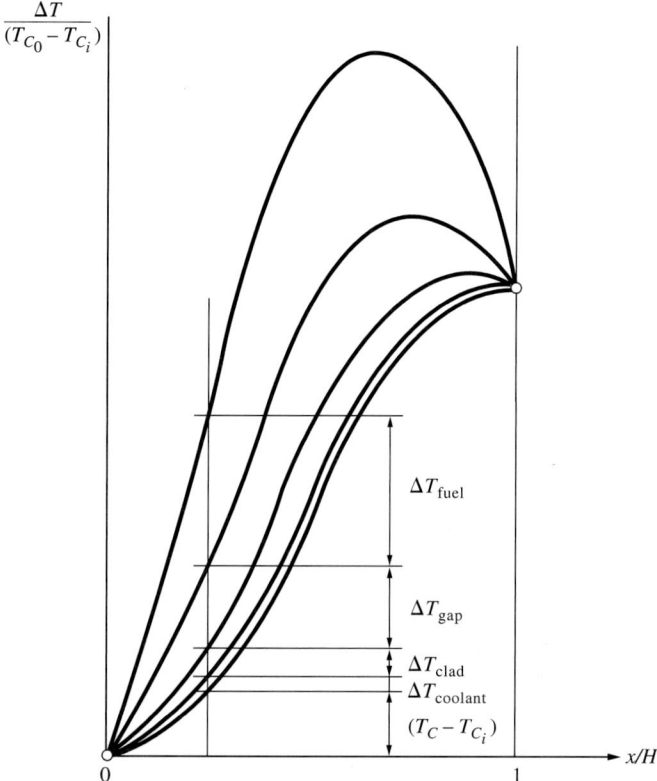

Figure 2.52 Temperature distribution in a fuel element.

■ REFERENCES

2.1 V. S. Arpacı, *Conduction Heat Transfer*. Addison-Wesley, Reading, MA 1966.

2.2 V. S. Arpacı and P. S. Larsen, *Convection Heat Transfer*. Prentice-Hall, Englewood Cliffs, NJ, 1984.

2.3 H. Fenech and W. M. Rohsenow, Chapter 16, "Heat Transfer," in *The Technology of Nuclear Reactor Safety*, T.J. Thompson and J.G. Beckerley (eds.). MIT Press, 1973.

2.4 J. J. Duderstadt and L. J. Hamilton. *Nuclear Reactor Analysis*. Wiley, New York, 1976.

2.5 M. M. El-Wakil, *Nuclear Energy Conversion*. Intext Press, New York, 1971.

2.6 M. G. Cooper, B. B. Mikic, and M. M. Yovanovich, "Thermal Contact Conductances," *Int. J. Heat Mass Trans.*, 12, 279, 1969.

2.7 K. A. Gardner, "Efficiency of Extended Surfaces," *Trans. ASME*, 67, 621, 1945.

2.8 Steam:, *Its Generation and Use*, Babcock & Wilcox, New York, 1978.

110 Chap. 2 Steady Conduction

2.8 COMPUTER PROGRAM APPENDIX

```
C-----------------------------------
C     EX2-1.F (START)
C-----------------------------------
 PROGRAM MAIN
 IMPLICIT REAL*8 (A-H,K-Z)
 PI=4*ATAN(1.)
 WRITE(*,*) 'EXAMPLE 2.1....'
C--------------------------------------------------
C     INPUT DATA
C--------------------------------------------------
 WRITE(*,*) 'INPUT THE FOLLOWING DATA...'
 WRITE(*,*) 'T_INFTY: C'
 READ(*,*) TINFTY
 WRITE(*,*) 'T_M: C'
 READ(*,*) TM
 WRITE(*,*) 'L_1: cm'
 READ(*,*) L1
 WRITE(*,*) 'L_2: cm'
 READ(*,*) L2
     WRITE(*,*) 'K_1: W/m.K'
 READ(*,*) K1
     WRITE(*,*) 'K_2: W/m.K'
 READ(*,*) K2
 WRITE(*,*) 'H_1: W/m^2.K'
 READ(*,*) H1
 WRITE(*,*) 'H_2: W/m^2.K'
 READ(*,*) H2
C--------------------------------------------------
C     UNIT CONVERSION
C--------------------------------------------------
 L1=0.01*L1
 L2=0.01*L2
C--------------------------------------------------
C     CALCULATION
C--------------------------------------------------
 R1=L1/K1+1/H1
 R2=L2/K2+1/H2
 RT=1/(1/R1+1/R2)
 Q=(TM-TINFTY)/RT
C--------------------------------------------------
C     ANSWER
C--------------------------------------------------
 WRITE(*,*) 'RADIANT HEAT FLUX NEEDED FOR THIS PROCESS IS'
 WRITE(*,*) Q/1000,' kW/m^2'
 STOP
 END
C-----------------------------------
C     EX2-1.F (END)
C-----------------------------------
C-----------------------------------
C     EX2-2.F (START)
C-----------------------------------
 PROGRAM MAIN
 IMPLICIT REAL*8 (A-H,K-Z)
 PI=4*ATAN(1.)
 WRITE(*,*) 'EXAMPLE 2.2....'
C--------------------------------------------------
C     INPUT DATA
```

```
C-----------------------------------------------------
      WRITE(*,*) 'INPUT THE FOLLOWING DATA...'
      WRITE(*,*) 'R_1: cm'
      READ(*,*) R1
      WRITE(*,*) 'R_2: cm'
      READ(*,*) R2
      WRITE(*,*) 'H_I: W/m^2.K'
      READ(*,*) HI
      WRITE(*,*) 'H_O: W/m^2.K'
      READ(*,*) HO
         WRITE(*,*) 'K: W/m.K'
      READ(*,*) K
      WRITE(*,*) 'T_INFTY: C'
      READ(*,*) TINFTY
         WRITE(*,*) 'K_FIBER: W/m.K'
      READ(*,*) KFIBER
      WRITE(*,*) 'L_FIBER: mm'
      READ(*,*) LFIBER
C-----------------------------------------------------
C     UNIT CONVERSION
C-----------------------------------------------------
      R1=0.01*R1
      R2=0.01*R2
      LFIBER=0.001*LFIBER
C-----------------------------------------------------
C     CALCULATION
C-----------------------------------------------------
      RI=1/(HI*2*PI*R1)
      R1=(R2-R1)/(K*PI*(R2+R1))
      RO=1/(HO*2*PI*R2)
      Q1=(120-20)/(RI+R1+RO)
      R3=R2+LFIBER
      R2=(R3-R2)/(KFIBER*PI*(R3+R2))
      RO=1/(HO*2*PI*R3)
      Q2=(120-20)/(R2+RO)
C-----------------------------------------------------
C     ANSWER
C-----------------------------------------------------
      WRITE(*,*) 'HEAT LOSS PER UNIT LENGTH OF BARE PIPE IS'
      WRITE(*,*) Q1,' W/m'
      WRITE(*,*) 'HEAT LOSS PER UNIT LENGTH OF INSULATED PIPE IS'
      WRITE(*,*) Q2,' W/m'
      STOP
      END
C---------------------------------------
C     EX2-1.F (END)
C---------------------------------------
C---------------------------------------
C     EX2-13.F (START)
C---------------------------------------
      PROGRAM MAIN
      IMPLICIT REAL*8 (A-H,K-Z)
      PI=4*ATAN(1.)
      WRITE(*,*) 'EXAMPLE 2.13....'
C---------------------------------------
C     INPUT DATA
```

```fortran
C------------------------------------------------
      WRITE(*,*) 'INPUT THE FOLLOWING DATA...'
      WRITE(*,*) 'V: m/s'
      READ(*,*) V
      WRITE(*,*) 'T_I: C'
      READ(*,*) TI
      WRITE(*,*) 'D_1: cm'
      READ(*,*) D1
      WRITE(*,*) 'D_2: cm'
      READ(*,*) D2
      WRITE(*,*) 'L: m'
      READ(*,*) L
      WRITE(*,*) 'T_INFTY: C'
      READ(*,*) TINFTY
      WRITE(*,*) 'H_I: W/m^2.K'
      READ(*,*) HI
      WRITE(*,*) 'H_O: W/m^2.K'
      READ(*,*) HO
         WRITE(*,*) 'K: W/m.K'
      READ(*,*) K
      WRITE(*,*) 'DENSITY (kg/m^3) OF WATER AT   ',TI+273,' K'
      READ(*,*) RHO
         WRITE(*,*) 'SPECIFIC HEAT (J/kg.K) OF WATER AT   ',TI+273,' K'
      READ(*,*) CP
C------------------------------------------------
C     UNIT CONVERSION
C------------------------------------------------
      D1=0.01*D1
      D2=0.01*D2
C------------------------------------------------
C     CALCULATION
C------------------------------------------------
      P=PI*D2
      A=PI*D1**2/4
      TE=TINFTY+(TI-TINFTY)*EXP(-HO*P*L/(RHO*CP*A*V))
C------------------------------------------------
C     ANSWER
C------------------------------------------------
      WRITE(*,*) 'EXIT TEMPERATURE OF THE WATER IS'
      WRITE(*,*) TE,' C'
      STOP
      END
C----------------------------------------
C     EX2-13.F (END)
C----------------------------------------
C----------------------------------------
C     EX2-14.F (START)
C----------------------------------------
      PROGRAM MAIN
      IMPLICIT REAL*8 (A-H,K-Z)
      PI=4*ATAN(1.)
      WRITE(*,*) 'EXAMPLE 2.14....'
C------------------------------------------------
C     INPUT DATA
```

```
C-----------------------------------------------
      WRITE(*,*) 'INPUT THE FOLLOWING DATA...'
      WRITE(*,*) 'V: m/s'
      READ(*,*) V
      WRITE(*,*) 'T_I: C'
      READ(*,*) TI
      WRITE(*,*) 'D_I: cm'
      READ(*,*) DI
      WRITE(*,*) 'L: m'
      READ(*,*) L
      WRITE(*,*) 'T_E: C (SATURATION TEMPERATURE OF WATER AT 2 ATM)'
      READ(*,*) TE
      WRITE(*,*) 'LATENT HEAT (kJ/kg) OF WATER AT T_E'
      READ(*,*) HFG
      WRITE(*,*) 'AVERAGE DENSITY (kg/m^3) OF WATER'
      READ(*,*) RHO
         WRITE(*,*) 'AVERAGE SPECIFIC HEAT (kJ/kg.K) OF WATER'
      READ(*,*) CP
C-----------------------------------------------
C     UNIT CONVERSION
C-----------------------------------------------
      DI=0.01*DI
      HFG=1000*HFG
      CP=1000*CP
C-----------------------------------------------
C     CALCULATION
C-----------------------------------------------
      AI=PI*DI**2/4
      DOTM=RHO*AI*V
      DT=TE-TI
      DOTWE=DOTM*(CP*DT+HFG)
      DOTWEO=DOTM*CP*DT
      LO=L*DOTWEO/DOTWE
C-----------------------------------------------
C     ANSWER
C-----------------------------------------------
      WRITE(*,*) 'POWER NEEDED TO EVAPORATE WATER COMPLETELY IS'
      WRITE(*,*) DOTWE/1000,' kW'
      WRITE(*,*) 'FRACTION OF PIPE WITH WATER STARTING TO EVAPORATE'
      WRITE(*,*) LO,' m'
      STOP
      END
C-----------------------------------
C     EX2-14.F (END)
C-----------------------------------
```

EXERCISES

2.1 A heat flux q maintains the surfaces of a plate 3 cm thick at temperatures T_1 and T_2. Consider the following two cases:

(a) $q = 25{,}000$ W/m^2, $T_1 = 200\,°\text{C}$, $T_2 = 100\,°\text{C}$

(b) $q = 30{,}000$ W/m^2, $T_1 = 300\,°\text{C}$, $T_2 = 200\,°\text{C}$

Evaluate the thermal conductivity of the slab from each case. What is your conclusion? Establish an approximate expression for the conductivity.

114 Chap. 2 Steady Conduction

2.2 The cross section of the wall of a wooden house is shown in Fig. 2P–1. To save heating costs, the outside wood is to be replaced with a brick wall. Find the reduction in heat loss from the house. *Data*: Conductivities for drywall, insulation, wood, and brick are $k_d = 0.2$, $k_i = 0.1$, $k_w = 0.2$, and $k_b = 0.7$ W/m·K, the inside and outside coefficients of heat transfer are $h_i = 10$, and $h_o = 30$ W/m²·K.

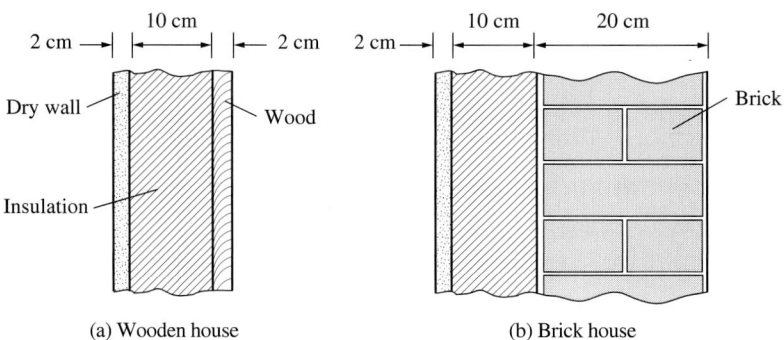

Figure 2P–1: (a) Wooden house, (b) brick house.

2.3 A family, realizing that they could not afford the cost of a brick wall, recently covered their wooden house with cork 2 cm thick. The conductivity of cork is $k_c = 0.04$ W/m·K. Compute the reduction in the heat loss from the wooden house of the preceding problem.

2.4 The interior of a refrigerator, having inside dimensions of 0.4 by 0.4 m base area and 1.2 m height, is to be maintained at 4 °C. The walls of the refrigerator are constructed of two mild-steel sheets 4 mm thick with 6 cm of glass-wool insulation between them. The inside and outside coefficients of heat transfer are 10 and 15 W/m²·K, respectively. Evaluate the rate at which heat must be removed from the interior to maintain the specified temperature in a kitchen at 25 °C. What will the temperature at the outer surface of the wall be? Assuming a 30% efficiency for the compressor, estimate the power need for the refrigerator.

2.5 Molten lead at 430 °C is stored in a welded stainless steel tank with walls 0.6 cm thick and insulated on the outside with 2.5 cm of 85% magnesia insulation. The melting point of lead is 327 °C. The heat transfer coefficient between the insulation and the air may be taken as 10 W/m²·K, and that at the interface between liquid lead and frozen lead may be taken as 20 W/m²·K. The ambient air temperature is 20 °C. Estimate the steady thickness of the layer of lead which will freeze on the inside surfaces of the tank. What is your conclusion?

2.6 In a manufacturing operation, a sheet of plastic 1 cm thick is to be glued to a sheet of cork board 3 cm thick (Fig. 2P–2). To affect a bond, the glue is to be maintained at a temperature of 30 °C for a considerable period of time. This is accomplished by a source of radiant heat, applied uniformly over the surface of the plastic. The exposed sides of the cork and the plastic have a heat transfer coefficient by convection of 10 W/m²·K, and the room temperature during the operation is 25 °C. Estimate the rate at which heat must be supplied to the surface of the plastic to obtain the required temperature at the interface. The thermal resistance of the glue may be neglected. The thermal conductivities of plastic and of cork are 2.3 and 0.042 W/m·K, respectively. Draw the thermal circuit for the system.

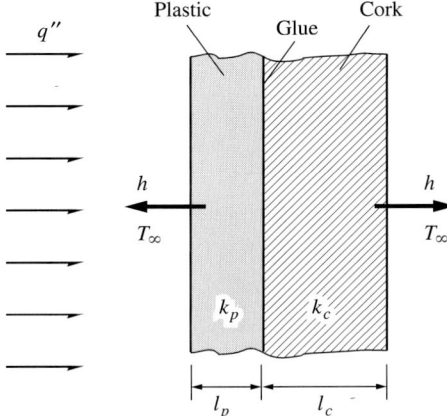

Figure 2P–2

2.7 Neglecting the effect of curvature, assume that an industrial brake system may be simulated by a flat plate (brake drum) moving on a composite plate (brake shoe) with a constant velocity V (Fig. 2P–3). The constant and uniform interface pressure is p. The coefficient of dry friction is μ, the ambient temperature is T_∞, and the heat transfer coefficients are h_1, h_2. The thermal conductivities and the thicknesses of the plates are k_1, k_2, and ℓ_1, ℓ_2, respectively. **(a)** Find the heat transfer to the drum and to the shoe. **(b)** Find the maximum temperature of the brake. **(c)** Draw the analogous electrical circuit for part (a).

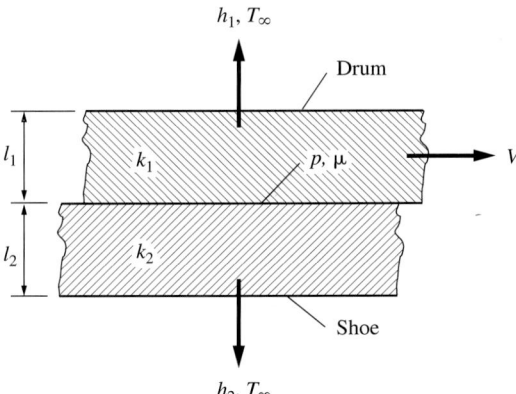

Figure 2P–3

2.8 A vertical boiler pipe of inner radius R_i and outer radius R_0 is subjected to a uniform heat flux q'' (Fig. 2P–4). A water flow evaporates through this pipe. The inside heat transfer coefficient h is very large. Neglecting axial conduction in the pipe, **(a)** evaluate the local temperature difference between the inner and outer surfaces of the pipe, **(b)** evaluate the local difference between the bulk temperature of the water and the inner surface temperature of the pipe. *Data*: Pipe diameters $D_0 = 12$ cm and $D_i = 10$ cm. Furnace temperature $T_g = 1400\,°$C. Evaporating water pressure in the pipe $p = 10$ MPa. Thermal conductivity of steel pipe $k = 20$ W/m·K.

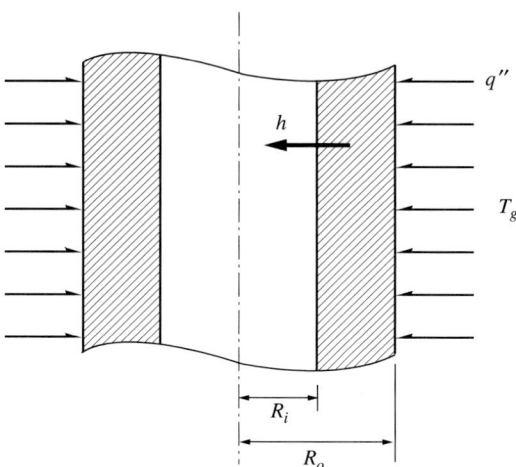

Figure 2P–4

2.9 A section of a carbon steel pipe 5 m long (12 cm OD and 10 cm ID) is insulated with 20-cm-OD 85% magnesia insulation. Saturated steam at 25 psi is admitted at one end, condensation occurs within the pipe, and saturated liquid is removed at the other end by a steam trap (Fig. 2P–5).

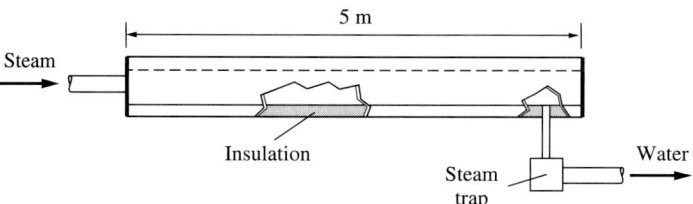

Figure 2P–5

It is desired to calculate the capacity of this steam trap in kg per hour of liquid flow, assuming that a constant pressure of 175 kPa is maintained in the pipe while heat is transferred to an ambient at $T_\infty = 25\ °C$. The end effects may be neglected. The heat transfer coefficient on the inside of the pipe is 5,000 W/m²·K and that on the outside of the insulation is 10 W/m²·K. The following properties of steam may be used:

p	T	v_f	v_g	h_f	h_g
175 kPa	116°C	0.001057 m³/kg	1.0036 m³/kg	487 kJ/kg	2,700.6 kJ/kg

2.10 A tube with 5 cm OD is maintained at $-30\ °C$ by a refrigerant boiling inside the tube. The tube is placed in water, and ice forms on it. The water temperature is 10 °C and the heat transfer coefficient between the ice and the liquid water is 60 W/m²·K. Estimate the thickness of the layer of ice on the tube at steady state. The conductivity of ice is $k = 0.2$ W/m·K.

2.11 A brake system consists of a hollow cylinder and a sleeve of negligible thickness (Fig. 2P–6). The relative angular velocity, pressure, and coefficient of dry friction between the cylinder and sleeve are ω, p, and μ, respectively. Find **(a)** the steady interface temperature, **(b)** the radius R corresponding to the lowest interface temperature.

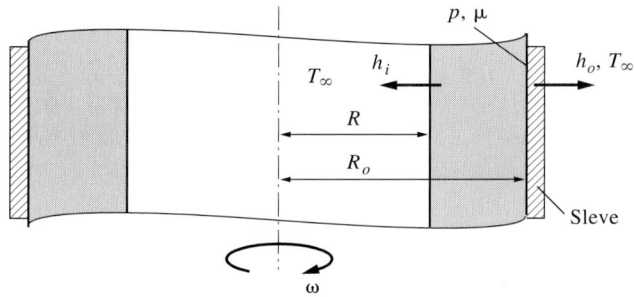

Figure 2P–6

2.12 Electrical power $\dot{W}_e$ is generated in a hollow sphere of radii R_i and R_0 (Fig. 2P–7). Under steady conditions, evaluate **(a)** the temperature drop across the thickness of sphere, **(b)** the outer surface temperature of the sphere. *Data:* $R_i = 5$ cm, $R_0 = 8$ cm, $\dot{W}_e = 750$ W, $T_\infty = 20\,°\text{C}$, $k = 20$ W/m·K, $h = 10$ W/m^2·K.

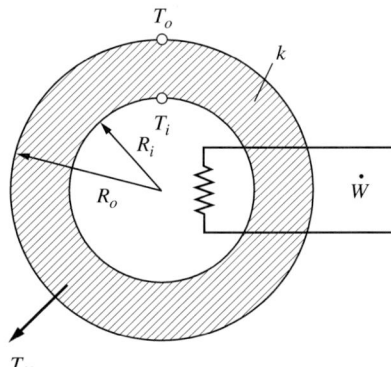

Figure 2P–7

2.13 A spherical liquid oxygen (LOX) tank has a diameter of 3 m. The boiling point of LOX is $-183\,°\text{C}$. Estimate the thickness of insulation which will reduce the boil-off rate in the steady state to no more than 0.005 kg/s. The heat of vaporization of LOX is 214 kJ/kg. The conductivity of the insulation is $k = 0.01$ W/m·K. The coefficient of outside heat transfer is h = 10 W/m^2·K, and the ambient temperature is $T_\infty = 20\,°\text{C}$.

2.14 The shield of a nuclear reactor can be idealized by a large flat plate 25 cm thick having a thermal conductivity of 4 W/m·K. Radiation from the interior of the reactor penetrates the shield and generates energy in the shield which decreases exponentially from a value of 200 kW/m^3 at the inner surface to a value of 20 kW/m^3 at a distance of 12.5 cm from the interior surface. For the case where the exterior surface is kept at 40 °C by forced convection, determine the temperature at the inner surface of the shield.

2.15 A 12-gage (2 mm diameter) wire with a resistance of 0.05 ohms/m carries a 15-ampere current and is insulated with rubber (k = 0.013 W/m·K). The heat transfer coefficient for the outer surface is 10 W/m²·K. Consider the wire temperature to be uniform and the ambient temperature to be 25 °C. Find (a) the thickness of insulation which will result in the lowest operating temperature of the wire, (b) the temperature of the outer surface of the rubber insulation.

2.16 Consider a fin 2ℓ long. One-half of the fin is peripherally insulated. The end temperature T_w of this half is specified. The other half transfers heat with a coefficient h to an ambient at temperature $T_\infty < T_w$. The end of this half is insulated (Fig. 2P–8). Find the temperature of the middle plane and of the insulated end of the fin.

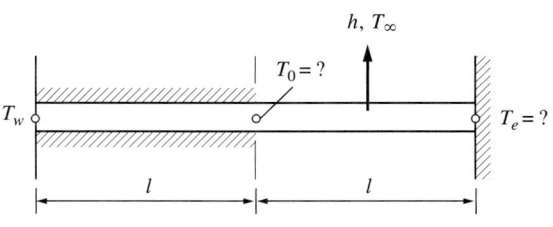

Figure 2P–8

2.17 Consider a stainless steel rod 2.5 cm in diameter and 1 m long mounted between two supports maintained at 50 °C. The rod is in air at temperature $T_\infty = 300$ °C. The thermal conductivity of the rod is $k = 20$ W/m·K. For two values of the transfer coefficient corresponding to $h = 5$ and 50 W/m²·K, evaluate (a) the temperature of the rod half way between the ends, (b) the rate of heat transfer from the air to the rod.

2.18 Consider a stainless steel turbine blade with height $h = 5$ cm, periphery $P = 8$ cm, and cross-sectional area $A = 2.5 \times 10^{-4}$ m². The conductivity of the blade is $k = 20$ W/m·K. The base temperature of the blade is $T_w = 600$ °C. The blade is exposed, as a part of a steam or a gas turbine, to superheated steam or hot gases at a temperature $T_\infty = 1,000$ °C. The heat transfer coefficient may be assumed to be 300 and 200 W/m²·K, respectively. Find the tip temperature of and the heat transfer to the blade for each case.

2.19 A wall separates a gas from a liquid. The temperature of the gas, T_g, is different from that of the fluid, T_ℓ. It is proposed to increase the heat transfer between the gas and the liquid by adding fins to either the gas-side or the liquid side. To which side must the fins be added for the best result? *Data:* Brass wall, $k = 100$ W/m·K, rectangular brass fins, 1 mm thick, 2.5 cm long, and spaced 2 mm apart. The gas-side and the liquid-side coefficients of heat transfer are $h_g = 10$ and $h_\ell = 1,000$ W/m²·K, respectively.

2.20 A solid rod ($2\ell = 20$ cm) with cross-sectional area A and perimeter P is insulated peripherally (Fig. 2P–9). Uniform internal energy $u''' = 500$ kW/m³ is generated in the rod. The thermal conductivity of the rod is $k = 50$ W/m·K. The end temperatures of the rod are $T_1 = 0$ °C and $T_2 = 100$ °C. **(a)** Find the temperature of the rod at $x = \ell$. **(b)** Sketch the temperature distribution of the rod.

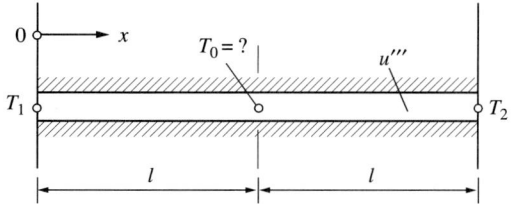

Figure 2P–9

2.21 Consider an infinitely long fin. Internal energy u''' is steadily generated in a part of the fin 2ℓ long (Fig. 2P–10). The entire fin transfers heat with a coefficient h to an ambient at temperature T_∞. **(a)** Find the temperature distribution within the fin. **(b)** Resolve the problem for a fin $2(\ell + L)$ long with insulated ends. The internal energy continues to be generated in the 2ℓ long central part of the fin.

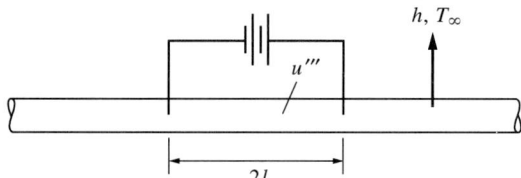

Figure 2P–10

2.22 Uniform internal energy u''' is generated in a semi-infinite rod of cross-sectional area A and periphery P (Fig. 2P–11). The peripheral and tip heat transfer coefficients are the same, say h. The ambient temperature is T_∞. If heat loss from the tip is neglected, the temperature of the rod is uniform, $T - T_\infty = u'''A/hP$. Find the distance δ over which the effect of heat loss from the tip is appreciable.

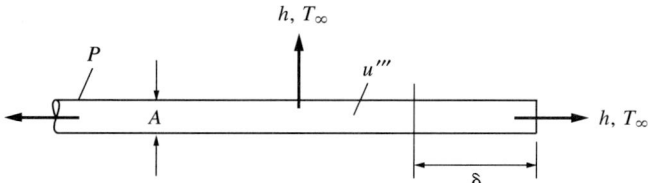

Figure 2P–11

2.23 In the process of constructing a thermocouple, two thin wires of different thermal conductivity are soldered with a heat flux q'' applied locally. Express q'' in terms of the melting temperature of the solder. *Data:* Wire diameter $d = 1$ mm, ambient air temperature $T_\infty = 20\ °C$, melting temperature $T_m = 260\ °C$, conductivity of the wire $k_1 = 20$, $k_2 = 60$ W/m·K, heat transfer coefficient $h = 10$ W/m²·K.

2.24 Electrical energy u''' is generated uniformly in an infinite rod (Fig. 2P–12). The upper half of the rod transfers heat with a coefficient h_1 to an ambient, while the lower half with coefficient $h_2\ (> h_1)$ exchanges heat with another ambient. Both ambients have the same temperature, say T_∞. Find the steady temperature of the rod. Reconsider the problem for two ambients at different temperatures, say T_1 and T_2. Find the relation between T_1 and T_2 such that the rod temperature remains approximately constant.

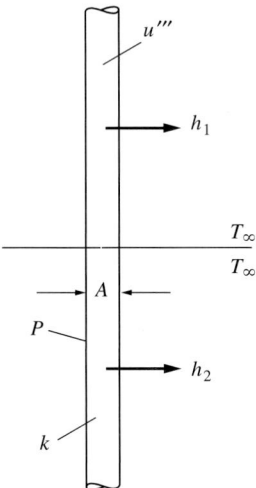

Figure 2P–12

2.25 Reconsider Ex. 2.12. Resolve the problem assuming (a) that the heat loss from the part of the rod in air is negligible, (b) a finite rod with both ends insulated.

2.26 A fluid flows steadily with velocity V through a thick wall pipe of length ℓ (Fig. 2P–13). The inlet temperature of the pipe is T_0. The effect of axial conduction is negligible. Under steady conditions, find (a) the outlet temperature of the fluid, (b) the axial variation in the inside and outside temperatures of the pipe wall. (c) Now, let the pipe be at the focus of a concentrating collector so that it receives uniform peripheral heat flux q'' from the sun. Resolve (a) and (b).

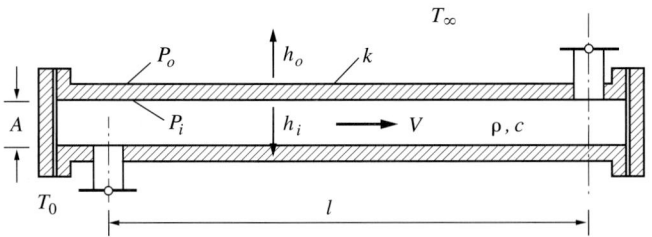

Figure 2P–13

2.27 In an industrial drawing process, a metal sheet of thickness δ steadily moves on rollers. This metal sheet is subjected to uniform heat flux q'' (Fig. 2P–14). (a) Neglecting the effect of axial conduction, find the steady axial temperature of the sheet. (b) A specific temperature, say T_ℓ, is desired at location $x = \ell$. Determine the angular velocity of the rollers required to achieve this temperature.

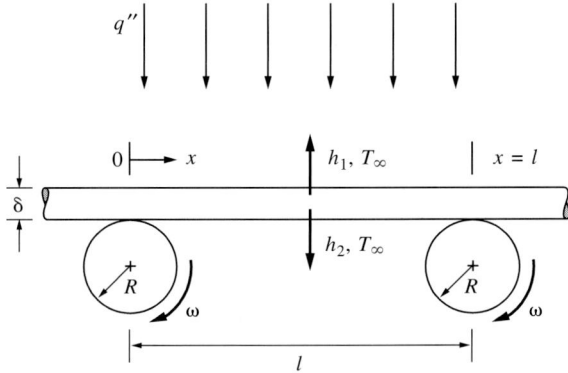

Figure 2P–14

2.28 An electrical conductor 2 cm in diameter fits closely inside a long hollow cylinder having an inside diameter of 2 cm and an outside diameter of 6 cm. The material from which the cylinder is made has a temperature-dependent thermal conductivity (Fig. 2P–15). The outer surface of the cylinder is maintained at 160 °C, and the electrical conductor dissipates 500 W per m of length. Compute the temperature at the inside surface of the hollow cylinder.

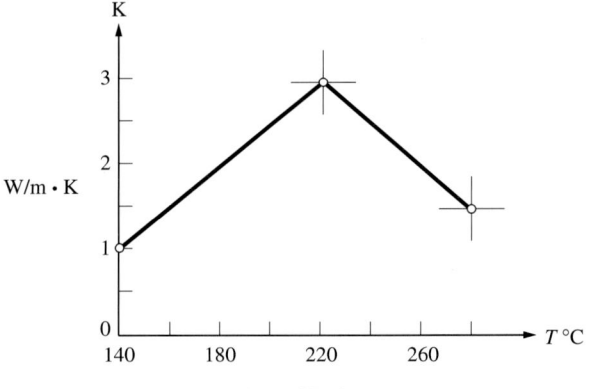

Figure 2P–15

2.29 A liquid is heated by flowing through a heat exchanger made of a heating rod co-axially placed in a closed cylindrical shell (Fig. 2P–16). The effect of axial conduction is negligible. Find the steady axial temperature distribution of the system. Repeat the problem including the effect of **(a)** the radial temperature distribution of the rod, **(b)** a large heat transfer coefficient.

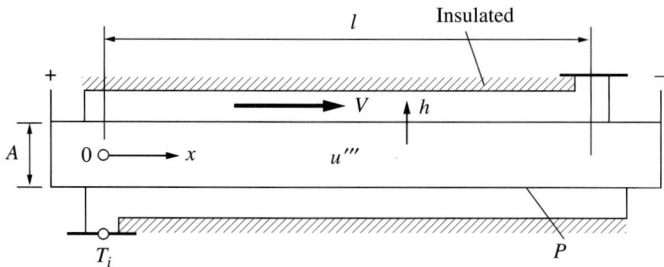

Figure 2P–16

2.30 A liquid enters the bottom of a large insulated pipe of diameter D and length L at a temperature T_∞, flows slowly upward with velocity V, and then spills over the top in a thin layer (Fig. 2P–17). A radiation heat flux q_{R_0} is incident on the upper surface of the liquid and is absorbed within the liquid such that the flux at any point within the liquid is given by

$$q''_R = q''_{R_0} e^{-\gamma x}.$$

The heat transfer between the upper surface of the liquid and the ambient is negligible. **(a)** Find the temperature of the liquid spilling over the top of the container. **(b)** Determine analytically an expression for the heat transfer coefficient at the underside of the upper liquid surface.

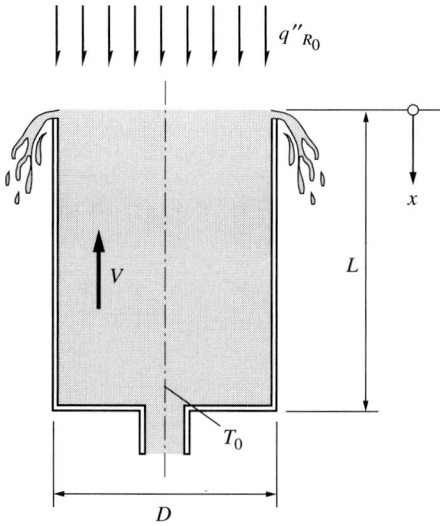

Figure 2P–17

2.31 In an experiment designed to measure the power output from a combustion process, two cylindrical shells of diameters D_i and D_o and length ℓ have been placed concentrically around a flame (Fig. 2P–18). This process yields an axisymmetric heat flux on the walls of the inside cylinder of the form

$$q'' = q_0 x(L - x); \quad q'' \text{ in } [\text{W/m}^2], \quad q_0 \text{ in } [\text{W/m}^4].$$

Water flows between the cylinders with inlet temperature T_i and velocity U. The water temperature at the exit is measured to be T_e. The outer cylinder is insulated externally. **(a)** Determine the total power output from the flames. **(b)** Determine q_0.

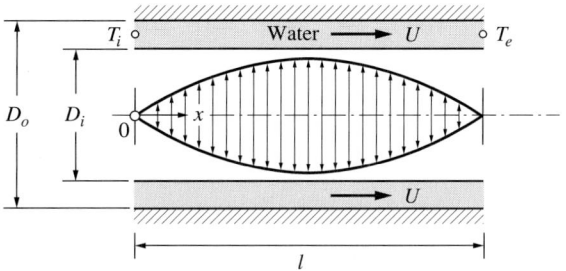

Figure 2P–18

2.32 A rotary heat exchanger can be idealized as a sleeve of radius R, width ℓ, and thickness δ rotating steadily with angular frequency ω (Fig. 2P–19). One-half of the sleeve is in an ambient at temperature T_1; the second half is in another ambient at temperature T_2. Find the temperature distribution of and the heat transfer from the exchanger.

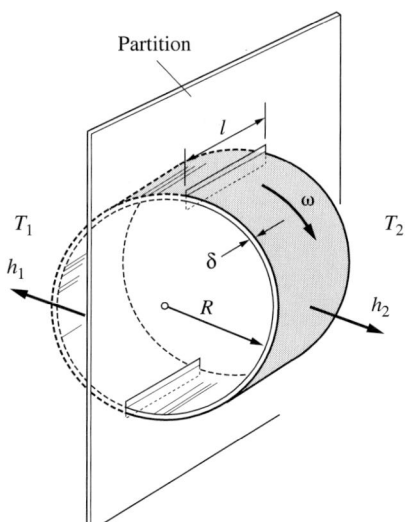

Figure 2P–19

2.33 An electrically heated screen is placed across a steady flow streaming with velocity U_∞ (Fig. 2P–20). *Including the effect of axial conduction* and assuming specified **(a)** screen temperature T_w or **(b)** power supply to the screen P, determine the steady temperature distribution within the flow.

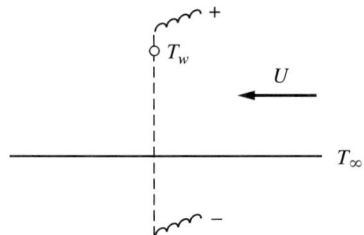

Figure 2P–20

2.34 Reconsider the second key problem of convection. Find the temperature of the fluid *including the effect of axial conduction.*

2.35 An electrical wire will be coated in a continuous process (Fig. 2P–21). The wire is pulled with a uniform velocity V while it is heated over a length L by a peripheral heat flux q''. The wire goes through a coating powder of very low conductivity. **(a)** Sketch the wire temperature in the three regions shown in the figure. **(b)** Find the wire temperature in these regions.

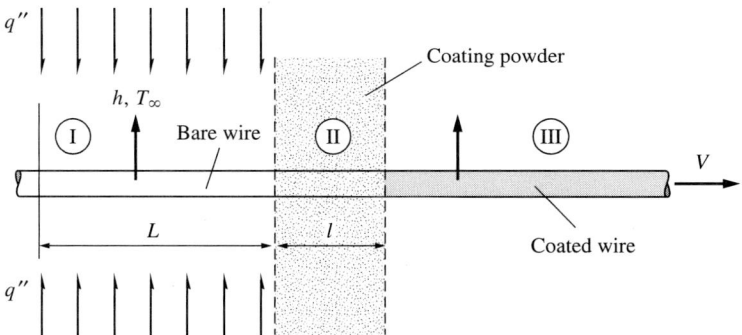

Figure 2P–21

2.36 An insulated thin-walled vessel initially contains saturated water vapor at temperature T_s (Fig. 2P–22). The external surface of the bottom of the vessel is exposed to the surrounding air at a temperature T_∞ ($T_\infty < T_s$). Determine the instantaneous thickness $X(t)$ of the condensate.

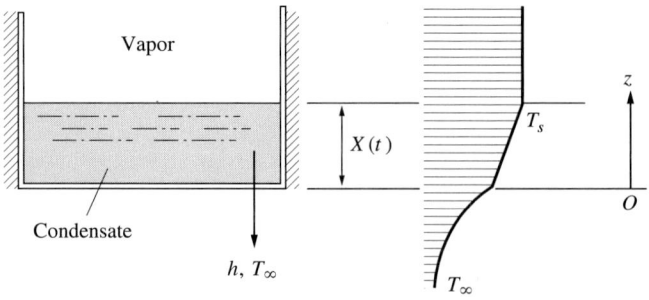

Figure 2P–22

CHAPTER 3

UNSTEADY/STEADY, MULTIDIMENSIONAL CONDUCTION

In Chapter 2 we learned how to interpret the spatial temperature distribution of steady problems in terms of the Biot number (recall Fig. 2.15). Here we extend this interpretation to the spatial temperature distribution of unsteady problems. Consider a solid transferring heat unsteadily with a coefficient h to an ambient at temperature T_∞. Let the instantaneous surface temperature be T_w, and the temperature of the solid relative to this temperature be ΔT in the neighborhood of the surface (Fig. 3.1).

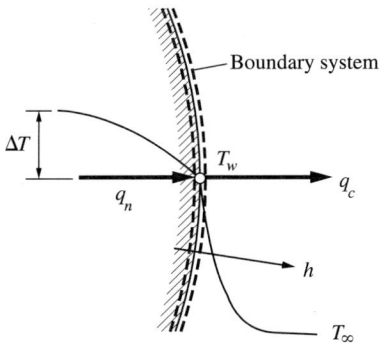

Figure 3.1 First law for a boundary system.

> In this and the following chapters we continue to employ the five steps of formulation but no longer make explicit reference to each step.

For a boundary system, the first law of thermodynamics reduces to

$$q_n = q_c$$

which, in terms of Fourier's law and Newton's law, and a characteristic length ℓ for the solid, may be dimensionally expressed as

$$k \frac{\Delta T}{\ell} \sim h(T_w - T_\infty),$$

or, in terms of the Biot number, as

$$\frac{h\ell}{k} = Bi \sim \frac{\Delta T}{(T_w - T_\infty)}. \tag{3.1}$$

For a small Biot number, say 0.1,

$$\Delta T \ll (T_w - T_\infty),$$

and the temperature drop across the solid can be neglected when compared with the temperature drop in the ambient. Consequently, for unsteady problems,

$$\boxed{Bi \leq 0.1}$$

implies a **spatially lumped** instantaneous temperature for the solid (the same criterion may also be used for steady problems). Since the elimination of spacewise temperature variation considerably simplifies unsteady problems, we examine first these (lumped) problems.

3.1 LUMPED PROBLEMS ($Bi \leq 0.1$)

An illustrative example for these problems is the cooling of a small metal ball (or billet) in a constant-temperature (large) bath. Let the ball, heated to a uniform temperature T_0, be quenched into the bath at temperature T_∞.

Consider the entire ball to be a lumped system. The first law of thermodynamics, Eq. (1.10), applied to this system yields (Fig. 3.2),

$$\frac{dU}{dt} = -q_c A, \tag{3.2}$$

where U is proportional to the total energy and A is the surface area of the ball. Now we may relate U and T by referring to thermodynamics.

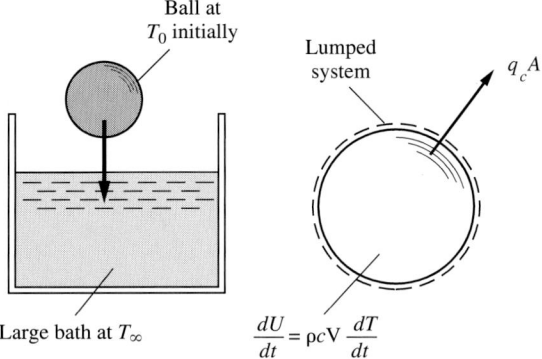

Figure 3.2 An illustrative example for lumped problems.

First, we have, by definition,

$$dU = m\,du = \rho V\,du,$$

ρ and V being the density and volume of the ball. Next, we may recall from thermodynamics that $u(v, T)$ and $h(p, T)$ for a **pure substance** and write

$$du = \left(\frac{\partial u}{\partial T}\right)_v dT + \left(\frac{\partial u}{\partial v}\right)_T dv$$

and

$$dh = \left(\frac{\partial h}{\partial T}\right)_p dT + \left(\frac{\partial h}{\partial p}\right)_T dp.$$

Then, in terms of the definitions of specific heat at constant volume and at constant pressure,

$$c_v = \left(\frac{\partial u}{\partial T}\right)_v, \quad c_p = \left(\frac{\partial h}{\partial T}\right)_p,$$

we have, for a constant-volume process,

$$du = c_v dT,$$

and, for a constant-pressure process,

$$dh = c_p dT.$$

Also, from the definition of enthalpy,

$$h = u + pv,$$

we get, for a constant-volume and a constant-pressure process,

$$dh = du$$

or

$$c_p = c_v = c.$$

That is why we usually refer to the specific heat for solids and liquids without any reference to volume or pressure [a fact already used in the development leading to Eqs. (2.148) and (2.149)]. Thus,

$$dU = \rho V \, du = \rho c V \, dT,$$

and Eq. (3.2) may be rearranged as

$$\rho c V \frac{dT}{dt} = -q_c A. \tag{3.3}$$

Finally, Newton's law, inserted into Eq. (3.3), gives the governing equation,

$$\rho c V \frac{dT}{dt} + hA(T - T_\infty) = 0 \tag{3.4}$$

subject to the initial condition

$$T(0) = T_0. \tag{3.5}$$

The solution of Eq. (3.4) satisfying Eq. (3.5) is[1]

$$\frac{T(t) - T_\infty}{T_0 - T_\infty} = \exp\left(-\frac{hA}{\rho c V} t\right) \tag{3.6}$$

or

$$\frac{T(t) - T_\infty}{T_0 - T_\infty} = \exp\left(-\frac{t}{\tau}\right), \tag{3.7}$$

where

$$\tau = \rho c V / hA. \tag{3.8}$$

A practical measure for the cooling time of the lumped ball, obtained from the tangent of Eq. (3.6) at $t = 0$ by letting

$$\frac{T(\tau) - T_\infty}{T_0 - T_\infty} = 1 - \left(\frac{hA}{\rho c V}\right) \tau = 0,$$

is called the **time constant**[2] (Fig. 3.3).

---

[1] The solution of Eq. (3.4) is obtained in a manner identical to that of Eq. (2.156).

[2] Also, on dimensional grounds, Eq. (3.4) readily gives $\rho c V \frac{T}{\tau} \sim hAT$, or $\tau \sim \rho c V / hA$.

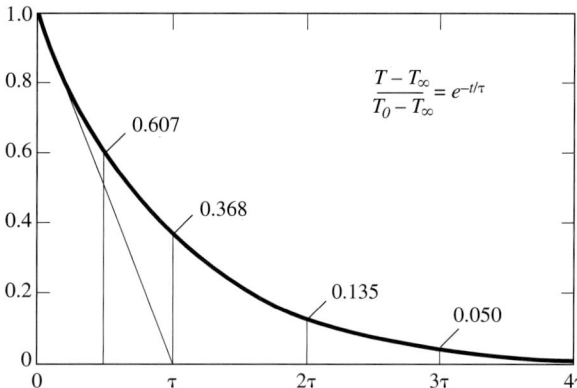

Figure 3.3 The time constant.

A better approximation for the **cooling time**, say t_0, may be obtained from an integral approach in a manner similar to the concept of penetration depth considered in Ex. 2.11, the cooling time being approximately the time the ball takes to reach temperature T_∞. Now integrate Eq. (3.4) over the cooling time,

$$\rho c V T \Big|_{T_0}^{T_\infty} = -hA \int_0^{t_0} (T - T_\infty) dt, \tag{3.9}$$

and select (as sketched in Fig. 3.4) an approximate temperature profile satisfying the apparent physics of the problem, that is,

$$T(0) = T_0, \quad T(t_0) = T_\infty \quad \text{and} \quad \frac{dT(t_0)}{dt} = 0.$$

In terms of polynomials, for example, the parabola $\theta = C_0 + C_1 t + C_2 t^2$ that satisfies these conditions yields

$$\frac{T(t) - T_\infty}{T_0 - T_\infty} = \left(1 - \frac{t}{t_0}\right)^2. \tag{3.10}$$

Inserting Eq. (3.10) into Eq. (3.9) gives

$$\rho c V (T_\infty - T_0) = hA(T_0 - T_\infty)\left(-\frac{1}{3}t_0\right)$$

or

$$t_0 = 3(\rho c V / hA), \tag{3.11}$$

or, in view of

$$\frac{V}{A} = \frac{\pi D^3/6}{\pi D^2} = \frac{1}{6}D,$$

$$t_0 = \frac{1}{2}(\rho c D / h). \tag{3.12}$$

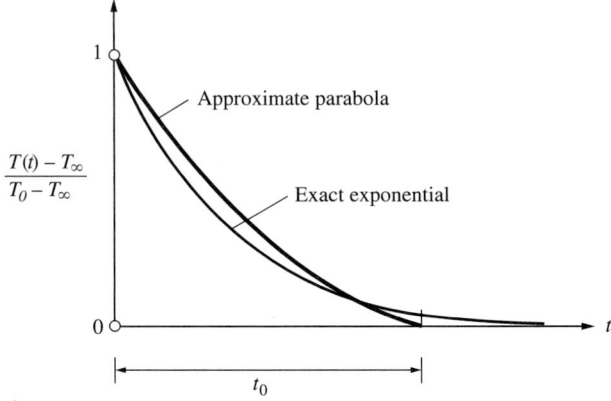

Figure 3.4 Exact and approximate temperatures.

Now, consider a steel or copper ball 4 cm in diameter to be dropped into a stagnant gas, oil, water, or liquid metal bath. From Appendix B1,

$$\rho_s \cong 8{,}000 \text{ kg/m}^3, \quad c_s \cong 500 \text{ J/kg·K}, \quad k_s \cong 10 \text{ W/m·K},$$
$$\rho_c \cong 9{,}000 \text{ kg/m}^3, \quad c_c \cong 400 \text{ J/kg·K}, \quad k_c \cong 400 \text{ W/m·K},$$

Also assume $\rho_s c_s \cong \rho_c c_c$, and refer to Table 1.2 for the heat transfer coefficients. The cooling time of these balls is given in Table 3.1. Clearly, since $Bi > 0.1$ for three of the steel cases and one copper case, the cooling time needs to be evaluated from a distributed solution in those cases.

Table 3.1 The cooling time of different balls

Fluid	h $\left[\dfrac{W}{m^2 \cdot K}\right]$	$Bi = \dfrac{hR}{k}$ [−]		t_0 [s]
		Steel	Copper	Steel and Copper
Gas	10	2×10^{-2}	5×10^{-4}	8,000
Oil	100	2×10^{-1}	5×10^{-3}	800
Water	1,000	2	5×10^{-2}	80
Liquid metal	5,000	20	5×10^{-1}	8

In Table 3–1 we used the radius of the sphere as the characteristic length. For axially symmetric cooling (or heating), the radius of a sphere or a long cylinder is the obvious choice for this length. For irregular bodies,

$$L = 2 - 3(V/A) \qquad (3.13)$$

may be used as a characteristic length, where 2 and 3 denote the number of finite dimensions. Equation (3.13) yields $L = R$ for both sphere and cylinder. For cooling

EXAMPLE 3.1

A thermocouple is needed to record a transient thermal process in which a fluid temperature rises from $T_0 = 20\,°C$ to $T_\infty = 1400\,°C$ (or drops from T_∞ to T_0) over a time interval of $t_d = 100$ ms. The heat transfer coefficient is $h \cong 6{,}000\,W/m^2 \cdot K$. We wish to determine the characteristics (type and size) of the thermocouple.

The thermocouple characteristics can be determined with the knowledge gained in the foregoing illustrative example. The heating or cooling response time of the thermocouple needs to be an order of magnitude smaller than t_d, that is,

$$t_0 = 3\tau \ll t_d,$$

say,

$$t_0 \cong 0.1 t_d. \tag{3.14}$$

Assuming the thermocouple to be a spherical bead, we have from Eqs. (3.12) and (3.14)

$$\frac{1}{2}(\rho c D / h) = 0.1 t_d$$

or

$$D = 0.20(h t_d / \rho c). \tag{3.15}$$

If the thermocouple were to be assumed cylindrical, noting $V/A = D/4$, we get from the combination of Eqs. (3.11) and (3.14),

$$\frac{3}{4}(\rho c D / h) = 0.1 t_d$$

or

$$D = 0.13(h t_d / \rho c). \tag{3.16}$$

For thermocouple wires,

$$\rho c = (3 \text{ to } 4) \times 10^6 \,J/\,m^3 \cdot K. \tag{3.17}$$

Then, the combination of Eqs. (3.15)–(3.17) gives

$$D[m] = (0.33 \text{ to } 0.67) \times 10^{-7} \left[\frac{m^3 \cdot K}{J}\right] h \left[\frac{W}{m^2 \cdot K}\right] t_d[s]. \tag{3.18}$$

For a disturbance time $t_d = 100$ ms in a hot fluid flow with $h \cong 6000\,W/m^2 \cdot K$,

$$D = 0.02 - 0.04 \text{ mm}, \tag{3.19}$$

which is a measure for the diameter of the thermocouple wire. For a temperature range of $20 - 1400\,°C$, Table 3.2 suggests,[3] for example, platinum (10 or 13)% rhodium-platinum as the thermocouple material. ◆

---

[3] For further details (environmental limitations, average sensitivities, etc.) see Ref. 2.

Table 3.2 Reference [2]

Thermocouple material	Temperature range °C
Copper-Constantan	~200–350
Iron-Constantan	0–750
Chromel-Constantan	~200–900
Chromel-Alumel	~200–1250
Platinum 10% Rhodium-Platinum	0–1450
Platinum 13% Rhodium-Platinum	0–1450
Tungsten-Tungsten 26% Rhenium	0–2320

EXAMPLE 3.2

Reconsider Ex. 1.2 for an uninsulated wire. We wish to determine the unsteady temperature of the wire.

For the lumped wire, the first law of thermodynamics, now combined with Newton's law, yields the governing equation,

$$\rho c V \frac{dT}{dt} = -hA(T - T_\infty) + u''' V \qquad (3.20)$$

subject to

$$T(0) = T_\infty. \qquad (3.21)$$

Here the generation of electrical power per unit volume, u''', and the total resistance of the wire are related by

$$u''' V = V^{*2}/R.$$

In terms of

$$\Theta = T - T_\infty - \frac{u''' V}{hA},$$

Eq. (3.20) reduces to

$$\frac{d\Theta}{dt} + \left(\frac{hA}{\rho c V}\right)\Theta = 0 \qquad (3.22)$$

subject to

$$\Theta(0) = -\frac{u''' V}{hA}. \qquad (3.23)$$

The solution of Eq. (3.22),

$$\Theta = C \exp\left[-\left(\frac{hA}{\rho c V}\right)t\right],$$

becomes with Eq. (3.23),

$$\Theta = -\frac{u''' V}{hA} \exp\left[-\left(\frac{hA}{\rho c V}\right) t\right],$$

or, in terms of T and τ,

$$\frac{T - T_\infty}{u''' V / hA} = 1 - e^{-t/\tau}. \tag{3.24}$$

◆

EXAMPLE 3.3

Consider the drum-shoe brake system of a car (Fig. 3.5). The density, heat capacity, and thickness of the brake shoes and drum are (ρ_1, ρ_2), (c_1, c_2), and (δ_1, δ_2), respectively. The initial temperature of the system is uniform, say T_∞. During a stopping time t_s, the pressure p between the shoes and drum, as well as the coefficient of dry friction μ, are assumed constant and the angular velocity $\omega(t)$ is specified. We wish to determine the unsteady temperature of the brake system.

Assume that Biot numbers for the inner and outer surfaces satisfy the conditions $h_1 \delta_1 / k_1 < 0.1$ and $h_2 \delta_2 / k_2 < 0.1$, and lumped conditions prevail. Also, in view of the fact that δ_1 and δ_2 are small compared with the mean radius R of the brake system, neglect the effect of curvature. For the lumped system shown in Fig. 3.6, the first law of thermodynamics combined with Newton's law for the outer and inner surfaces results in the governing equation for the brake system,

$$(\rho_1 c_1 \delta_1 + \rho_2 c_2 \delta_2) \frac{dT}{dt} = \mu p V(t) - (h_1 + h_2)(T - T_\infty) \tag{3.25}$$

subject to initial condition

$$T(0) = T_\infty. \tag{3.26}$$

Here T denotes the common lumped temperature of the drum and shoes, $V(t) = \omega(t) R$ the tangential velocity, and $\mu p V$ the friction power per unit area. Note that, for the system chosen, this power term should be interpreted as an internal energy generation. (Study the formulation of

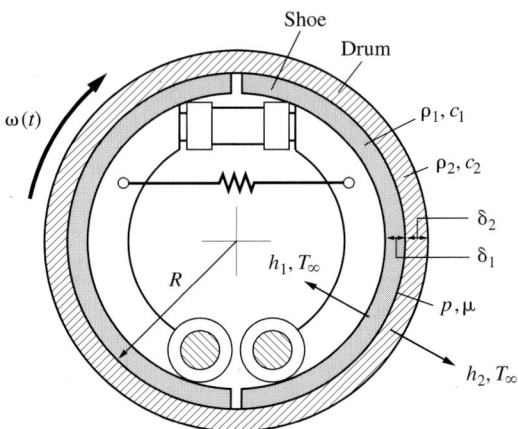

Figure 3.5 Brake system of a car.

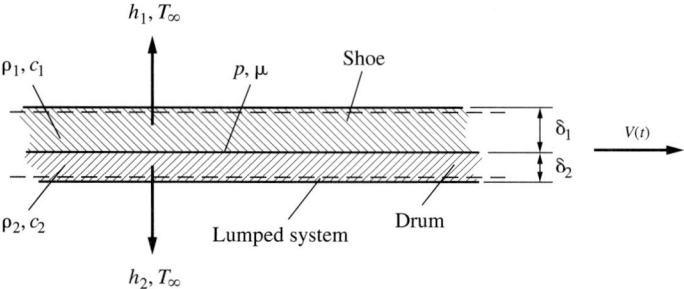

Figure 3.6 Model for the brake system.

the present problem in terms of two systems, one for the drum, the other for the shoes. Interpret the friction power in terms of these systems.)

Introducing $\theta = T - T_\infty$, $m = (h_1 + h_2)/(\rho_1 c_1 \delta_1 + \rho_2 c_2 \delta_2)$, and $n = \mu p/(\rho_1 c_1 \delta_1 + \rho_2 c_2 \delta_2)$, Eqs. (3.25) and (3.26) may be rearranged as

$$\frac{d\theta}{dt} + m\theta = nV(t), \quad \theta(0) = 0. \tag{3.27}$$

The general solution for the homogeneous part of Eq. (3.27) is

$$\theta(t) = C_1 e^{-mt}. \tag{3.28}$$

Now, following the method of the **variation of parameters**,[4] assume C_1 to be time dependent and insert Eq. (3.28) into Eq. (3.27). The result, after some rearrangement, is

$$\frac{dC_1}{dt} = nV(t)e^{mt}. \tag{3.29}$$

The integration of Eq. (3.29), for an arbitrary $V(t)$, gives

$$C_1(t) = C_2 + n \int_0^t V(t^*)e^{mt^*} dt^*, \tag{3.30}$$

where C_2 is a constant. Inserting Eq. (3.30) into Eq. (3.28), we have

$$\theta(t) = C_2 e^{-mt} + n \int_0^t V(t^*)e^{-m(t-t^*)} dt^*. \tag{3.31}$$

Writing Eq. (3.31) for the initial condition, Eq. (3.27), multiplying the result by $\exp(-mt)$ and subtracting it from Eq. (3.31) yields the unsteady temperature of the brake system for an arbitrary velocity,

$$\theta(t) = n \int_0^t V(t^*)e^{-m(t-t^*)} dt^*. \tag{3.32}$$

The integration of this equation for a particular variation in velocity presents no difficulty. For a linear decrease in velocity over a stopping time t_s,

$$V(t) = V_0 \left(1 - \frac{t}{t_s}\right), \tag{3.33}$$

---

[4] See a text on differential equations.

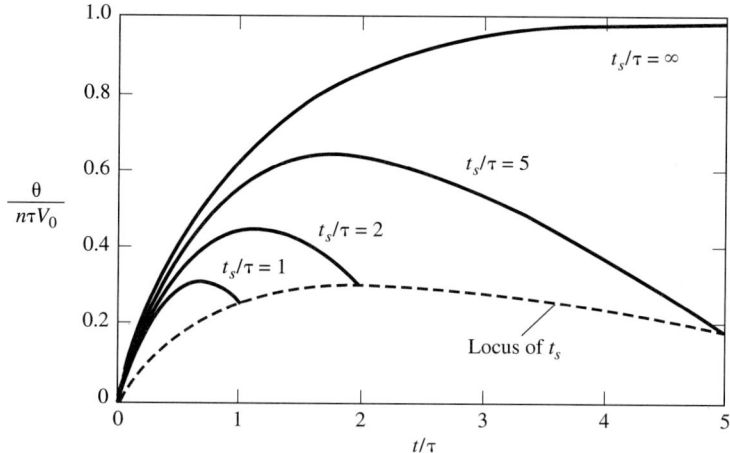

Figure 3.7 Unsteady brake temperature.

where V_0 is the initial value of the velocity. In terms of this velocity, Eq. (3.32) may be written as

$$\theta(t) = nV_0 \int_0^t \left(1 - \frac{t^*}{t_s}\right) e^{-m(t-t^*)} dt^*.$$

The integration of the first term of the integrand directly and the second term by parts, or the use of a text on integral tables, or the selection of appropriate functions for the nonhomogeneous part yields the unsteady temperature of the brake system for a linearly decreasing velocity,

$$\frac{\theta(t)}{n\tau V_0} = \left\{ (1 - e^{-t/\tau}) - \frac{\tau}{t_s}\left[\frac{t}{\tau} - (1 - e^{-t/\tau})\right] \right\}, \qquad (3.34)$$

where $\tau = m^{-1}$ and the terms in brackets show the effect of the decrease in velocity. Equation (3.34) for various τ/t_s is plotted in Fig. 3.7. Note that the brake velocity becomes zero at $t = t_s$. Afterwards the brake system cools down. This is a new problem with $V = 0$, and h_1, h_2 are different (usually smaller) from the original heat transfer coefficients.

◆

EXAMPLE 3.4 ⊕

Reconsider the illustrative example involving a small metal ball to be dropped into a constant-temperature bath. Let the bath volume be small now and its temperature no longer remain constant during the cooling of the ball. Also, for simplicity, let the bath be insulated. We wish to determine the unsteady temperature of the ball and that of the bath.

Assume that the Biot number for the ball is small and that the bath is slowly stirred. Consequently, both the ball and bath temperatures are instantaneously uniform. For the lumped ball and the lumped bath, the first law of thermodynamics combined with Newton's law yields the governing equations,

$$\frac{dT_1}{dt} = -m_1(T_1 - T_2), \qquad (3.35)$$

$$\frac{dT_2}{dt} = m_2(T_1 - T_2), \tag{3.36}$$

subject to

$$T_1(0) = T_0, \quad T_2(0) = T_\infty. \tag{3.37}$$

Here subscripts 1 and 2 refer respectively to the ball and bath, $m_1 = hA/\rho_1 c_1 V_1$ and $m_2 = hA/\rho_2 c_2 V_2$.

In terms of operator $D \equiv d/dt$, Eqs. (3.35) and (3.36) may be rearranged as

$$(D + m_1)T_1 - m_1 T_2 = 0, \tag{3.38}$$

$$-m_2 T_1 + (D + m_2)T_2 = 0. \tag{3.39}$$

The solution of this set, either by the theory of determinants,

$$\begin{bmatrix} (D + m_1) & -m_1 \\ -m_2 & (D + m_2) \end{bmatrix} \begin{bmatrix} T_1 \\ T_2 \end{bmatrix} = 0,$$

or, by the usual method of elimination [solving, for example, T_1 from Eq. (3.39) in terms of T_2 and inserting the result into Eq. (3.38)], yields

$$\left[D^2 + (m_1 + m_2)D\right] T_{1,2} = 0,$$

or

$$\frac{d^2 T_{1,2}}{dt^2} + (m_1 + m_2)\frac{dT_{1,2}}{dt} = 0. \tag{3.40}$$

The solution of Eq. (3.40) is

$$T_1 = A_1 + B_1 e^{-t/\tau}, \tag{3.41}$$

$$T_2 = A_2 + B_2 e^{-t/\tau}, \tag{3.42}$$

where $\tau = (m_1 + m_2)^{-1}$. Introducing these equations into Eq. (3.35) or (3.36), say into Eq. (3.35), and letting $\eta = m_2/m_1$, results in the identity

$$(1 + \eta)B_1 e^{-t/\tau} \equiv (A_1 - A_2) + (B_1 - B_2)e^{-t/\tau},$$

which can only be satisfied at each instant in time whenever

$$A_1 - A_2 = 0, \quad (1 + \eta)B_1 = B_1 - B_2$$

or

$$A_2 = A_1, \quad B_2 = -\eta B_1.$$

Introducing these into Eq. (3.42), we have

$$T_2 = A_1 - \eta B_1 e^{-t/\tau}. \tag{3.43}$$

Equations (3.41) and (3.43), in view of the initial conditions, Eq. (3.37), yield

$$T_0 = A_1 + B_1,$$

$$T_\infty = A_1 - \eta B_1.$$

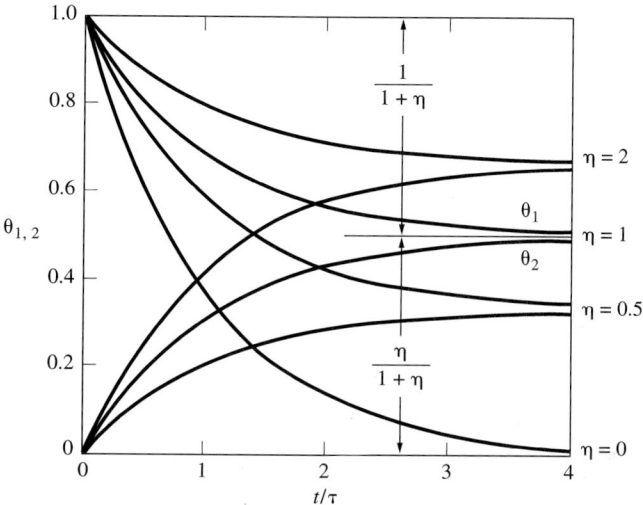

Figure 3.8 Ball and bath temperatures versus time.

Solving this algebraic set for A_1 and B_1, we get

$$A_1 = \frac{T_\infty + \eta T_0}{1 + \eta}, \quad B_1 = \frac{T_0 - T_\infty}{1 + \eta},$$

where, for convenience, A_1 may be arranged by the addition of $\pm \eta T_\infty/(1+\eta)$ as

$$A_1 = \frac{T_\infty + \eta T_0}{1 + \eta} \pm \frac{\eta T_\infty}{1 + \eta} = T_\infty + \eta \frac{T_0 - T_\infty}{1 + \eta}.$$

Inserting A_1 and B_1 into Eqs. (3.41) and (3.43) gives, after some rearrangement, the unsteady lumped temperature of the ball and that of the bath,

$$\frac{\theta_1(t)}{\theta_0} = \frac{1}{1 + \eta}(\eta + e^{-t/\tau}) \tag{3.44}$$

and

$$\frac{\theta_2(t)}{\theta_0} = \frac{\eta}{1 + \eta}(1 - e^{-t/\tau}). \tag{3.45}$$

where $\theta_{1,2} = T_{1,2} - T_\infty$ and $\theta_0 = T_0 - T_\infty$. The steady temperature of the system is

$$\frac{\theta_{1,2}}{\theta_0} = \frac{\eta}{1 + \eta}. \tag{3.46}$$

Clearly, as $\eta \to 0$, $\theta_2(t) \to 0$ (corresponding to a constant bath temperature of T_∞), and Eq. (3.44) is reduced to Eq. (3.7). Equations (3.44) and (3.45) versus t/τ for some values of η are plotted in Fig. 3.8.

◆

138 Chap. 3 Unsteady/steady, Multidimensional Conduction

Example 3.5

Reconsider Ex. 2–14. Let the uniform electrical power per unit volume u''' be suddenly generated within the pipe walls. Assume no evaporation in the water, and neglect the radial distribution of the wall temperature and the axial conduction both in the pipe walls and in the water. We wish to determine the unsteady wall and bulk water temperatures.

Consider a radially lumped and axially differential system and control volume for the pipe walls and the water flow, respectively (Fig. 3.9). The first law of thermodynamics applied to the system and control volume yields, after the use of Newton's law,

$$\rho_p c_p A_p \frac{\partial T_p}{\partial t} = -hP(T_p - T) + u''' A_p, \tag{3.47}$$

$$\rho c A \left(\frac{\partial T}{\partial t} + V \frac{\partial T}{\partial x} \right) = +hP(T_p - T), \tag{3.48}$$

where the subscript p refers to the pipe walls, A is the cross sectional area, P the inner perimeter of the pipe walls and V the water velocity. Introducing

$$m = \frac{hP}{\rho c A}, \quad m_p = \frac{hP}{\rho_p c_p A_p}, \quad n = \frac{u''' A_p}{\rho_p c_p A_p},$$

Eqs. (3.47) and (3.48) may be rearranged as

$$\frac{\partial T_p}{\partial t} + m_p(T_p - T) = n, \tag{3.49}$$

$$\frac{\partial T}{\partial t} + V \frac{\partial T}{\partial x} + m(T - T_p) = 0 \tag{3.50}$$

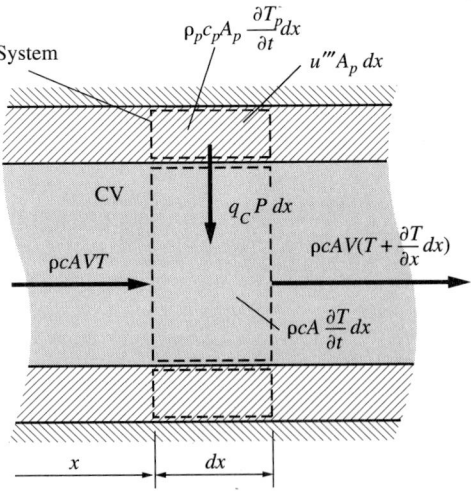

Figure 3.9 Pipe system and fluid control volume.

subject to
$$T_p(x, 0) = T(x, 0) = T(0, t) = T_i. \tag{3.51}$$

The complete solution of these (coupled, partial differential) equations is beyond the scope of this text.[5] Here we seek out a simple intuitive solution.

For example, consider the time interval
$$0 \leq t \leq t^* = \frac{x}{V}.$$

For $t \leq t^*$, or before the arrival of the inlet water to (any) location x, the physics of each location is independent of x. That is, $\partial T/\partial x \equiv 0$, and Eq. (3.50) reduces to

$$\frac{\partial T}{\partial t} + m(T - T_p) = 0, \; t \leq \frac{x}{V}. \tag{3.52}$$

The solution of Eqs. (3.49) and (3.52) subject to Eq. (3.51) can be found in a manner similar to that of Eqs. (3.35) and (3.36). The details of this solution are left to the reader. The result is

$$\frac{T_p(x, t) - T_i}{n/m(1 + \eta)^2} = \frac{t}{\tau} + \eta(1 - e^{-t/\tau}), \tag{3.53}$$

$$\frac{T(x, t) - T_i}{n/m(1 + \eta)^2} = \frac{t}{\tau} - (1 - e^{-t/\tau}), \tag{3.54}$$

where $\eta = m_p/m$ and $\tau = (m + m_p)^{-1}$. The wall and fluid temperatures versus time for various η are plotted in Fig. 3.10. As $\eta \to 0$ (that is, negligible heat transfer to the fluid), Eqs. (3.53) and (3.54) are reduced to

$$\frac{T_p(x, t) - T_i}{n/m} = \frac{t}{\tau}, \tag{3.55}$$

$$\frac{T(x, t) - T_i}{n/m} = \frac{t}{\tau} - (1 - e^{-t/\tau}) \tag{3.56}$$

which are valid only for small values of time.

For water ($\rho = 1{,}000$ kg/m^3, $c_p = 4{,}200$ J/kg·K) flowing steadily with velocity $V = 1$ m/s in a steel pipe of $1 -$ in. diameter ($D_i = 1.049$ in. $= 2.66$ cm, $D_0 = 1.315$ in. $= 3.34$ cm),

$$P = \pi D_i = \pi \times 2.66 = 8.36 \text{ cm},$$

$$A = \pi D_i^2/4 = \pi \times (2.66)^2/4 = 5.56 \text{ cm}^2,$$

$$A_p = \pi(D_0^2 - D_i^2)/4 = \pi \times (3.34^2 - 2.66^2)/4 = 3.2 \text{ cm}^2,$$

---

[5] See pp. 360–365 of *Conduction Heat Transfer* by Arpacı[1] for this solution.

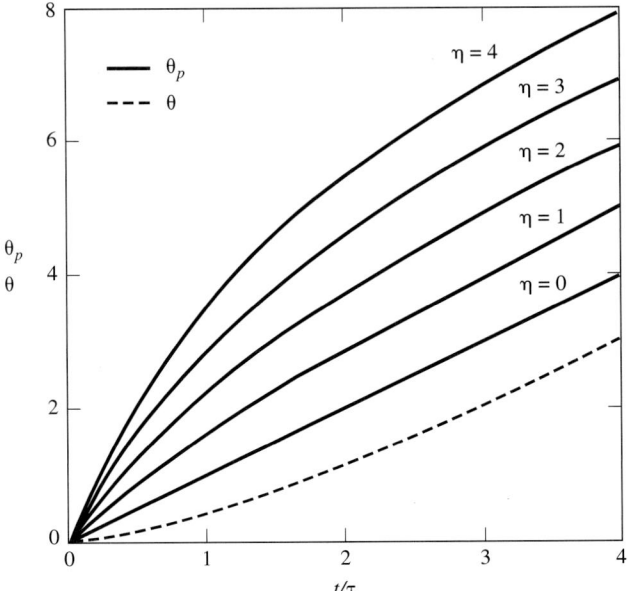

Figure 3.10 Unsteady temperatures in electrically heated flow in pipe.

and, assuming[6] $h \cong 10{,}000 \text{ W/m}^2 \cdot \text{K}$, we get

$$m = \frac{hP}{\rho c A} = \frac{10{,}000 \text{ W/m}^2\cdot\text{K} \times 8.36 \times 10^{-2} \text{ m}}{1{,}000 \text{ kg/m}^3 \times 4{,}200 \text{ J/kg}\cdot\text{K} \times 5.56 \times 10^{-4} \text{ m}^2} \cong 0.36 \text{ s}^{-1}$$

and

$$m_p = \frac{hP}{\rho_p c_p A_p} = \frac{10{,}000 \text{ W/m}^2\cdot\text{K} \times 8.36 \times 10^{-2} \text{ m}}{8{,}000 \text{ kg/m}^3 \times 500 \text{ J/kg}\cdot\text{K} \times 3.2 \times 10^{-4} \text{ m}^2} \cong 0.65 \text{ s}^{-1}$$

$$\eta = m_p/m = 0.65/0.36 = 1.81 \cong 2$$

and

$$\tau = (m + m_p)^{-1} = (0.36 + 0.65)^{-1} \cong 1 \text{ s}.$$

In Fig. 3.10, the unsteady temperature of the pipe walls and that of the water flow are respectively given by the solid line for $\eta = 2$ and the broken line. For any axial location, these temperatures are valid up to time $t = x/V$, or $t/\tau = x/V\tau$, which yields for $V = 1$ m/s and $\tau = 1$ s,

$$\frac{t}{\tau} = x. \qquad \blacklozenge$$

Before proceeding to unsteady distributed problems, we next consider a class of unsteady lumped problems periodically depending on time. These problems (lumped or distributed) find many practical applications.

---

[6] See Table 1.2.

3.2 PERIODIC PROBLEMS ⊕

In this chapter we have already classified unsteady problems with respect to their dependence on space (as lumped or distributed) and have so far studied the lumped problems. Now we may also classify these problems with respect to their dependence on time (as transient or periodic). Consequently

$$\text{An unsteady problem may be} \begin{cases} \text{spacewise} \begin{cases} \text{lumped,} & Bi < 0.1, \\ \text{distributed,} & Bi > 0.1, \end{cases} \\ \text{timewise} \begin{cases} \text{transient,} \\ \text{periodic,} \end{cases} \end{cases}$$

where each transient or periodic problem involves a **starting**, a **steady** and an **ending** time interval as shown in Fig. 3.11. In the preceding section we demonstrated the transient lumped case in terms of some examples. Here we illustrate the periodic case in terms of a lumped example. Periodic disturbances need not be harmonic. The linearity of conduction problems allows any periodic disturbance to be expressed in terms of its harmonics and reduces these problems to ones having harmonic disturbances.

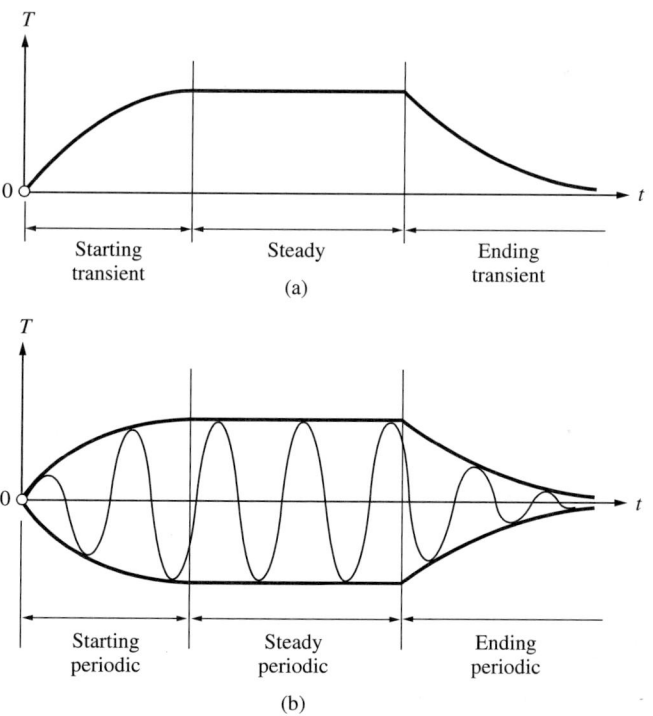

Figure 3.11 Transient and periodic variation.

EXAMPLE 3.6

Reconsider Ex. 3.2. Let the rate of electrical energy per unit volume of the wire,

$$u'''(t) = u'''_0 (1 + \cos \omega t),$$

be generated from the initial condition of the wire, $T(0) = T_\infty$. We wish to determine the temperature fluctuations in the wire.

The solution of the problem may be conveniently written as the superposition[7] of two problems, the first one being suddenly subjected to uniform u'''_0 and the second one to $u'''_0 \cos \omega t$. The first problem is identical to Ex. 3.2. Its solution is given then by Eq. (3.24). The formulation of the second problem, obtained from Eqs. (3.20) and (3.21) in terms of $\theta = T - T_\infty$, $m = hA/\rho c V$, and $n = u'''_0 / \rho c$, is

$$\frac{d\theta}{dt} + m\theta = n \cos \omega t \qquad (3.57)$$

subject to

$$\theta(0) = 0. \qquad (3.58)$$

Since Eq. (3.57) is a special form of Eq. (3.27), its solution may be directly written by letting $V(t^*) = \cos \omega t^*$. Thus

$$\theta(t) = n e^{-mt} \int_0^t e^{mt^*} \cos \omega t^* dt^*. \qquad (3.59)$$

For the integral of this equation, we usually refer to a text on integral tables (or employ integration by parts twice). Here we follow a novel third approach, which will prove convenient later.

Consider a **complementary** problem with oscillation $u'''(t) = u'''_0 \sin \omega t$. Let the temperature of this problem be $\theta^*(t)$. An inspection of the solution procedure for the **original** problem readily reveals the solution of the complementary problem,

$$\theta^*(t) = n e^{-mt} \int_0^t e^{mt^*} \sin \omega t^* dt^*. \qquad (3.60)$$

Multiplying this equation by the imaginary unit i of complex variables, adding the result to Eq. (3.59), and recalling Euler's formula

$$e^{i\beta} = \cos \beta + i \sin \beta,$$

we get the **complex temperature** ψ,

$$\psi(t) = \theta + i\theta^* = n e^{-mt} \int_0^t e^{(m+i\omega)t^*} dt^*, \qquad (3.61)$$

which may be readily integrated to give

$$\psi(t) = \frac{n}{m + i\omega} \left(e^{i\omega t} - e^{-mt} \right). \qquad (3.62)$$

Recalling the relation between the cartesian and polar forms of a complex number,

$$m + i\omega = \gamma e^{i\beta}, \qquad (3.63)$$

---

[7] Recall the use of the same concept in connection with a steady problem in Ex. 2.5.

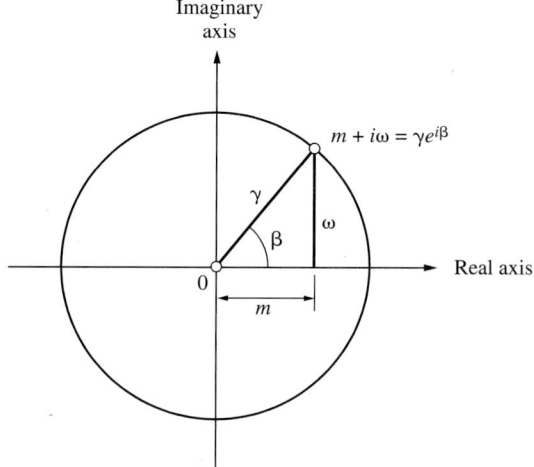

Figure 3.12 Cartesian and polar forms of a complex number.

where, according to Fig. 3.12,

$$\gamma = \sqrt{m^2 + \omega^2} \quad \text{and} \quad \tan \beta = \frac{\omega}{m},$$

Equation (3.62) may be conveniently rearranged in terms of Eq. (3.63) as

$$\psi(t) = \frac{n}{\gamma} \left[e^{i(\omega t - \beta)} - e^{-mt} e^{-i\beta} \right]. \tag{3.64}$$

The real part of this equation is the **starting periodic** solution of the **original** problem,

$$\theta(t) = \frac{n}{\gamma} \left[\cos(\omega t - \beta) - e^{-mt} \cos \beta \right],$$

or, explicitly,[8]

$$\theta(t) = \frac{n}{m^2 + \omega^2} (m \cos \omega t + \omega \sin \omega t - m e^{-mt}). \tag{3.65}$$

(The imaginary part of Eq. 3.64 is the starting periodic solution of the **complementary** problem.) As $t \to \infty$, the second term in brackets approaches zero, and Eq. (3.65) reduces to the **steady periodic** solution,

$$\theta(t) = \frac{n}{\gamma} \cos(\omega t - \beta),$$

or, explicitly,

$$\theta(t) = \frac{n}{m^2 + \omega^2} (m \cos \omega t + \omega \sin \omega t). \tag{3.66}$$

---

[8] The total solution is obtained by the superposition of Eqs. (3.24) and (3.65).

For distributed problems, the complete (starting or ending) periodic solution is difficult to obtain; in practical situations it may not even be needed. For example, stresses corresponding to the steady part of periodic temperatures are maximum in the cylinder walls of an engine. Therefore, a simple procedure which yields only the steady part of a periodic solution is quite important. This procedure may conveniently be described here in terms of the foregoing example.

Reconsider the governing equation of Ex. 3.6,

$$\frac{d\theta}{dt} + m\theta = n \cos \omega t. \tag{3.57}$$

Assume the governing equation for the complementary problem to be

$$\frac{d\theta^*}{dt} + m\theta^* = n \sin \omega t. \tag{3.67}$$

(Recall the only difference between the original and complementary problems: cosine oscillations of the former are replaced by sine oscillations in the latter or vice versa). Multiplying Eq. (3.67) by the imaginary unit i and adding the result to Eq. (3.57) yields, in terms of the complex temperature $\psi = \theta + i\theta^*$,

$$\frac{d\psi}{dt} + m\psi = n e^{i\omega t}. \tag{3.68}$$

For **linear** problems, the input and response have the **same** harmonic variation. A steady periodic solution must then have the form

$$\psi(t) = \phi e^{i\omega t}, \tag{3.69}$$

where ϕ is a parameter to be determined (for distributed problems ϕ is a function of space). This solution, being only valid as $t \to \infty$, need not satisfy the initial condition stated by Eq. (3.58). Inserting Eq. (3.69) into Eq. (3.68), recalling Eq. (3.63), and rearranging give

$$\phi = \frac{n}{m + i\omega} = \frac{n}{\gamma} e^{-i\beta},$$

or, in view of Eq. (3.69),

$$\psi(t) = \frac{n}{\gamma} e^{i(\omega t - \beta)}, \tag{3.70}$$

which is the steady periodic part of Eq. (3.64). The real part of Eq. (3.70), being Eq. (3.66), or Eq. (3.65) for $t \to \infty$, is the steady periodic solution of the original problem. The imaginary part of Eq. (3.70) is the steady periodic solution of the complementary problem. ◆

Having completed our discussion of unsteady lumped problems, we proceed now to unsteady distributed problems.

3.3 DISTRIBUTED PROBLEMS ($Bi > 0.1$). DIFFERENTIAL FORMULATION

In Chapter 2 we studied steady, one-dimensional conduction. In many realistic problems, however, conduction is two or even three dimensional. Examples are composite walls involving parallel paths (Fig. 2.6), extended surfaces with a transversally large Biot number, corners of a room (Fig. 3.13), etc. Accordingly, in this section we proceed

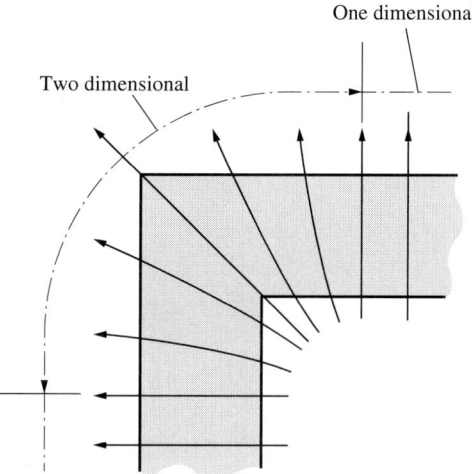

Figure 3.13 Two-dimensional corner.

to a study of unsteady multidimensional conduction. For convenience, multidimensional concepts are illustrated first in terms of two-dimensional problems and later are extended to three-dimensional problems.

Consider a two-dimensional geometry, say, a long cylinder of arbitrary cross section [Fig. 3.14(a)]. Let the rate of internal energy generation per unit volume be u'''. Neglect the axial temperature variation. We wish to formulate the unsteady, two-dimensional conduction for the cross section of this cylinder.

The first law of thermodynamics, Eq. (1.10), applied to the two-dimensional differential system shown in Fig. 3.14(a), and interpreted in terms of Fig. 3.14(b), gives (per unit length of the cylinder)

$$\rho dxdy \frac{\partial u}{\partial t} = +q_x dy - \left(q_x + \frac{\partial q_x}{\partial x}dx\right)dy + q_y dx - \left(q_y + \frac{\partial q_y}{\partial y}dy\right)dx + u''' dxdy,$$

or, after relating u to T in a manner similar to the development in Section 3.1 for the rate of internal energy of a lumped system, and dividing each term by $dxdy$,

$$\rho c \frac{\partial T}{\partial t} = -\frac{\partial q_x}{\partial x} - \frac{\partial q_y}{\partial y} + u'''. \qquad (3.71)$$

The two-dimensional Fourier's law of conduction,

$$q_x = -k\frac{\partial T}{\partial x}, \quad q_y = -k\frac{\partial T}{\partial y}, \qquad (3.72)$$

inserted into Eq. (3.71) yields, after the third dimension is included by inspection, the governing equation for unsteady, three-dimensional conduction in cartesian coordinates,

$$\rho c \frac{\partial T}{\partial t} = \frac{\partial}{\partial x}\left(k\frac{\partial T}{\partial x}\right) + \frac{\partial}{\partial y}\left(k\frac{\partial T}{\partial y}\right) + \frac{\partial}{\partial z}\left(k\frac{\partial T}{\partial z}\right) + u'''. \qquad (3.73)$$

For a uniform thermal conductivity, recalling the definition of thermal diffusivity $\alpha = k/\rho c$, Eq. (3.73) may be reduced to

$$\frac{1}{\alpha}\frac{\partial T}{\partial t} = \frac{\partial^2 T}{\partial x^2} + \frac{\partial^2 T}{\partial y^2} + \frac{\partial^2 T}{\partial z^2} + \frac{u'''}{k}, \qquad (3.74)$$

or, in terms of the Laplacian operator ∇^2,

$$\frac{1}{\alpha}\frac{\partial T}{\partial t} = \nabla^2 T + \frac{u'''}{k}, \qquad (3.75)$$

where

$$\nabla^2 \equiv \frac{\partial^2}{\partial x^2} + \frac{\partial^2}{\partial y^2} + \frac{\partial^2}{\partial z^2}. \qquad (3.76)$$

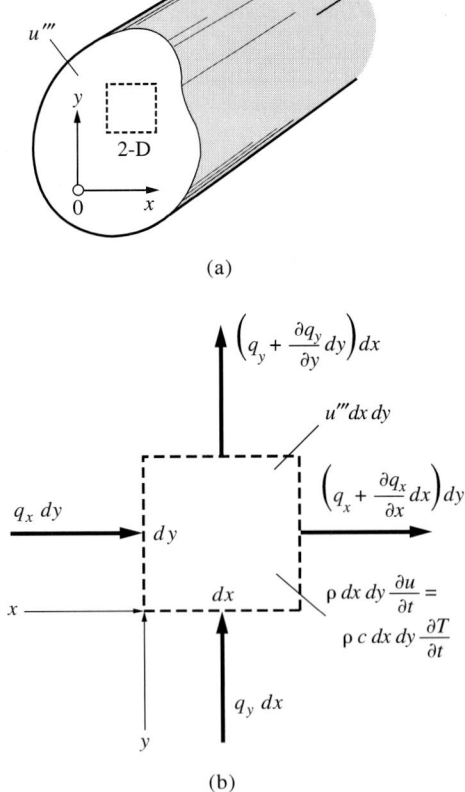

Figure 3.14 First law for a two-dimensional system.

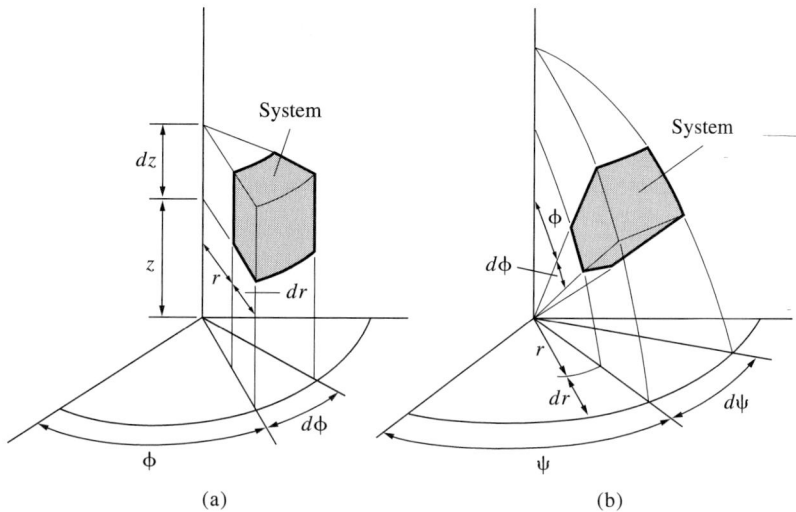

Figure 3.15 Cylindrical and spherical systems.

Many problems of conduction, however, may be more conveniently handled in terms of a cylindrical or spherical coordinate system. The Laplacian for a three-dimensional cylinder [Fig. 3.15(a)] is

$$\nabla^2 \equiv \frac{\partial}{\partial r}\left(r\frac{\partial}{\partial r}\right) + \frac{1}{r^2}\frac{\partial^2}{\partial \phi^2} + r\frac{\partial^2}{\partial z^2}, \qquad (3.77)$$

and for a three-dimensional sphere [Fig. 3.15(b)] is

$$\nabla^2 \equiv \frac{1}{r^2}\frac{\partial}{\partial r}\left(r^2\frac{\partial}{\partial r}\right) + \frac{1}{r^2 \sin\phi}\frac{\partial}{\partial \phi}\left(\sin\phi\frac{\partial}{\partial \phi}\right) + \frac{1}{r^2 \sin^2\phi}\frac{\partial^2}{\partial \psi^2}. \qquad (3.78)$$

For an application of the foregoing general considerations, reconsider the flat plate (key problem) of Section 2.3. Let the uniform internal energy u''' be suddenly generated and thereafter held constant in the plate which has a uniform initial temperature T_∞.

The governing equation for a variable conductivity, obtained from the one-dimensional form of Eq. (3.73), is

$$\rho c \frac{\partial T}{\partial t} = \frac{\partial}{\partial x}\left(k\frac{\partial T}{\partial x}\right) + u''', \qquad (3.79)$$

which is the unsteady form of Eq. (2.47). The governing equation for a uniform conductivity, obtained from the one-dimensional form of Eq. (3.74), is

$$\frac{1}{\alpha}\frac{\partial T}{\partial t} = \frac{\partial^2 T}{\partial x^2} + \frac{u'''}{k}, \qquad (3.80)$$

which is the unsteady form of Eq. (2.48). The initial condition and boundary conditions to be satisfied by Eq. (3.79) or Eq. (3.80) are

$$T(x, 0) = T_\infty, \qquad (3.81)$$

$$\frac{\partial T(0, t)}{\partial x} = 0, \qquad (3.82)$$

and[9]

$$-k\frac{\partial T(\ell, t)}{\partial x} = h\left[T(\ell, t) - T_\infty\right]. \qquad (3.83)$$

The solution of this formulation, including the foregoing partial differential equation, is beyond the scope of this book.

Having learned the general formulation of unsteady, three-dimensional problems and its illustration in terms of the foregoing one-dimensional example, we proceed now to a simplified formulation of these problems.

Consider a flat plate of thickness δ (Fig. 3.16). Let the temperature distribution in the plate be time dependent. The upward and downward heat transfer coefficients respectively are h_1 and h_2, and the Biot numbers based on these coefficients allow a transversally lumped formulation.

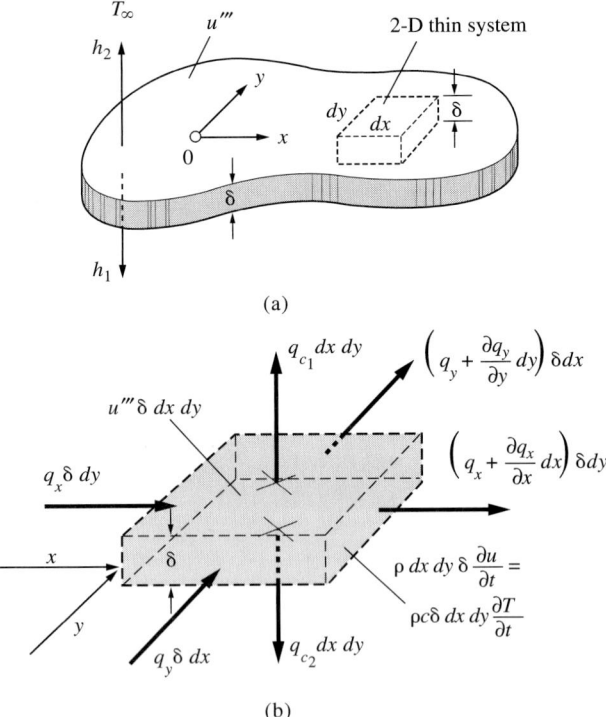

Figure 3.16 First law for a thin plate.

---

[9] Equations (3.82) and (3.83) are the unsteady forms of Eqs. (2.49) and (2.50), respectively.

The first law applied to the transversally lumped, otherwise differential system shown in Fig. 3.16(a), and interpreted in terms of Fig. 3.16(b), yields

$$\rho\delta\frac{\partial u}{\partial t} = -\delta\frac{\partial q_x}{\partial x} - \delta\frac{\partial q_y}{\partial y} - q_{c1} - q_{c2} + u'''\delta. \quad (3.84)$$

Relating u to T, employing two-dimensional Fourier's law and Newton's law for upward and downward convection, Eq. (3.84) may be rearranged to give

$$\frac{1}{\alpha}\frac{\partial T}{\partial t} = \frac{\partial^2 T}{\partial x^2} + \frac{\partial^2 T}{\partial y^2} - \left(\frac{h_1 + h_2}{k\delta}\right)(T - T_\infty) + \frac{u'''}{k}. \quad (3.85)$$

For an application of this formulation, reconsider Ex. 2.9. Let the fin have a uniform initial temperature T_∞, and let the base temperature be suddenly raised to temperature T_0 and held constant thereafter.

We have from Eq. (3.85), after eliminating the energy generation and the y dependence of temperature, and introducing $m^2 = (h_1 + h_2)/k\delta$,

$$\frac{1}{\alpha}\frac{\partial T}{\partial t} = \frac{\partial^2 T}{\partial x^2} - m^2(T - T_\infty) \quad (3.86)$$

subject to the initial condition

$$T(x, 0) = T_\infty, \quad (3.87)$$

and boundary conditions

$$T(0, t) = T_0, \quad \lim_{x \to \infty} T(x, t) = T_\infty. \quad (3.88)$$

As we learned in this chapter, the formulation of unsteady distributed problems leads to partial differential equations. The solution of these equations is much more involved than that of ordinary differential equations. Among the techniques available, the analytical and computational methods are most frequently referred to. Exact analytical methods such as separation of variables and transform calculus are beyond the scope of the text. However, the method of complex temperature and the use of charts based on exact analytical solutions, being useful for some practical problems, are respectively discussed in Sections 3.4 and 3.6. Among approximate analytical methods, the integral method, already introduced in Sections 2.4 and 3.1, is further discussed in Section 3.5. The analog solution technique is also briefly treated in Section 3.7.

3.4 STEADY PERIODIC SOLUTION ⊕

In Section 3.2 we focused on the unsteady solution and its steady part for periodic lumped problems. We learned then the practical importance of steady periodic solutions and, in terms of the method of complex temperature, an easy way of obtaining only the steady part of periodic solutions. In this section we apply the method of complex temperature to the steady periodic solution of distributed problems.[10] Since the method equally applies to both lumped and distributed problems, we proceed directly to an illustrative example for distributed problems.

EXAMPLE 3.7

Consider a semi-infinite solid whose surface temperature is oscillating as $\theta_0 \cos \omega t$ relative to an ambient at temperature T_∞. We wish to determine the steady periodic temperature of this solid.

Since the solution of interest is to be valid only for large values of time, there is no need for an initial condition. Then, in terms of $\theta = T - T_\infty$, the formulations of the original and complementary problems are, respectively,

$$\frac{1}{\alpha}\frac{\partial \theta}{\partial t} = \frac{\partial^2 \theta}{\partial x^2}, \qquad \frac{1}{\alpha}\frac{\partial \theta^*}{\partial t} = \frac{\partial^2 \theta^*}{\partial x^2}, \tag{3.89}$$

$$\theta(0, t) = \theta_0 \cos \omega t, \qquad \theta^*(0, t) = \theta_0 \sin \omega t \tag{3.90}$$

$$\theta(\infty, t) = 0, \qquad \theta^*(\infty, t) = 0. \tag{3.91}$$

Now, the complementary problem multiplied by i and added to the original problem yields the problem satisfied by the complex temperature, $\psi = \theta + i\theta^*$,

$$\frac{1}{\alpha}\frac{\partial \psi}{\partial t} = \frac{\partial^2 \psi}{\partial x^2}, \tag{3.92}$$

$$\theta(0, t) = \theta_0 e^{i\omega t}, \tag{3.93}$$

$$\theta(\infty, t) = 0. \tag{3.94}$$

The steady periodic solution,

$$\psi(x, t) = \phi(x) e^{i\omega t}, \tag{3.95}$$

with ϕ now depending on x, introduced into Eqs. (3.92)-(3.94), gives

$$\frac{d^2 \phi}{dx^2} - \left(\frac{i\omega}{\alpha}\right) \phi = 0, \tag{3.96}$$

$$\phi(0) = \theta_0, \tag{3.97}$$

$$\phi(\infty) = 0. \tag{3.98}$$

The solution of Eq. (3.96) subject to Eqs. (3.97) and (3.98) is

$$\frac{\phi}{\theta_0} = \exp\left[-\left(\frac{i\omega}{\alpha}\right)^{1/2} x\right]. \tag{3.99}$$

---

[10] The unsteady solution of these problems (see, for example, Ex. 7.28 of Reference 1) is beyond the scope of this text.

Finally, inserting Eq. (3.99) into Eq. (3.95), and noting that $i^{1/2} = (1+i)/2^{1/2}$, we get the complex temperature,

$$\frac{\psi(x,t)}{\theta_0} = \exp\left[-\left(\frac{\omega}{2\alpha}\right)^{1/2} x\right] \exp\left\{i\left[\omega t - \left(\frac{\omega}{2\alpha}\right)^{1/2} x\right]\right\}, \tag{3.100}$$

and from its real part, the steady periodic temperature of the original problem,

$$\frac{\theta(x,t)}{\theta_0} = \exp\left[-\left(\frac{\omega}{2\alpha}\right)^{1/2} x\right] \cos\left[\omega t - \left(\frac{\omega}{2\alpha}\right)^{1/2} x\right]. \tag{3.101}$$

This equation is perspectively sketched in Fig. 3.17. At a given depth x, the temperature oscillates with amplitude $\exp\left[-(\omega/2\alpha)^{1/2} x\right]$ relative to the amplitude of the disturbance and has the time lag $x/(2\alpha\omega)^{1/2}$ relative to the disturbance.

The penetration depth of this disturbance, obtained in terms of an approximate profile $(x/\delta)^2$, is

$$\int_0^\delta \left(\frac{x}{\delta}\right)^2 dx = \int_0^\infty \exp\left[-\left(\frac{\omega}{2\alpha}\right)^{1/2} x\right] dx,$$

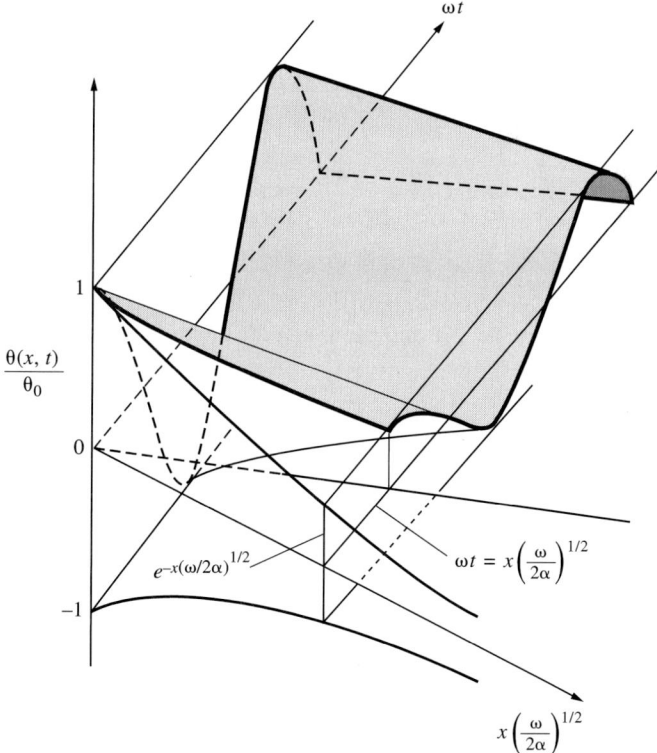

Figure 3.17 Amplitude and phase of temperature oscillations.

152 Chap. 3 Unsteady/steady, Multidimensional Conduction

or,

$$\delta \cong 4\sqrt{\frac{\alpha}{\omega}}. \qquad (3.102)$$

For a reciprocating engine (with $n = 100$ to $4{,}000$ rpm) the approximate range of ω in view of $\omega = 2\pi n/60 \sim n/10$ is 10 to 400 s^{-1}. Assuming $\alpha \cong 20 \times 10^{-6}$ for a cast iron cylinder block, we have $\delta = 5\text{–}1$ mm. ◆

3.5 INTEGRAL FORMULATION. APPROXIMATE SOLUTION ⊕

The integral formulation is a convenient tool for obtaining approximate solutions and, consequently, is suitable to problems having complicated exact solutions and is indispensable for complex problems having no exact solution. This formulation is obtained by integrating the differential form of the general laws over the entire geometry of a given problem. Next, an approximate profile is constructed for the dependent variable, say the temperature, involving a product of one-dimensional functions, each depending on one of the independent variables. Leaving one of these functions unknown and to be determined by the integral formulation, the other functions are written in terms of polynomials satisfying the conditions (initial or boundary) of problem. So far we have employed the integral formulation in connection with the penetration depth (of heat) in a steady fin problem (Section 2.4.1) and the cooling time of a lumped ball (Section 3.1). Here, we extend the use of this formulation to unsteady distributed problems in terms of a couple of illustrative examples.

EXAMPLE 3.8

Consider a flat plate of thickness 2ℓ. For a suddenly generated internal energy u''' we wish to obtain the integral formulation and its solution by approximate profiles.

Noting that the problem is symmetric relative to the midplane (equivalent to an insulated surface) and following the five steps of formulation, we apply the first law of thermodynamics to the system shown in Fig. 3.18 and, assuming the energy is generated electrically, get

$$\frac{dU}{dt} = -q_{x=\ell}A + U_e.$$

Evaluating dU/dt and U_e by integrating the corresponding differential terms (Fig. 3.18), we obtain

$$\frac{d}{dt}\int_0^\ell \rho c T\, dx = -q_{x=\ell} + \int_0^\ell u'''(x)\, dx, \qquad (3.103)$$

where from Fourier's law,

$$q_{x=\ell} = -k\left(\frac{\partial T}{\partial x}\right)_{x=\ell}. \qquad (3.104)$$

Inserting Eq. (3.104) into Eq. (3.103), we obtain the governing integral equation,

$$\frac{d}{dt}\int_0^\ell \rho c T\, dx = k\left(\frac{\partial T}{\partial x}\right)_{x=\ell} + \int_0^\ell u'''(x)\, dx. \qquad (3.105)$$

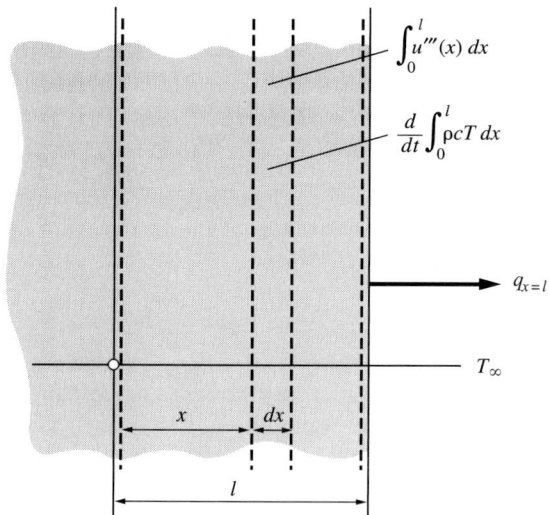

Figure 3.18 First law for an integral system.

The initial and boundary conditions are identical to those of the differential formulation. Hence Eq. (3.105), subject to Eqs. (3.81)-(3.83), completes the integral formulation of the problem.

An **approach** for an approximate solution is to assume the plate temperature to be a product of one-dimensional functions,

$$T(x,t) - T_\infty = X(x)\tau(t). \tag{3.106}$$

If an unsteady temperature distribution resembles its ultimate steady limit, we may select $X(x)$ to be the steady solution of the problem. Thus, in view of Eq. (2.59), an approximate unsteady solution may be written as

$$T(x,t) - T_\infty = \frac{u'''\ell^2}{2k}\left[1 - \left(\frac{x}{\ell}\right)^2 + \frac{2}{B}\right]\tau(t). \tag{3.107}$$

Inserting Eq. (3.107) into Eq. (3.105), and integrating the latter, we obtain the differential equation

$$\frac{d\tau}{dt} + \frac{3\alpha}{\ell^2}\left(\frac{B}{B+3}\right)(\tau - 1) = 0, \tag{3.108}$$

subject to the initial condition

$$\tau(0) = 0. \tag{3.109}$$

The solution of Eq. (3.108), first satisfied by Eq. (3.109), next introduced into Eq. (3.107), gives the approximate solution

$$\frac{T(x,t) - T_\infty}{u'''\ell^2/k} = \frac{1}{2}\left[1 - \left(\frac{x}{\ell}\right)^2 + \frac{2}{B}\right]\left[1 - \exp\left(-\frac{3\alpha}{\ell^2}\frac{B}{B+3}t\right)\right]. \tag{3.110}$$

(What happens to Eq. 3.110 as $B \to \infty$? Recall Fig. 2.15.) ◆

EXAMPLE 3.9

Reconsider a semi-infinite plate. Let the initial temperature of the plate be uniform, say T_0, and the surface temperature be suddenly changed to T_∞. We wish to develop the integral formulation of this problem and its solution by approximate profiles.

This problem can be best described in terms of the concept of **penetration depth** of heat, $\delta(t)$, shown in Fig. 3.19. The first law of thermodynamics applied to the **control volume** of thickness δ yields

$$\frac{dU}{dt} = \rho c A \frac{d\delta}{dt} T_0 - q_{x=\delta} A. \tag{3.111}$$

Note that the moving penetration depth carries an enthalpy flow into the expanding control volume. The coordinate axis measured from the moving penetration depth proves convenient in the construction of approximate profiles. Evaluating the internal energy of the control volume by integrating the differential internal energy, Eq. (3.111) becomes

$$\frac{d}{dt} \int_0^{\delta(t)} \rho c T \, dx = \rho c T_0 \frac{d\delta}{dt} - q_{x=\delta},$$

or, alternatively,

$$\frac{d}{dt} \int_0^{\delta(t)} \rho c (T - T_0) \, dx = -q_{x=\delta}. \tag{3.112}$$

Fourier's law,

$$q_{x=\delta} = -k \left(\frac{\partial T}{\partial x}\right)_{x=\delta},$$

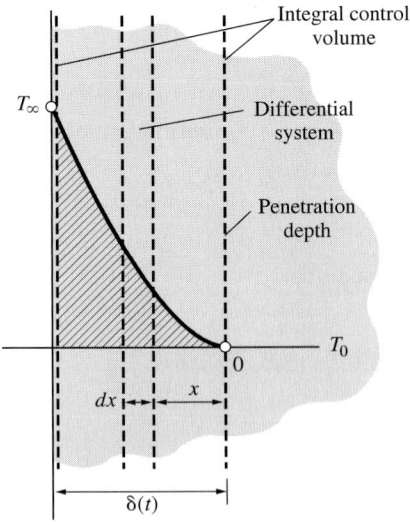

Figure 3.19 Penetration depth.

inserted into Eq. (3.112) yields the governing integral equation,

$$\frac{d}{dt}\int_0^{\delta(t)} \rho c(T - T_0)dx = k\left(\frac{\partial T}{\partial x}\right)_{x=\delta}. \qquad (3.113)$$

The initial and boundary conditions are

$$T(x, 0) = T_0; \quad T(\delta, t) = T_\infty, \quad \frac{\partial T(0, t)}{\partial x} = 0. \qquad (3.114)$$

The steady solution of the problem (corresponding to $t \to \infty$) is $T = T_\infty$. Any scaling of the steady solution does not appear to resemble the unsteady temperature distribution. This fact suggests a **second approach** for the construction of approximate profiles, one that requires the selection of polynomials satisfying the boundary conditions. For example, a parabola satisfying the boundary conditions of Eq. (3.114) gives

$$T(x, t) - T_0 = (T_\infty - T_0)\left(\frac{x}{\delta}\right)^2. \qquad (3.115)$$

Introducing Eq. (3.115) into Eq. (3.113), and integrating the latter with the assumption of constant properties, we obtain

$$\frac{1}{3}\rho c \frac{d\delta}{dt} = 2\frac{k}{\delta},$$

or,

$$d\delta^2 = 12\alpha dt \qquad (3.116)$$

subject to

$$\delta(0) = 0. \qquad (3.117)$$

The solution of Eq. (3.116) which satisfies Eq. (3.117) is the unsteady penetration depth,

$$\delta = \sqrt{12\alpha t}. \qquad (3.118)$$

Now, the time required for heat to penetrate to a distance ℓ in a solid can be readily evaluated from Eq. (3.118) as[11]

$$t_0 = \ell^2/12\alpha. \qquad (3.119)$$

The penetration time of heat to $\ell = 4$ cm in steel and copper are shown in Table 3.3 (based on the numerical values used in Section 3.1). For convenience, the numbers in the last column are somewhat rounded. Copper conducts heat much better than steel, thus has a much shorter penetration time.

Table 3.3 Penetration time for steel and copper

Solid	k [W/m·K]	ρ [kg/m^3]	c [J/kg·K]	α [m^2/s]	t_0 [s]
Steel	10	8,000	500	2.5×10^{-6}	50
Copper	400	9,000	400	1.1×10^{-4}	1

---

[11] The type of boundary condition affects the numerical constant of Eq. (3.119).

Inserting Eq. (3.118) into Eq. (3.116), we have the approximate unsteady temperature

$$\frac{T(x,t) - T_0}{T_\infty - T_0} = \frac{1}{12}\left(\frac{x^2}{\alpha t}\right). \qquad (3.120)$$

◆

3.6 CHARTED EXACT SOLUTIONS

In the preceding section we studied the formulation of unsteady distributed problems and indicated the somewhat involved nature of their solutions. This section is devoted to the use of charts obtained from these solutions without actually working out the solutions.

3.6.1 Flat Plate (Key Problem)

Consider an infinitely long flat plate of thickness 2ℓ, initially having a uniform temperature, say T_0. From this condition, the plate is suddenly immersed in a fluid at constant temperature T_∞. The plate exchanges heat with the fluid with a heat transfer coefficient h.

The governing equation, obtained from the one-dimensional form of Eq. (3.80) in the absence of any energy generation, is

$$\frac{1}{\alpha}\frac{\partial T}{\partial t} = \frac{\partial^2 T}{\partial x^2} \qquad (3.121)$$

subject to initial condition

$$T(x, 0) = T_0 \qquad (3.122)$$

and the boundary conditions relative to the x coordinate measured from the middle plane,

$$\frac{\partial T(0, t)}{\partial x} = 0, \quad -k\frac{\partial T(\ell, t)}{\partial x} = h[T(\ell, t) - T_\infty]. \qquad (3.123)$$

The exact analytical solution of this problem is (see, for example, Ex. 5.3 of Reference 1)

$$\frac{T - T_\infty}{T_0 - T_\infty} = 2\sum_{n=1}^{\infty}\left(\frac{\sin\mu_n}{\mu_n + \sin\mu_n \cos\mu_n}\right)e^{-\mu_n^2 Fo}\cos\mu_n\xi, \qquad (3.124)$$

where $\xi = x/\ell$, $Fo = \alpha t/\ell^2$ (Fourier number), μ_n are the roots of $\mu_n \sin\mu_n = Bi \cos\mu_n$, and $Bi = h\ell/k$. As can be readily seen from Eq. (3.124), or from the nondimensionalization of the foregoing formulation, the dimensionless temperature depends on the dimensionless distance, ξ, dimensionless time, Fo, and the ratio of internal resistance to external resistance, $(\ell/k)/(1/h) = h\ell/k = Bi$,

$$\frac{T - T_\infty}{T_0 - T_\infty} = f(\xi, Fo, Bi).$$

In Fig. 3.20 this temperature is plotted against Fo for the middle plane ($\xi = 0$) and surface ($\xi = 1$) of the plate, with $1/Bi$ as a parameter.[12]

A useful quantity may be the heat transfer Q between half the plate and the ambient over a time interval $(0, t)$. This heat transfer relative to the initial internal energy of the plate, $Q_0 = \rho c A \ell (T_0 - T_\infty)$, A being one surface area, is

$$\frac{Q}{Q_0} = \frac{-kA \int_0^t \frac{\partial}{\partial x}(T - T_\infty)\Big|_{x=\ell} dt}{\rho c A \ell (T_0 - T_\infty)},$$

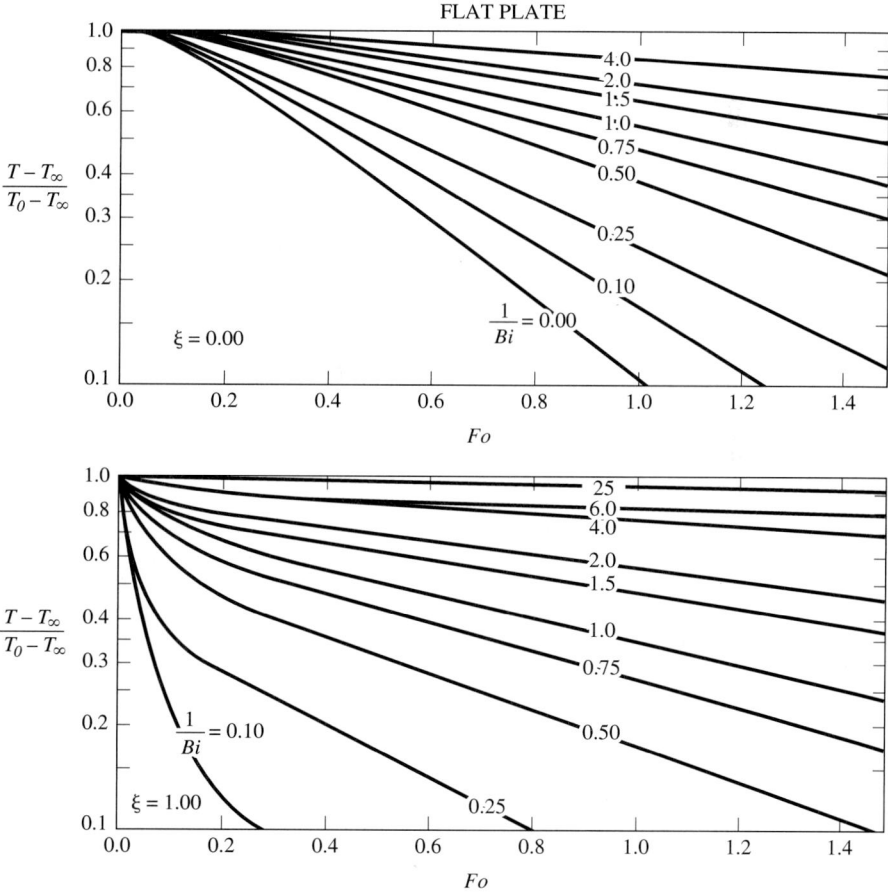

Figure 3.20 Middle plane and surface temperatures.

---

[12] Figures 3.20, 3.22, 3.24, 3.26, 3.27 from Boelter et al. [5] (see Appendix D for more detailed charts from Heisler [8])

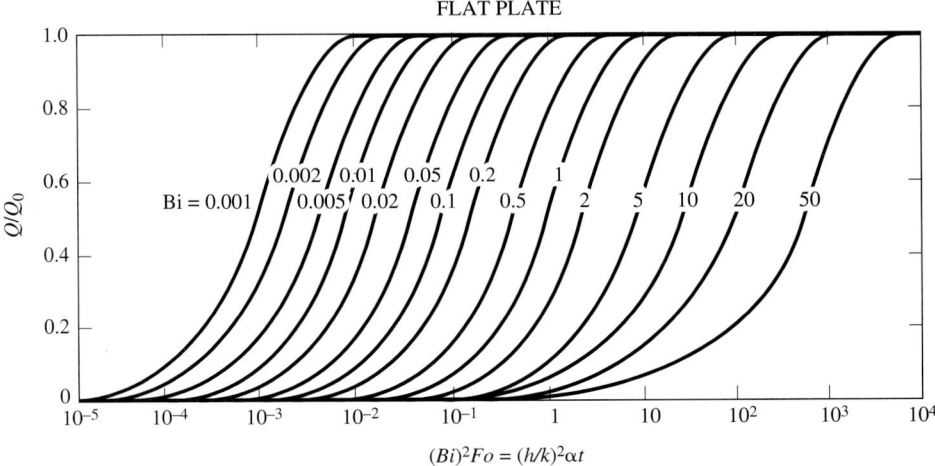

Figure 3.21 Heat transfer over interval $(0, t)$.

or, in terms of Eq. (3.124),

$$\frac{Q}{Q_0} = 2 \sum_{n=1}^{\infty} \frac{\sin^2 \mu_n}{\mu_n(\mu_n + \sin \mu_n \cos \mu_n)} \left(1 - e^{-\mu_n^2 Fo}\right). \quad (3.125)$$

In Fig. 3.21 this equation is plotted against $(Bi)^2 Fo$ for some values of Bi.[13] Equation (3.125) could also be plotted as a function of Bi and Fo. The difference in these arrangements is mainly due to the fact that Figs. 3.20 and 3.21 are taken from two separate sources. Note, however, that both parameters in the (Bi, Fo) combination depend on ℓ, while only one parameter in the $[Bi, (Bi)^2 Fo]$ combination depends on ℓ. The following examples illustrate the use of these charts.

EXAMPLE 3.10[14]

A steel plate of thickness $2\ell = 5$ cm, heated to a uniform initial temperature $T_0 = 550\,°\text{C}$ for a heat treatment process, is plunged into a water bath at constant temperature $T_\infty = 50\,°\text{C}$. The density, specific heat, and thermal conductivity of the plate are $\rho = 8 \times 10^3$ kg/m^3, $c_p = 420$ J/kg·K, and $k = 17$ W/m·K, respectively. The heat transfer coefficient is $h = 340$ W/m^2·K.

We wish to determine the middle plane and surface temperatures of the plate and the heat loss from the plate after two minutes.

The needed numerical values for the dimensionless numbers are

$$Bi = \frac{h\ell}{k} = \frac{340 \times (0.05/2)}{17} = 1/2, \quad \frac{1}{Bi} = 2$$

$$\left(\alpha = \frac{k}{\rho c_p} = \frac{17}{8 \times 10^3 \times 420} = 5.06 \times 10^{-6}\text{ m}^2/\text{s} \cong 5.0 \times 10^{-6}\text{ m}^2/\text{s}\right),$$

---

[13] Figures 3.21, 3.23, 3.25 from H. Grober, Erk, and Grigull [6].

[14] The FORTRAN program EX3–10.F is listed in the appendix of this chapter.

$$Fo = \frac{\alpha t}{\ell^2} = \frac{5.0 \times 10^{-6} \times 120}{(0.05/2)^2} = 0.96 \cong 1, \quad Bi^2 Fo \cong \left(\frac{1}{2}\right)^2 \times 1 = 0.25.$$

Then, from the chart of Fig. 3.20 for $\xi = 0$,

$$\frac{T - T_\infty}{T_0 - T_\infty} \cong 0.72,$$

which gives

$$T_{\text{mid}} = (550 - 50) \times 0.72 + 50 = 410 \,°\text{C}.$$

From the chart of the same figure for $\xi = 1$,

$$\frac{T - T_\infty}{T_0 - T_\infty} \cong 0.56,$$

which gives

$$T_{\text{surf}} = (550 - 50) \times 0.56 + 50 = 330 \,°\text{C}.$$

By the computer program provided, the interested reader may compute Bi and Fo for a parametric study of T_{mid} and T_{surf} of some steel and iron plates plunged into different liquids.

Also, from Fig. 3.21, for $Bi = 1/2$ and $Bi^2 Fo \cong 0.25$,

$$\frac{Q}{Q_0} \cong 0.30.$$

Then,

$$\frac{Q_0}{A} = \rho c_p \ell (T_0 - T_\infty) = 8 \times 10^3 \times 420 \times \left(\frac{0.05}{2}\right) \times 500 = 42{,}000 \text{ kJ/m}^2,$$

$$\frac{Q}{A} \cong 0.3 \times 42{,}000 = 12{,}600 \text{ kJ/m}^2 \quad \text{lost from each surface.}$$

◆

One-dimensional charts find an important application in the solution of multidimensional problems. It can be shown (see, for example, Section 5.2 of Ref. 1) that the dimensionless unsteady temperature of an infinitely long rod of rectangular cross section $2\ell \times 2L$ may be expressed as the product of the dimensionless temperature of an infinite flat plate of thickness 2ℓ times the dimensionless temperature of an infinite flat plate of thickness $2L$,

$$\left(\frac{T - T_\infty}{T_0 - T_\infty}\right)_{2\ell \times 2L, \text{rod}} = \left(\frac{T - T_\infty}{T_0 - T_\infty}\right)_{2\ell, \text{plate}} \times \left(\frac{T - T_\infty}{T_0 - T_\infty}\right)_{2L, \text{plate}}. \quad (3.126)$$

Also, the dimensionless unsteady temperature of a rod of finite height $2H$ and of rectangular cross section $2\ell \times 2L$ may be written as

$$\left(\frac{T - T_\infty}{T_0 - T_\infty}\right)_{2\ell \times 2L \times 2H, \text{rod}} = \left(\frac{T - T_\infty}{T_0 - T_\infty}\right)_{2\ell, \text{plate}} \times \left(\frac{T - T_\infty}{T_0 - T_\infty}\right)_{2L, \text{plate}} \times \left(\frac{T - T_\infty}{T_0 - T_\infty}\right)_{2H, \text{plate}}. \quad (3.127)$$

EXAMPLE 3.11

Reconsider the preceding example for a long rod with square cross section (5 cm × 5 cm).

For the axis of the rod,

$$\left(\frac{T - T_\infty}{T_0 - T_\infty}\right)_{\xi=0,\xi=0} = 0.72 \times 0.72 \cong 0.52,$$

which gives

$$T_{\text{axis}} = (550 - 50) \times 0.52 + 50 = 310\,°\text{C}.$$

For the middle of each surface,

$$\left(\frac{T - T_\infty}{T_0 - T_\infty}\right)_{\xi=0,\xi=1} = 0.72 \times 0.56 \cong 0.40$$

which gives

$$T_{\text{middle}} = (550 - 50) \times 0.40 + 50 = 250\,°\text{C}.$$

For a corner,

$$\left(\frac{T - T_\infty}{T_0 - T_\infty}\right)_{\xi=1,\xi=1} = 0.56 \times 0.56 \cong 0.31$$

which gives

$$T_{\text{corner}} = (550 - 50) \times 0.31 + 50 = 205\,°\text{C}.$$

◆

EXAMPLE 3.12

Reconsider Ex. 3.10 for a cube (5 cm × 5 cm × 5 cm).

For the center of the cube,

$$\left(\frac{T - T_\infty}{T_0 - T_\infty}\right)_{\xi=0,\xi=0,\xi=0} = 0.72 \times 0.72 \times 0.72 \cong 0.37,$$

which gives

$$T_{\text{center}} = (550 - 50) \times 0.37 + 50 = 235\,°\text{C}.$$

For the corner of the cube,

$$\left(\frac{T - T_\infty}{T_0 - T_\infty}\right)_{\xi=1,\xi=1,\xi=1} = 0.56 \times 0.56 \times 0.56 \cong 0.18,$$

which gives

$$T_{\text{corner}} = (550 - 50) \times 0.18 + 50 = 140\,°\text{C}.$$

◆

3.6.2 Solid Cylinder (Key Problem)

Reconsider the flat plate (key problem) now for an infinitely long cylinder of diameter $2R$. The exact analytical solution of this problem (see, for example, Ex. 5.8 of Reference 1) may be implicitly expressed as

$$\frac{T - T_\infty}{T_0 - T_\infty} = f(\rho, Fo, Bi), \tag{3.128}$$

where $\rho = r/R$, $Fo = \alpha t/R^2$, and $Bi = hR/k$. In Fig. 3.22 this temperature is plotted against Fo for the center ($\rho = 0$) and surface ($\rho = 1$) of the cylinder, with $1/Bi$ as a parameter.

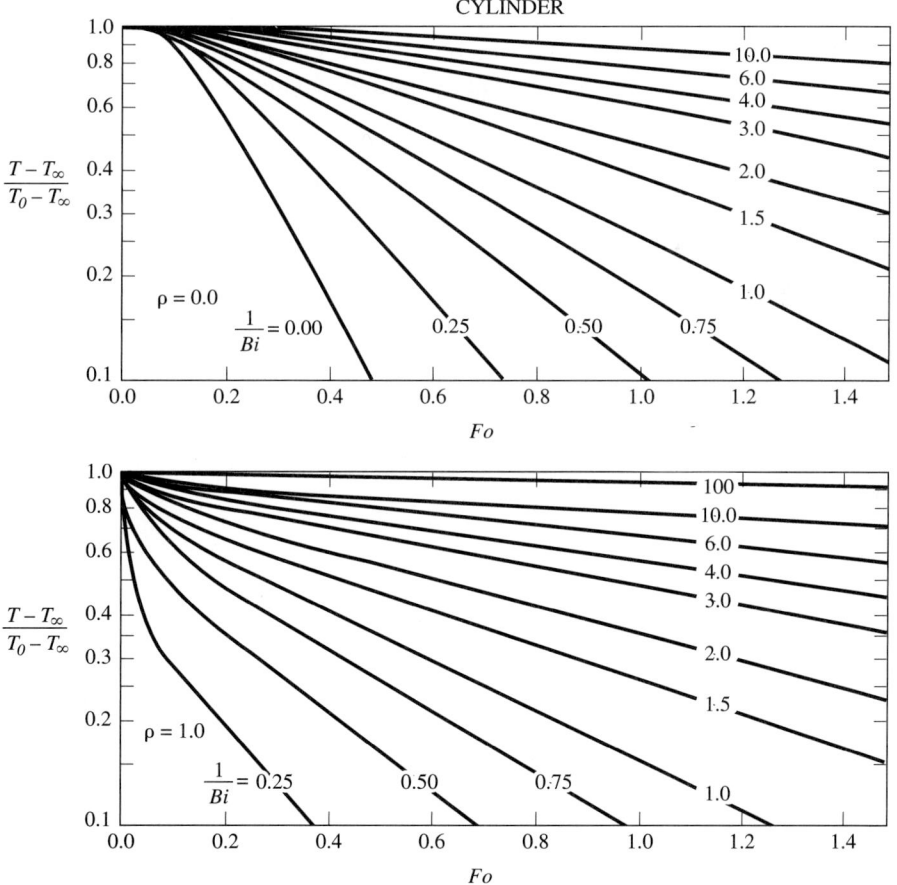

Figure 3.22 Centerline and surface temperatures.

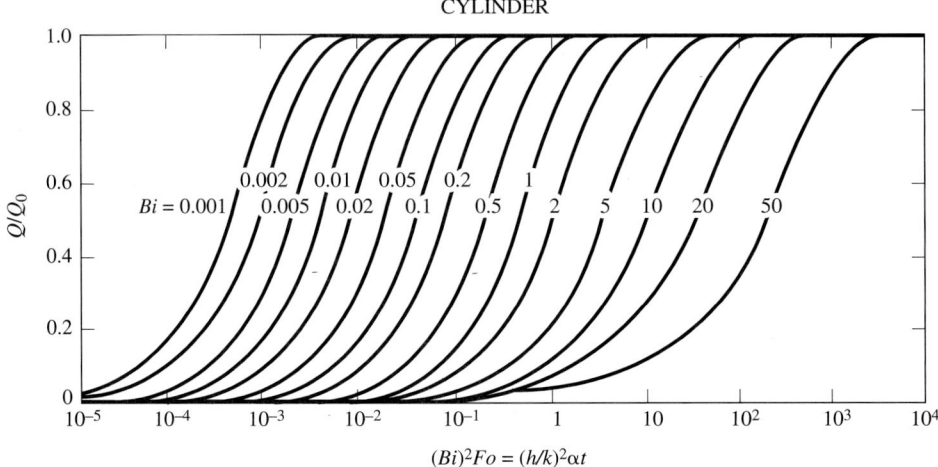

Figure 3.23 Heat transfer over interval $(0, t)$.

The heat transfer Q over a time interval $(0, t)$ and per unit length of this cylinder relative to the initial internal energy of the cylinder per unit length,

$$Q_0 = \rho c \pi R^2 (T_0 - T_\infty),$$

is plotted against $(Bi)^2 Fo$ in Fig. 3.23 for some values of Bi.

The one-dimensional charts for a flat plate and an infinitely long cylinder can be used to determine the temperature of a cylinder of finite length. It can be shown that the dimensionless unsteady temperature of a solid cylinder of height $2H$ and diameter $2R$ may be expressed as the product of the dimensionless temperature of a flat plate of thickness $2H$ and the temperature of an infinitely long cylinder of diameter $2R$,

$$\left(\frac{T - T_\infty}{T_0 - T_\infty}\right)_{2H \times 2R, \text{finite cylinder}} = \left(\frac{T - T_\infty}{T_0 - T_\infty}\right)_{2H, \text{plate}} \times \left(\frac{T - T_\infty}{T_0 - T_\infty}\right)_{2R, \text{cylinder}}. \tag{3.129}$$

EXAMPLE 3.13

Reconsider Ex. 3.10 for an infinitely long cylinder of diameter $2R = 5$ cm.

For $\alpha t / R^2 \cong 1$ and $Bi = hR/k = 0.5$, from Fig. 3.22 for $\rho = 0$,

$$\left(\frac{T - T_\infty}{T_0 - T_\infty}\right)_{\rho=0} \cong 0.47,$$

which gives the axis temperature,

$$T_{\text{axis}} = (550 - 50) \times 0.47 + 50 = 285\,°\text{C}.$$

From Fig. 3.22 for $\rho = 1$,

$$\left(\frac{T - T_\infty}{T_0 - T_\infty}\right)_{\rho=1} \cong 0.37,$$

which gives the surface temperature,

$$T_{\text{surface}} = (550 - 50) \times 0.37 + 50 = 235\,°\text{C}.$$

Compare these temperatures with those obtained in Ex. 3.10. ◆

Example 3.14

Reconsider Ex. 3.10 for a finite solid cylinder of height $2H = 5$ cm and diameter $2R = 5$ cm.

We have already obtained, for a flat plate of thickness $2H = 5$ cm,

$$\left(\frac{T - T_\infty}{T_0 - T_\infty}\right)_{\xi=0} \cong 0.72 \quad \text{and} \quad \left(\frac{T - T_\infty}{T_0 - T_\infty}\right)_{\xi=1} \cong 0.56$$

and for an infinitely long cylinder of diameter $2R = 5$ cm,

$$\left(\frac{T - T_\infty}{T_0 - T_\infty}\right)_{\rho=0} \cong 0.47 \quad \text{and} \quad \left(\frac{T - T_\infty}{T_0 - T_\infty}\right)_{\rho=1} \cong 0.37.$$

The hottest spot of this cylinder is at location ($\xi = 0$, $\rho = 0$). Then, for this spot,

$$\left(\frac{T - T_\infty}{T_0 - T_\infty}\right)_{\xi=0,\rho=0} \cong 0.72 \times 0.47 \cong 0.34,$$

which gives

$$T_{\text{hottest}} = (550 - 50) \times 0.34 + 50 = 220\,°\text{C}.$$

For the coldest spot of the cylinder,

$$\left(\frac{T - T_\infty}{T_0 - T_\infty}\right)_{\xi=1,\rho=1} \cong 0.56 \times 0.37 \cong 0.21,$$

which gives

$$T_{\text{coldest}} = (550 - 50) \times 0.21 + 50 = 155\,°\text{C}.$$

◆

3.6.3 Solid Sphere (Key Problem)

Reconsider the flat plate (key problem) now for solid sphere of diameter $2R$. The exact analytical solution of this problem (see, for example, Ex. 5.10 of Reference 1) is parametrically identical to Eq. (3.128). In Fig. 3.24 the unsteady temperature of the sphere is plotted against Fo for the center ($\rho = 0$) and surface ($\rho = 1$) of the sphere, with Bi as parameter, where $Fo = \alpha t/R^2$ and $Bi = hR/k$.

164 Chap. 3 Unsteady/steady, Multidimensional Conduction

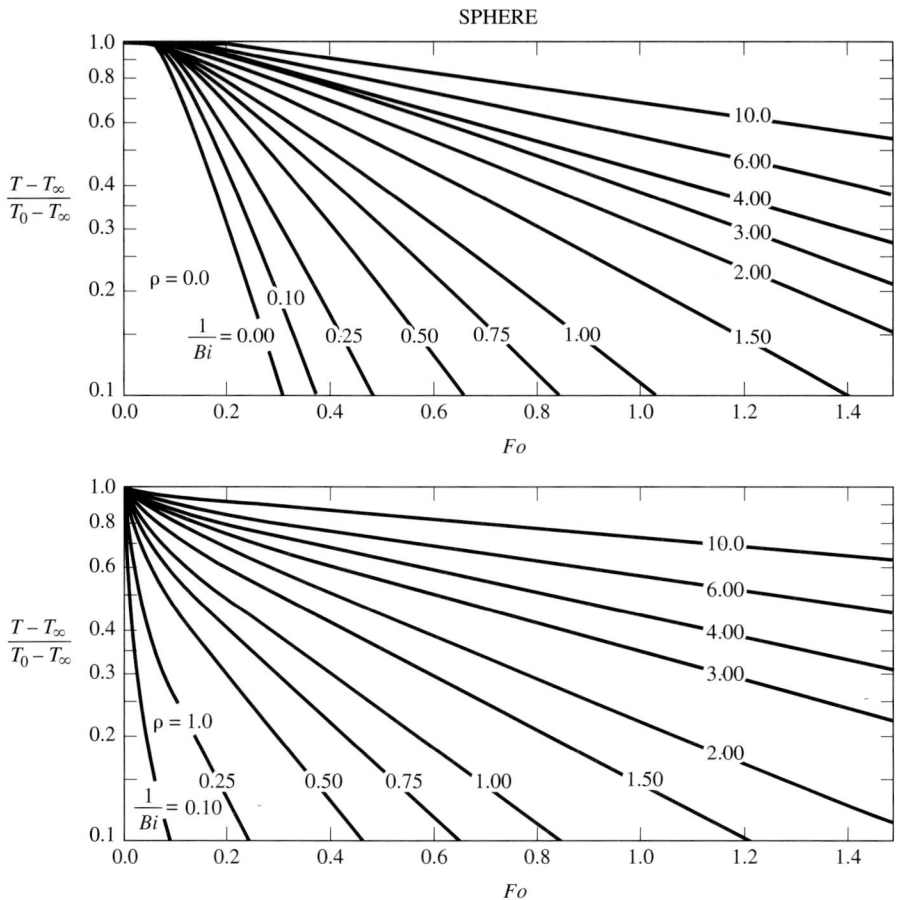

Figure 3.24 Centerline and surface temperatures.

The heat transfer Q from the sphere over a time interval $(0, t)$, relative to the initial internal energy of the sphere,

$$Q_0 = \rho c \frac{4}{3}\pi R^3 (T_0 - T_\infty),$$

is plotted against $(Bi)^2 Fo$ in Fig. 3.25 for some values of Bi.

EXAMPLE 3.15

Reconsider Ex. 3.10 for a solid sphere of diameter $2R = 5$ cm.

From Fig. 3.24 for $\rho = 0$,

$$\left(\frac{T - T_\infty}{T_0 - T_\infty}\right)_{\rho=0} \cong 0.30,$$

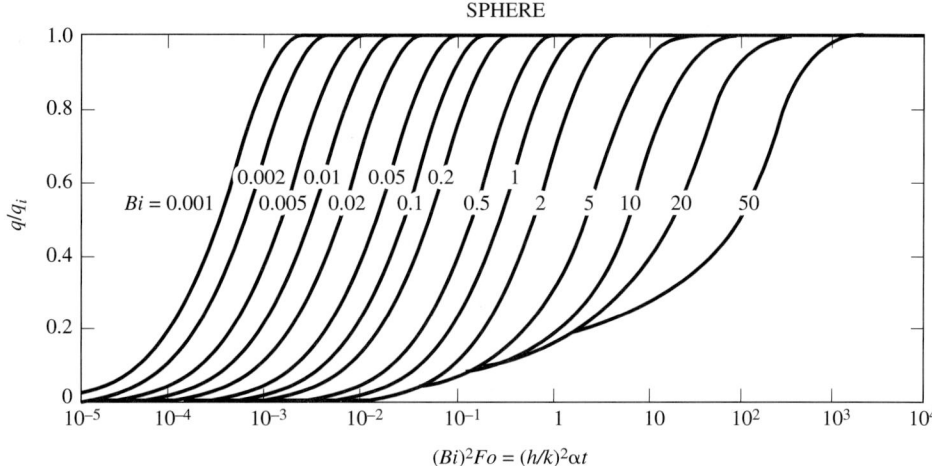

Figure 3.25 Heat transfer over interval $(0, t)$.

which gives

$$T_{\text{center}} = (550 - 50) \times 0.30 + 50 = 200\,°\text{C}.$$

From Fig. 3.24 for $\rho = 1$,

$$\left(\frac{T - T_\infty}{T_0 - T_\infty}\right)_{\rho=1} \cong 0.23,$$

which gives the surface temperature,

$$T_{\text{surface}} = (550 - 50) \times 0.23 + 50 = 165\,°\text{C}.$$

Compare these temperatures with the hottest and coldest temperatures of the cube of Ex. 3.12 and the finite cylinder of Ex. 3.13. Justify the results with physical reasoning. ◆

The foregoing considerations for finite geometry are now extended to semi-infinite geometry.

3.6.4 Semi-infinite Plate

Consider a semi-infinite plate at a uniform initial temperature T_0. From this condition, the ambient temperature is suddenly changed to a temperature T_∞. The heat transfer coefficient is h.

The unsteady temperature of the plate corresponding to an infinite heat transfer coefficient is

$$\left(\frac{T - T_\infty}{T_0 - T_\infty}\right)_x = \text{erfc}\left(\frac{x}{2\sqrt{\alpha t}}\right), \qquad (3.130)$$

and that corresponding to a finite heat transfer coefficient is

$$\left(\frac{T - T_\infty}{T_0 - T_\infty}\right)_x = \text{erfc}\left(\frac{x}{2\sqrt{\alpha t}}\right) - \exp\left[\frac{hx}{k} + \left(\frac{h}{k}\right)^2 \alpha t\right] \text{erfc}\left[\frac{x}{2\sqrt{\alpha t}} + \frac{h}{k\sqrt{\alpha t}}\right], \tag{3.131}$$

where erfc is the complementary error function [see, for example, Chapter 7 of Reference 1 for an exact method of solution leading to Equations (3.130) and (3.131)]. In Fig. 3.26, Eq. (3.130) is plotted against $x/(\sqrt{\alpha t})$. In Fig. 3.27(a),(b), Eq. (3.131) is plotted against different values of the local Biot modulus, $Bi_x = hx/k$, with $(h/k)^2 \alpha t$ as a parameter.

By referring to Eqs. (3.130) and (3.131), we may write the temperature of two- and three-dimensional corners in the form

$$\left(\frac{T - T_\infty}{T_0 - T_\infty}\right)_{x,y} = \left(\frac{T - T_\infty}{T_0 - T_\infty}\right)_x \times \left(\frac{T - T_\infty}{T_0 - T_\infty}\right)_y \tag{3.132}$$

and

$$\left(\frac{T - T_\infty}{T_0 - T_\infty}\right)_{x,y,z} = \left(\frac{T - T_\infty}{T_0 - T_\infty}\right)_x \times \left(\frac{T - T_\infty}{T_0 - T_\infty}\right)_y \times \left(\frac{T - T_\infty}{T_0 - T_\infty}\right)_z. \tag{3.133}$$

Similarly, the temperature of a semi-infinite rod of diameter $2R$ may be obtained from

$$\left(\frac{T - T_\infty}{T_0 - T_\infty}\right)_{2R,z} = \left(\frac{T - T_\infty}{T_0 - T_\infty}\right)_{2R} \times \left(\frac{T - T_\infty}{T_0 - T_\infty}\right)_z, \tag{3.134}$$

where $[(T - T_\infty)/(T_0 - T_\infty)]_{2R}$ is given by Eq. (3.128), and is plotted in Figure 3.22.

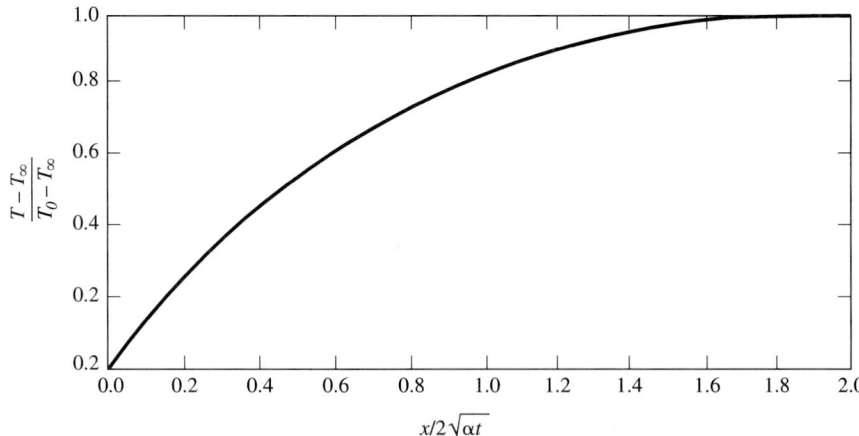

Figure 3.26 Semi-infinite solid with infinite h.

Sec. 3.6 Charted Exact Solutions 167

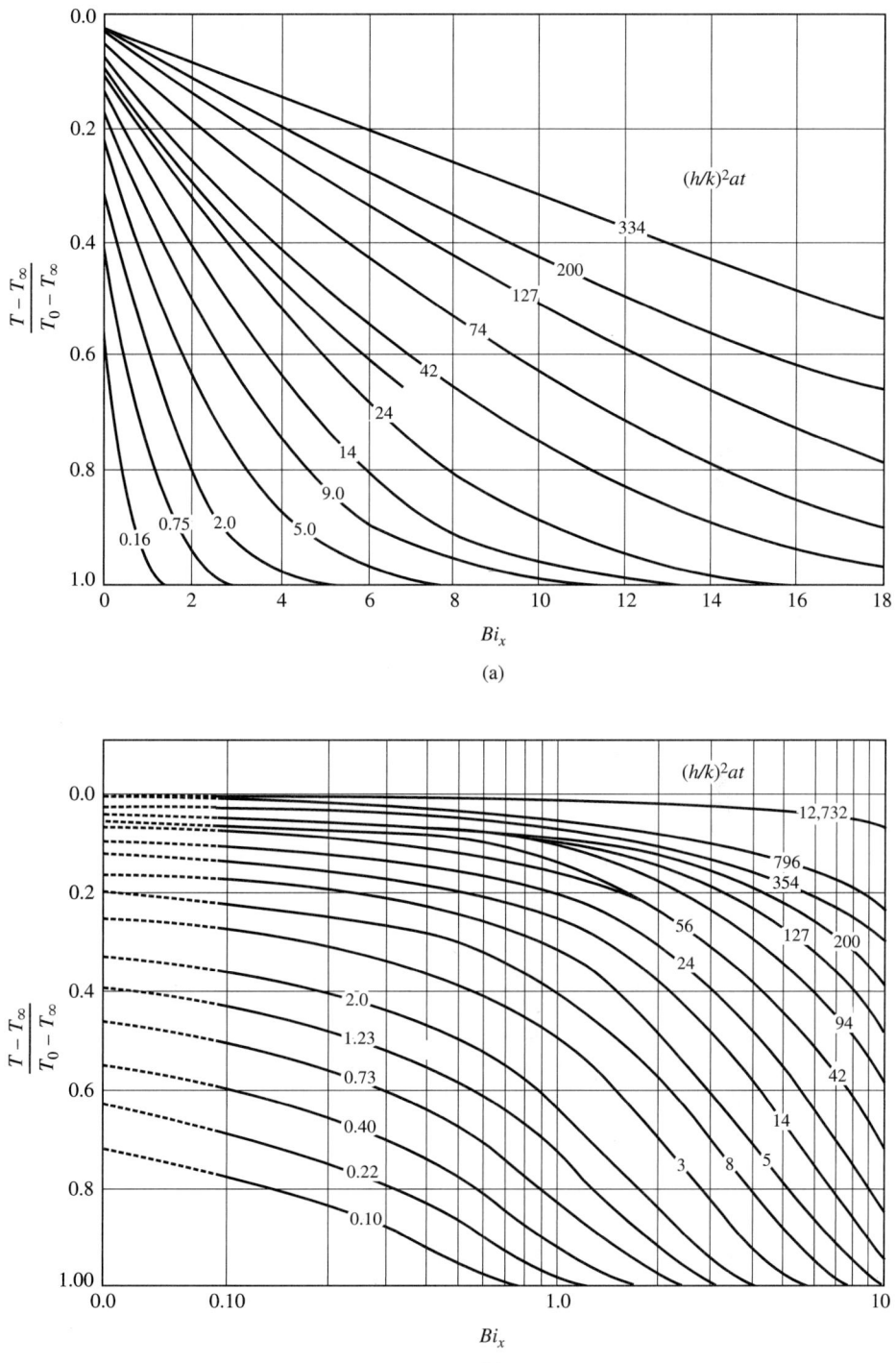

Figure 3.27 Semi-infinite solid with finite h. (a) Linear Bi_x, (b) logarithmic Bi_x.

3.7 MIXED (DIFFERENTIAL-DIFFERENCE) FORMULATION. ANALOG SOLUTION

As we have seen in the preceding sections, the solution of unsteady conduction problems is, in general, not mathematically simple, and one must usually resort to a number of solution methods to evaluate the unsteady temperature distribution. We have also learned how to obtain solutions by using the available charts for a class of analytical results. In Chapter 4 we will explore the use of numerical computations to evaluate multidimensional and unsteady conduction problems. These computations require approximate difference formulations to represent time and spatial derivatives. Actually there exists a third and hybrid (analog) method that allows us to evaluate the temperature distribution in a conduction problem by using a timewise differential and spacewise difference formulation. This method utilizes electrical circuits to represent unsteady conduction problems. The circuits are selected in such a way that the voltages (representing temperatures) obey the same differential equations as the temperature.

An analog solution of conduction problems requires a circuit capable of performing multiplication by a constant, addition, integration, and differentiation. Electric circuits that can accomplish these operations are called passive circuits if they include only fixed resistors and capacitors and possibly inductors and transformers. They are called active circuits if, like electronic amplifiers, they involve additional elements drawing energy from an external source. All passive circuit elements suffer from loading errors. In general, these elements operate correctly only if the impedance at the output of the circuit element is very high. Since active circuit elements such as electronic amplifiers have input impedances often in excess of several megaohms, the coupling of a passive circuit element with an amplifier eliminates the loading error on the element. Of course, the use of amplifiers increases both the first cost and the operating cost of the analog circuit. However, this expense may be well justified by the increased accuracy of the solution. For most general-purpose analog computation utilize the high-gain DC amplifier, which is discussed in the next section. In general, computations based on AC analog computations are less expensive to build and require simpler auxiliaries. The phase shifts in AC circuits, however, tend to make this type of computer less accurate than DC computers.

3.7.1 Active Circuit Elements. High-Gain DC Amplifiers

The most commonly employed active elements are high-gain DC amplifiers, function generators, and function multiplier-dividers. We shall consider only amplifiers.

The design of amplifiers and the function of their components are beyond the scope of the text. Interested readers may consult the references at the end of this chapter. Here we picture the amplifier simply as a black box (Fig. 3.28) and discuss only those amplifier characteristics conceptually important for analog computer solutions. Designating the input and output potentials relative to the ground by e_i and e_0, respectively, we first define the gain A of the of amplifier by the ratio[15]

$$A = -\frac{e_0}{e_i}.$$

---

[15] The sign of this ratio will be discussed in connection with Fig. 3.29.

Sec. 3.7 Mixed (Differential-Difference) Formulation. Analog Solution

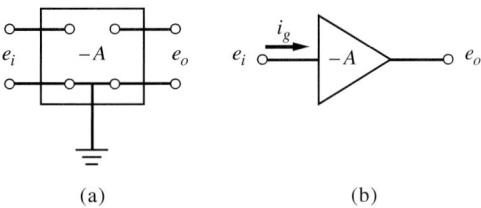

Figure 3.28 Amplifier. (a) Schematic, (b) symbolic.

The basic requirement of the amplifier is to have as high a gain as possible. The practical range of this gain is 10^4-10^8. Assuming that, for the purpose of recording, the output potential is to be kept between ± 100 volts, we obtain a maximum possible value of 0.01 volts for the input potential. Since this potential is usually connected to a grid which has an extremely high impedance, the input current is very small. Hence the three basic properties required from the amplifier as an analog component are: (1) a very high gain A, (2) approximately zero input potential e_i, (3) negligibly small input current i_g. There are other important properties required from the amplifier, such as linear output over a wide range, flat frequency response, low noise level, etc. These, however, are means for improved accuracy, rather than basic requirements for the amplifier.

Let us now consider a circuit consisting of a high-gain DC amplifier, an input resistor R_1, and a feedback resistor R_0 (Fig. 3.29). A potential, say e_1, is applied to the input of the amplifier through the resistor R_1. Since $i_g = 0$ according to property (3), Kirchoff's law applied to the input of the amplifier gives

$$i_1 + i_0 = 0.$$

Then, relating currents to potentials by Ohm's law, and noting that $e_i = 0$ because of property (2), we obtain

$$\frac{e_1}{R_1} + \frac{e_0}{R_0} = 0,$$

which may be rearranged in the form

$$e_0 = -\left(\frac{R_0}{R_1}\right) e_1. \tag{3.135}$$

Thus we have shown that the circuit element given by Fig. 3.29 is a multiplier. Clearly, the same circuit can be used as a sign changer by taking the ratio R_0/R_1 equal to unity.

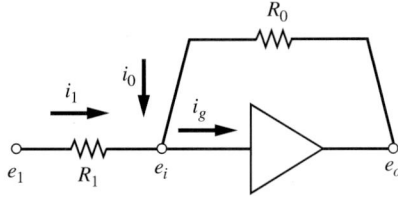

Figure 3.29 Multiplier.

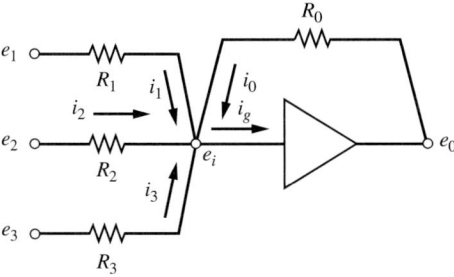

Figure 3.30 Adder.

Note that in analog computer applications the DC amplifier always includes a feedback circuit, as exemplified by Fig. 3.29. It is for this reason that the amplifier gain A has to be negative. If it were positive, an increase at the output potential would increase the input potential, which, in turn, would further increase the output potential, causing instability. The sign of amplifier gain is made negative by using an odd number of amplification stages. This is a problem of electronics, however, and is not our concern. Note, moreover, that this sign has nothing to do with that of Eq. (3.135).

Let us now reconsider the circuit element given by Fig. 3.29 and replace the input resistor by three parallel resistors, as shown in Fig. 3.30. Again, Kirchoff's law applied to the input of the amplifier in the light of property (3), and rearranged by Ohm's law and property (2), gives

$$\frac{e_1}{R_1} + \frac{e_2}{R_2} + \frac{e_3}{R_3} + \frac{e_0}{R_0} = 0,$$

which yields

$$e_0 = -\left(\frac{R_0}{R_1}e_1 + \frac{R_0}{R_2}e_2 + \frac{R_0}{R_3}e_3\right). \tag{3.136}$$

We learn from Eq. (3.136) that the circuit element of Fig. 3.30 can be used as an adder, as clearly seen from the special case $R_0/R_1 = R_0/R_2 = R_0/R_3 = 1$.

We now return once more to the circuit element given by Fig. 3.29, this time replacing the feedback element by a capacitor C as indicated in Fig. 3.31. Kirchoff's law applied to the input of the amplifier and rearranged by Ohm's and Coulomb's laws then gives

$$\frac{e}{R} + C\frac{de_0}{dt} = 0,$$

which, by integration, may be rearranged in the form

$$e_0 = -\frac{1}{RC}\int_0^t e\,d\tau + e_0(0), \tag{3.137}$$

Sec. 3.7 Mixed (Differential-Difference) Formulation. Analog Solution

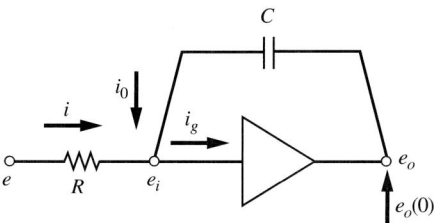

Figure 3.31 Integrator.

where $e_0(0)$ is the initial condition, the potential to which the capacitor is charged at $t = 0$. Equation (3.137) indicates that the circuit element of Fig. 3.31 can be employed as an integrator.

Clearly, a DC amplifier may also be used as a differentiator. However, whenever possible this operation should be avoided because of noise. So far we have seen the use of the DC amplifier as a multiplier, a sign changer, an adder, and an integrator.[16] This background is sufficient for solving unsteady lumped problems, which are illustrated next.

EXAMPLE 3.16

Reconsider the lumped brake system of Ex. 3.3 for a sudden jump in unit velocity.

Letting $V(t) = 1$ in Eq. (3.27), the temperature difference $\Theta = T - T_\infty$ satisfies

$$\frac{d\theta}{dt} + m\theta = n, \quad \theta(0) = 0. \tag{3.138}$$

We wish to build an electrical circuit such that a voltage E (representing θ) obeys a similar differential equation. A time derivative is obtained through the use of a capacitor, and a term directly related to voltage is obtained using a resistor. The analog formulation corresponding to Eq. (3.138) obtained applying Kirchoff's current law to point 0 (zero potential) is

$$C\frac{dE}{dt} + \frac{E}{R} + \frac{(-E_0)}{R_0} = 0. \tag{3.139}$$

An active circuit which satisfies Eq. (3.139) at the inlet node of the DC amplifier provides the required analog solution (Fig. 3.32). The values of R_0, R, and C are selected by considering the proportionality of m to $1/RC$ and that of n to E_0/R_0C. Note that E_0 is a variable source proportional to $V(t)$. ◆

---

[16] The two other important active circuit elements, function generators and function multiplier-dividers, require a somewhat lengthy and special treatment and will not be considered here. The interested reader may consult the references cited at the end of this chapter.

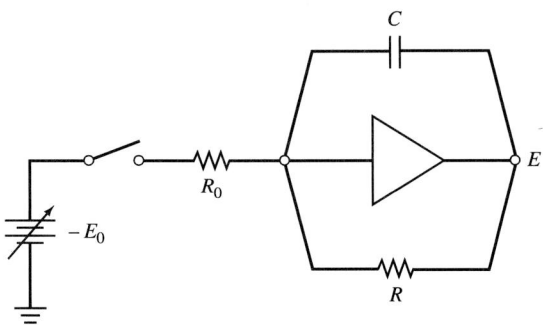

Figure 3.32 Electrical circuit.

Example 3.17

Reconsider Ex.3.5 and its lumped formulation

$$\frac{dT_1}{dt} = -m_1(T_1 - T_2), \tag{3.35}$$

$$\frac{dT_2}{dt} = m_2(T_1 - T_2), \tag{3.36}$$

subject to

$$T_1(0) = T_0, \quad T_2(0) = T_\infty. \tag{3.37}$$

We wish to solve this problem in terms of electrical analogy.

Simulating T_1 by E and T_2 by e, we find that the corresponding analog equations are

$$C_1\frac{dE}{dt} + \frac{E}{R_1} + \frac{-e}{R_1} = 0, \tag{3.140}$$

$$C_2\frac{de}{dt} + \frac{e}{R_2} + \frac{-E}{R_2} = 0, \tag{3.141}$$

$$E(0) = E_0, \quad e(0) = E_\infty. \tag{3.142}$$

Although the first two terms of Eqs. (3.140) and (3.141) are easily satisfied at the input of two DC amplifiers, the third term of each equation requires the minus value of the potential associated with the first two terms of the other equation. Thus the use of another amplifier as a sign changer at the output of each of the first two amplifiers is needed. However, by simply changing the sign of Eq. (3.140) or (3.141), we can eliminate the amplifiers employed as sign changers. Multiplying one of these equations, say the latter, by -1 gives

$$C_2\frac{d(-e)}{dt} + \frac{(-e)}{R_2} + \frac{E}{R_2} = 0, \tag{3.143}$$

Now we need only E and $-e$. In fact, the circuit indicated in Fig. 3.33 satisfies Eqs. (3.140) and (3.143). Note that capacitors must initially be loaded according to Eqs. (3.142). Also, the values of R_1, R_2, C_1, and C_2 must be selected such that m_1 is proportional to $1/R_1C_1$ and m_2 to $1/R_2C_2$. ◆

Sec. 3.7 Mixed (Differential-Difference) Formulation. Analog Solution 173

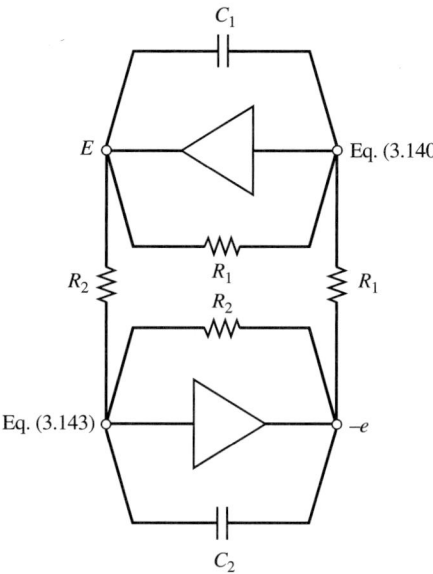

Figure 3.33 Example 3.17

Having learned the analog solution of unsteady lumped problems, we proceed now to the analog solution of unsteady distributed problems. The analog computer, although capable of a continuous integration in time, is silent to any space integration. For example, consider the unsteady, one-dimensional governing equation for conduction,

$$\frac{1}{\alpha}\frac{\partial T}{\partial t} = \frac{\partial^2 T}{\partial x^2}. \tag{3.144}$$

Explicitly, this equation is differential both in time and space. Since the analog computer can handle only the timewise integration of Eq. (3.144), we have to use other means for its spacewise integration. For example, we may replace the space differential with a difference approximation. We have then a mixed (timewise differential and spacewise difference) formulation which follows. A difference formulation involves the selection of a given set of spatial locations (nodes) and approximation of derivatives using linear interpolation. Here we somewhat loosely use this formulation and save a rigorous treatment to the next chapter.

Consider the finite difference system shown in Fig. 3.34. Let T_i represent the temperature at node i and $q_{i-1,i}$ represent the heat flux from node $i-1$ to node i. The first law applied to this system,

$$\rho c (A \Delta x) \frac{dT_i}{dt} = q_{i-1,i} A + q_{i+1,i} A,$$

rearranged by the help of Fourier's law, now written in difference forms

$$q_{i-1,i} = k \frac{T_{i-1} - T_i}{\Delta x}, \quad q_{i+1,i} = k \frac{T_{i+1} - T_i}{\Delta x},$$

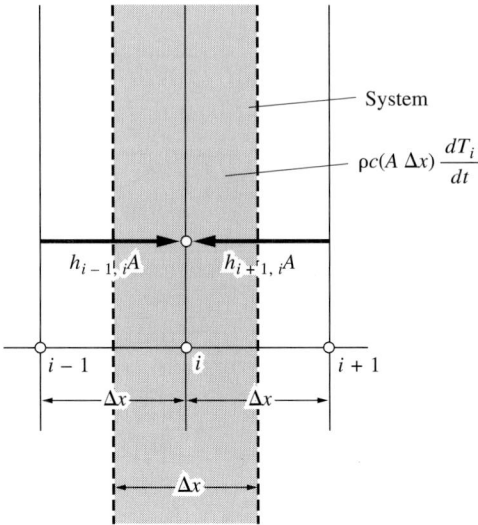

Figure 3.34 First law for a one-dimensional difference system.

gives the governing mixed equation for node i,

$$\frac{(\Delta x)^2}{\alpha} \frac{dT_i}{dt} = T_{i+1} + T_{i-1} - 2T_i, \qquad (3.145)$$

where $\alpha = k/\rho c$. The analog of this equation is

$$C \frac{d(-E_i)}{dt} + \frac{E_{i+1}}{R} + \frac{E_{i-1}}{R} + \frac{(-E_i)}{R/2} = 0. \qquad (3.146)$$

The circuit element shown in Fig. 3.35 satisfies this equation. Extension to two-dimensional geometry (Fig. 3.36), to three-dimensional geometry, boundary conditions, and to other physics presents no difficulty.

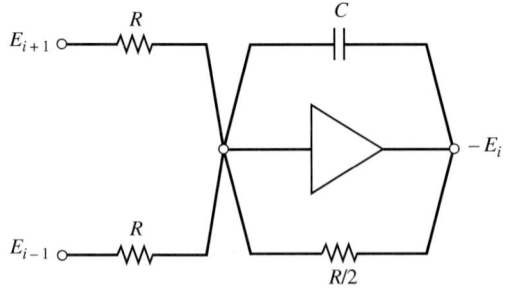

Figure 3.35 Unsteady, one-dimensional conduction.

Sec. 3.7 Mixed (Differential-Difference) Formulation. Analog Solution

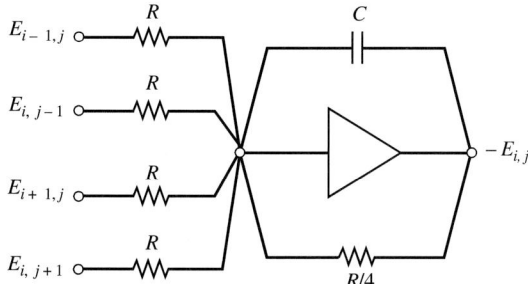

Figure 3.36 Unsteady, two-dimensional conduction.

Example 3.18

Consider a flat plate of thickness ℓ, initially in an ambient at uniform temperature T_∞. From this condition, one surface is subjected to uniform flux q''. Heat is transferred from the other surface with coefficient h.

For a demonstration rather than an accurate solution, consider a coarse subdivision based on four equal parts. For the typical inner node we may use the circuit element in Fig. 3.35; however, we need to develop the circuit elements corresponding to the boundary nodes.

In terms of Fig. 3.37, the differential-difference formulation of the boundary associated with the imposed heat flux is

$$\frac{(\Delta x)^2}{2\alpha} \frac{dT_0}{dt} = T_1 - T_0 + \frac{q''\Delta x}{k}. \tag{3.147}$$

Since $(\Delta x)^2/\alpha$ has been made proportional to RC in connection with the inner nodes, the analog of Eq. (3.147) gives

$$\frac{E_1}{R/2} + \frac{E''}{R/2} + \frac{(-E_0)}{R/2} + C\frac{d(-E_0)}{dt} = 0, \tag{3.148}$$

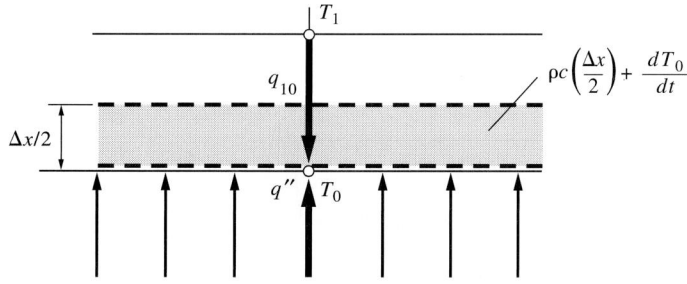

Figure 3.37 Heat-flux boundary.

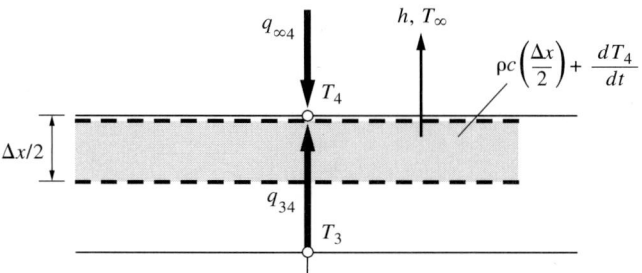

Figure 3.38 Convection boundary.

where E'' is proportional to $q''\Delta x/k$. The circuit element that satisfies Eq. (3.148) is given in Fig. 3.39.

In terms of Fig. 3.38, the differential-difference formulation of the boundary that transfers heat to the ambient is

$$\frac{(\Delta x)^2}{2\alpha}\frac{dT_4}{dt} = T_3 - T_4 + \Delta B(T_\infty - T_4), \tag{3.149}$$

where $\Delta B = h\Delta x/k$ is a cell Biot number. The analog of Eq. (3.149) yields

$$\frac{E_3}{R/2} + \frac{E_\infty}{R/2\Delta B} + \frac{-E_4}{R/2(1+\Delta B)} + C\frac{d(-E_4)}{dt} = 0, \tag{3.150}$$

where E_∞ is proportional to T_∞. The circuit element indicated in Fig. 3.40 satisfies Eq. (3.150). If the numerical value of ΔB requires a fraction of an available resistance, any necessary adjustment can be made by means of a potentiometer.

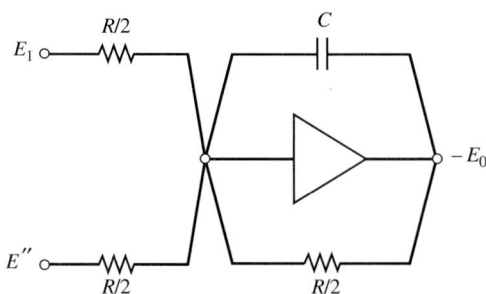

Figure 3.39 Boundary with specified flux.

When the appropriate circuit elements are assigned to nodes, the existence of both plus and minus potentials requires the use of additional amplifiers as sign changers. However, these amplifiers can be eliminated, as in the preceding example, by simply multiplying the potentials of alternate circuit elements by -1. The analog solution thus obtained is shown in Fig. 3.41. The circuit operates as follows: initially the input and output potentials have the same value E_∞, which may conveniently be assumed zero; then the input potential is suddenly changed to E''. ◆

Here we conclude our study of the analog computer by the following general remarks:

Units commonly employed in practice are the megaohm (10^6 ohms) for resistors and the microfarad (10^{-6} farad) for capacitors. The product RC is then expressed in

$$RC = \text{ohm} \times \text{farad} = \text{second}.$$

Variables of conduction and related analog solutions have limitations on their possible magnitudes and rates of change with time; hence in practice the corresponding thermal and analog variables must be related by amplitude- and time-scale factors. More specifically, an amplitude-scale factor is needed because the computer variable, the output potential of a DC amplifier, is limited, often to ± 100 volts; to measure a temperature variation over a range greater than $\pm 100\,°C$, we would need a scale factor between potentials and temperatures. Similarly, the use of a time-scale factor becomes necessary when the transient time of a problem is too short or too long; for short transients the components of the computer and of the recording equipment may not function adequately; for long transients amplifier drift and capacitor leakage may introduce appreciable errors. Furthermore, for reasons of economy, we often prefer to use the computer for a shorter time than the actual time. An extensive treatment of scale factors may be found in Chapter 3 of Ref. 7.

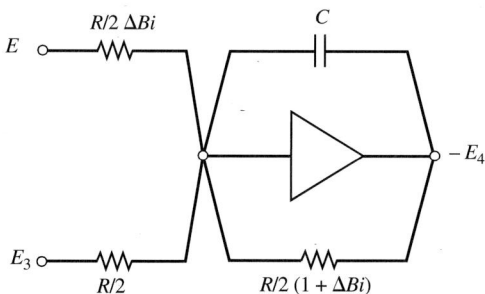

Figure 3.40 Convective boundary.

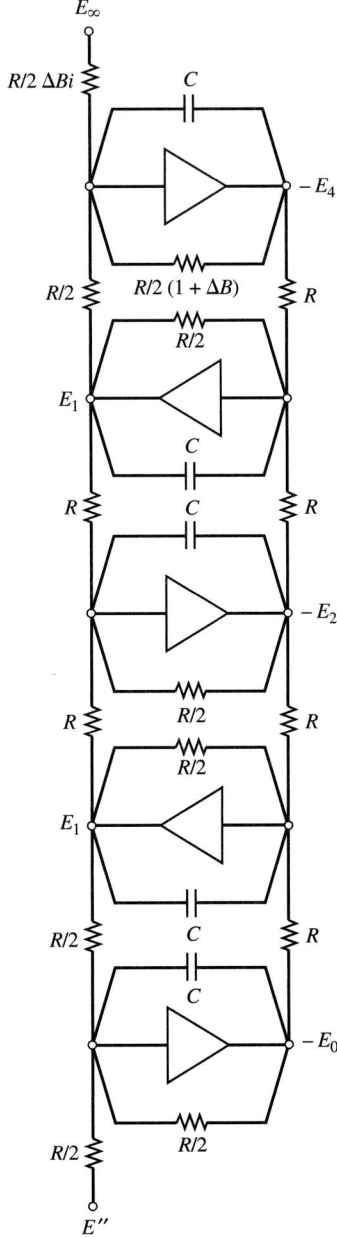

Figure 3.41 Example 3.18.

■ REFERENCES

3.1 V. S. Arpacı, *Conduction Heat Transfer*, Addison-Wesley, Reading, MA 1966.

3.2 Omega Engineering, *Temperature Measurement Handbook and Encyclopedia*, Omega Eng., Inc., CT, 1987.

3.3 H. D. Baker, E. A. Ryder, and N. H. Baker, *Temperature Measurement in Engineering*, Vol.2, Omega Eng., Inc., CT, 1975.

3.4 P. J. Schneider, *Temperature Response Charts*, John Wiley & Sons, New York, 1963.

3.5 L. M. K. Boelter, V. H. Cherry, H. A. Johnson, and R. C. Martinelli, *Heat Transfer Notes*, McGraw-Hill, New York, 1965.

3.6 H. Grober, S. Erk, and U. Grigull, *Fundamentals of Heat Transfer*, McGraw-Hill, New York, 1961.

3.7 R. Tomovic and W. J. Karplus, *High-Speed Analog Computers*, John Wiley & Sons, New York, 1962.

3.8 M. P. Heisler, "Temperature Charts for Induction and Constant Tempersture Heating," Trans. ASME, 1969, 227, 1947.

3.8 APPENDIX

```
C-----------------------------------
C     EX3-10.F (START)
C-----------------------------------
      PROGRAM MAIN
      IMPLICIT REAL*8 (A-H,K-Z)
      PI=4*ATAN(1.)
      WRITE(*,*) 'EXAMPLE 3.10....'
C--------------------------------------------------
C     INPUT DATA
C--------------------------------------------------
      WRITE(*,*) 'INPUT THE FOLLOWING DATA...'
      WRITE(*,*) '2L: cm'
      READ(*,*) L
      WRITE(*,*) 'T_0: C'
      READ(*,*) T0
      WRITE(*,*) 'T_INFTY: C'
      READ(*,*) TINFTY
      WRITE(*,*) 'RHO: kg/m^3'
      READ(*,*) RHO
           WRITE(*,*) 'C_P: J/kg.K'
      READ(*,*) CP
           WRITE(*,*) 'K: W/m.K'
      READ(*,*) K
      WRITE(*,*) 'H: W/m^2.K'
      READ(*,*) H
      WRITE(*,*) 'T: min'
      READ(*,*) T
C--------------------------------------------------
C     UNIT CONVERSION
C--------------------------------------------------
      L=0.01*L/2
      T=60*T
C--------------------------------------------------
C     CALCULATION
C--------------------------------------------------
      BI=H*L/K
      ALPHA=K/(RHO*CP)
      FO=ALPHA*T/L**2
C--------------------------------------------------
C     CHART READINGS
```

```
C------------------------------------------------
      WRITE(*,*) 'INPUT (T-T_INFTY)/(T_0-T_INFTY) READING FIG. 3.20'
      WRITE(*,*) 'WITH 1/BI=',1/BI,' AND FO=',FO,'    AT XI=0'
      READ(*,*) READ1
      WRITE(*,*) 'INPUT (T-T_INFTY)/(T_0-T_INFTY) READING FIG. 3.20'
      WRITE(*,*) 'WITH 1/BI=',1/BI,' AND FO=',FO,'    AT XI=1'
      READ(*,*) READ2
      WRITE(*,*) 'INPUT Q/Q_0 BY LOOKING UP FIG. 3.21'
      WRITE(*,*) 'WITH BI=',BI,' AND BI^2*FO=',BI**2*FO
      READ(*,*) READ3
      WRITE(*,*)
C------------------------------------------------
C     CALCULATION
C------------------------------------------------
      TMID=(T0-TINFTY)*READ1+TINFTY
      TSURF=(T0-TINFTY)*READ2+TINFTY
      Q0=RHO*CP*L*(T0-TINFTY)
      Q=READ3*Q0
C------------------------------------------------
C     ANSWER
C------------------------------------------------
      WRITE(*,*) 'TEMPERATURE AT THE MIDDLE PLANE IS'
      WRITE(*,*) TMID,' C'
      WRITE(*,*) 'TEMPERATURE AT THE SURFACE IS'
      WRITE(*,*) TSURF,' C'
      WRITE(*,*) 'HEAT LOSS FROM THE PLATE AFTER ',T,' MIN IS'
      WRITE(*,*) Q/1000,' kJ/m^2 EACH SURFACE'
      STOP
      END
C------------------------------------
C     EX3-10.F (END)
C------------------------------------
```

■ EXERCISES

3.1 Two square plates are to be glued together by means of a radiant heat source, q'', acting on one plate (Fig. 3P–1). The glue becomes effective at a temperature T_m. The ambient temperature is T_∞. It may be assumed that δ_1 and δ_2 are small, k_1 and k_2 are large, so that **lumped** conditions prevail. Find the time required to obtain a bond between the plates.

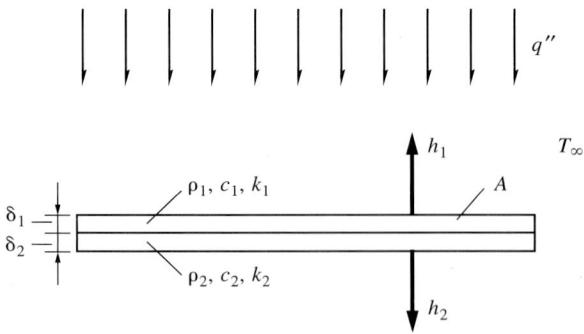

Figure 3P–1

3.2 A liquid metal flowing with a uniform velocity V through a pipe of periphery P and cross section A is electrically heated at the rate of u''' over a length ℓ of the pipe (Fig. 3P–2). The inlet temperature of the liquid is equal to the ambient temperature T_∞. The outside heat transfer coefficient is h. **Neglecting the effect of axial conduction:**.

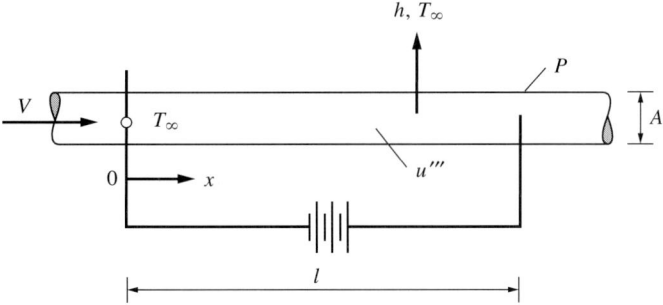

Figure 3P–2

(a) Find the **steady** axial temperature distribution of the liquid metal.

(b) The flow of the liquid metal is suddenly stopped because of a pump failure. Find the **unsteady** temperature of the liquid metal.

3.3 Nuclear power per unit volume

$$u''' = u_0''' \cos \pi \left(\frac{x}{2H} \right)$$

is steadily generated in a rod of height $2H$ (Fig. 3P–3). Let the power generation be suddenly stopped. Neglect the effect of axial conduction and the heat loss from the ends. Determine the temperature decay in the rod.

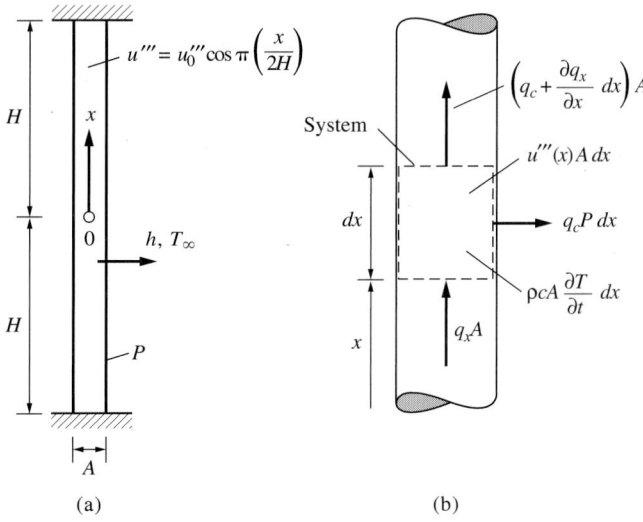

Figure 3P–3

3.4 Consider a flat plate steadily oscillating with velocity $V_0 \sin \omega t$ on top of another flat plate (Fig. 3P–4). The pressure and dry friction coefficient between the plate are p and μ, respectively. Find the steady periodic temperature fluctuations in the system.

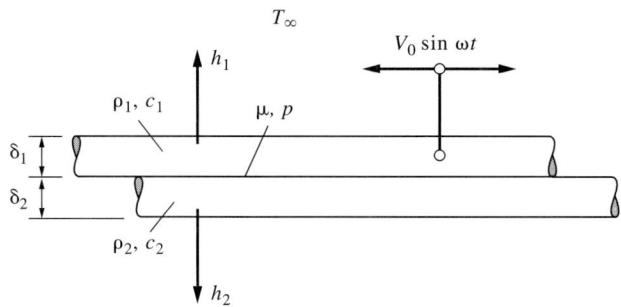

Figure 3P–4

3.5 Reconsider Ex. 3.9 for a suddenly applied heat flux q''.

3.6 Extend the steady thermal length discussed in Section 2.4.1 to an unsteady length resulting from a suddenly changed base temperature.

3.7 A 2-kg household "iron" is made of aluminum and has a 500-W heating element. The ambient is at 20 °C, and the surface area of the iron is 0.05 m^2, while the heat transfer coefficient may be assumed constant at 10 W/m$^2 \cdot$K. How long will it take the iron to reach 100 °C after being turned on?

3.8 A fireproof safe is to be constructed. Its walls consist of two 2-mm steel sheets with a layer of asbestos board between them. Using the chart for a slab, estimate the thickness of asbestos required to give 1 hr of fire protection on the basis that, for an outside temperature of 800 °C, the inside temperature is not to rise above 120 °C during this period. The heat transfer coefficient at the exterior surface is 25 W/m$^2 \cdot$K.

3.9 Reconsider Ex. 3.15 after replacing steel with (a) aluminum, (b) copper. What conclusions can you make?

3.10 A large steel plate (with a thermal diffusivity of 1.3×10^{-5} m^2/s and a thermal conductivity of 40 W/m·K) is 3 cm thick. At zero time it begins to receive heat on one side at the rate of 100 kW/m^2 while the other side is exposed to a fluid at -20 °C with a heat transfer coefficient of 500 W/m$^2 \cdot$K. If the initial temperature of the entire plate is 40 °C, determine the temperature at the midplane after 5 sec.

3.11 A cylindrical mild steel billet, 3 cm OD by 10 cm long, initially at 500 °C, is plunged into a large container filled with oil at 100 °C. The average heat transfer coefficient between the steel and oil is 500 W/m$^2 \cdot$K. Find the time required to cool (a) the center and (b) the surface of the billet to 250 °C. Assume that (i) the billet is lumped, (ii) the billet is not lumped, (iii) the billet is very long. Comment on the comparison of these results.

3.12 Consider a thermopile infrared detector having an active surface area A and initially at the ambient temperature T_∞ (Fig. 3P–5). The active sensing element (or, active junction, equivalently) of the detector consists of two very thin layers ($\delta_1 + \delta_2 \sim 10^{-5}$ cm) of bismuth and tellurium formed by vacuum deposition onto a film of Mylar. The upper surface is suddenly subjected to a heat flux of q'' [W/m^2]. The heat transferred from the detector to a sink at a temperature T_{sink} via a connection may be approximated as $\dot{Q} = \dot{Q}_0(T - T_{\text{sink}})$ [W], where $\dot{Q}_0$ is a known quantity and T is the temperature of the detector active junction.

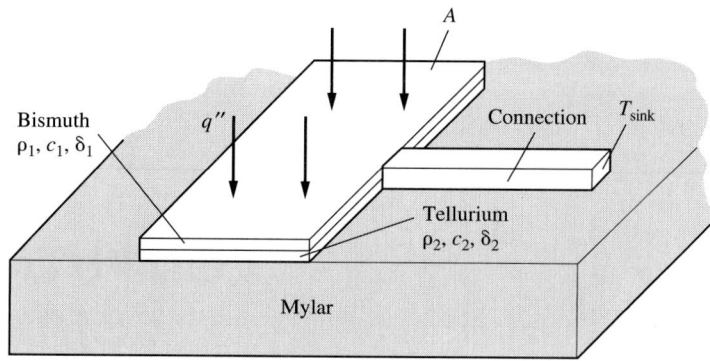

Figure 3P–5

For a **lumped** detector active junction (bismuth and tellurium layers of thickness δ_1 and δ_2, respectively, and surface area A), neglecting the conductive loss to the Mylar as well as the convective and radiative losses to the ambient, determine (a) the steady temperature, (b) the unsteady temperature, and (c) the time constant of the thermopile.

3.13 Reconsider Ex. 3.3. Devise an electric circuit for linearly decreasing energy function of the brake. What is the analog solution of this problem?

3.14 Reconsider Ex. 3.7. Devise an electric circuit for harmonically oscillating energy generation. What is the analog solution of this problem?

3.15 Draw the analog circuit element for unsteady extended surfaces.

CHAPTER 4

COMPUTATIONAL CONDUCTION

So far we have learned the formulation of steady one-dimensional problems which lead to ordinary differential equations, and the formulation of unsteady one-dimensional and (steady and unsteady) multidimensional problems which lead to partial differential equations. In the formulation of multidimensional problems we simply extended the one-dimensional development. However, the solution methods available for partial differential equations are not usually obtained by extending those suitable to ordinary differential equations. They need to be developed separately. Among the methods available for partial differential equations are a variety of **analytical** and **numerical** techniques. The former apply only to problems involving regular geometry, which may not be appropriate for practical and/or technological problems; the latter apply to irregular as well as regular geometries and they are suited to these problems. In this text we shall consider only the numerical methods. The most important feature of these methods is the **discrete formulation**.

4.1 DISCRETE FORMULATION

There are a number of approaches to discrete Formulation; some are illustrated here in terms of a steady one-dimensional fin problem (recall Ex. 2.9). For an infinitely long fin, with a specified base temperature T_0, transferring heat with a coefficient h to an ambient at temperature T_∞, an exact solution for temperature is

$$\frac{\theta(x)}{\theta_0} = e^{-mx}, \qquad (4.1)$$

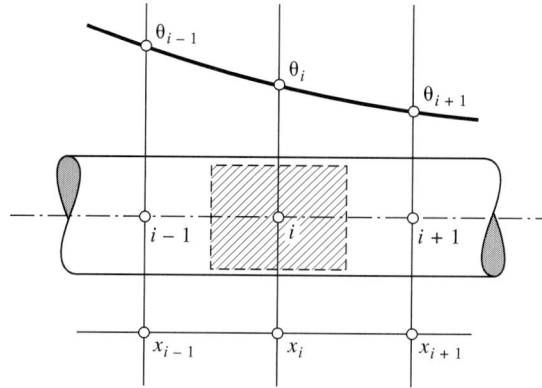

Figure 4.1 Discrete system.

where $\theta = T - T_\infty$ and $m = (hP/kA)^{1/2}$. Now, for a discrete formulation, subdivide the fin into a number (say n) of equal (or unequal) intervals (grids, cells) of length $\Delta x = x_{i+1} - x_i$, $i = 1, 2, \ldots, n$, and assign a discrete temperature θ_i to each grid point, hereafter called node, located at $x_i = (i-1)\Delta x$, $i = 1, 2, \ldots, n+1$ (Fig. 4.1). The numerical solution is based on determining these discrete temperatures. Here, we need not consider the whole problem yet and will confine our attention only to an element (difference system or control volume) involving three inner nodes (i.e., nodes way from boundaries). We limit our discussion hereafter to equal intervals except for Section 4.2.1, where the extension to uneven intervals will be explored.

4.1.1 Exact Discrete Formulation

The analytical solution [Eq. (4.1)] written for each of the three nodes at x_{i-1}, x_i, and x_{i+1} gives

$$\theta_{i-1} = \theta_0 e^{-mx_{i-1}}, \quad \theta_i = \theta_0 e^{-mx_i}, \quad \theta_{i+1} = \theta_0 e^{-mx_{i+1}}.$$

After letting $x_{i-1} = x_i - \Delta x$ and $x_{i+1} = x_i + \Delta x$, the sum

$$\theta_{i-1} + \theta_{i+1} = 2\theta_0 \frac{e^{-m(x_i - \Delta x)} + e^{-m(x_i + \Delta x)}}{2}$$

$$= 2\theta_0 e^{-mx_i} \frac{e^{m\Delta x} + e^{-m\Delta x}}{2}$$

$$= 2\theta_i \cosh m\Delta x$$

yields the **exact discrete** formulation relating three nodal temperatures as

$$\theta_{i-1} - (2\cosh m\Delta x)\theta_i + \theta_{i+1} = 0,$$

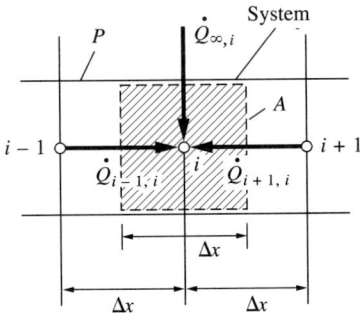

Figure 4.2 First law (difference system).

or, in terms of the **cell Biot number**,

$$B = (m\Delta x)^2 \equiv \frac{hP}{kA}(\Delta x)^2 \qquad (4.2)$$

as

$$\theta_{i-1} - (2\cosh\sqrt{B})\theta_i + \theta_{i+1} = 0. \qquad (4.3)$$

We shall use Eq. (4.3) as a base for the following approximate discrete formulations.

4.1.2 Finite-Difference/Finite-Volume Formulation

The **five steps of formulation** considered in the preceding chapters for **differential formulation** are now applied to a **finite-difference formulation**:

Step 1. Assign a difference system with length Δx to each node, as shown in Fig. 4.1. Let the system be centered around the node.

Step 2. State the first law, Eq. (1.16), in terms of the system assigned to each node. Since all terms of Eq. (1.16) other than the **net** heat flux are absent, the first law reduces to a heat balance. For node i, we have in terms of Fig. 4.2,

$$\dot{Q}_{i-1,i} + \dot{Q}_{i+1,i} + \dot{Q}_{\infty,i} = 0, \qquad (4.4)$$

where $\dot{Q}_{i-1,i}$ represents the heat flux from node $i-1$ to i and $\dot{Q}_{\infty,i}$ represents the convection heat transfer with the ambient. Note that the order of subscripts indicates the direction of the heat fluxes, all arbitrarily assumed to be toward the system, so that we conveniently have the same (plus) sign for each flux. The numerical values obtained at the end determine the actual direction of each flux.

Step 3. For the system around node i, express Fourier's law:

$$\dot{Q}_{i-1,i} = kA\frac{T_{i-1} - T_i}{\Delta x}, \quad \dot{Q}_{i+1,i} = kA\frac{T_{i+1} - T_i}{\Delta x}, \qquad (4.5)$$

and Newton's law

$$\dot{Q}_{\infty,i} = hP\Delta x(T_\infty - T_i). \qquad (4.6)$$

Step 4. Obtain the governing (finite-difference) equation by inserting Eqs. (4.5) and (4.6) into Eq. (4.4),

$$kA\frac{T_{i-1} - T_i}{\Delta x} + kA\frac{T_{i+1} - T_i}{\Delta x} + hP\Delta x(T_\infty - T_i) = 0,$$

which may be rearranged, multiplying each term by $\Delta x/kA$, and in view of Eq. (4.2), as

$$T_{i-1} - (2 + B)T_i + T_{i+1} + BT_\infty = 0,$$

or, in terms of $\theta = T - T_\infty$, as

$$\theta_{i-1} - (2 + B)\theta_i + \theta_{i+1} = 0. \tag{4.7}$$

Equation (4.7) applies to all inner nodes of the fin. Note that the difference between Eqs. (4.3) and (4.7) is confined to the coefficient of θ_i. The Maclaurin expansion of

$$\cosh\sqrt{B} = 1 + \frac{1}{2}B + \frac{1}{24}B^2 + \cdots$$

indicates that the coefficient of θ_i in the finite-difference formulation corresponds to a truncated Maclaurin expansion of the exact coefficient. Boundary conditions and Step 5 for finite-difference formulations are discussed later. Next, we investigate this difference by another discrete formulation.

Suppose we repeat the foregoing development by following an integral formulation in terms of a piecewise linear profile shown in Fig. 4.3. The differential equation governing the temperature distribution is given by Eq. (2.112). Integrating Eq. (2.112) from $x_{i-1/2} = x_i - \Delta x/2$ to $x_{i+1/2} = x_i + \Delta x/2$ gives

$$\left.\frac{d\theta}{dx}\right|_{x_i-\Delta x/2}^{x_i+\Delta x/2} - m^2 \int_{x_i-\Delta x/2}^{x_i+\Delta x/2} \theta(x)dx = 0, \tag{4.8}$$

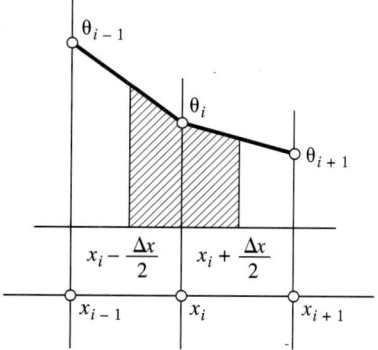

Figure 4.3 Piece-wise linear temperature profile.

or

$$\frac{\theta_{i+1} - \theta_i}{\Delta x} - \frac{\theta_i - \theta_{i-1}}{\Delta x} - \frac{m^2}{2}\left[\left(\frac{\theta_{i-1} + \theta_i}{2} + \theta_i\right) + \left(\theta_i + \frac{\theta_i + \theta_{i+1}}{2}\right)\right]\frac{\Delta x}{2} = 0,$$

which, in view of Eq. (4.2), leads to a **finite-volume** formulation:

$$\theta_{i-1} - 2\left(\frac{8 + 3B}{8 - B}\right)\theta_i + \theta_{i+1} = 0. \tag{4.9}$$

Note that an expansion of

$$\frac{8 + 3B}{8 - B} = 1 + \frac{1}{2}B + \frac{1}{16}B^2 + \cdots$$

reproduces the first two terms of the exact coefficient expansion.

A final method, the so-called finite-element formulation, requiring a variational formulation or the Galerkin method of weighted residuals, is beyond the scope of the text (see, for example, Arpacı and Larsen 1984, ch. 7).

Table 4.1 compares the results of the foregoing discrete formulations for a typical inner node. Since Eqs. (4.3), (4.7), and (4.9) differ only in the coefficient of θ_i, these coefficients, their percent error relative to the exact coefficient, and their series expansions for small values of B are given in Table 4.1. Figure 4.4 shows the coefficients for larger values of B. As the number of grids $n(= \ell/\Delta x)$ is increased, the cell Biot number B decreases, and all formulations approach the exact one.

As demonstrated by the foregoing example, the finite difference is the simplest of the difference formulations, and it will be employed throughout this chapter. We demonstrate the numerical solution of a finite-difference formulation in the following example.

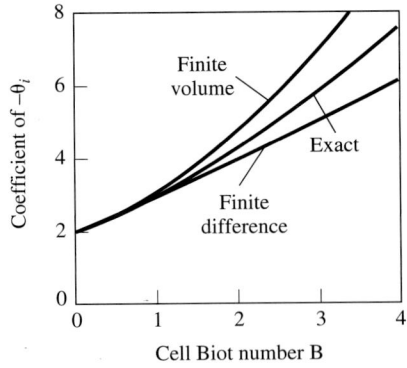

Figure 4.4 Coefficient of $-\theta_i$ versus cell Biot number for different discretizations.

Table 4.1 Coefficients of θ_i for internal nodes.

Procedure and Profile	Equation Number	Coefficient	Expansion for small B	Percent Error for $B = 1$
Exact	4.3	$2\cosh\sqrt{B}$	$2 + B + B^2/12 + \cdots$	0.0
Finite difference	4.6	$2 + B$	$2 + B$	2.79
Finite volume	4.8	$2\frac{8+3B}{8-B}$	$2 + B + B^2/8 + \cdots$	1.54

Example 4.1

In terms of Ex. 2.10 (a cylindrical fin of finite length ℓ, with a specified base temperature and an insulated tip) with $T(0) = T_1 = 120\,°C$, $T_\infty = 20\,°C$, $\ell = 10$ cm, $D = 1$ cm, $k = 20$ W/m·K, $h = 250$ W/m²·K, and the fin divided into $N = 5$ elements of equal size with $N + 1 = 6$ nodes (Fig. 4.5), we wish to determine numerically the temperature distribution and heat transfer rate and to comment on the accuracy.

The finite-difference formulation for inner nodes (2, 3, 4, 5) is given by Eq. (4.7). A similar formulation for the boundary nodes needs to be developed. Here, the temperature of the base is specified and only the insulated tip (node 6) needs to be analyzed. Note that the finite-difference formulation of a boundary condition is related to a finite-difference volume. Accordingly, an application of the first four steps of our formulation procedure to the boundary (difference) system of length $\Delta x/2$ shown in Fig. 4.6 gives

$$kA\frac{T_5 - T_6}{\Delta x} + hP\frac{\Delta x}{2}(T_\infty - T_6) = 0, \quad (4.10)$$

which may be rearranged as

$$T_5 - \frac{1}{2}(2 + B)T_6 + \frac{1}{2}BT_\infty = 0 \quad (4.11)$$

or, in terms of θ,

$$\theta_5 - \frac{1}{2}(2 + B)\theta_6 = 0. \quad (4.12)$$

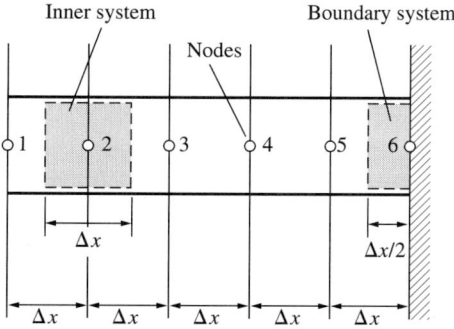

Figure 4.5 Finite-difference systems.

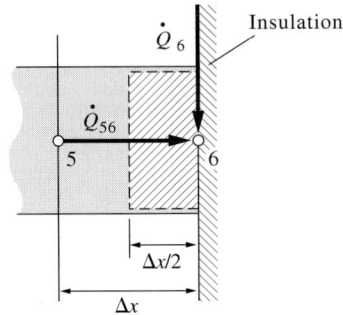

Figure 4.6
Boundary-difference system.

Since θ_1 is already specified, no finite-difference equation is written for node 1. Table 4.2 shows the finite-difference equations for nodes 1 through 6. By expressing Table 4.2 in terms of a matrix, the finite-difference formulation of the fin becomes

$$\begin{bmatrix} -(2+B) & 1 & 0 & 0 & 0 \\ 1 & -(2+B) & 1 & 0 & 0 \\ 0 & 1 & -(2+B) & 1 & 0 \\ 0 & 0 & 1 & -(2+B) & 1 \\ 0 & 0 & 0 & 1 & -(2+B)/2 \end{bmatrix} \begin{bmatrix} \theta_2 \\ \theta_3 \\ \theta_4 \\ \theta_5 \\ \theta_6 \end{bmatrix} = \begin{bmatrix} -\theta_1 \\ 0 \\ 0 \\ 0 \\ 0 \end{bmatrix}, \quad (4.13)$$

where the "known" temperatures are placed in the righthand side. This leads to a matrix equation of the form $AX = C$, which can then be solved for X.

For the given data, the grid size is

$$\Delta x = \frac{\ell}{N} = \frac{10 \text{ cm}}{5} = 2 \text{ cm},$$

Table 4.2 Finite-difference equations

	Node Number	Finite-Difference Equation
Boundary	1	θ_1 specified
	2	$-(2+B)\theta_2 + \theta_3 = -\theta_1$
	3	$\theta_2 - (2+B)\theta_3 + \theta_4 = 0$
Internal	4	$\theta_3 - (2+B)\theta_4 + \theta_5 = 0$
	5	$\theta_4 - (2+B)\theta_5 + \theta_6 = 0$
Boundary	6	$\theta_5 - \frac{1}{2}(2+B)\theta_6 = 0$

and, noting $P/A = 4/D$ for a circular fin, the cell Biot number is

$$B = \frac{4 \times 250 \text{ W/m}^2\cdot\text{K}}{20 \text{ W/m}\cdot\text{K} \times 0.01\text{m}}(0.02\text{m})^2 = 2, \tag{4.14}$$

and the base boundary condition is

$$\theta_1 = T_1 - T_\infty = 120\,°\text{C} - 20\,°\text{C} = 100\,°\text{C}.$$

Equation (4.13) now becomes

$$\begin{bmatrix} -4 & 1 & 0 & 0 & 0 \\ 1 & -4 & 1 & 0 & 0 \\ 0 & 1 & -4 & 1 & 0 \\ 0 & 0 & 1 & -4 & 1 \\ 0 & 0 & 0 & 1 & -2 \end{bmatrix} \begin{bmatrix} \theta_2 \\ \theta_3 \\ \theta_4 \\ \theta_5 \\ \theta_6 \end{bmatrix} = \begin{bmatrix} -100 \\ 0 \\ 0 \\ 0 \\ 0 \end{bmatrix}. \tag{4.15}$$

This matrix equation can be easily solved using a mathematical software package such as Matlab or Maple.

One may wish to solve the matrix manually rather than utilizing a software. A matrix of this form (because of the fact that the coefficient matrix has a nonzero main diagonal as well as two nonzero neighboring upper and lower diagonals) is called **tridiagonal**. A technique for efficiently solving a tridiagonal system of linear algebraic equations is due to Thomas (1949). This particular algorithm has a wide variety of applications and is explained here.

Divide the first row of Eq. (4.15) by 4 (the absolute value of the first diagonal element) and add to row 2. The resulting matrix is

$$\begin{bmatrix} -1 & 0.25 & 0 & 0 & 0 \\ 0 & -3.75 & 1 & 0 & 0 \\ 0 & 1 & -4 & 1 & 0 \\ 0 & 0 & 1 & -4 & 1 \\ 0 & 0 & 0 & 1 & -2 \end{bmatrix} \begin{bmatrix} \theta_2 \\ \theta_3 \\ \theta_4 \\ \theta_5 \\ \theta_6 \end{bmatrix} = \begin{bmatrix} -25 \\ -25 \\ 0 \\ 0 \\ 0 \end{bmatrix}. \tag{4.16}$$

Next, divide the second row of Eq. (4.16) by 3.75 (the absolute value of the second diagonal element) and add to row 3:

$$\begin{bmatrix} -1 & 0.25 & 0 & 0 & 0 \\ 0 & -1 & 0.26667 & 0 & 0 \\ 0 & 0 & -3.73333 & 1 & 0 \\ 0 & 0 & 1 & -4 & 1 \\ 0 & 0 & 0 & 1 & -2 \end{bmatrix} \begin{bmatrix} \theta_2 \\ \theta_3 \\ \theta_4 \\ \theta_5 \\ \theta_6 \end{bmatrix} = \begin{bmatrix} -25 \\ -6.6667 \\ -6.6667 \\ 0 \\ 0 \end{bmatrix}. \tag{4.17}$$

In a similar manner, divide the third row of Eq. (4.17) by 3.73333 and add to row 4:

$$\begin{bmatrix} -1 & 0.25 & 0 & 0 & 0 \\ 0 & -1 & 0.26667 & 0 & 0 \\ 0 & 0 & -1 & 0.26786 & 0 \\ 0 & 0 & 0 & -3.73214 & 1 \\ 0 & 0 & 0 & 1 & -2 \end{bmatrix} \begin{bmatrix} \theta_2 \\ \theta_3 \\ \theta_4 \\ \theta_5 \\ \theta_6 \end{bmatrix} = \begin{bmatrix} -25 \\ -6.6667 \\ -1.7857 \\ -1.7857 \\ 0 \end{bmatrix}. \tag{4.18}$$

Finally, divide the fourth row of Eq. (4.18) by 3.73214 and add to row 5:

$$\begin{bmatrix} -1 & 0.25 & 0 & 0 & 0 \\ 0 & -1 & 0.26667 & 0 & 0 \\ 0 & 0 & -1 & 0.26786 & 0 \\ 0 & 0 & 0 & -1 & 0.26794 \\ 0 & 0 & 0 & 0 & -1.73206 \end{bmatrix} \begin{bmatrix} \theta_2 \\ \theta_3 \\ \theta_4 \\ \theta_5 \\ \theta_6 \end{bmatrix} = \begin{bmatrix} -25 \\ -6.6667 \\ -1.7857 \\ -0.4785 \\ -0.4785 \end{bmatrix}. \quad (4.19)$$

The last row of Eq. (4.19) readily gives

$$\theta_6 = \frac{-0.4785}{-1.73206} = 0.276.$$

Then, a backward substitution gives the other unknowns. From the fourth row of Eq. (4.19),

$$\theta_5 = 0.26794\theta_6 + 0.4785 = 0.26794 \times 0.276 + 0.4785 = 0.552;$$

from the third row,

$$\theta_4 = 0.26794\theta_5 + 1.7857 = 0.26794 \times 0.552 + 1.7857 = 1.93;$$

from the second row,

$$\theta_3 = 0.26667\theta_4 + 6.6667 = 0.26667 \times 1.93 + 6.6667 = 7.18;$$

and, finally, from the first row,

$$\theta_2 = 0.25\theta_3 + 25 = 0.25 \times 7.18 + 25 = 26.8.$$

Having found the discrete solution θ_i, $i = 1, 2, \ldots, 6$, we can now calculate the total heat transfer from the fin. Consider the difference system for base node 1 as shown in Fig. 4.7. The first four steps of our formulation procedure applied to this system yield

$$q_1 A + kA \frac{T_2 - T_1}{\Delta x} + hP \frac{\Delta x}{2}(T_\infty - T_1) = 0,$$

which may be arranged, by dividing with A and in view of $\theta = T - T_\infty$, as

$$q_1 = \frac{k}{\Delta x} \left[\frac{1}{2}(2 + B)\theta_1 - \theta_2 \right], \quad (4.20)$$

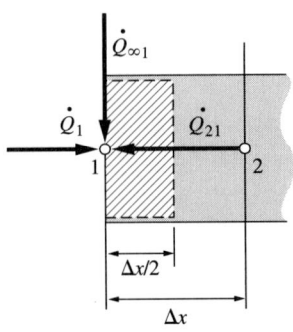

Figure 4.7 Boundary-difference system.

where q_1 is the heat flux at the base. Inserting θ_1 and θ_2 into Eq. (4.20), we get

$$q_1 = \frac{20\text{W/m·K}}{0.02\text{m}}\left[\frac{1}{2}(4)100 - 26.80\right] °\text{C}$$

or

$$q_1 = 173{,}200 \text{ W/m}^2.$$

The FORTRAN program EX4–1.F is listed in the Appendix for the numerical procedure discussed here.

The analytical (exact) solution for the temperature distribution is given by Eq. (2.128):

$$\frac{\theta}{\theta_0} = \frac{\cosh m(\ell - x)}{\cosh m\ell} \tag{2.128}$$

for x measured from the base, and the heat flux by Eq. (2.129),

$$q_1 = \theta_0 \left(\frac{hPk}{A}\right)^{1/2} \tanh m\ell. \tag{2.129}$$

A comparison of the numerical and exact solutions in Table 4.3 shows a large discrepancy between the solutions, especially as one moves away from the base grid point. In the next example, accuracy is improved by using a smaller grid size. ◆

Table 4.3 Numerical vs. exact solution ($B = 2$)

Node Number	X [m]	Exact Solution	Numerical Solution	Error %
1	.00	100.00	100.00	.00
2	.02	24.31	26.80	10.22
3	.04	5.91	7.18	21.49
4	.06	1.44	1.93	34.10
5	.08	.37	.55	49.32
6	.10	.17	.28	62.63
Heat Flux at the base		.1414E+06	.1732E+06	22.47

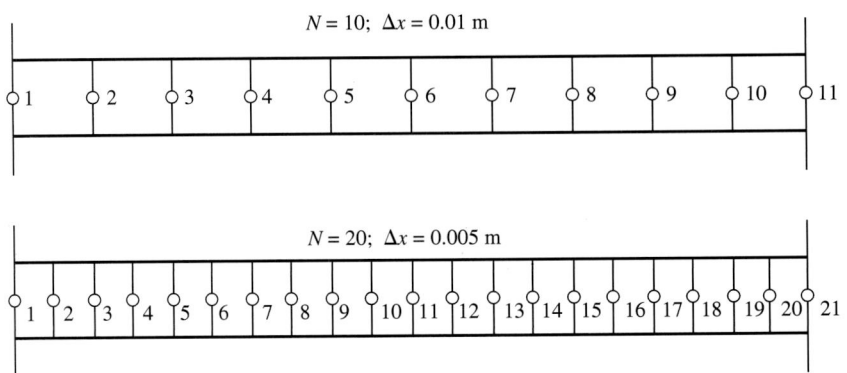

Figure 4.8 Finite-difference systems.

EXAMPLE 4.2

Reconsider Ex. 4.1 with $N = 10$, i.e., $\Delta x = 0.01$ m and $N = 20$, i.e., $\Delta x = 0.005$ m (Fig. 4.8). We wish to determine the effect of the grid size.

In view of Eq. (4.14), for $\Delta x = 0.01$ m,

$$B = 0.5,$$

and for $\Delta x = 0.005$ m,

$$B = 0.125,$$

and the rest of the parameters of Ex. 4.1 remain the same. We can utilize the same matrix manipulation through Eqs. (4.15) to (4.19). However, it becomes tedious to handle the mathematics manually when N is large. The computer program EX4–2.F, listed in the Appendix, solves the problem. The comparison between the exact and numerical solutions is summarized in Table 4.4. As to be expected, the accuracy is significantly improved when a smaller B corresponding to a smaller grid size is utilized. Other numerical schemes such as finite volume, or a higher-order approximation, may be employed to improve the accuracy further. ◆

4.2 MULTIDIMENSIONAL FORMULATION

Reconsider the infinitely long cylinder of Section 3.3 (Fig. 3.14). Subdivide the cross section of the cylinder by straight lines separated both vertically and horizontally $\Delta \ell$ distance apart, forming the so-called **square network** [Fig. 4.9(a)]. Significant differences in temperature gradients ($\partial T/\partial x$ and $\partial T/\partial y$) or geometry may dictate the use of a **rectangular** network instead (i.e., the use of $\Delta x \neq \Delta y$).

The first law applied to the two-dimensional difference system shown in Fig. 4.9(a), and interpreted in terms of Fig. 4.9(b), gives per-unit thickness of the cylinder:

$$\dot{Q}_{10} + \dot{Q}_{20} + \dot{Q}_{30} + \dot{Q}_{40} + u'''(\Delta \ell)^2 = 0. \qquad (4.21)$$

Table 4.4 Numerical vs. exact solution ($B = 0.5$ and 0.125).

Node Number	X [m]	Exact Solution	Numerical Solution	Error %
1	.00	100.00	100.00	.00
2	.01	49.31	50.00	1.41
3	.02	24.31	25.00	2.83
4	.03	11.99	12.50	4.28
5	.04	5.91	6.25	5.75
6	.15	2.92	3.13	7.24
7	.16	1.44	1.57	8.78
8	.17	.72	.79	10.40
9	.18	.37	.42	12.17
10	.19	.21	.24	14.01
11	.10	.17	.20	14.98
Heat Flux at the base		$.1414E+06$	$.1500E+06$	6.07
1	.00	100.00	100.00	.00
2	.005	70.22	70.35	.18
3	.01	49.31	49.49	.36
4	.015	34.62	34.81	.55
5	.02	24.31	24.49	.73
6	.025	17.07	17.23	.91
7	.03	11.99	12.12	1.10
8	.035	6.42	8.53	1.28
9	.04	5.91	6.00	1.46
10	.045	4.15	4.22	1.65
11	.05	2.92	2.97	1.84
12	.055	2.05	2.09	2.02
13	.06	1.44	1.47	2.21
14	.065	1.02	1.04	2.41
15	.07	.72	.74	2.61
16	.075	.51	.53	2.81
17	.08	.37	.38	3.03
18	.085	.27	.28	3.26
19	.09	.21	.22	3.47
20	.095	.18	.19	3.64
21	.10	.17	.18	3.70
Heat Flux at the base		$.1414E+06$	$.1436E+06$	1.55

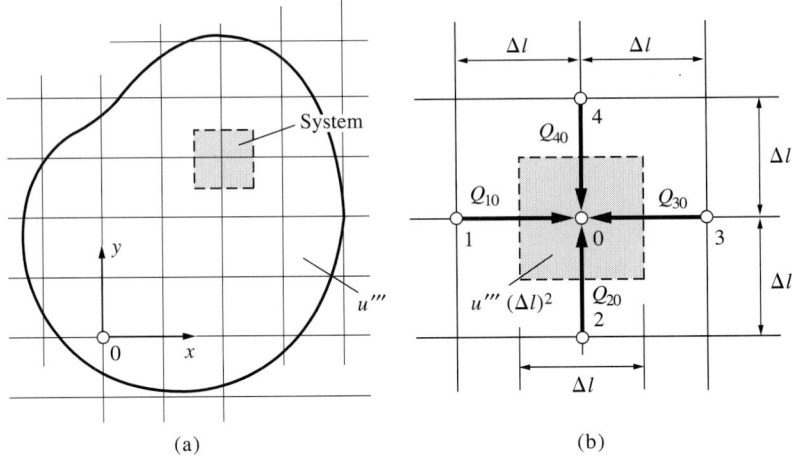

Figure 4.9 The first law for a two-dimensional difference system.

Fourier's law appropriate for the unit thickness of Fig. 4.9,

$$\dot{Q}_{10} = k\Delta\ell \frac{T_1 - T_0}{\Delta\ell}, \quad \dot{Q}_{20} = k\Delta\ell \frac{T_2 - T_0}{\Delta\ell}, \quad \dot{Q}_{30} = k\Delta\ell \frac{T_3 - T_0}{\Delta\ell}, \quad \dot{Q}_{40} = k\Delta\ell \frac{T_4 - T_0}{\Delta\ell}, \tag{4.22}$$

inserted into Eq. (4.21) yields, after dividing each term by k, the governing equation for node 0,

$$T_1 + T_2 + T_3 + T_4 - 4T_0 + \frac{u'''(\Delta\ell)^2}{k} = 0 \tag{4.23}$$

or

$$T_0 = \frac{T_1 + T_2 + T_3 + T_4}{4} + \frac{u'''(\Delta\ell)^2}{4k}, \tag{4.24}$$

Eq. (4.24) being convenient for an iteration method. Note that Eqs. (4.23) and (4.24) apply equally to all **inner** nodes.

For the statement of boundary conditions we begin with regular boundaries and consider the finite-difference systems shown in Fig. 4.10(a) for a **corner node** and a **side node**. We consider three types of boundary conditions: specified temperature, specified heat flux, and heat transfer to ambient.

For boundaries with a specified temperature distribution there is no need to consider corner or side nodes. Governing equations for neighboring inner nodes involve the specified temperature of the boundaries. For boundaries subject to a specified heat flux, the first four steps of formulation give, for the corner node [Fig. 4.10(b)],

$$k\frac{\Delta\ell}{2}\frac{T_1 - T_0}{\Delta\ell} + k\frac{\Delta\ell}{2}\frac{T_2 - T_0}{\Delta\ell} + (q_1'' + q_2'')\frac{\Delta\ell}{2} = 0,$$

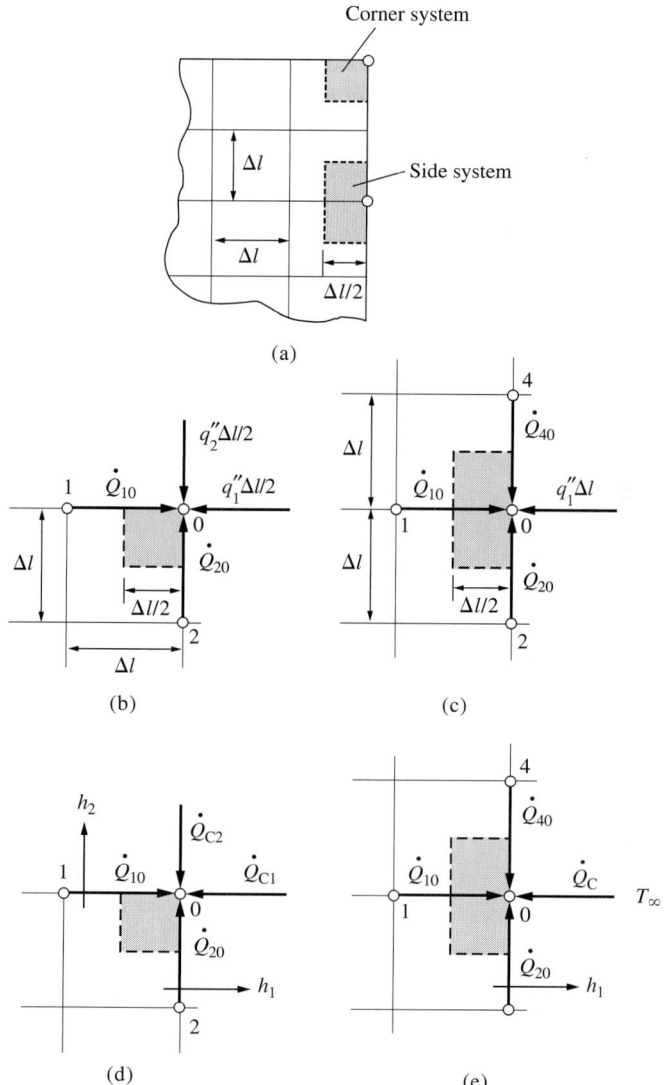

Figure 4.10 The first law for a corner and a side system.

which may be rearranged as

$$\frac{1}{2}(T_1 + T_2) - T_0 + (q_1'' + q_2'')\frac{\Delta \ell}{2k} = 0, \qquad (4.25)$$

and, for the side node [Fig. 4.10(c)],

$$k\Delta\ell \frac{T_1 - T_0}{\Delta\ell} + k\frac{\Delta\ell}{2}\frac{T_2 - T_0}{\Delta\ell} + k\frac{\Delta\ell}{2}\frac{T_4 - T_0}{\Delta\ell} + q_1''\Delta\ell = 0,$$

which may be rearranged as

$$T_1 + \frac{1}{2}(T_2 + T_4) - 2T_0 + \frac{q_1'' \Delta \ell}{k} = 0. \quad (4.26)$$

Clearly, for q_1'' and/or $q_2'' \to 0$, Eqs. (4.25) and (4.26) give boundary conditions for an **insulated** corner or side node.

For boundaries transferring heat with a specified coefficient h, the four steps of formulation give, for the corner node [Fig. 4.10(d)],

$$k\frac{\Delta\ell}{2}\frac{T_1 - T_0}{\Delta\ell} + k\frac{\Delta\ell}{2}\frac{T_2 - T_0}{\Delta\ell} + (h_1 + h_2)\frac{\Delta\ell}{2}(T_\infty - T_0) = 0,$$

which may be rearranged as

$$\frac{1}{2}(T_1 + T_2) - \left[1 + (h_1 + h_2)\frac{\Delta\ell}{2k}\right]T_0 + (h_1 + h_2)\frac{\Delta\ell}{2k}T_\infty = 0, \quad (4.27)$$

and, for the side node [Fig. 4.10(e)],

$$k\Delta\ell\frac{T_1 - T_0}{\Delta\ell} + k\frac{\Delta\ell}{2}\frac{T_2 - T_0}{\Delta\ell} + k\frac{\Delta\ell}{2}\frac{T_4 - T_0}{\Delta\ell} + h_1\Delta\ell(T_\infty - T_0) = 0,$$

which may be rearranged as

$$T_1 + \frac{1}{2}(T_2 + T_4) - \left(2 + \frac{h_1\Delta\ell}{k}\right)T_0 + \frac{h_1\Delta\ell}{k}T_\infty = 0. \quad (4.28)$$

What happens as h_1 and/or $h_2 \to 0$? Note that Eqs. (4.27) and (4.28) may conveniently be written in terms of $\theta = T - T_\infty$ as

$$\frac{1}{2}(\theta_1 + \theta_2) - \left[1 + (h_1 + h_2)\frac{\Delta\ell}{2k}\right]\theta_0 = 0 \quad (4.29)$$

and

$$\theta_1 + \frac{1}{2}(\theta_2 + \theta_4) - \left(2 + \frac{h_1\Delta\ell}{k}\right)\theta_0 = 0. \quad (4.30)$$

Next, we consider irregular boundaries such as turbine blades. These boundaries may be approximated in terms of, say, a cartesian coordinate system [Fig. 4.11(a)]. A more accurate formulation, however, requires the construction of modified difference equations. The four steps applied to the difference system shown in Fig. 4.11(b) give

$$\frac{2}{\xi(1+\xi)}T_1' + \frac{2}{1+\eta}T_2 + \frac{2}{1+\xi}T_3 + \frac{2}{\eta(1+\eta)}T_4' - 2\left(\frac{1}{\xi} + \frac{1}{\eta}\right)T_0 + \frac{u'''(\Delta\ell)^2}{k} = 0. \quad (4.31)$$

As $\xi \to 1$ and $\eta \to 1$, Eq. (4.31) reduces to Eq. (4.24), as expected.

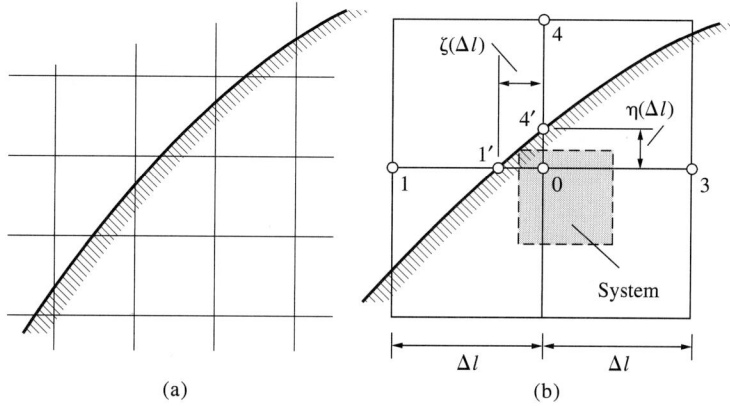

Figure 4.11 Irregular boundary.

For a simplified finite-difference formulation of steady, multidimensional problems, reconsider the flat plate of thickness δ (Fig. 3.16). The four steps of formulation, interpreted now in terms of the two-dimensional finite difference system shown in Fig. 4.12 yield

$$k\delta\Delta\ell\frac{T_1 - T_0}{\Delta\ell} + k\delta\Delta\ell\frac{T_2 - T_0}{\Delta\ell} + k\delta\Delta\ell\frac{T_3 - T_0}{\Delta\ell} + k\delta\Delta\ell\frac{T_4 - T_0}{\Delta\ell}$$
$$+ (h_1 + h_2)(\Delta\ell)^2(T_\infty - T_0) + u'''\delta(\Delta\ell)^2 = 0,$$

which may be rearranged as

$$T_1 + T_2 + T_3 + T_4 - \left[4 + (h_1 + h_2)\frac{(\Delta\ell)^2}{k\delta}\right]T_0 + (h_1 + h_2)\frac{(\Delta\ell)^2}{k\delta}T_\infty + \frac{u'''(\Delta\ell)^2}{k} = 0 \tag{4.32}$$

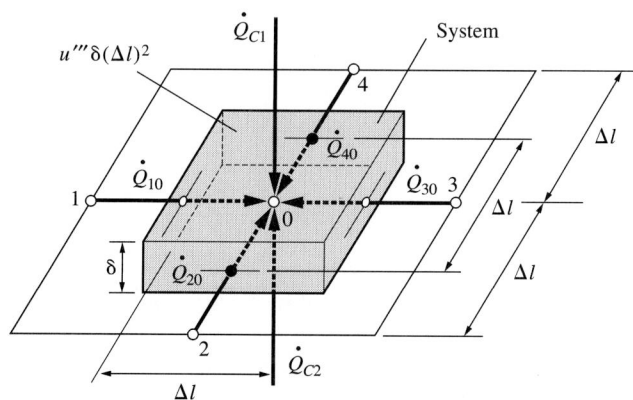

Figure 4.12 The first law for a two-dimensional difference system for a plate.

or, in terms of θ,

$$\theta_1 + \theta_2 + \theta_3 + \theta_4 - \left[4 + (h_1 + h_2)\frac{(\Delta\ell)^2}{k\delta}\right]\theta_0 + \frac{u'''(\Delta\ell)^2}{k} = 0.$$

As $(h_1 + h_2) \to 0$, Eq. (4.32) reduces to Eq. (4.24), as expected. Boundary conditions associated with Eq. (4.32) will not be elaborated here because of space considerations. Having studied the finite-difference formulation of steady, multidimensional problems, we illustrate now a numerical solution of this formulation in terms of an iteration method.

EXAMPLE 4.3

Three surfaces of a square rod are kept at $0\,°C$ and the remaining surface at $100\,°C$. We wish to find the temperature distribution and the effect of the grid size on this rod.

Let us consider the network shown in Fig. 4.13. Our primary objective is to demonstrate the selection of an initial set of estimates for iteration rather than accuracy. Because of the horizontal symmetry of the problem, $T_{2,2} = T_{4,2}$, $T_{2,3} = T_{4,3}$, $T_{2,4} = T_{4,4}$, we need to consider only the temperature of the six locations indicated in the figure by solid circles. In view of Eq. (4.24) with $u''' = 0$, the corresponding difference equations are

$$T_{3,2} = (100 + T_{4,2} + T_{3,3} + T_{2,2})/4 = 25 + (2T_{4,2} + T_{3,3})/4,$$

$$T_{3,3} = (T_{3,2} + T_{4,3} + T_{3,4} + T_{2,3})/4 = (T_{3,2} + 2T_{4,3} + T_{3,4})/4,$$

$$T_{3,4} = (T_{3,3} + T_{4,4} + 0 + T_{2,4})/4 = (T_{3,3} + 2T_{4,4})/4,$$

$$T_{4,2} = (100 + 0 + T_{4,3} + T_{3,2})/4 = 25 + (T_{4,3} + T_{3,2})/4,$$

$$T_{4,3} = (T_{4,2} + 0 + T_{4,4} + T_{3,3})/4 = (T_{4,2} + T_{4,4} + T_{3,3})/4,$$

$$T_{4,4} = (T_{4,3} + 0 + 0 + T_{3,4})/4 = (T_{4,3} + T_{3,4})/4;$$

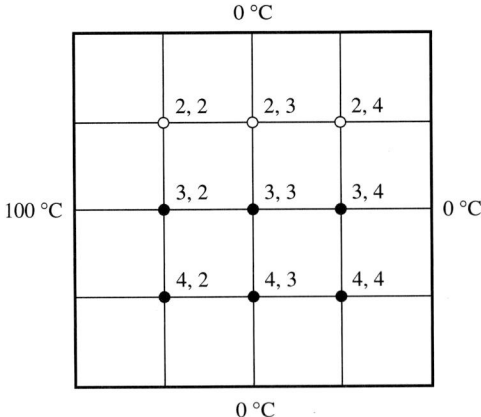

Figure 4.13 Square rod.

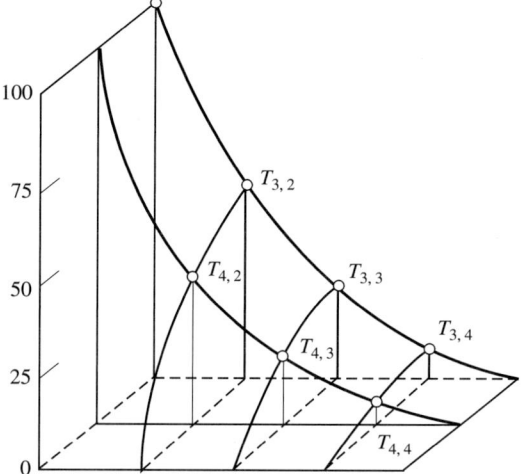

Figure 4.14 Temperatures of the square rod.

where the first and second subscripts identify the x and y locations of a node, respectively. Now, we may start by guessing the temperature of locations (3, 2), (3, 3), (3, 4), (4, 2), (4, 3) and (4, 4). The exact temperature of node (3, 3) is $T_{3,3} = 25\,°C$ (note that if four sides have the same temperature, say 100 °C, then $T_{3,3} = 100\,°C$) (Fig. 4.14). A linear interpolation gives $T_{3,2} \cong 62\,°C$, $T_{3,4} \cong 12\,°C$, $T_{4,2} \cong 31\,°C$, $T_{4,3} \cong 12\,°C$, and $T_{4,4} \cong 6\,°C$. The iterations are given in Table 4.5. For the interested reader a FORTRAN program, EX4–3.F, is listed in the Appendix for this iteration.

Table 4.5 Iteration table for temperature distribution of two-dimensional rod

Iteration Step	$T_{3,2}$	$T_{3,3}$	$T_{3,4}$	$T_{4,2}$	$T_{4,3}$	$T_{4,4}$
Guess	62.00	25.00	12.00	31.00	12.00	6.00
1	49.88	23.12	9.20	40.47	17.40	6.65
2	51.43	24.06	9.39	42.21	18.23	6.91
3	52.17	24.53	9.59	42.60	18.51	7.03
4	52.44	24.77	9.70	42.74	18.63	7.08
5	52.56	24.88	9.76	42.80	18.69	7.11
6	52.62	24.94	9.79	42.83	18.72	7.13
7	52.65	24.97	9.81	42.84	18.74	7.14
8	52.66	24.99	9.81	42.85	18.74	7.14
9	52.67	24.99	9.82	42.85	18.75	7.14
10	52.67	25.00	9.82	42.86	18.75	7.14

The foregoing problem, which has an exact analytical solution, was used here for illustration of a computational solution. However, the problem of the next example has no analytical solution. Its computational solution is indispensable. ◆

EXAMPLE 4.4

The inner and outer surfaces of a chimney are kept at 100 °C and 0 °C, respectively. The ratio between the outer and inner sides is 4 to 1. We wish to find the temperature distribution in the chimney.

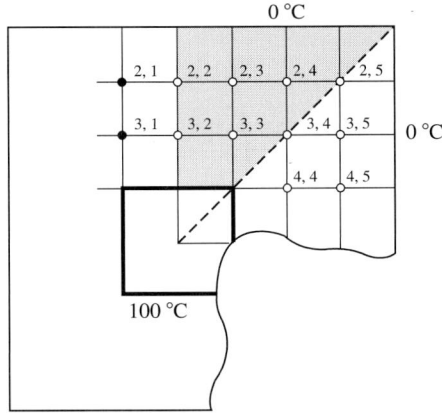

Figure 4.15 Chimney.

Because of the thermal symmetry, we need to consider only one-eighth of the chimney (Fig. 4.15) which leads to

$$T_{2,1} = T_{2,3}, \quad T_{3,1} = T_{3,3}, \quad T_{3,5} = T_{2,4}, \quad T_{4,4} = T_{3,3}, \quad T_{4,5} = T_{2,3}.$$

In view of Fig. 4.9 and Eq. (4.24) with $u''' = 0$, the appropriate iterative equation is

$$T_{i,j} = \frac{T_{i-1,j} + T_{i,j-1} + T_{i+1,j} + T_{i,j+1}}{4},$$

for $i = 2, 3$ and $j = 2, 3, 4, 5$. Next we guess the node temperatures. Again, following a linear interpolation, we assume $T_{3,2} = T_{3,4} = T_{3,3} \cong 66$ °C and $T_{2,2} = T_{2,5} = T_{2,3} = T_{2,4} \cong 33$ °C. The FORTRAN program EX4–4.F listed in the Appendix is utilized to solve this problem, and the iteration results are shown in Table 4.6. ◆

4.2.1 Nonuniform Grid Spacing ⊕

The methods discussed in the preceding sections apply also to nonuniform grids. A nonuniform grid spacing may yield a more accurate solution than a uniform grid for the same number of grid points, provided the grid is strained to be "problem fitted" in some manner. The derivation of difference operators is based on the Taylor-series expansions for $T_{i+1} = T(x_i + \Delta x_i)$ and $T_{i-1} = T(x_i - \Delta x_{i-1})$, where $\Delta x_i / \Delta x_{i-1} = s_i$ is a measure of the **local straining** of the grid (Fig. 4.16). In general, s_i is a function of location, but for a geometric progression s_i is a constant. For notational convenience let $s_i = s$ and $\Delta x_{i-1} = \Delta$. Then, replacing Δx_i by $s\Delta$ and Δx_{i-1} by Δ in the truncated Taylor expansion of T_{i+1} and T_{i-1} gives

$$T_{i+1} = T_i + s\Delta \left(\frac{dT}{dx}\right)_i + \frac{(s\Delta)^2}{2}\left(\frac{d^2T}{dx^2}\right)_i, \tag{4.33}$$

$$T_{i-1} = T_i - \Delta \left(\frac{dT}{dx}\right)_i + \frac{(\Delta)^2}{2}\left(\frac{d^2T}{dx^2}\right)_i, \tag{4.34}$$

Table 4.6 Iteration table for temperature distribution of two-dimensional Chimney

Iteration Step	$T_{2,2}$	$T_{2,3}$	$T_{2,4}$	$T_{2,5}$	$T_{3,2}$	$T_{3,3}$	$T_{3,4}$
Guess	33.00	33.00	33.00	33.00	66.00	66.00	66.00
1	33.00	33.00	33.00	8.25	66.25	66.31	41.33
2	33.06	33.09	20.67	10.33	66.42	60.21	41.96
3	33.15	28.51	20.20	10.38	63.39	58.47	40.05
4	30.10	27.19	19.41	10.07	61.76	57.25	39.00
5	29.04	26.42	18.87	9.57	60.88	56.58	38.02
6	28.43	25.97	18.39	9.27	60.40	56.10	37.44
7	28.08	25.64	18.09	9.09	60.07	55.79	37.06
8	27.84	25.43	17.90	8.98	59.85	55.59	36.83
9	27.68	25.29	17.77	8.91	59.71	55.46	36.67
10	27.57	25.20	17.70	8.86	59.62	55.37	36.57
11	27.51	25.14	17.64	8.83	59.56	55.32	36.50
12	27.46	25.11	17.61	8.81	59.53	55.28	36.46
13	27.43	25.08	17.59	8.80	59.50	55.26	36.43
14	27.42	25.07	17.57	8.79	59.48	55.25	36.42
15	27.40	25.06	17.57	8.78	59.47	55.24	36.41
16	27.40	25.05	17.56	8.78	59.47	55.23	36.40
17	27.39	25.05	17.56	8.78	59.46	55.23	36.39
18	27.39	25.04	17.55	8.78	59.46	55.22	36.39
19	27.39	25.04	17.55	8.78	59.46	55.22	36.39
20	27.39	25.04	17.55	8.78	59.46	55.22	36.39
21	27.38	25.04	17.55	8.78	59.46	55.22	36.39

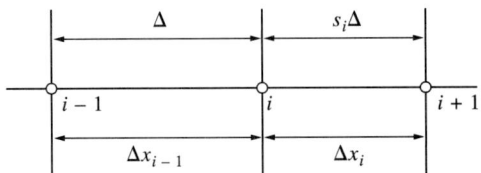

Figure 4.16 Nonuniform grid spacing.

and multiplying Eq. (4.34) by s^2 and subtracting it from Eq. (4.33) yields the second-order central difference for the first derivative at node i,

$$\left(\frac{dT}{dx}\right)_i = \frac{T_{i+1} + (s^2 - 1)T_i - s^2 T_{i-1}}{s(s + 1)\Delta}. \tag{4.35}$$

Next, the sum of Eq. (4.34) multiplied by s and Eq. (4.33) gives the second-order central-difference form of the second derivative at node i,

$$\left(\frac{d^2T}{dx^2}\right)_i = \frac{T_{i+1} - (s + 1)T_i + s T_{i-1}}{s(s + 1)\Delta^2/2}. \tag{4.36}$$

Some of the basic difference formulas now have lower accuracy due to straining, suggesting the need to limit the magnitude of the straining. The use of polynomial interpolation for boundary conditions follows the previous development but usually leads to more elaborate expressions.

EXAMPLE 4.5

Consider a square rod with boundaries at $0\,°C$ and a circular hole inside which is maintained at a temperature of $100\,°C$ as shown in Fig. 4.17. We wish to determine the steady temperature distribution.

The governing differential equation of this problem,

$$\frac{d^2T}{dx^2} + \frac{d^2T}{dy^2} = 0,$$

in view of Eq. (4.36) and Fig. 4.17, leads to

$$\frac{T_{i+1,j} - (s_i + 1)T_{i,j} + s_i T_{i-1,j}}{s_i(s_i + 1)\Delta_i^2/2} + \frac{T_{i,j+1} - (s_j + 1)T_{i,j} + s_j T_{i,j-1}}{s_j(s_j + 1)\Delta_j^2/2} = 0$$

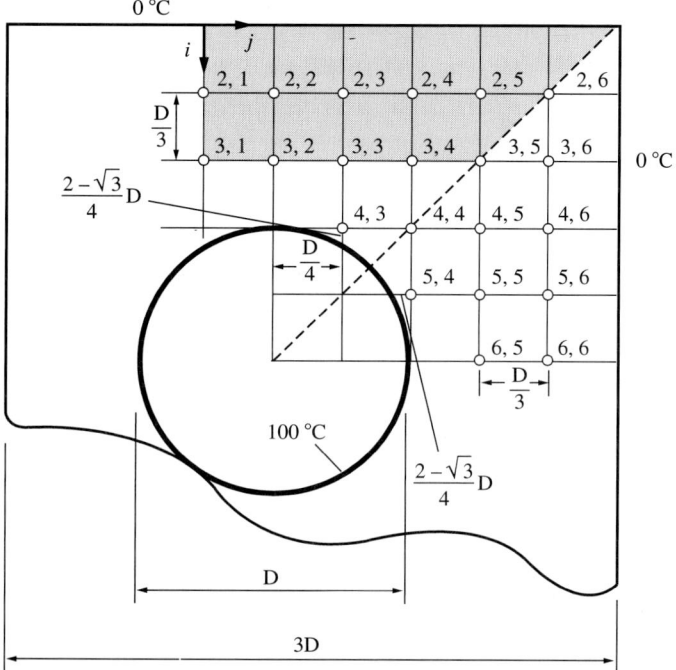

Figure 4.17 Square rod with a circular hole.

or

$$T_{i,j}\left[\frac{1}{s_i\Delta_i^2}+\frac{1}{s_j\Delta_j^2}\right]=\frac{T_{i+1,j}+s_iT_{i-1,j}}{s_i(s_i+1)\Delta_i^2}+\frac{T_{i,j+1}+s_jT_{i,j-1}}{s_j(s_j+1)\Delta_j^2}=0,$$

where $i = 2, 3, 4$ and $j = 2, \ldots, 6$. Note that

$$T_{2,1}=T_{2,3},\quad T_{3,1}=T_{3,3},\quad T_{5,4}=T_{4,3},\quad T_{5,5}=T_{3,3},\quad T_{5,6}=T_{2,3}.$$

With an initial temperature guess $T_{i,j}=0$, the FORTRAN program EX4–5.F solves the problem. Iteration results are shown in Table. 4.7. ◆

So far we have demonstrated how to utilize numerical techniques to solve two-dimensional steady conduction problems. We now proceed to problems involving enthalpy flow.

Table 4.7 Iteration table for temperature distribution of two-dimensional rod

Iteration Step	$T_{2,2}$	$T_{2,3}$	$T_{2,4}$	$T_{2,5}$	$T_{2,6}$	$T_{3,2}$	$T_{3,3}$	$T_{3,4}$	$T_{3,5}$	$T_{4,3}$	$T_{4,4}$
Guess	.	.	.	.	.	.	.	.	.	.	.
1	.00	.00	.00	.00	.08	18.00	5.76	1.88	.47	75.22	21.89
2	3.24	2.07	1.80	.41	.26	22.27	21.64	12.10	4.41	80.06	47.99
3	5.34	5.95	4.64	2.33	1.08	32.81	29.85	22.10	9.90	84.51	54.65
4	9.71	9.97	8.56	4.89	2.25	38.86	36.51	27.89	14.63	86.20	59.46
5	13.37	13.59	11.61	7.12	3.37	43.77	40.90	32.17	18.20	87.38	62.37
6	16.58	16.38	13.99	8.89	4.29	47.16	44.06	35.21	20.95	88.15	64.51
7	18.97	18.48	15.75	10.25	5.03	49.61	46.34	37.46	23.04	88.71	66.04
8	20.76	20.02	17.08	11.29	5.61	51.39	48.00	39.13	24.63	89.12	67.18
9	22.07	21.17	18.06	12.07	6.05	52.70	49.24	40.37	25.81	89.42	68.03
10	23.03	22.01	18.80	13.66	6.38	53.66	50.15	41.30	26.70	89.65	68.65
11	23.75	22.64	19.34	13.10	6.62	54.37	50.83	41.99	27.36	89.81	69.12
12	24.28	23.11	19.75	13.43	6.80	54.90	51.33	42.50	27.85	89.94	69.46
13	24.67	23.45	20.06	13.68	6.94	55.29	51.70	42.89	28.21	90.03	69.72
14	24.96	23.71	20.28	13.86	7.04	55.58	51.98	43.17	28.48	90.09	69.91
15	25.18	23.91	20.45	13.99	7.12	55.80	52.19	43.38	28.68	90.15	70.05
16	25.34	24.05	20.57	14.09	7.17	55.96	52.34	43.54	28.83	90.18	70.16
17	25.46	24.15	20.67	14.17	7.21	56.08	52.46	43.65	28.94	90.21	70.23
18	25.55	24.23	20.74	14.22	7.24	56.17	52.54	43.74	29.02	90.23	70.29
19	25.62	24.29	20.79	14.26	7.27	56.24	52.61	43.80	29.08	90.25	70.34
20	25.67	24.34	20.83	14.29	7.28	56.29	52.65	43.85	29.13	90.26	70.37
21	25.71	24.37	20.85	14.32	7.30	56.33	52.69	43.89	29.16	90.27	70.39
22	25.73	24.39	20.88	14.33	7.31	56.35	52.72	43.91	29.19	90.27	70.41
23	25.75	24.41	20.89	14.35	7.31	56.37	52.73	43.93	29.21	90.28	70.42
24	25.77	24.42	20.90	14.36	7.32	56.39	52.75	43.95	29.22	90.28	70.43
25	25.78	24.43	20.91	14.36	7.32	56.40	52.76	43.96	29.23	90.28	70.44
26	25.79	24.44	20.92	14.37	7.33	56.41	52.77	43.97	29.24	90.29	70.45
27	25.80	24.45	20.92	14.37	7.33	56.41	52.77	43.97	29.24	90.29	70.45
28	25.80	24.45	20.93	14.37	7.33	56.42	52.78	43.98	29.25	90.29	70.45
29	25.80	24.45	20.93	14.38	7.33	56.42	52.78	43.98	29.25	90.29	70.45
30	25.81	24.46	20.93	14.38	7.33	56.42	52.78	43.98	29.25	90.29	70.46
31	25.81	24.46	20.93	14.38	7.33	56.43	52.79	43.98	29.26	90.29	70.46
32	25.81	24.46	20.93	14.38	7.33	56.43	52.79	43.98	29.26	90.29	70.46
33	25.81	24.46	20.93	14.38	7.33	56.43	52.79	43.99	29.26	90.29	70.46
34	25.81	24.46	20.93	14.38	7.33	56.43	52.79	43.99	29.26	90.29	70.46

4.2.2 Effect of Enthalpy Flow ⊕

Reconsider the formulation of the key problems discussed in Section 2.5,

$$\frac{d^2\theta}{dx^2} - \frac{V}{\alpha}\frac{d\theta}{dx} - \frac{hP\theta}{kA} = 0, \qquad (4.37)$$

with convective heat loss to the ambient [recall Eq. (2.151)], and

$$\frac{d^2\theta}{dx^2} - \frac{V}{\alpha}\frac{d\theta}{dx} - \frac{q''P}{kA} = 0, \qquad (4.38)$$

with a specified heat flux [recall Eq. (2.161)]. In Section 4.1 we discussed the finite-difference formulation of the conduction and source terms of these equations. Here, we consider the convective terms by using the upstream (upwind) finite-difference. For $V > 0$, the temperature at node i is influenced by the upstream temperature at node $i - 1$. Then the finite-difference formulation for Eq. (4.37) is

$$\frac{\theta_{i-1} - 2\theta_i + \theta_{i+1}}{(\Delta x)^2} - \frac{V(\theta_i - \theta_{i-1})}{\alpha \Delta x} - \frac{hP\theta_i}{kA} = 0, \qquad (4.39)$$

and for Eq. (4.38) is

$$\frac{\theta_{i-1} - 2\theta_i + \theta_{i+1}}{(\Delta x)^2} - \frac{V(\theta_i - \theta_{i-1})}{\alpha \Delta x} - \frac{q''P}{kA} = 0. \qquad (4.40)$$

The next example demonstrates the procedure for a numerical solution of this problem.

EXAMPLE 4.6

Consider the problem involving a combination of the two key problems as shown in Fig. 4.18. We wish to determine the steady temperature distribution.

The governing differential formulation in terms of $\theta = T - T_\infty$ is

$$\frac{d^2\theta}{dx^2} - \frac{V}{\alpha}\frac{d\theta}{dx} + \frac{4q''}{kD} = 0, \qquad 0 < x < \ell,$$

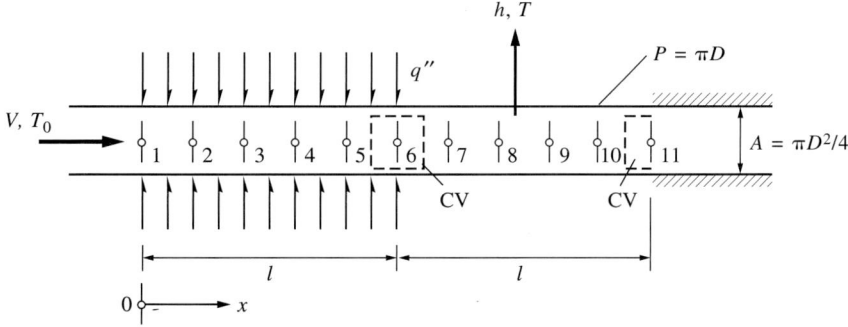

Figure 4.18 A moving circular rod.

and

$$\frac{d^2\theta}{dx^2} - \frac{V}{\alpha}\frac{d\theta}{dx} - \frac{4h\theta}{kD} = 0, \quad \ell < x < 2\ell,$$

subject to boundary conditions

$$\theta(0) = \theta_0, \quad \frac{d\theta}{dx}(x = 2\ell) = 0.$$

Rearranging Eq. (4.40) for the first domain in terms of $Pe = V\Delta x/\alpha$, we obtain

$$\theta_{i-1} - \left(\frac{2 + Pe}{1 + Pe}\right)\theta_i + \frac{1}{1 + Pe}\theta_{i+1} = -\frac{4q''\Delta x^2}{kD(1 + Pe)},$$

for $i = 2, 3, 4, 5$, and rearranging Eq. (4.39) for the second domain,

$$\theta_{i-1} - \left(\frac{2 + Pe + \dfrac{4h\Delta x^2}{kD}}{1 + Pe}\right)\theta_i + \frac{1}{1 + Pe}\theta_{i+1} = 0,$$

for $i = 7, 8, 9, 10$. For the interface between two domains ($i = 6$), from the first law

$$\theta_{i-1} - \left(\frac{2 + Pe + \dfrac{2h\Delta x^2}{kD}}{1 + Pe}\right)\theta_i + \frac{1}{1 + Pe}\theta_{i+1} = -\frac{2q''\Delta x^2}{kD(1 + Pe)}.$$

The boundary conditions take the form of

$$\theta_1 = \theta_0, \quad \theta_{10} - \left(\frac{2 + Pe + \dfrac{2h\Delta x^2}{kD}}{1 + Pe}\right) - \frac{1}{1 + Pe}\theta_{11} = 0.$$

The matrix representation of this problem is then

$$\begin{bmatrix} -M & N & 0 & 0 & 0 & 0 & 0 & 0 & 0 & 0 \\ 1 & -M & N & 0 & 0 & 0 & 0 & 0 & 0 & 0 \\ 0 & 1 & -M & N & 0 & 0 & 0 & 0 & 0 & 0 \\ 0 & 0 & 1 & -M & N & 0 & 0 & 0 & 0 & 0 \\ 0 & 0 & 0 & 1 & -M' & N & 0 & 0 & 0 & 0 \\ 0 & 0 & 0 & 0 & 1 & -M'' & N & 0 & 0 & 0 \\ 0 & 0 & 0 & 0 & 0 & 1 & -M'' & N & 0 & 0 \\ 0 & 0 & 0 & 0 & 0 & 0 & 1 & -M'' & N & 0 \\ 0 & 0 & 0 & 0 & 0 & 0 & 0 & 1 & -M'' & N \\ 0 & 0 & 0 & 0 & 0 & 0 & 0 & 0 & 1 & -(M' - N) \end{bmatrix} \begin{bmatrix} \theta_2 \\ \theta_3 \\ \theta_4 \\ \theta_5 \\ \theta_6 \\ \theta_7 \\ \theta_8 \\ \theta_9 \\ \theta_{10} \\ \theta_{11} \end{bmatrix} = \begin{bmatrix} -\theta_1 - D' \\ -D' \\ -D' \\ -D' \\ -D'/2 \\ 0 \\ 0 \\ 0 \\ 0 \\ 0 \end{bmatrix},$$

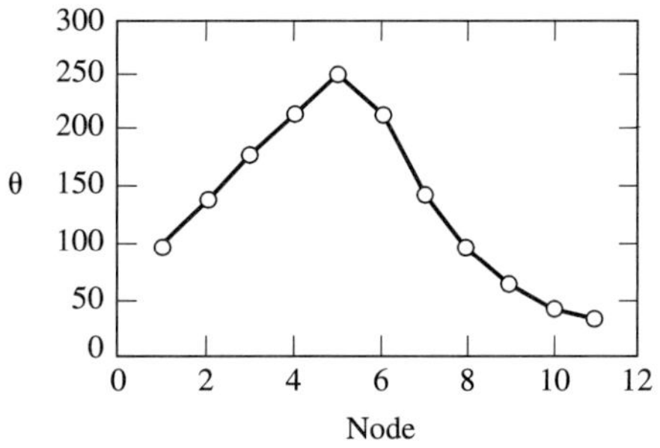

Figure 4.19 Temperatures of the moving rod.

where

$$N = \frac{1}{1 + Pe}, \quad D' = \frac{4q'' \Delta x^2}{kD(1 + Pe)},$$

$$M = \frac{2 + Pe}{1 + Pe}, \quad M' = \frac{2 + Pe + \frac{2h\Delta x^2}{kD}}{1 + Pe}, \quad M'' = \frac{2 + Pe + \frac{4h\Delta x^2}{kD}}{1 + Pe}.$$

The node temperatures are obtained by solving this matrix equation. For an illustration, let $\theta_0 = 100\,°C$, $\alpha = 10^{-4}\,m^2/s$, $V = 0.01\,m/s$, $\Delta x = 0.1\,m$, $h = 250\,W/m^2 \cdot K$, $k = 20\,W/m \cdot K$, $q'' = 20{,}000\,W/m^2$, $D = 0.1\,m$; then

$$Pe = 10, \quad N = \frac{1}{11}, \quad D' = \frac{400}{11}, \quad M = \frac{12}{11}, \quad M' = \frac{17}{11}, \quad M'' = 2.$$

The steady temperature distribution is shown in Fig. 4.19. For the interested reader, the FORTRAN program EX4–6.F listed in the Appendix solves the problem.

◆

4.3 TRUNCATION ERROR

So far the inductive approach followed in this chapter, which is based on the five steps of formulation, directly leads to the discrete formulation of a given problem. However, it does not provide information on the accuracy of this formulation. In this section we deal with the error involved with discrete formulations, which is usually called the truncation error.

Consider two Taylor series expansions of $T(x)$ near x_i, yielding at $x_i + \Delta x$ and $x_i - \Delta x$, respectively,

$$T_{i+1} \equiv T(x_i + \Delta x) = T_i + \Delta x \left(\frac{dT}{dx}\right)_i + \frac{(\Delta x)^2}{2!}\left(\frac{d^2 T}{dx^2}\right)_i + \frac{(\Delta x)^3}{3!}\left(\frac{d^3 T}{dx^3}\right)_i + \cdots,$$

(4.41)

$$T_{i-1} \equiv T(x_i - \Delta x) = T_i - \Delta x \left(\frac{dT}{dx}\right)_i + \frac{(\Delta x)^2}{2!}\left(\frac{d^2 T}{dx^2}\right)_i - \frac{(\Delta x)^3}{3!}\left(\frac{d^3 T}{dx^3}\right)_i + \cdots. \quad (4.42)$$

From Eq. (4.41) we obtain for node i at x_i the **first-order forward difference**,

$$\left(\frac{dT}{dx}\right)_i \cong \frac{T_{i+1} - T_i}{\Delta x} + O(\Delta x), \quad (4.43)$$

from Eq. (4.42) the **first-order backward difference**,

$$\left(\frac{dT}{dx}\right)_i \cong \frac{T_i - T_{i-1}}{\Delta x} + O(\Delta x), \quad (4.44)$$

from the difference between Eqs. (4.41) and (4.42) the **second-order central difference**,

$$\left(\frac{dT}{dx}\right)_i \cong \frac{T_{i+1} - T_{i-1}}{2\Delta x} + O(\Delta x)^2, \quad (4.45)$$

and, finally, from the sum of Eqs. (4.41) and (4.42) the **second-order central difference**,

$$\left(\frac{d^2 T}{dx^2}\right)_i \cong \frac{T_{i-1} - 2T_i + T_{i+1}}{(\Delta x)^2} + O(\Delta x)^2. \quad (4.46)$$

Other linear combinations of Taylor series expansions, or the use of manipulations such as substitutions in addition, yield other difference operators.

The foregoing results are of the form

$$D = FD + TE, \quad (4.47)$$

The derivative (D) being approximated by the finite-difference operator (FD) to within a **truncation error** (TE) (or, discretization error). The foregoing mathematical consideration provides an estimate of the accuracy of the discretization of the difference operators. It shows that TE is of the order of $(\Delta x)^2$ for the central difference, but only $O(\Delta x)$ for the forward and backward difference operators of first order. Equations (4.41) and (4.42) involve 2 or 3 nodes around node i at x_i, leading to 2- and 3-point difference operators. Considering additional Taylor series expansions extending to nodes $i + 2$ and $i - 2$ etc., located at $x_i + 2\Delta x$ and $x_i - 2\Delta x$, etc., respectively, one may derive 4- and 5-point difference formulas with associated truncation errors. Results summarized in Table 4.8 show that a TE of $O(\Delta x)^4$ can be achieved in this manner. The penalty for this increased accuracy is the increased complexity of the coefficient matrix of the resulting system of equations.

We demonstrate now the truncation error in terms of the fin example of Section 4.1. Using Eq. (4.46) for $d^2\theta/dx^2$ and θ_i for θ, we obtain

$$\underbrace{\frac{d^2\theta}{dx^2} - m^2\theta}_{\text{ODE}} = \underbrace{\frac{\theta_{i-1} - 2\theta_i + \theta_{i+1}}{(\Delta x)^2} - m^2\theta_i}_{\text{FDA}} + \underbrace{O(\Delta x)^2}_{\text{TE}}. \quad (4.48)$$

Table 4.8 2-, 3-, 4- and 5-point difference formulas

Number of Points	Difference formula	Difference Type
2	$\dfrac{dT_i}{dx} = \dfrac{T_{i+1}-T_i}{\Delta x} + O(\Delta x)$	Forward
	$\dfrac{dT_i}{dx} = \dfrac{T_i-T_{i-1}}{\Delta x} + O(\Delta x)$	Backward
3	$\dfrac{dT_i}{dx} = \dfrac{-3T_i+4T_{i+1}-T_{i+2}}{2\Delta x} + O(\Delta x)^2$	Forward
	$\dfrac{dT_i}{dx} = \dfrac{3T_i-4T_{i-1}+T_{i-2}}{2\Delta x} + O(\Delta x)^2$	Backward
	$\dfrac{dT_i}{dx} = \dfrac{T_{i+1}-T_{i-1}}{2\Delta x} + O(\Delta x)^2$	Center
	$\dfrac{d^2T_i}{dx^2} = \dfrac{T_{i+2}-2T_{i+1}+T_i}{(\Delta x)^2} + O(\Delta x)$	Forward
	$\dfrac{d^2T_i}{dx^2} = \dfrac{T_i-2T_{i-1}+T_{i-2}}{(\Delta x)^2} + O(\Delta x)$	Backward
	$\dfrac{d^2T_i}{dx^2} = \dfrac{T_{i+1}-2T_i+T_{i-1}}{(\Delta x)^2} + O(\Delta x)^2$	Center
4	$\dfrac{dT_i}{dx} = \dfrac{-T_{i+2}+8T_{i+1}-8T_{i-1}+T_{i-2}}{12(\Delta x)} + O(\Delta x)^4$	Center
	$\dfrac{d^2T_i}{dx^2} = \dfrac{-T_{i+3}+4T_{i+2}-5T_{i+1}+2T_i}{(\Delta x)^2} + O(\Delta x)^2$	Forward
	$\dfrac{d^2T_i}{dx^2} = \dfrac{2T_i-5T_{i-1}+4T_{i-2}-T_{i-3}}{(\Delta x)^2} + O(\Delta x)^2$	Backward
5	$\dfrac{d^2T_i}{dx^2} = \dfrac{-T_{i+2}+16T_{i+1}-30T_i+16T_{i-1}-T_{i-2}}{12(\Delta x)^2} + O(\Delta x)^4$	Center

In solving the finite-difference approximation, we let FDA = 0 and, in fact, do not solve the differential equation, but rather the difference between the ODE and TE. For example, in Table 4.3, the deviation of the numerical results from those of the exact solution is caused by the truncation error, since Δx is not small enough to eliminate the effect of the truncated terms.

As illustrated in the foregoing discussion, the numerical solution is expected to approach the exact solution as we use more refined discretizations. In reality, however, computer solutions may be somewhat limited in accuracy by the number of digits employed by the processor. This restriction associated with rounding to a finite number of digits results in **round-off errors**. These errors increase with the number of arithmetic operations required to produce a solution, hence with the number of grid points. The latter depends on the size of the problem and the degree to which the discretization is refined. The ultimate accuracy of a numerical solution can therefore be expected to be achieved in a trade-off between truncation errors and round-off errors. The effect of

4.4 UNSTEADY CONDUCTION

4.4.1 Explicit Finite-Difference Formulation

As a simple illustration, we wish to develop the unsteady one-dimensional finite-difference formulation for a flat plate of thickness ℓ having an initial temperature difference $\Delta T = T_1 - T_2$ between its surfaces. Assume the surface of the plate with temperature T is suddenly insulated.

Applying the first law of thermodynamics to the system shown in Fig. 4.20 (a uniform grid is assumed), and relating the result to temperatures by means of the Fourier law, we obtain

$$\rho c \Delta x \frac{T_i^{n+1} - T_i^n}{\Delta t} = k \frac{T_{i-1}^n - T_i^n}{\Delta x} + k \frac{T_{i+1}^n - T_i^n}{\Delta x}, \tag{4.49}$$

and, after some arrangement,

$$T_i^{n+1} - T_i^n = F\left(T_{i-1}^n - 2T_i^n + T_{i+1}^n\right) \tag{4.50}$$

or

$$T_i^{n+1} = F\left[T_{i-1}^n + (1/F - 2)T_i^n + T_{i+1}^n\right], \tag{4.51}$$

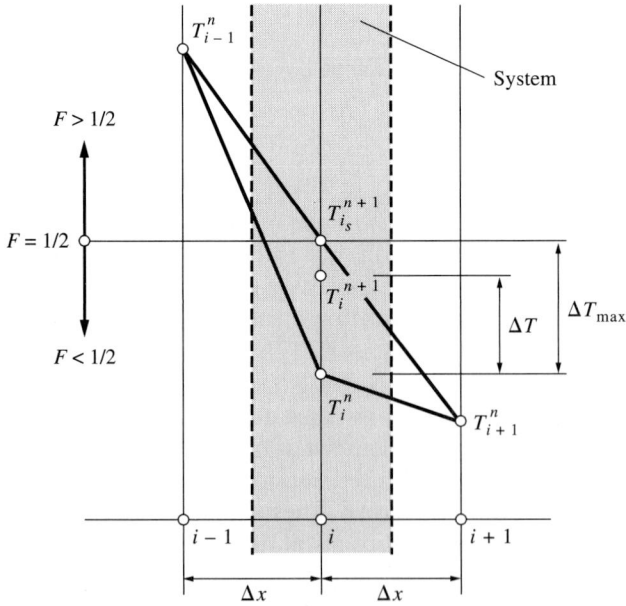

Figure 4.20 Finite-difference approximation.

where, as before, subscripts denote locations, the newly introduced superscripts denote time intervals—that is, $T_i^n = T(x_i, t_n)$, $T_{i+1}^n = T(x_i + \Delta x, t_n)$, $T_i^{n+1} = T(x_i, t_n + \Delta t)$—,

$$F = \frac{\alpha \Delta t}{(\Delta x)^2} \qquad (4.52)$$

is the **cell Fourier number** and $\alpha = k/\rho c$ the thermal diffusivity. Clearly, choosing some values for the intervals Δx and Δt, and referring to a table of properties, we can evaluate F. With known F, as well as the temperatures from a time step t_n, T_i^n, $i = 1, \ldots, N + 1$, the temperature at location i for subsequent time $n + 1$ is obtained from Eq. (4.51). This is an **explicit** time **marching** scheme which allows direct calculation of the unknowns at each time step from the preceding step. However, under certain conditions the scheme may be mathematically unstable, which we consider next.

4.4.2 Stability of Explicit Scheme

Let us see what happens when we carry out the calculations for two slightly different values of F, say 5/11 and 5/9. This corresponds to slightly different values for the time intervals if the plate is divided into equal space steps (grids), say 10, for both cases. Inspection of these figures reveals that although the results for $F = 5/11$ are quite satisfactory, those for $F = 5/9$ are affected by some sort of accumulated and amplified error, the so-called **instability**. This error has nothing to do with the truncation or round-off error, which can be made negligibly small in the present example; it is, rather, a property of difference equations, and it increases with successively smaller values of Δx unless Δt is also reduced accordingly. We therefore learn the important fact that the **explicit difference formulation of an unsteady problem is never complete without the statement of a stability criterion.** Formal but somewhat lengthy mathematical proofs may be found in the literature (see, for example, Anderson et al., 1984). Here we give a simple physical argument based on the fact that the steady one-dimensional temperature distribution in a plate of constant thermal conductivity is a straight line. Consequently, the maximum attainable by temperature T_i^{n+1} is its steady limit $T_{i,s}^{n+1}$. From the steady limit of Eq. (4.50) which is independent of the initial temperature T_i^n, which corresponds to $F = 1/2$, we have then

$$T_{i,s}^{n+1} = \frac{1}{2}\left(T_{i-1}^n + T_{i+1}^n\right), \qquad (4.53)$$

or, by subtracting T_i^n from both sides,

$$T_{i,s}^{n+1} - T_i^n = \frac{1}{2}\left(T_{i-1}^n - 2T_i^n + T_{i+1}^n\right). \qquad (4.54)$$

Note that the steady value may also be obtained when the lefthand side of Eq. (4.50) is set to zero. From the ratio of Eqs. (4.50) and (4.54)

$$\frac{T_i^{n+1} - T_i^n}{T_{i,s}^{n+1} - T_i^n} = 2F, \qquad (4.55)$$

which, in view of

$$T_i^{n+1} - T_i^n \le T_{i,s}^{n+1} - T_i^n$$

(see Fig. 4.20), gives the **stability criterion** as

$$\boxed{F \le \frac{1}{2}}.\qquad(4.56)$$

The reason for the selecting of 5/9 and 5/11 for F in Figs. 4.21 and 4.22 now becomes clear. That is, the case of $F > 1/2$, which causes instability in the difference equations, violates the physics of the problem by leading to a temperature that exceeds its steady value.[1] This stability criterion can also be obtained by requiring that the coefficient of T_i^n in Eq. (4.51) be positive.

Consider next the effect of **enthalpy flow** on the stability at inner nodes. The governing equation now becomes

$$\frac{1}{\alpha}\frac{\partial \theta}{\partial t} = \frac{\partial^2 \theta}{\partial x^2} - \frac{V}{\alpha}\frac{\partial \theta}{\partial x},\qquad(4.57)$$

where $\theta = T - T_\infty$. Using forward difference in time, the first-order backward difference for convection, and the second-order central difference for conduction, the

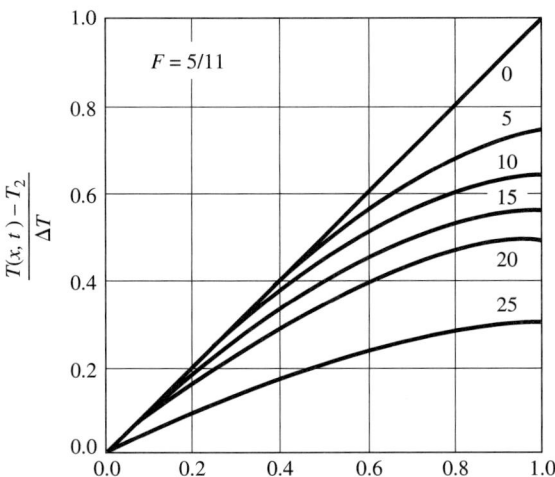

Figure 4.21 Stable solution.

---

[1] Note that the steady limit, which corresponds to $F = 1/2$ and yields temperature $T_{i,s}^{n+1}$ given by Eq. (4.53), readily admits a graphical interpretation as follows. The temperature change T_i^{n+1} at location i in the time interval $(n+1)\Delta t - n\Delta t$ is the arithmetic mean of the neighboring temperatures T_{i-1}^n and T_{i+1}^n taken from the previous time step; thus the intersection of the straight line connecting T_{i-1}^n and T_{i+1}^n with the isothermal plane of location i gives $T_{i,s}^{n+1}$ (Fig. 4.20). This is the basis of the **Binder-Schmidt graphical method**.

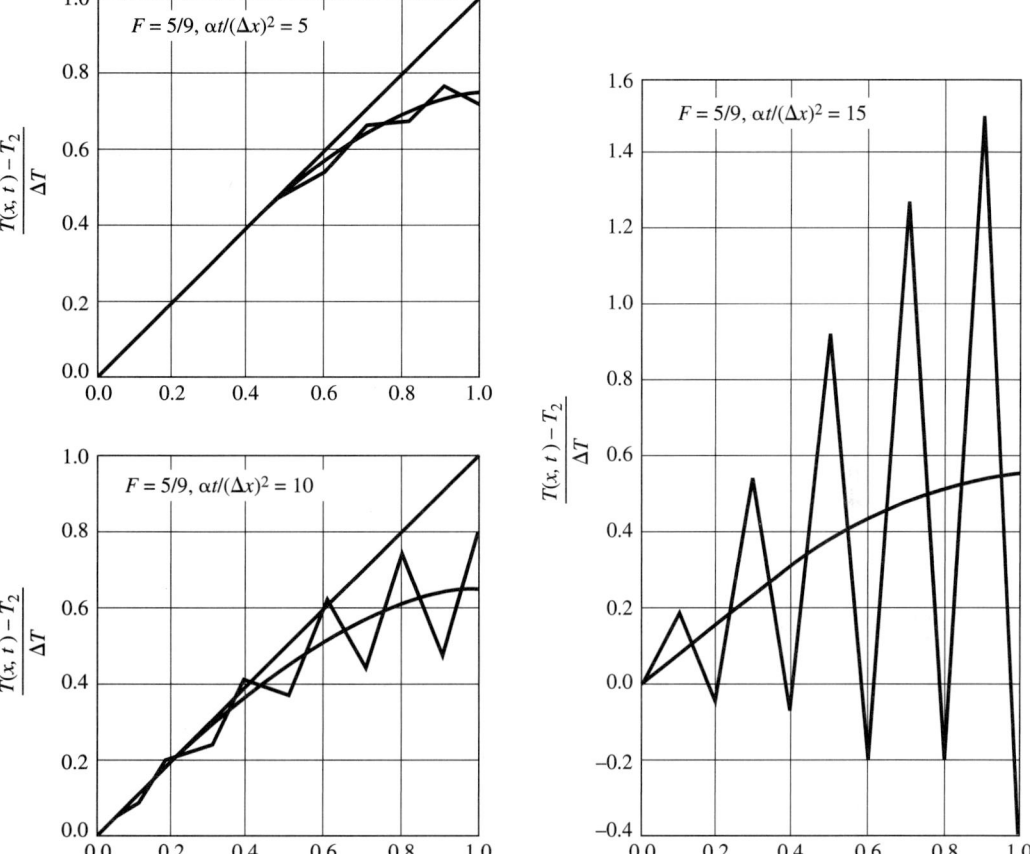

Figure 4.22 Unstable solutions.

finite-difference form of this equation becomes

$$\frac{1}{\alpha}\frac{\theta_i^{n+1} - \theta_i^n}{\Delta t} = \frac{\theta_{i-1}^n - 2\theta_i^n + \theta_{i+1}^n}{(\Delta x)^2} - \frac{V}{\alpha}\frac{\theta_i^n - \theta_{i-1}^n}{\Delta x} = 0 \qquad (4.58)$$

or

$$\theta_i^{n+1} - \theta_i^n = F\left[(1 + P)\theta_{i-1}^n - (2 + P)\theta_i^n + \theta_{i+1}^n\right], \qquad (4.59)$$

where

$$\boxed{P = V\Delta x/\alpha} \qquad (4.60)$$

is the Peclet number for an inner cell. Following the physical approach θ_i^{n+1} must be bounded between θ_i^n and the steady solution obtained by setting the lefthand side of Eq. (4.59) to zero,

$$0 = (1 + P)\theta_{i-1}^n - (2 + P)\theta_{i,s}^{n+1} + \theta_{i+1}^n,$$

which may be rearranged as

$$\theta_{i,s}^{n+1} - \theta_i^n = \frac{(1+P)\theta_{i-1}^n - (2+P)\theta_i^n + \theta_{i+1}^n}{2+P}. \quad (4.61)$$

The ratio of Eqs. (4.59) and (4.61) yields the stability criterion

$$0 < \frac{\theta_i^{n+1} - \theta_i^n}{\theta_{i,s}^{n+1} - \theta_i^n} = F(2+P) \leq 1,$$

or

$$F \leq \frac{1}{2+P}. \quad (4.62)$$

After rearranging Eq. (4.59) for θ_i^{n+1}, the same criterion can again be recovered by setting the coefficient of θ_i^n positive. We learn from this result that the presence of enthalpy flow further restricts the size of the time step for a given spatial discretization. A constant source term, say u'''/k, however, has no effect on the stability of the numerical scheme.

Finally, let us consider a **boundary node**-for example, the boundary of a flat plate-now subject to convective heat transfer to an ambient at temperature T_∞. The discretized balance of thermal energy and the Fourier law of conduction applied to a $\Delta x/2$-thick boundary difference system (Fig. 4.23) yields for $\theta = T - T_\infty$

$$\rho c A \left(\frac{\Delta x}{2}\right) \frac{\theta_1^{n+1} - \theta_1^n}{\Delta t} = +kA \frac{\theta_2^n - \theta_1^n}{\Delta x} - hA\theta_1^n \quad (4.63)$$

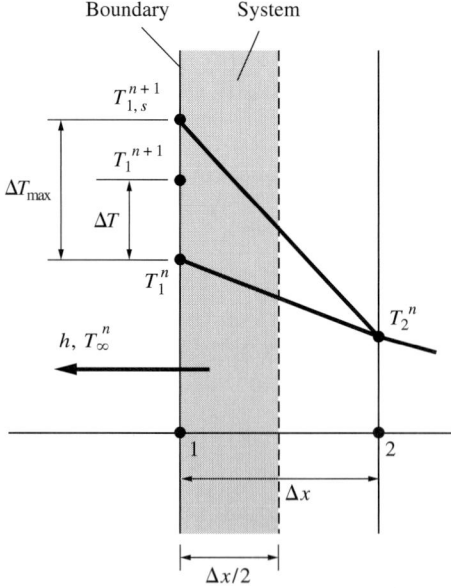

Figure 4.23 Discretized balance of boundary thermal energy.

or

$$\theta_1^{n+1} - \theta_1^n = 2F\left[\theta_2^n - (1+B)\theta_1^n\right], \tag{4.64}$$

or

$$\theta_1^{n+1} = 2F\left\{\theta_2^n + [1/2F - (1+B)]\theta_1^n\right\}, \tag{4.65}$$

where

$$\boxed{B = h\Delta x/k} \tag{4.66}$$

denotes the **boundary-cell Biot number**.[2] For the limiting steady solution which is independent of the initial temperature θ_1^n and corresponds to $1/2F = (1+B)$, we have

$$\theta_{1,s}^{n+1} = \frac{\theta_2^n}{1+B}, \tag{4.67}$$

or, by subtracting θ_1^n from both sides,

$$\theta_{1,s}^{n+1} - \theta_1^n = \frac{\theta_2^n - 1 + B\theta_1^n}{1+B}. \tag{4.68}$$

From the ratio of Eqs. (4.64) and (4.68),

$$\frac{\theta_1^{n+1} - \theta_1^n}{\theta_{1,s}^{n+1} - \theta_1^n} = 2F(1+B),$$

which, in view of the fact that

$$\theta_1^{n+1} - \theta_1^n \leq \theta_{1,s}^{n+1} - \theta_1^n$$

(see Fig. 4.20), yields the stability criterion,

$$F \leq \frac{1}{2(1+B)}, \tag{4.69}$$

which is more restrictive relative to that of Eq. (4.56) for inner nodes. Since the stability criteria for inner and boundary nodes are different, the more restrictive boundary criterion must be selected to ensure overall stability. It may readily be shown that the stability for a boundary node for which temperature is specified is not more restrictive than that of an inner node.

---

[2] Note different definitions for B in Eqs. (4.2) and (4.66).

4.4.3 Truncation Error of Explicit Scheme

With the Taylor expansions given by Eqs. (4.41) and (4.42) and similar expansions in time, we employ the **forward difference in time** and the **central difference in space** to get

$$\underbrace{\frac{\partial T}{\partial t} - \alpha \frac{\partial^2 T}{\partial x^2}}_{\text{PDE}} = \underbrace{\frac{T_i^{n+1} - T_i^n}{\Delta t} - \alpha \frac{T_{i-1}^n - 2T_i^n + T_{i+1}^n}{(\Delta x)^2}}_{\text{FDA}}$$

$$\underbrace{- \frac{1}{2}\left(\frac{\partial^2 T}{\partial x^2}\right)_i^n \Delta t + \frac{\alpha}{12}\left(\frac{\partial^4 T}{\partial x^4}\right)_i^n (\Delta x)^2 + O(\Delta t)^2 + O(\Delta x)^4}_{\text{TE}}. \quad (4.70)$$

Equation (4.70) indicates that the governing partial differential equation (PDE) equals the finite-difference approximation (FDA) to within a truncation error (TE) which is of $O[\Delta t, (\Delta x)^2]$. Clearly, the first-order forward difference has only first-order accuracy (recall Eq. 4.43). However, differentiating $\partial T/\partial t = \alpha \partial^2 T/\partial x^2$ twice with respect to x and once with respect to t, respectively, shows that the first two terms of TE can be combined into one,

$$\text{TE} = (1 - 6F)\frac{\alpha}{12}\left(\frac{\partial^4 T}{\partial x^4}\right)_i^n (\Delta x)^2 + O(\Delta t)^2 + O(\Delta x)^4. \quad (4.71)$$

Thus, for the particular choice of $F = 1/6$ the explicit scheme has TE of $O[(\Delta t)^2, (\Delta x)^4]$, which assures a higher accuracy. However, the choice $\Delta t = (\Delta x)^2/(6\alpha)$ implies time steps too small to be of practical interest in most cases. Irrespective of the size of the time step relative to that of the space step, TE $\to 0$, hence FDA $\to$ PDE, as $\Delta t \to 0$ and $\Delta x \to 0$, which shows the discrete formulation to be **consistent**.

EXAMPLE 4.7

A flat plate with thickness $\ell = 10$ cm and diffusivity $\alpha = 1 \times 10^{-4}$ m^2/s is initially at temperature of 0 °C. One surface of the plate is kept at a temperature of 0 °C while the other is suddenly raised to 100 °C. We wish to determine the transient temperature distribution within the plate by using an explicit numerical scheme.

In this example, the grid system of Ex. 4.1 is employed, i.e., $N = 5$ and $\Delta x = 0.02$ m. In view of Eqs. (4.51) and (4.56), the time step is specified as

$$\Delta t \leq \frac{(\Delta x)^2}{2\alpha} = 2 \text{ s}.$$

Here, we let $\Delta t = 1$ s, i.e., $F = 0.25$, to ensure the stability. The FORTRAN program EX4–7.F solves this problem. Table 4.9 gives the iteration results. ◆

Table 4.9 Iteration table for transient temperature distribution (explicit)

Step	Time	T_1	T_2	T_3	T_4	T_5	T_6
1	1.0	100.00	25.00	.00	.00	.00	.00
2	2.0	100.00	37.50	6.25	.00	.00	.00
3	3.0	100.00	45.31	12.50	1.56	.00	.00
4	4.0	100.00	50.78	17.97	3.91	.39	.00
5	5.0	100.00	54.88	22.66	6.54	1.17	.00
6	6.0	100.00	58.11	26.68	9.23	2.22	.00
7	7.0	100.00	60.72	30.18	11.84	3.42	.00
8	8.0	100.00	62.91	33.23	14.32	4.67	.00
9	9.0	100.00	64.76	35.92	16.63	5.91	.00
10	10.0	100.00	66.36	38.31	18.78	7.12	.00
11	11.0	100.00	67.76	40.44	20.74	8.25	.00
12	12.0	100.00	68.99	42.34	22.54	9.31	.00
13	13.0	100.00	70.08	44.06	24.19	10.29	.00
14	14.0	100.00	71.05	45.59	25.68	11.19	.00
15	15.0	100.00	71.93	46.98	27.04	12.02	.00
16	16.0	100.00	72.71	48.23	28.27	12.77	.00
17	17.0	100.00	73.41	49.36	29.38	13.45	.00
18	18.0	100.00	74.05	50.38	30.39	14.07	.00
19	19.0	100.00	74.62	51.30	31.31	14.63	.00
20	20.0	100.00	75.13	52.13	32.14	15.14	.00
21	21.0	100.00	75.60	52.88	32.89	15.61	.00
22	22.0	100.00	76.02	53.56	33.57	16.03	.00
23	23.0	100.00	76.40	54.18	34.18	16.40	.00
24	24.0	100.00	76.75	54.73	34.74	16.75	.00
25	25.0	100.00	77.06	55.24	35.24	17.06	.00
26	26.0	100.00	77.34	55.69	35.69	17.34	.00
27	27.0	100.00	77.59	56.10	36.10	17.59	.00
28	28.0	100.00	77.82	56.48	36.48	17.82	.00
29	29.0	100.00	78.03	56.81	36.81	18.03	.00
30	30.0	100.00	78.22	57.12	37.12	18.22	.00
31	31.0	100.00	78.39	57.39	37.39	18.39	.00
32	32.0	100.00	78.54	57.64	37.64	18.54	.00
33	33.0	100.00	78.68	57.87	37.87	18.68	.00
34	34.0	100.00	78.81	58.07	38.07	18.81	.00
35	35.0	100.00	78.92	58.25	38.25	18.92	.00
36	36.0	100.00	79.02	58.42	38.42	19.02	.00
37	37.0	100.00	79.12	58.57	38.57	19.12	.00
38	38.0	100.00	79.20	58.71	38.71	19.20	.00
39	39.0	100.00	79.28	58.83	38.83	19.28	.00
40	40.0	100.00	79.35	58.94	38.94	19.35	.00

4.4.4 Implicit Scheme

So far, we have considered the explicit scheme of finite-difference formulations and its stability criterion for an illustrative example. The use of the explicit scheme becomes somewhat cumbersome when a rather small Δx is selected to eliminate the truncation error for accuracy. The Δt allowed then by the stability criterion may be so small that an enormous amount of calculations may be required. We now intend to eliminate this difficulty by giving different forms to the equations resulting from the finite-difference formulation. Let us take the case of one-dimensional conduction in unsteady problems, for which we obtained the difference equation given by Eq. (4.50). Consider a formulation of the problem in terms of **backward** rather than **forward** differences in time. That is, decrease the time from $t_{n+1} = (n+1)\Delta t$ to $t_n = n\Delta t$. Thus we obtain

$$T_i^{n+1} - T_i^n = F\left(T_{i-1}^{n+1} - 2T_i^{n+1} + T_{i+1}^{n+1}\right). \tag{4.72}$$

Subtracting from this relation its initial steady value given by Eq. (4.53) yields

$$T_i^{n+1} - T_i^n = F\left[\left(T_{i-1}^{n+1} - T_{i-1}^n\right) + 2\left(T_i^{n+1} - T_i^n\right) + \left(T_{i+1}^{n+1} - T_{i+1}^n\right)\right]. \tag{4.73}$$

Since all differences in Eq. (4.73) are positive according to Fig. 4.20,[3] this equation is satisfied under all circumstances. Therefore, it is **unconditionally stable**. We obtain this stability, however, at the cost of a new algebraic complexity. Recall Eq. (4.50), in which all temperatures except T_i^{n+1} are known, and the latter is obtained by solution of the equation. By contrast, in Eq. (4.72) only T_i^n is known, and the application of this equation to the nodes yields a tridiagonal matrix which then can be solved by the method discussed in Ex. 4.1.

EXAMPLE 4.8

Reconsider Ex. 4.7. We wish to determine the transient temperature distribution of the flat plate by using an implicit numerical scheme.

In view of Eq. (4.72), we can write a matrix representation for each inner node as

$$[C][T]^{n+1} = [T]^n = [D] \tag{4.74}$$

or, explicitly,

$$\begin{bmatrix} -(2+F^{-1}) & 1 & 0 & 0 \\ 1 & -(2+F^{-1}) & 1 & 0 \\ 0 & 1 & -(2+F^{-1}) & 1 \\ 0 & 0 & 1 & -(2+F^{-1}) \end{bmatrix} \begin{bmatrix} T_2^{n+1} \\ T_3^{n+1} \\ T_4^{n+1} \\ T_5^{n+1} \end{bmatrix} = \begin{bmatrix} -100 - T_2^n F^{-1} \\ -T_3^n F^{-1} \\ -T_4^n F^{-1} \\ -T_5^n F^{-1} \end{bmatrix}$$

(4.75)

This tridiagonal matrix is identical in form to Eq. (4.13) of Ex. 4.1 and can be solved in a similar manner. The FORTRAN program EX4–8.F listed in the Appendix solves the problem. Table 4.10 shows the iteration results. ◆

---

[3] The only other possibility is a downward concave temperature distribution for which all differences are negative, and Eq. (4.73) is again unconditionally satisfied.

Sec. 4.4 Unsteady Conduction

Table 4.10 Iteration table for transient temperature distribution (implicit)

Step	Time	T_1	T_2	T_3	T_4	T_5	T_6
1	1.0	100.00	17.16	2.94	.50	.08	.00
2	2.0	100.00	29.29	7.11	1.57	.32	.00
3	3.0	100.00	38.13	11.61	3.11	.73	.00
4	4.0	100.00	44.76	16.03	4.96	1.31	.00
5	5.0	100.00	49.86	20.16	7.01	2.04	.00
6	6.0	100.00	53.90	23.95	9.14	2.89	.00
7	7.0	100.00	57.16	27.38	11.29	3.81	.00
8	8.0	100.00	59.85	30.46	13.40	4.77	.00
9	9.0	100.00	62.11	33.23	15.43	5.75	.00
10	10.0	100.00	64.02	35.72	17.36	6.73	.00
11	11.0	100.00	65.67	37.95	19.18	7.68	.00
12	12.0	100.00	67.11	39.97	20.88	8.60	.00
13	13.0	100.00	68.37	41.78	22.47	9.48	.00
14	14.0	100.00	69.49	43.43	23.93	10.31	.00
15	15.0	100.00	70.48	44.91	25.29	11.09	.00
16	16.0	100.00	71.36	46.26	26.54	11.81	.00
17	17.0	100.00	72.15	47.48	27.69	12.49	.00
18	18.0	100.00	72.87	48.59	28.74	13.12	.00
19	19.0	100.00	73.51	49.59	29.71	13.70	.00
20	20.0	100.00	74.09	50.51	30.60	14.23	.00
21	21.0	100.00	74.62	51.35	31.41	14.72	.00
22	22.0	100.00	75.10	52.10	32.15	15.17	.00
23	23.0	100.00	75.53	52.80	32.83	15.59	.00
24	24.0	100.00	75.93	53.43	33.45	15.97	.00
25	25.0	100.00	76.28	54.00	34.02	16.32	.00
26	26.0	100.00	76.61	54.53	34.54	16.63	.00
27	27.0	100.00	76.91	55.01	35.02	16.93	.00
28	28.0	100.00	77.18	55.44	35.45	17.19	.00
29	29.0	100.00	77.43	55.84	35.85	17.44	.00
30	30.0	100.00	77.65	56.20	36.21	17.66	.00
31	31.0	100.00	77.86	56.53	36.54	17.86	.00
32	32.0	100.00	78.04	56.84	36.84	18.05	.00
33	33.0	100.00	78.21	57.11	37.11	18.22	.00
34	34.0	100.00	78.37	57.36	37.37	18.37	.00
35	35.0	100.00	78.51	57.59	37.60	18.51	.00
36	36.0	100.00	78.64	57.80	37.80	18.64	.00
37	37.0	100.00	78.76	58.00	38.00	18.76	.00
38	38.0	100.00	78.87	53.17	38.17	18.87	.00
39	39.0	100.00	78.97	58.33	38.33	18.97	.00
40	40.0	100.00	79.06	58.48	38.48	19.06	.00

4.4.5 Crank-Nicolson Method

Although there is only one explicit formulation for a given problem, the implicit formulation may be written in a number of forms. For example, we may express the righthand side of Eq. (4.72) as

$$T_i^{n+1} - T_i^n = \frac{F}{2}\left[\left(T_{i-1}^{n+1} - 2T_i^{n+1} + T_{i+1}^{n+1}\right) + \left(T_{i-1}^n - 2T_i^n + T_{i+1}^n\right)\right], \quad (4.76)$$

which is known as the **Crank-Nicolson scheme**. The reader is referred to Anderson et al. (1984) for further discussion on this aspect.

Note that unsteady problems are initial-value problems. The solution of their explicit difference formulation is trivial; it amounts to the evaluation of the temperature change at each node and at each time step in terms of the temperature of the same node and the neighboring nodes taken from the preceding time step; the procedure is carried out step by step to subsequent intervals. The solution of the implicit finite-difference formulation reduces at each node and time step to an algebraic equation involving the temperature of the same node and the neighboring nodes expressed in terms of the next time interval; thus we obtain a set of algebraic equations to be solved by a direct or iteration method as discussed in Section 4.2. Note that steady problems, in general, are boundary-value problems, but those having motion and negligible conduction in a given direction are conceptually initial-value problems and can be solved in a manner similar to unsteady problems.

EXAMPLE 4.9

Reconsider Ex. 4.7. We now wish to determine the transient temperature distribution of the flat plate by using the Crank-Nicolson implicit scheme.

In view of Eq. (4.76), the proper difference equation for each inner node is

$$T_{i-1}^{n+1} - 2(1 + F^{-1})T_i^{n+1} + T_{i+1}^{n+1} = -T_{i-1}^n + 2(1 - F^{-1})T_i^n - T_{i+1}^n \equiv D_i,$$

which has the same form as Eq. (4.74). This set of equations can be solved in the same way as described in the previous example. The FORTRAN program EX4–9.F solves the problem. Table. 4.11 shows the iteration results.

◆

So far, we have employed three different numerical schemes (explicit, implicit, Crank-Nicolson) to solve a one-dimensional unsteady conduction problem. Pros and cons for these schemes are:

1. The explicit scheme is very easy to implement in a computer program, must satisfy a restrictive stability criterion, and is first-order accurate in time.
2. Writing a computer program for an implicit scheme requires some matrix manipulations and is somewhat involved. It is also first-order accurate in time but unconditionally stable.
3. The Crank-Nicolson scheme improves the accuracy of the implicit scheme to second-order accuracy in time but still requires the usual matrix manipulations of that scheme.

Table 4.11 Iteration table for transient temperature distribution (Crank-Nicolson)

Step	Time	T_1	T_2	T_3	T_4	T_5	T_6
1	1.0	100.00	20.20	2.04	.21	.02	.00
2	2.0	100.00	32.99	6.67	1.01	.13	.00
3	3.0	100.00	41.64	11.82	2.51	.43	.00
4	4.0	100.00	47.84	16.74	4.50	.96	.00
5	5.0	100.00	52.50	21.20	6.76	1.70	.00
6	6.0	100.00	56.14	25.17	9.13	2.61	.00
7	7.0	100.00	59.07	28.69	11.49	3.63	.00
8	8.0	100.00	61.49	31.79	13.77	4.70	.00
9	9.0	100.00	63.53	34.55	15.95	5.79	.00
10	10.0	100.00	65.27	37.00	17.99	6.87	.00
11	11.0	100.00	66.78	39.20	19.89	7.91	.00
12	12.0	100.00	68.11	41.16	21.65	8.90	.00
13	13.0	100.00	69.27	42.93	23.27	9.83	.00
14	14.0	100.00	70.31	44.52	24.76	10.70	.00
15	15.0	100.00	71.23	45.95	26.13	11.51	.00
16	16.0	100.00	72.06	47.25	27.37	12.26	.00
17	17.0	100.00	72.80	48.43	28.51	12.94	.00
18	18.0	100.00	73.47	49.49	29.55	13.57	.00
19	19.0	100.00	74.08	50.45	30.50	14.15	.00
20	20.0	100.00	74.62	51.33	31.30	14.67	.00
21	21.0	100.00	75.12	52.12	32.14	15.15	.00
22	22.0	100.00	75.57	52.84	32.86	15.59	.00
23	23.0	100.00	75.97	53.50	33.51	15.99	.00
24	24.0	100.00	76.34	54.09	34.10	16.30	.00
25	25.0	100.00	76.68	54.63	34.63	16.69	.00
26	26.0	100.00	76.98	55.12	35.12	16.99	.00
27	27.0	100.00	77.26	55.56	35.57	17.26	.00
28	28.0	100.00	77.51	55.97	35.97	17.51	.00
29	29.0	100.00	77.73	56.34	36.34	17.74	.00
30	30.0	100.00	77.94	56.67	36.67	17.94	.00
31	31.0	100.00	78.13	56.97	36.97	18.13	.00
32	32.0	100.00	78.30	57.25	37.25	18.30	.00
33	33.0	100.00	78.45	57.50	37.50	18.46	.00
34	34.0	100.00	78.60	57.73	37.73	18.60	.00
35	35.0	100.00	78.72	57.94	37.94	18.72	.00
36	36.0	100.00	78.84	58.12	38.12	18.84	.00
37	37.0	100.00	78.95	58.29	38.29	18.95	.00
38	38.0	100.00	79.04	53.45	38.45	19.04	.00
39	39.0	100.00	79.13	58.59	38.59	19.13	.00
40	40.0	100.00	79.21	58.72	38.72	19.21	.00

4.5 EULER'S METHOD

The timewise integration of lumped problems leading to ordinary differential equation(s) needs a treatment different from the timewise integration of distributed problems involving partial differential equation(s) considered in Section 4.4. Here, we demonstrate this integration in terms of Euler's method applied to Exs. 4.10 and 4.11.

Reconsider the lumped brake problem of Ex. 3.3 with

$$\frac{d\theta}{dt} + m\theta = nV_0\left(1 - \frac{t}{t_s}\right) \qquad (4.77)$$

subject to the initial condition

$$\theta(0) = 0,$$

m, n, V_0, and t_s being known constants, and $\theta = T - T_\infty$. Rearranging Eq. (4.77) by leaving the derivative term alone on the left-hand side,

$$\frac{d\theta}{dt} = -m\theta + nV_0\left(1 - \frac{t}{t_s}\right), \qquad (4.78)$$

which may be expressed in a general form as

$$\boxed{\frac{d\theta}{dt} = f(t, \theta)}, \qquad (4.79)$$

where

$$f \equiv f(t, \theta) = -m\theta + nV_0\left(1 - \frac{t}{t_s}\right). \qquad (4.80)$$

Now, consider an instant i where the solution is known and introduce a time-step size

$$\tau = \Delta t = t_{i+1} - t_i. \qquad (4.81)$$

Inserting known t_i and θ_i into Eq. (4.80) gives $f(t_i, \theta_i)$. In terms of $f(t_i, \theta_i)$, prescribed time-step size τ, and Eq. (4.79), we determine θ_{i+1} at t_{i+1} from Taylor's expansion,

$$\theta_{i+1} = \theta_i + \tau \left.\frac{d\theta}{dt}\right|_i + \frac{\tau^2}{2} \left.\frac{d^2\theta}{dt^2}\right|_i + O(\tau^3), \qquad (4.82)$$

or, in view of Eq. (4.79), from

$$\theta_{i+1} = \theta_i + \tau f(t_i, \theta_i) + \frac{\tau^2}{2} \left.\frac{df}{dt}\right|_i + O(\tau^3). \qquad (4.83)$$

Neglecting $O(\tau^2)$ and higher-order terms, Eq. (4.83) is usually approximated by

$$\theta_{i+1} \cong \theta_i + \tau f(t_i, \theta_i), \qquad (4.84)$$

which is known as **Euler's method**. In the following example we use this method to solve Eq. (3.27).

EXAMPLE 4.10

Reconsider Ex. 3.3. Let

$$\delta_1 = 0.4 \text{ cm}, \quad \delta_2 = 0.5 \text{ cm}, \quad \mu = 0.3, \quad p = 10 \text{ kPa}, \quad V_0 = 25 \text{ m/s}, \quad t_s = 10 \text{ s},$$
$$\rho_1 = 4{,}000 \text{ kg/m}^3, \quad c_1 = 600 \text{ J/kg·K}, \quad \rho_2 = 8{,}000 \text{ kg/m}^3, \quad c_2 = 450 \text{ J/kg·K},$$
$$h_1 = h_2 = 30 \text{ W/m}^2\text{·K}, \quad k_1 = 0.2 \text{ W/m·K}, \quad k_2 = 15 \text{ W/m·K}.$$

We wish to numerically determine the unsteady temperature of the brake and compare the numerical solution with the analytical solution.

From the foregoing data we have, for

$$m = \frac{h_1 + h_2}{\rho_1 c_1 \delta_1 + \rho_2 c_2 \delta_2},$$

$$m = \frac{60 \text{ W/m}^2\text{·K}}{(4{,}000 \times 600 \times 0.4 \times 10^{-2} + 8{,}000 \times 450 \times 0.5 \times 10^{-2}) \text{ J/m}^2\text{·K}}$$

$$= \frac{60}{3 \times 10^4} = 2 \times 10^{-3} \text{ s}^{-1},$$

and, for

$$n = \frac{\mu p}{\rho_1 c_1 \delta_1 + \rho_2 c_2 \delta_2},$$

$$n = \frac{0.3 \times 10^4 \text{ N/m}^3}{3 \times 10^4 \text{ J/m}^2\text{·K}} = 0.1 \text{ K/m}.$$

Then, from Eq. (4.84), we have

$$\theta_{i+1} = \theta_i + \tau \left[-2 \times 10^{-3} \theta_i + 2.5 \left(1 - \frac{t_i}{10} \right) \right].$$

For a time interval of 10 s we pick $\tau = \Delta t = 0.1$ s. The time history of θ is shown in Fig. 4.24 and compared with the exact solution, Eq. (3.34). For the interested reader, the FORTRAN program EX4–10.F for this problem is listed in the Appendix. ◆

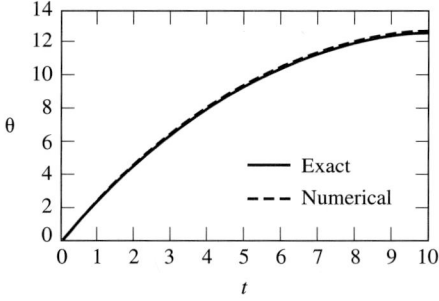

Figure 4.24 Transient temperature.

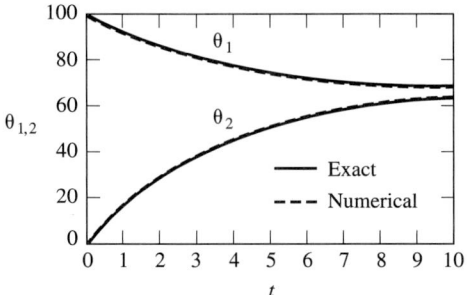

Figure 4.25 Temperatures of the ball and the bath.

Example 4.11

Reconsider Ex. 3.4. We wish to simultaneously determine the unsteady lumped ball and bath temperatures.

The governing equations of this problem are [recall Eqs. (3.35)-(3.37)]

$$\frac{d\theta_1}{dt} = -m_1(\theta_1 - \theta_2), \quad \theta_1(0) = \theta_0,$$

$$\frac{d\theta_2}{dt} = m_2(\theta_1 - \theta_2), \quad \theta_2(0) = 0,$$

and the exact solutions are given by Eqs. (3.44) and (3.45). Here, let $m_1 = 0.1$, $m_2 = 0.2$, $\theta_0 = T_0 - T_\infty = 100$, $\eta = m_2/m_1 = 2$, $\tau = 1/(m_1 + m_2) = 10/3$, and $\tau = \Delta t = 0.1$. The difference equations are

$$(\theta_1)_{i+1} = (\theta_1)_i - \tau m_1(\theta_1 - \theta_2)_i$$

and

$$(\theta_2)_{i+1} = (\theta_2)_i + \tau m_2(\theta_1 - \theta_2)_i.$$

The numerical solution is shown in Fig. 4.25 against the exact solution. For the interested reader, the FORTRAN program EX4–11.F of this problem is listed in the Appendix. ◆

4.6 CONCLUDING REMARKS

In this chapter we introduced numerical techniques to solve two-dimensional steady/unsteady conduction, one-dimensional convection, and the time integration of lumped problems which are discussed and solved analytically in Chapters 2 and 3. An extensive reference list is provided at the end of this chapter for interested readers regarding accuracy, stability, efficiency, and geometry concerns. Some FORTRAN programs listed in the Appendix are useful for students not familiar with computer languages. These programs are easy to modify for the solution of homework problems of this chapter.

REFERENCES

4.1 M. Abramowitz, and I. A. Stegun, *Handbook of Mathematical Functions*. Dover, New York, 1972.
4.2 V. S. Arpacı, *Conduction Heat Transfer*. Addison-Wesley, Reading, MA, 1966.
4.3 V. S. Arpacı, and P. S. Larsen, *Convection Heat Transfer*, Prentice-Hall, Englewood Cliffs, NJ, 1985.
4.4 R. L. Burden, J. D. Faires, and A. C. Reynolds, *Numerical Analysis*. PWS, Boston, MA, 1981.
4.5 B. Carnahan, H. A. Luther, and J. O. Wilkes, *Applied Numerical Methods*. Wiley, New York, 1969.
4.6 B. Carnahan, and J. O. Wilkes, *Fortran 77 with MTS and the IBM-PC*. The University of Michigan, Ann Arbor, 1984a.
4.7 B. Carnahan, and J. O. Wilkes, *The IBM Personal Computers and the Michigan Terminal System*. The University of Michigan, Ann Arbor, 1984b.
4.8 G. Dahlquist, and A. Bjorck, *Numerical Methods*. Prentice-Hall, Englewood Cliffs, NJ, 1974.
4.9 A. D. Gosman, B. E. Launder, and G. J. Reece, *Computer-Aided Engineering, Heat Transfer and Fluid Flow*. Ellis Horwood/Wiley, New York, 1985.
4.10 IMSL FORTRAN 77 Library (User's manual), Houston, TX, 1982.
4.11 Y. Jaluria, and K. E. Torrance, *Computational Heat Transfer*. Hemisphere, Washington, DC, 1986.
4.12 W. J. Minkowycz, E. M. Sparrow, G. E. Schneider, and R. H. Pletcher, *Handbook of Numerical Heat Transfer*. Wiley, New York, 1988.
4.13 G. E. Myers, *Analytical Methods in Conduction Heat Transfer*. McGraw-Hill, New York, 1971.
4.14 S. V. Patankar, *Numerical Heat Transfer and Fluid Flow*. Hemisphere, Washington, DC, 1980.
4.15 R. Peyret, and T. D. Taylor, *Computational Methods for Fluid Flow*. Springer-Verlag, New York, 1985.
4.16 R. D. Richtmyer, and K. W. Morton, *Difference Methods for Initial-value Problems*. 2d ed. Interscience/Wiley, New York, 1967.
4.17 P. J. Roache, *Computational Fluid Dynamics*. Hermosa, Albuquerque, NM, 1985.
4.18 D. B. Spalding, *Int. J. Num. Methods Eng.*, 4, 551, 1972.
4.19 D. A. Anderson, J. C. Tannehill, and R. H. Pletcher, *Computational Fluid Mechanics and Heat Transfer*. Hemisphere/ McGraw-Hill, New York, 1984
4.20 J. Boyd, *Spectral Methods-Lecture Notes for AOS 555*, 2 vols. The University of Michigan, Ann Arbor, 1986.
4.21 D. Gottlieb, and S. A. Orszag, *Numerical Analysis of Spectral Methods: Theory and Applications*. J. W. Arrowsmith, Bristol, England, 1981.
4.22 L. Fox, and I. B. Parker, *Chebyshev Polynomials in Numerical Analysis*. Oxford University, London, 1968
4.23 S. A. Orszag, and M. Israeli, *Ann. Rev. Fluid Mech.*, 6, 281, 1974.
4.24 A. J. Baker, *Finite Element Computational Fluid Mechanics*. Hemisphere/McGraw-Hill, New York, 1983.
4.25 N. Kikuchi, *Finite Element Methods in Mechanics*. Cambridge University, 1985.

228 Chap. 4 Computational Conduction

4.7 COMPUTER PROGRAM APPENDIX

```
C-----------------------------------
C      EX4-1.F (START)
C-----------------------------------
       PROGRAM MAIN
       IMPLICIT REAL*8 (A-H,O-Z)
       DIMENSION C(6,6),T(6),D(6)

       DO 10 J=1,6
       T(J)=0.
       D(J)=0.
       DO 20 I=1,6
       C(I,J)=0.
20     CONTINUE
10     CONTINUE

       DO 30 I=2,5
       C(I,I)=-4.
       C(I,I-1)=1.
       C(I,I+1)=1.
30     CONTINUE
       C(5,6)=1.
       C(6,6)=-2.
       D(2)=-100.

       DO 40 I=2,5
       C(I,I+1)=C(I,I+1)/(-C(I,I))
       D(I)=D(I)/(-C(I,I))
       C(I+1,I+1)=C(I+1,I+1)+C(I,I+1)
       D(I+1)=D(I+1)+D(I)
40     CONTINUE

       T(1)=100.
       T(6)=D(6)/C(6,6)
       DO 50 I=5,2,-1
       T(I)=C(I,I+1)*T(I+1)-D(I)
50     CONTINUE

       WRITE(*,700)
700    FORMAT(2X,'Node Number',2X,'X [m]',2X,'Exact Solution',2X,
      +'Numerical Solution',2X,'Error %')
       DO 60 I=1,6
       X=(I-1.)*0.02
       TEXACT=100.*COSH(-SQRT(5000.)*(0.1-X))/COSH(-SQRT(5000.)*0.1)
       ERR=ABS(T(I)-TEXACT)/TEXACT*100.
       WRITE(*,710) I,X,TEXACT,T(I),ERR
710    FORMAT(5X,I3,6X,F5.2,5X,F6.2,12X,F6.2,8X,F6.2)
60     CONTINUE

       Q1=1000.*(2.*T(1)-T(2))
       Q1EXACT=T(1)*SQRT(4.*250.*20/0.01)*TANH(SQRT(5000.)*0.1)
       ERR=ABS(Q1-Q1EXACT)/Q1EXACT*100.
       WRITE(*,720) Q1EXACT,Q1,ERR
720    FORMAT(2X,'Heat flux',/,2X,'at the base',8X,E12.4,4X,E12.4,
      +7X,F6.2)
       STOP
       END
C-----------------------------------
C      EX4-1.F (END)
C-----------------------------------
C-----------------------------------
C      EX4-2.F  (START)
C-----------------------------------
       PROGRAM MAIN
       IMPLICIT REAL*8 (A-H,O-Z)
       PARAMETER (NMAX=101)
       DIMENSION C(NMAX,NMAX),T(NMAX),D(NMAX)
```

```
              N=11
              B=0.5
              DO 10 J=1,N
              T(J)=0.
              D(J)=0.
              DO 20 I=1,N
              C(I,J)=0.
20            CONTINUE
10            CONTINUE
              DO 30 I=2,N-1
              C(I,I)=-(2.+B)
              C(I,I-1)=1.
              C(I,I+1)=1.
30            CONTINUE
              C(N-1,N)=1.
              C(N,N)=-(2.+B)/2.
              D(2)=-100.
              DO 40 I=2,N-1
              C(I,I+1)=C(I,I+1)/(-C(I,I))
              D(I)=D(I)/(-C(I,I))
              C(I+1,I+1)=C(I+1,I+1)+C(I,I+1)
              D(I+1)=D(I+1)+D(I)
40            CONTINUE
              T(1)=100.
              T(N)=D(N)/C(N,N)
              DO 50 I=N-1,2,-1
              T(I)=C(I,I+1)*T(I+1)-D(I)
50            CONTINUE
              WRITE(*,700)
700           FORMAT(2X,'Node Number',2X,'X [m]',2X,'Exact Solution',2X,
             +'Numerical Solution',2X,'Error %')
              DO 60 I=1,N
              X=(I-1.)/(N-1.)*0.1
              TEXACT=100.*COSH(-SQRT(5000.)*(0.1-X))/COSH(-SQRT(5000.)*0.1)
              ERR=ABS(T(I)-TEXACT)/TEXACT*100.
              WRITE(*,710) I,X,TEXACT,T(I),ERR
710           FORMAT(5X,I3,6X,F5.2,5X,F6.2,12X,F6.2,8X,F6.2)
60            CONTINUE
              Q1=20.*10.*(N-1)*(0.5*(2+B)*T(1)-T(2))
              Q1EXACT=T(1)*SQRT(4.*250.*20/0.01)*TANH(SQRT(5000.)*0.1)
              ERR=ABS(Q1-Q1EXACT)/Q1EXACT*100.
              WRITE(*,720) Q1EXACT,Q1,ERR
720           FORMAT(2X,'Heat flux',/,2X,'at the base',8X,E12.4,4X,E12.4,
             +7X,F6.2)
              STOP
              END
C-----------------------------------
C      EX4-2.F  (END)
C-----------------------------------
C-----------------------------------
C      EX4-3.F  (START)
C-----------------------------------
              PROGRAM MAIN
              IMPLICIT REAL*8 (A-H,O-Z)
              DIMENSION T(5,5)
              DO 10 J=1,5
              DO 10 I=1,5
              T(I,J)=0.
10            CONTINUE
              DO 20 I=1,5
              T(I,1)=100.
```

```
 20     CONTINUE
        T(3,3)=25.
        T(3,2)=62.
        T(3,4)=12.
        T(4,2)=31.
        T(4,3)=12.
        T(4,4)=6.
        T(2,2)=T(4,2)
        T(2,3)=T(4,3)
        T(2,4)=T(4,4)

        DO 30 ITER=1,100
        ICONV=1
        DO 40 J=2,4
        DO 40 I=2,4
        TOLD=T(I,J)
        T(I,J)=(T(I,J-1)+T(I+1,J)+T(I,J+1)+T(I-1,J))/4.
        ERR=ABS(T(I,J)-TOLD)
        IF(ERR.GT.0.001) ICONV=0
 40     CONTINUE
        WRITE(*,700) ITER,((T(I,J),J=2,4),I=3,4)
 700    FORMAT(5X,I3,5X,6F8.2)
        IF(ICONV.EQ.1) GOTO 900
 30     CONTINUE
 900    STOP
        END
C------------------------------------
C       EX4-3.F   (END)
C------------------------------------
C------------------------------------
C       EX4-4.F   (START)
C------------------------------------
        PROGRAM MAIN
        IMPLICIT REAL*8 (A-H,O-Z)
        DIMENSION T(4,6)

        DO 10 J=1,6
        DO 10 I=1,4
        T(I,J)=0.
 10     CONTINUE

        T(4,2)=100.
        T(4,3)=100.
        T(3,2)=66.
        T(3,3)=66.
        T(3,4)=66.
        T(2,2)=33.
        T(2,3)=33.
        T(2,4)=33.
        T(2,5)=33.

        DO 20 ITER=1,100
        ICONV=1
        T(2,1)=T(2,3)
        T(3,1)=T(3,3)
        T(4,5)=T(2,3)
        T(4,4)=T(3,3)
        DO 40 J=2,5
        DO 40 I=2,3
        TOLD=T(I,J)
        T(I,J)=(T(I,J-1)+T(I+1,J)+T(I,J+1)+T(I-1,J))/4.
        ERR=ABS(T(I,J)-TOLD)
        IF(ERR.GT.0.001) ICONV=0
 40     CONTINUE
        WRITE(*,700) ITER,((T(I,J),J=2,5),I=2,3)
 700    FORMAT(3X,I3,3X,8F7.2)
        IF(ICONV.EQ.1) GOTO 900
```

```
      20    CONTINUE
      900   STOP
            END
C----------------------------------
C     EX4-4.F  (END)
C----------------------------------
C----------------------------------
C     EX4-5.F  (START)
C----------------------------------
            PROGRAM MAIN
            IMPLICIT REAL*8 (A-H,O-Z)
            DIMENSION T(5,6),X(6),Y(7),DI(5,6),SI(5,6),DJ(5,6),SJ(5,6)
            D=1.
            DO 10 J=1,6
            DO 10 I=1,5
            T(I,J)=0.
      10    CONTINUE
            DO 20 I=1,4
            X(I)=(I-1.)*D/3.
      20    CONTINUE
            DO 30 I=5,6
            X(I)=X(4)+(I-4.)*D/4.
      30    CONTINUE
            DO 40 J=1,4
            Y(J)=(J-1.)*D/4.
      40    CONTINUE
            DO 50 J=5,7
            Y(J)=Y(4)+(J-4.)*D/3.
      50    CONTINUE
            DO 60 J=2,6
            DO 60 I=2,4
            DI(I,J)=X(I)-X(I-1)
            SI(I,J)=(X(I+1)-X(I))/(X(I)-X(I-1))
            DJ(I,J)=Y(J)-Y(J-1)
            SJ(I,J)=(Y(J+1)-Y(J))/(Y(J)-Y(J-1))
      60    CONTINUE
            SI(4,3)=3.*(2.-SQRT(3.))/4.
            T(4,2)=100.
            T(5,3)=100.
            DO 70 ITER=1,100
            ICONV=1
            T(2,1)=T(2,3)
            T(3,1)=T(3,3)
            T(5,4)=T(4,3)
            T(5,5)=T(3,3)
            T(5,6)=T(2,3)
            DO 80 J=2,6
            DO 80 I=2,4
            IF((I.EQ.4).AND.(J.EQ.2)) GOTO 80
            TOLD=T(I,J)
            C0=1./(SI(I,J)*DI(I,J)**2)+1./(SJ(I,J)*DJ(I,J)**2)
            C1=(T(I+1,J)+SI(I,J)*T(I-1,J))/(SI(I,J)*(SI(I,J)+1.)*DI(I,J)**2)
            C2=(T(I,J+1)+SJ(I,J)*T(I,J-1))/(SJ(I,J)*(SJ(I,J)+1.)*DJ(I,J)**2)
            T(I,J)=(C1+C2)/C0
            ERR=ABS(T(I,J)-TOLD)
            IF(ERR.GT.0.001) ICONV=0
      80    CONTINUE
            WRITE(*,700) ITER,(T(2,J),J=2,6),(T(3,J),J=2,5),(T(4,J),J=3,4)
      700   FORMAT(3X,I3,3X,11F6.2)
            IF(ICONV.EQ.1) GOTO 900
      70    CONTINUE
```

```
900     STOP
        END
C----------------------------------
C       EX4-5.F  (END)
C----------------------------------
C----------------------------------
C       EX4-6.F (START)
C----------------------------------
        PROGRAM MAIN
        IMPLICIT REAL*8 (A-H,O-Z)
        REAL M,M1,M2,N,D1
        DIMENSION C(11,11),D(11),T(11)

        PE=10.
        N=1./(1+PE)
        M=(2.+PE)/(1.+PE)
        M1=(2.+PE+2.5)/(1.+PE)
        M2=(2.+PE+5.)/(1.+PE)
        D1=400./(1.+PE)

        DO 10 J=1,11
        D(J)=0.
        T(J)=0.
        DO 20 I=1,11
        C(I,J)=0.
20      CONTINUE
10      CONTINUE
        T(1)=100.

        DO 30 I=2,10
        IF(I.LE.5) THEN
             C(I,I)=-M
        ELSEIF(I.GE.7) THEN
             C(I,I)=-M2
        ELSE
             C(I,I)=-M1
        ENDIF
        C(I,I-1)=1.
        C(I,I+1)=N
30      CONTINUE
        C(11,11)=-(M1-N)
        C(11,10)=1.

        D(2)=-T(1)-D1
        DO 50 I=3,5
        D(I)=-D1
50      CONTINUE
        D(6)=-D1/2.

        DO 60 I=2,10
        C(I,I+1)=C(I,I+1)/(-C(I,I))
        D(I)=D(I)/(-C(I,I))
        C(I+1,I+1)=C(I+1,I+1)+C(I,I+1)
        D(I+1)=D(I+1)+D(I)
60      CONTINUE

        T(11)=D(11)/C(11,11)
        DO 70 I=10,2,-1
        T(I)=C(I,I+1)*T(I+1)-D(I)
70      CONTINUE
40      CONTINUE

        WRITE(*,*) ' NODE',' TEMPERATURE'
        DO 80 I=1,11
        WRITE(*,700) I,T(I)
80      CONTINUE
700     FORMAT(2X,I4,F10.4)
```

```
      STOP
      END
C------------------------------------
C     EX4-6.F (END)
C------------------------------------
C------------------------------------
C     EX4-7.F (START)
C------------------------------------
      PROGRAM MAIN
      IMPLICIT REAL*8 (A-H,O-Z)
      DIMENSION TN(6),TN1(6)

      DT=1.
      F=0.25

      DO 10 J=1,6
      TN(J)=0.
10    CONTINUE
      TN(1)=100.

      DO 20 N=1,50
      TN1(1)=100.
      TN1(6)=0.
      DO 30 I=2,5
      TN1(I)=F*(TN(I-1)+(1./F-2.)*TN(I)+TN(I+1))
30    CONTINUE
      T=N*DT
      WRITE(*,700) N,T,(TN1(I),I=1,6)
700   FORMAT(2X,I4,2X,F6.1,2X,6F7.2)
      DO 40 I=1,6
      TN(I)=TN1(I)
40    CONTINUE
20    CONTINUE

      STOP
      END
C------------------------------------
C     EX4-7.F (END)
C------------------------------------
C------------------------------------
C     EX4-8.F (START)
C------------------------------------
      PROGRAM MAIN
      IMPLICIT REAL*8 (A-H,O-Z)
      DIMENSION C(6,6),D(6),TN(6),TN1(6)

      DT=1.
      F=0.25

      DO 10 J=1,6
      TN(J)=0.
      TN1(J)=0.
      DO 20 I=1,6
      C(I,J)=0.
20    CONTINUE
10    CONTINUE
      TN(1)=100.
      TN1(1)=100.

      DO 40 N=1,50

      DO 30 I=2,5
      C(I,I)=-(2.+1./F)
      C(I,I-1)=1.
      C(I,I+1)=1.
30    CONTINUE

      D(2)=-TN(1)-TN(2)/F
      DO 50 I=3,5
      D(I)=-TN(I)/F
```

```
 50     CONTINUE
        DO 60 I=2,4
        C(I,I+1)=C(I,I+1)/(-C(I,I))
        D(I)=D(I)/(-C(I,I))
        C(I+1,I+1)=C(I+1,I+1)+C(I,I+1)
        D(I+1)=D(I+1)+D(I)
 60     CONTINUE
        TN1(5)=D(5)/C(5,5)
        DO 70 I=4,2,-1
        TN1(I)=C(I,I+1)*TN1(I+1)-D(I)
 70     CONTINUE
        T=N*DT
        WRITE(*,700) N,T,(TN1(I),I=1,6)
 700    FORMAT(2X,I4,2X,F6.1,2X,6F7.2)
        DO 80 I=1,6
        TN(I)=TN1(I)
 80     CONTINUE
 40     CONTINUE
        STOP
        END
C-----------------------------------
C       EX4-8.F (END)
C-----------------------------------
C-----------------------------------
C       EX4-9.F (START)
C-----------------------------------
        PROGRAM MAIN
        IMPLICIT REAL*8 (A-H,O-Z)
        DIMENSION C(6,6),D(6),TN(6),TN1(6)

        DT=1.
        F=0.25

        DO 10 J=1,6
        TN(J)=0.
        TN1(J)=0.
        DO 20 I=1,6
        C(I,J)=0.
 20     CONTINUE
 10     CONTINUE
        TN(1)=100.
        TN1(1)=100.

        DO 40 N=1,50

        DO 30 I=2,5
        C(I,I)=-(2.+2./F)
        C(I,I-1)=1.
        C(I,I+1)=1.
 30     CONTINUE
        D(2)=-2.*TN(1)+2.*(1.-1./F)*TN(2)-TN(3)
        DO 50 I=3,4
        D(I)=-TN(I-1)+2.*(1.-1./F)*TN(I)-TN(I+1)
 50     CONTINUE
        D(5)=-TN(4)+2.*(1.-1./F)*TN(5)

        DO 60 I=2,4
        C(I,I+1)=C(I,I+1)/(-C(I,I))
        D(I)=D(I)/(-C(I,I))
        C(I+1,I+1)=C(I+1,I+1)+C(I,I+1)
        D(I+1)=D(I+1)+D(I)
 60     CONTINUE
        TN1(5)=D(5)/C(5,5)
        DO 70 I=4,2,-1
        TN1(I)=C(I,I+1)*TN1(I+1)-D(I)
```

```
        70      CONTINUE
                T=N*DT
                WRITE(*,700) N,T,(TN1(I),I=1,6)
        700     FORMAT(2X,I4,2X,F6.1,2X,6F7.2)
                DO 80 I=1,6
                TN(I)=TN1(I)
        80      CONTINUE
        40      CONTINUE
                STOP
                END
C-----------------------------------
C       EX4-9.F (END)
C-----------------------------------
C-----------------------------------
C       EX4-10.F (START)
C-----------------------------------
                PROGRAM MAIN
                IMPLICIT REAL*8 (A-H,O-Z)
                DIMENSION T(101)

                T(1)=0.
                H=0.1

                WRITE(*,700)
        700     FORMAT(' TIME    NUMERICAL    EXACT')

                DO 10 I=1,100
                TIME=(I-1.)*H
                F=-0.002*T(I)+2.5*(1.-TIME/10.)
                T(I+1)=T(I)+H*F
                TEXACT=1250.*(1.-EXP(-TIME/500))-62500.*
                +(TIME/500-(1.-EXP(-TIME/500)))
                WRITE(*,710) TIME,T(I),TEXACT
        10      CONTINUE
        710     FORMAT(1X,F4.1,2F10.2)
                STOP
                END
C-----------------------------------
C       EX4-10.F (END)
C-----------------------------------
C-----------------------------------
C       EX4-11.F (START)
C-----------------------------------
                PROGRAM MAIN
                IMPLICIT REAL*8 (A-H,O-Z)
                REAL M1,M2,ETA,TAU
                DIMENSION T1(101),T2(101)

                M1=0.1
                M2=0.2
                T0=100.
                ETA=M2/M1
                TAU=1./(M1+M2)
                T1(1)=T0
                T2(1)=0.
                H=0.1

                WRITE(*,700)
        700     FORMAT(' TIME   NUMERICAL    EXACT    NUMERICAL    EXACT')
```

```
            DO 10 I=1,100
            TIME=(I-1.)*H
            F1=-M1*(T1(I)-T2(I))
            F2=M2*(T1(I)-T2(I))
            T1(I+1)=T1(I)+H*F1
            T2(I+1)=T2(I)+H*F2
            TEXACT1=T0/(1.+ETA)*(ETA+EXP(-TIME/TAU))
            TEXACT2=T0*ETA/(1.+ETA)*(1.-EXP(-TIME/TAU))
            WRITE(*,710) TIME,T1(I),TEXACT1,T2(I),TEXACT2
     10     CONTINUE
     710    FORMAT(1X,F4.1,4F10.2)
            STOP
            END
C-------------------------------------
C          EX4-11.F (END)
C-------------------------------------
```

EXERCISES

4.1 Reconsider Ex. 4.3. Use a parabolic interpolation for the initial temperatures. Comment on the number of iteration steps.

4.2 Determine the effect of a finite heat transfer coefficient on Ex. 4.4. Use $h = 10$ W/m$^2 \cdot$ K for both inner and outer surface.

4.3 Consider a bar of square cross section. Write the finite-difference formulation of the problem for the network shown in Fig. 4P–1. Solve these equations by (a) iteration, (b) direct elimination. *Data:* $u''' = 10^6$ W/m^3, $L = 20$ mm, $k = 20$ W/m·K, $h = 700$ W/m$^2 \cdot$K, $T_\infty = 25\,°$C.

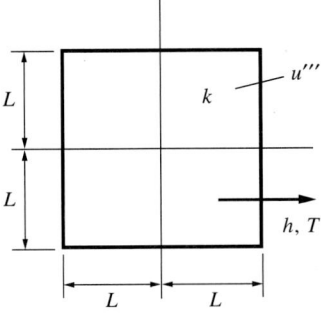

Figure 4P–1

4.4 The steady temperature distribution of a rod of square cross section is desired using a finite-difference formulation and an iteration solution (Fig. 4P–2). Employ the largest possible network, but do not lump the whole cross section. *Data:* $q'' = 40{,}000$ W/m^2, $L = 50$ mm, $k = 20$ W/m·K, $h = 1{,}500$ W/m$^2 \cdot$K, $T_\infty = 0\,°$C.

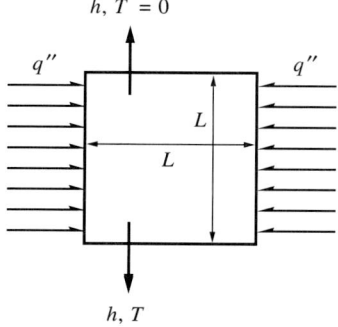

Figure 4P–2

4.5 The steady temperature distribution of a square plate of thickness δ in an ambient at $0\,°\mathrm{C}$ and with side boundary temperatures of $0\,°\mathrm{C}$ and $100\,°\mathrm{C}$ is desired using a finite-difference formulation and an iteration solution (Fig. 4P–3). Employ the largest possible network, but do not lump the whole cross section. *Data:* $L = 50$ mm, $\delta = 3$ mm, $k = 15$ W/m·K, $h_1 = 200$ W/m²·K, $h_2 = 40$ W/m²·K.

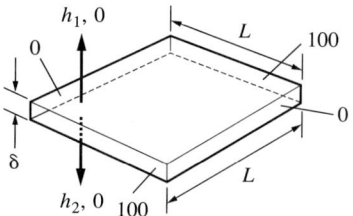

Figure 4P–3

4.6 The steady temperature distribution of a thin square plate is desired using a finite-difference formulation and an iteration solution (Fig. 4P–4). Employ a network finer than the coarsest possible network. *Data:* $L = 40$ mm, $\delta = 2$ mm, $k = 400$ W/m·K, $h = 1{,}000$ W/m²·K, $u''' = 10^7$ W/m³, $T_\infty = 20\,°\mathrm{C}$.

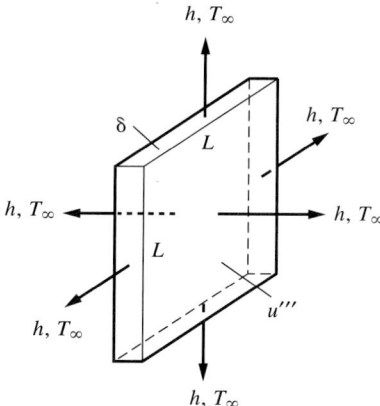

Figure 4P–4

4.7 The upper and lower surfaces of a triangular fin of length 90 mm transfer heat with a coefficient $h = 200$ W/m²·K to an ambient at temperature $T_\infty = 20\,°C$. The base temperature of the fin is maintained at $T_0 = 120\,°C$ (Fig. 4P–5). The thermal conductivity of the fin material is $k = 50$ W/m·K. Write the finite-difference formulation of the problem by considering a nodal spacing of 15 mm. Determine the tip temperature of and the heat transfer to the fin, and the efficiency of the fin, by the method of iteration.

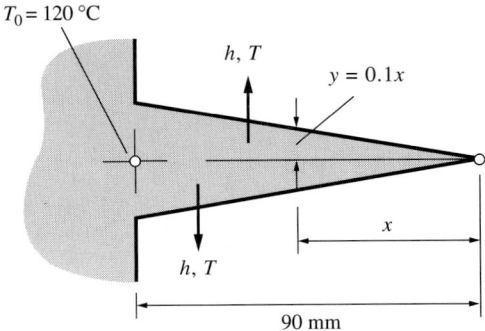

Figure 4P–5

4.8 Consider a rod with a trapezoidal cross section as shown in Fig. 4-P6 with the specified temperature boundary conditions. Determine the steady cross sectional temperature distribution.

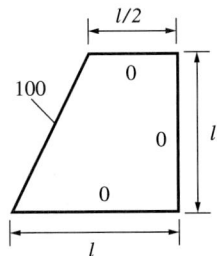

Figure 4P–6

4.9 Consider an infinitely long pipe with diameter $D = 0.2$ m. The upstream half of the pipe is insulated and the downstream half is subjected to a uniform heat flux $q'' = 10$ kW/m². A liquid metal ($\alpha = 5 \times 10^{-5}$ m²/s) with bulk velocity $V = 0.01$ m/s flows slowly through the pipe. Including the effect of axial conduction, determine the axial temperature distribution within the liquid metal.

4.10 Determine the stability criterion for the explicit scheme of an unsteady fin.

4.11 Reconsider Ex. 4.7. Assume that the surface that was kept at constant temperature of $0°C$ now transfers heat with a coefficient h to an ambient at $0\,°C$. Determine the transient temperature distribution within the plate for $h = 10, 100, 1{,}000$ W/m²·K by using an explicit scheme.

4.12 Repeat Prob. 4.11 by an implicit scheme.

4.13 Repeat Prob. 4.11 by the Crank-Nicolson scheme.

4.14 Consider a thin flat plate (of thickness $\delta_1 = 1$ cm, density $\rho_1 = 8,000$ kg/m^3, and specific heat $c_1 = 500$ J/kg·K) next to an ambient (of thickness $\delta_2 = 5$ cm, density $\rho_2 = 1,000$ kg/m^3, and specific heat $c_2 = 4,000$ J/kg·K) at an initial temperature $T_2 = 0\,°C$ with $h = 200$ W/m^2·K(Fig. 4P–7). One side of the plate and one side of the ambient are insulated. From this condition, electrical energy $u''' = 2$ kW/m^3 begins to be generated within the plate. Determine the unsteady lumped temperatures of the plate and the ambient.

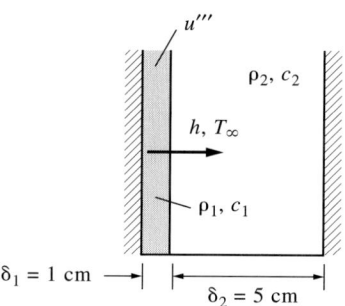

Figure 4P–7

CHAPTER 5

FOUNDATIONS OF CONVECTION

In Chapter 1 we distinguished between the three modes of heat transfer, and so far we have studied the conduction mode. From the definition of these modes, we introduced convection as being conduction (and/or radiation) in a moving medium (solid as well as fluid). However, solids are not suited for transport of energy, because they cannot be piped and branched while in motion. Also, radiation in a moving fluid, besides being rather involved, is less frequent in technological problems. It will be briefly introduced in Chapter 10. Hereafter in this text, unless otherwise specified, we assume **convection** to be **conduction in a moving fluid**.

The formulation of convection problems, to be outlined by including the effect of fluid motion into conduction, presents no real difficulties. Let us proceed to the formulation of these problems by the help of some conduction problems formulated in the preceding chapters. Recall the original conduction problem,

$$0 = \frac{d^2 T}{dx^2}. \tag{1.86}$$

In Chapter 2, we added energy generation,

$$0 = \frac{d^2 T}{dx^2} + \frac{u'''}{k}, \tag{2.37}$$

and the bulk motion (enthalpy flow) of an incompressible fluid (recall Section 2.5),

$$\frac{u}{\alpha}\frac{dT}{dx} = \frac{d^2T}{dx^2} + \frac{u'''}{k}, \tag{5.1}$$

u being the bulk velocity in the x direction and α the thermal diffusivity. In Chapter 3 we included two-dimensional conduction [recall Eq. (3.74)],

$$0 = \frac{\partial^2 T}{\partial x^2} + \frac{\partial^2 T}{\partial y^2} + \frac{u'''}{k}. \tag{5.2}$$

From Eqs. (5.1) and (5.2), and with two-dimensional motion, we have

$$\frac{1}{\alpha}\left(u\frac{\partial T}{\partial x} + v\frac{\partial T}{\partial y}\right) = \frac{\partial^2 T}{\partial x^2} + \frac{\partial^2 T}{\partial y^2} + \frac{u'''}{k}, \tag{5.3}$$

where (u, v) are the x and y components of the velocity V. In terms of the symbolic (vector) notation, and with extension to three dimensions, Eq. (5.3) may be generalized to

$$\frac{1}{\alpha}V\cdot\nabla T = \nabla^2 T + \frac{u'''}{k}. \tag{5.4}$$

Further extension to the unsteady case presents no difficulty but need not be considered here.

Suppose we now wish to determine the temperature of a fluid in motion from Eq. (5.4). First we need to know the local velocity of the fluid. It can be shown, in a manner similar to the development leading to Eq. (5.4), but starting with Newton's law of motion,

$$\frac{d}{dt}(mV) = F,$$

that the steady motion of an incompressible viscous fluid is governed by

$$\frac{1}{\nu}V\cdot\nabla V = \nabla^2 V + \frac{1}{\mu}(-\nabla p), \tag{5.5}$$

where ν and μ denote the kinematic and dynamic viscosities, respectively, and ∇p denotes the pressure gradient. The details of the development leading to Eq. (5.5) are not important for the present discussion. Note, however, the term-by-term similarity between Eqs. (5.4) and (5.5).

For a constant-property fluid, Eq. (5.5) is decoupled from Eq. (5.4) and is the only equation needed for velocity. Then, Eq. (5.4) in terms of the velocity obtained from Eq. (5.5) gives the steady temperature of an incompressible fluid in motion. Keep in mind, in addition to fluid temperatures, that convection studies are ultimately and more importantly concerned with heat transfer through a solid-fluid interface. In terms of a heat transfer coefficient h, this heat transfer is

$$q_C = h(T_w - T_\infty), \tag{5.6}$$

where T_w and T_∞ denote the interface and ambient temperatures, respectively. Also, expressing q_C by means of conduction in the fluid (Fig. 5.1), Eq. (5.6) may be written as

$$-k\left(\frac{\partial T}{\partial y}\right)_w = h(T_w - T_\infty), \tag{5.7}$$

or, in terms of a characteristic length ℓ for the fluid, as

$$Nu = \frac{h\ell}{k} = \frac{\partial}{\partial y^*}\left(\frac{T_w - T}{T_w - T_\infty}\right)_w, \quad y^* = \frac{y}{\ell}, \tag{5.8}$$

where k is the conductivity of the fluid and Nu is the **Nusselt number**.[1] Thus, the **convection heat transfer through an interface is the wall gradient of the dimensionless fluid temperature**. For all practical purposes, the distribution of the fluid temperature is confined to a narrow region (boundary layer) next to the wall. So, as we have done in Chapter 1 (recall Fig. 1.13), this wall gradient may closely be approximated by a thermal boundary layer,

$$Nu \sim \frac{\ell}{\delta}. \tag{1.59}$$

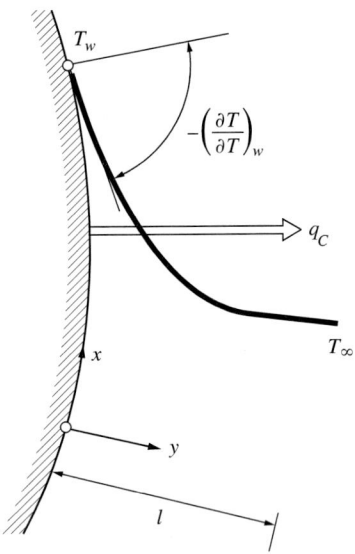

Figure 5.1 Illustration of the convection concept.

---

[1] Clearly, q_C may also be expressed by means of conduction in the solid, which leads to the definition of the Biot number [recall Eq. (3.1)]. Note the fundamental difference in the use of Eqs. (3.1) and (5.8). In conduction problems, h and T_∞ are given, and Eq. (3.1) is employed as a boundary condition. Because of their complexity, however, convection problems are usually solved in terms of simpler boundary conditions unrelated to h (such as specified temperature or heat flux), and Eq. (5.8) is utilized for the evaluation of h.

We now proceed to a brief review of the three methods available for the evaluation of convection heat transfer:

1. **Analytical solution of the fluid temperature distribution.** The wall gradient of this distribution gives the heat transfer coefficient. The exact analytical solution of convection problems is rather involved and is beyond the scope of this text. The concept of boundary layer (penetration depth) provides a convenient tool for approximate analytical solutions and will be considered in Sections 5.1 and 5.2.

2. **Analogy between heat and momentum transfer** The analogy coupled with wall-friction calculations or measurements provides the heat transfer coefficient. In Section 5.1 we develop this analogy for a particular problem which leads, in terms of the wall-friction coefficient of this problem, to the heat transfer coefficient.

3. **Dimensional analysis.** In the absence of an analytical solution or analogy between heat and momentum transfer, the (dimensionless) heat transfer coefficient may be obtained from the correlation of experimental data in terms of appropriate dimensionless numbers obtained from a dimensional analysis. In Section 5.3 we shall review the foundations of dimensional analysis in a manner particularly suited to heat transfer studies.

Actually, convection heat transfer in nature occurs in two different forms, the so-called **natural convection** and **forced convection**. A simple illustration is the motion of air around an ordinary light bulb. When the effect of temperature on the air density is taken into account, the air heated around the bulb gets lighter and rises relative to cold and denser air far from the bulb [Fig. 5.2(a)]. The **buoyant** force is the driving mechanism of natural convection. The most frequently encountered buoyancy is associated with gravity. Examples are the cooling of mechanical devices (such as heating, air conditioning, refrigeration equipment, etc.), the cooling of electrical systems (such as transmission lines, transistors, transformers, electric furnaces, etc.), and the comfort of humans and animals in a quiescent atmosphere. Other sources of natural convection are centrifugal forces which provide the internal cooling of turbine blades, inertial forces which affect cryogenic liquids in accelerating rockets, etc. Note that the rates of heat transfer by natural convection (resulting from gravity) are set by **buoyancy** and to a large extent are beyond our control. When heat transfer by natural convection is inadequate, fluid motion may be increased mechanically, say by a pump or fan [Fig. 5.2(b)]. Rates of heat transfer are then controlled largely by the power driving the **pump** or fan and consequently are within our control. This form of convection, known as forced convection, is of great technological importance and finds many applications. Examples are the cooling of gasoline and diesel engines, gas turbines, and various heat exchangers in conventional and nuclear power plants, etc.

Natural convection and forced convection, depending respectively on the magnitude of buoyancy and the power of the pump or fan, may be **laminar** or **turbulent**. As we know from fluid mechanics, the streamlines of a laminar flow behave in an orderly manner, while the streamlines of a turbulent flow fluctuate irregularly about a mean flow. In this chapter, we shall deal with laminar convection. Turbulent convection will be left to Chapter 6.

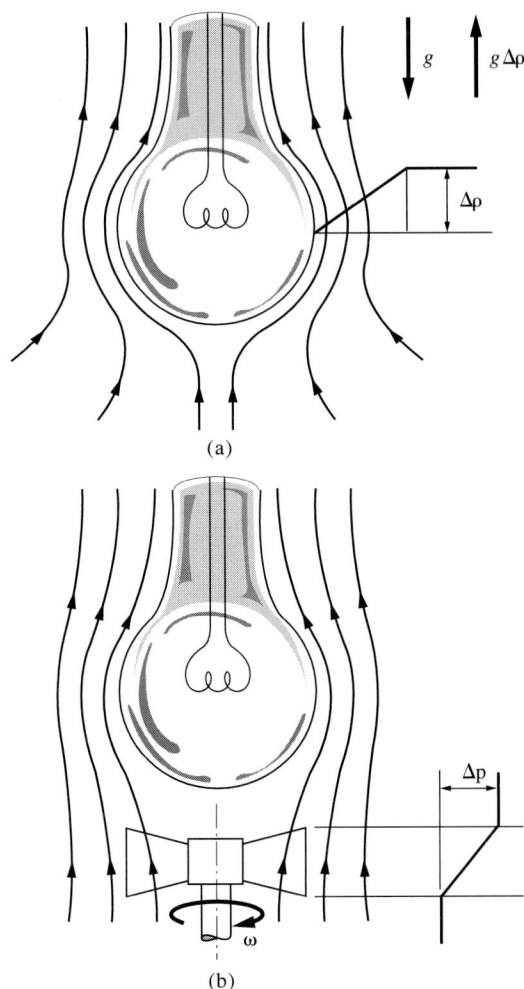

Figure 5.2 Natural and forced convection.

5.1 BOUNDARY-LAYER CONCEPT. LAMINAR FORCED CONVECTION ○

In Chapters 2 and 3 we have already introduced the concept of penetration depth for an approximate solution of conduction problems (recall Section 2.4.1, and Exs. 2.11 and 3.9). This concept, which we utilized to determine the steady or unsteady penetration depth of heat (or thermal boundary layer) in solids and stagnant fluids, actually applies to all diffusion processes, such as diffusion of momentum, mass, electricity, and neutrons, as well as diffusion (or conduction) of heat. It is a convenient tool for an approximate solution of conduction problems and is indispensable for convection problems, which are considerably more complicated than conduction problems.

Sec. 5.1 Boundary-Layer Concept. Laminar Forced Convection

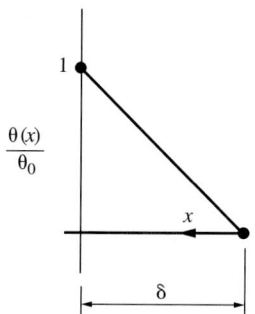

Figure 5.3 Linear temperature profile.

For example, in terms of a parabolic profile approximating the temperature distribution in an infinite fin, we obtained in Section 2.4.1

$$\delta = \sqrt{6}/m, \tag{2.125}$$

which is an illustration of constant penetration depth (or boundary layer). By definition, the effect of conduction is negligible beyond the penetration depth. This effect increases monotonically from naught at the edge of the penetration depth to a maximum at the base of fin. Now consider a further approximation, replacing the parabola with a linear profile (Fig. 5.3),

$$\frac{\theta(x)}{\theta_0} = \frac{x}{\delta}. \tag{5.9}$$

Inserting Eq. (5.9) into Eq. (2.123) results in

$$\frac{1}{\delta} = m^2 \frac{\delta}{2}$$

or

$$\delta = \sqrt{2}/m. \tag{5.10}$$

Clearly, except for a numerical constant depending on the assumed approximate profiles, Eqs. (2.125) and (5.10) for the penetration depths are identical.

Separately, on dimensional grounds, we have from the first law applied to a system of δ extent (Fig. 5.4),

$$kA\frac{\theta}{\delta} \sim h(P\delta)\theta \tag{5.11}$$

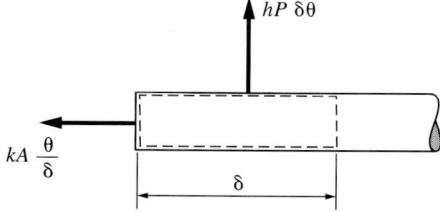

Figure 5.4 Schematic of the control volume.

or
$$\delta \sim 1/m, \tag{5.12}$$

which is silent to the numerical constant but otherwise identical to the preceding results.

For another illustration of the penetration-depth concept, using a parabolic profile, we obtained in Ex. 3.9

$$\delta = (12\alpha t)^{1/2},$$

which is an **unsteady** penetration depth in a semi-infinite plate. In terms of a linear profile, following the procedure used in Ex. 3.9, we get

$$\delta = (4\alpha t)^{1/2},$$

a result identical to Eq. (3.118) except for the numerical constant.

Again, on dimensional grounds, we have from the first law (applied this time to a control volume[2] of δ extent as shown in Fig. 5.5),

$$\rho c A \theta \frac{\delta}{t} \sim k A \frac{\theta}{\delta} \tag{5.13}$$

or
$$\delta \sim (\alpha t)^{1/2}, \tag{5.14}$$

which is silent to the numerical constant but otherwise identical to the results obtained in terms of the assumed approximate profiles.

The foregoing examples clearly demonstrate that **the boundary layer/penetration depth associated with a problem can be obtained from strict dimensional considerations except for the numerical constant involved with it. The numerical value of the constant requires the integration of the governing equation in terms of an approximate profile.**

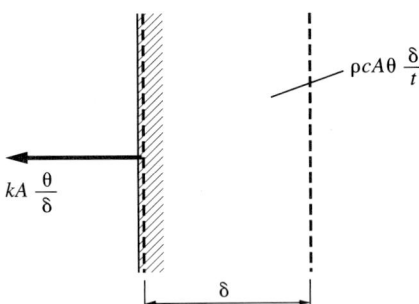

Figure 5.5 Schematic of the control volume.

---

[2] Note that mass flow $\rho A (d\delta/dt)$ into the control volume carries no enthalpy because of assumed $\theta_\infty = 0$.

Sec. 5.1 Boundary-Layer Concept. Laminar Forced Convection   247

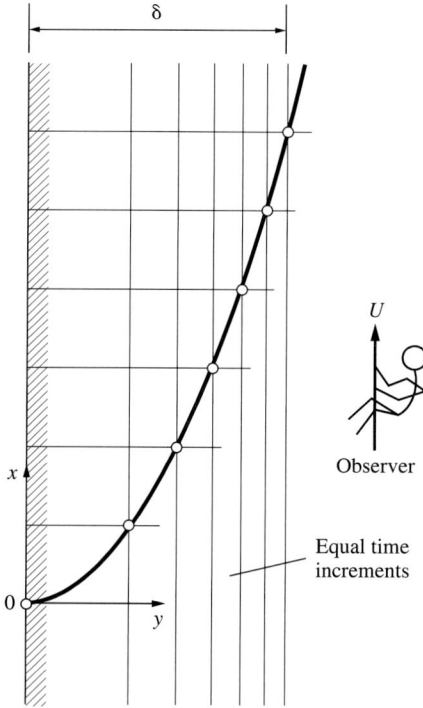

Figure 5.6 Thermal boundary layer relative to a moving observer.

In the present problem, an observer moving with constant velocity U parallel to the surface of the solid observes the boundary layer shown in Fig. 5.6. Replacing t of Eq. (3.118) with x/U,

$$\delta = (12\alpha x/U)^{1/2}, \tag{5.15}$$

which is a **spatial** boundary layer. Note that in terms of the $t = x/U$ transformation,

$$\frac{1}{\alpha}\frac{\partial T}{\partial t} = \frac{\partial^2 T}{\partial y^2} \tag{5.16}$$

becomes

$$\frac{U}{\alpha}\frac{\partial T}{\partial x} = \frac{\partial^2 T}{\partial y^2}, \tag{5.17}$$

by which the present problem may be interpreted as a semi-infinite solid at temperature T_∞ moving at constant velocity U relative to a wall kept at temperature T_w [Fig. 5.7(a)]. Also, from the first law applied to the control volume shown in Fig. 5.7(b), we have, on dimensional grounds,

$$0 \sim \rho c U \delta \left[T_\infty - \frac{1}{2}(T_w + T_\infty) \right] + q_w x$$

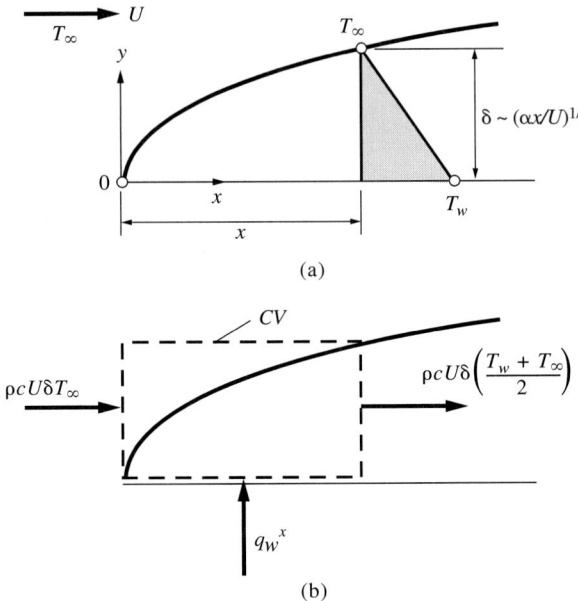

Figure 5.7 (a) Thermal boundary layer, (b) thermal energy balance.

or

$$\frac{1}{2}\rho c U \delta (T_w - T_\infty) \sim q_w x. \qquad (5.18)$$

However, an estimate on q_w is not readily available. Note that $q_w \to \infty$ as $x \to 0$ (why?) and

$$q_w \sim k\frac{T_w - T_\infty}{\delta} \qquad (5.19)$$

for any x far from $x = 0$. Inserting Eq. (5.19) into Eq. (5.18) gives

$$\frac{1}{2}\rho c U \delta (T_w - T_\infty) \sim kx\frac{T_w - T_\infty}{\delta} \qquad (5.20)$$

or

$$\delta \sim (\alpha x/U)^{1/2} \qquad (5.21)$$

or

$$\frac{x}{\delta} \sim Pe_x^{1/2} \qquad (5.22)$$

$Pe_x = Ux/\alpha$ being the local **Peclet number**.

Now, we wish to determine the local heat transfer from the boundary of the semi-infinite solid (or an inviscid fluid). In Chapter 1 we defined convection as

$$q_C = (q_K)_{\text{fluid boundaries}},$$

and, in terms of the local heat transfer coefficient[3] (Newton's cooling law),

$$q_C = h(T_w - T_\infty), \qquad (5.23)$$

we obtained

$$h(T_w - T_\infty) \sim k\frac{T_w - T_\infty}{\delta}$$

or

$$h \sim \frac{k}{\delta} \qquad (5.24)$$

or

$$\frac{hx}{k} = Nu_x \sim \frac{x}{\delta}, \qquad (5.25)$$

Nu_x being the **local Nusselt number** (or **dimensionless heat transfer coefficient**). In terms of Eq. (5.22) we have, from Eq. (5.25),

$$Nu_x \sim Pe_x^{1/2}. \qquad (5.26)$$

For the constant involved with Eqs. (5.22) and (5.26), actual temperature distributions are needed. In what follows is a first-order approximation in terms of assumed linear profiles.

EXAMPLE 5.1

Consider the control volume shown in Fig. 5.8.

The first law applied to this control volume gives

$$\rho c \int_0^\delta u(T - T_\infty)\,dy = \int_0^x q_w\,dx, \qquad (5.27)$$

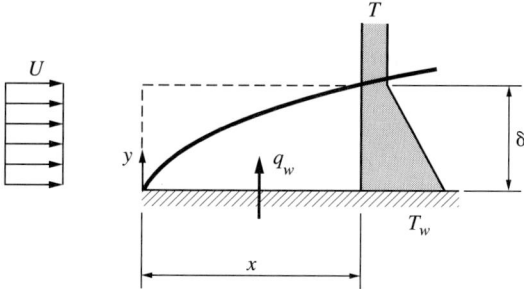

Figure 5.8 Schematic of the control volume.

---

[3] In Chapter 1 we overlooked the local nature of this coefficient.

which, in terms of the Fourier law, leads to the governing equation[4]

$$\int_0^\delta u(T - T_\infty)\,dy = -\alpha \int_0^x \left(\frac{\partial T}{\partial y}\right)_w dx. \qquad (5.28)$$

For a uniform U, Eq. (5.28) becomes

$$U \int_0^\delta (T - T_\infty)\,dy = -\alpha \int_0^x \left(\frac{\partial T}{\partial y}\right)_w dx. \qquad (5.29)$$

With a linear temperature profile shown in Fig. 5.8,

$$\frac{T - T_\infty}{T_w - T_\infty} = 1 - \frac{y}{\delta}, \qquad (5.30)$$

Eq. (5.29) yields

$$Nu_x = \frac{x}{\delta} = 0.5 Pe_x^{1/2}. \qquad (5.31)$$

Table 5.1 shows the numerical constants resulting from the use of higher-order profiles.

Table 5.1 Constants from various temperature profiles

Constants	Linear	Parabolic	Cubic	Exact
$Nu_x Pe_x^{-1/2}$	0.500	0.289	0.530	0.564

In Chapter 1 we learned the implicit relation

$$Nu = f(\text{Motion}).$$

Here, we obtain with Eq. (5.31) an explicit form of this relation. Note that the foregoing development leading to Eq. (5.26) is for (a semi-infinite solid moving with) uniform velocity U. As shown in Table 5.1, the use of higher-order profiles in general but not always gives more accurate results. These considerations are beyond the scope of this text (see, for example, Arpacı, 1966, or Arpacı and Larsen, 1984). ◆

---

[4] Integrated governing equations include also the boundary conditions.

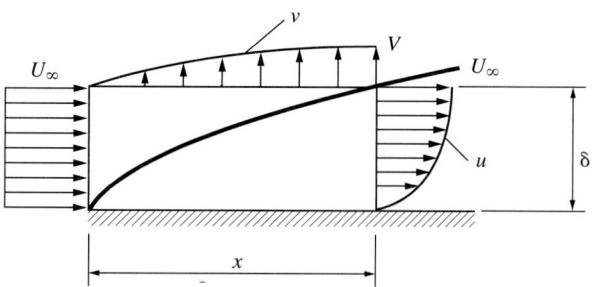

Figure 5.9 Actual velocity profiles.

In actual convection problems dealing with fluids, viscous forces retard the flow across a momentum boundary layer δ and bring the velocity down to zero on the boundary. Then, the flow becomes **two dimensional** because of **conservation of mass** (Fig. 5.9). The momentum boundary layer is usually different from the thermal boundary layer.[5] To evaluate the heat transfer in fluids near a boundary, first the velocity boundary layer needs to be determined.

For an incompressible fluid, the conservation of mass for the control volume shown in Fig. 5.9 gives, on dimensional grounds,

$$\rho U_\infty \delta \sim \frac{1}{2}\rho U_\infty \delta + \frac{1}{2}\rho V x. \tag{5.32}$$

The momentum balance for the same control volume yields

$$\rho U_\infty^2 \delta - \frac{1}{4}\rho U_\infty^2 \delta - \frac{1}{2}\rho U_\infty V x \delta \sim \tau_w x$$

or, in terms of Eq. (5.32),

$$\frac{1}{4}\rho U_\infty^2 \delta \sim \tau_w x. \tag{5.33}$$

Note that $\tau_w \to \infty$ as $x \to 0$ [recall the discussion leading to Eq. 5.19)] and assume for a Newtonian fluid

$$\tau_w \sim \mu \frac{U_\infty}{\delta} \tag{5.34}$$

for any x far from $x = 0$. Inserting Eq. (5.34) into Eq. (5.33) gives

$$\frac{1}{4}\rho U_\infty^2 \delta \sim \mu x \frac{U_\infty}{\delta} \tag{5.35}$$

or

$$\delta \sim (\nu x/U_\infty)^{1/2} \tag{5.36}$$

---

[5] When dealing with problems involving two boundary layers, δ is used for momentum and δ_θ for thermal energy.

or

$$\frac{x}{\delta} \sim Re_x^{1/2}, \tag{5.37}$$

$Re_x = U_\infty x/\nu$ being the local Reynolds number.

Here, the wall friction (resulting from the momentum flow retarded in the boundary layer) may be evaluated from

$$\frac{1}{2} f_x = \frac{\tau_w}{\rho U_\infty^2}, \tag{5.38}$$

f_x being the **local friction coefficient**. Then, in terms of Eq. (5.34),

$$\frac{1}{2} f_x \sim \frac{\nu}{U_\infty \delta} = \frac{\nu}{U_\infty x} \left(\frac{x}{\delta}\right) \tag{5.39}$$

which yields, in view of Eq. (5.37),

$$\frac{1}{2} f_x \sim \frac{1}{Re_x^{1/2}}. \tag{5.40}$$

For the constant involved with Eqs. (5.37) and (5.40), the actual velocity distribution is needed. In what follows is a first-order approximation based on linear profiles.

Example 5.2

Consider the control volume shown in Fig. 5.10.

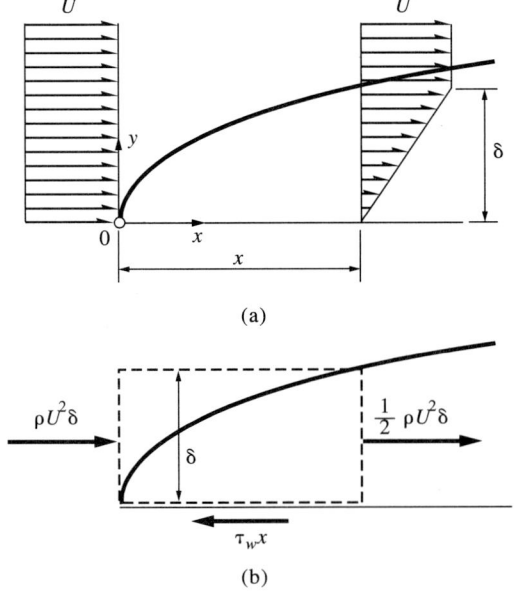

Figure 5.10 (a) Velocity boundary layer, (b) momentum balance.

The conservation of mass for this control volume gives

$$\rho U_\infty \delta = \rho \int_0^\delta u\, dy + \rho \int_0^x v\, dx. \tag{5.41}$$

The momentum balance for the same control volume yields

$$\rho U_\infty^2 \delta - \rho \int_0^\delta u^2\, dy - \rho U_\infty \int_0^x v\, dx = \int_0^x \tau_w\, dx \tag{5.42}$$

or, in terms of Eq. (5.41),

$$\rho \int_0^\delta u(U_\infty - u)\, dy = \int_0^x \tau_w\, dx. \tag{5.43}$$

For a Newtonian fluid, Eq. (5.43) leads to the governing equation.[6]

$$\int_0^\delta u(U_\infty - u)\, dy = \nu \int_0^x \left(\frac{\partial u}{\partial y}\right)_w dx. \tag{5.44}$$

In terms of a first-order linear profile shown in Fig. 5.10,

$$\frac{u}{U_\infty} = \frac{y}{\delta}, \tag{5.45}$$

Eq. (5.44) yields

$$\frac{U_\infty \delta}{6} = \nu \int_0^x \frac{dx}{\delta} \tag{5.46}$$

or, after differentiating with respect to x,

$$\delta \frac{d\delta}{dx} = 6\frac{\nu}{U_\infty} \tag{5.47}$$

which can be rearranged as

$$\frac{d\delta^2}{dx} = 12\frac{\nu}{U_\infty}. \tag{5.48}$$

Integration of this result readily gives

$$\delta = 3.464 \left(\frac{\nu x}{U_\infty}\right)^{1/2} \tag{5.49}$$

and

$$\frac{x}{\delta} = 0.289\, Re_x^{1/2}, \quad \frac{1}{2}f = \frac{0.289}{Re_x^{1/2}}. \tag{5.50}$$

Table 5.2 shows the numerical constants resulting from the use of higher-order profiles. ◆

---

[6] Integrated governing equations include also the boundary conditions.

Table 5.2 Constants from various velocity profiles.

Constants	Linear	Parabolic	Cubic	Exact
$\frac{1}{2} f Re_x^{1/2}$	0.289	0.365	0.323	0.332

To evaluate the heat transfer in fluids with different thermal and momentum boundary layers, we have to examine the velocity of the enthalpy flow. First, consider the case, $\delta > \delta_\theta$. Later we shall comment on the cases $\delta \sim \delta_\theta$ and $\delta \ll \delta_\theta$. Note from Fig. 5.11 that, assuming linear profiles for $\delta > \delta_\theta$, the velocity appropriate for the enthalpy flow is

$$U = U_\infty(\delta_\theta/\delta), \tag{5.51}$$

and the enthalpy flow of Eq. (5.20) in terms of this velocity yields

$$\delta_\theta^2(\delta_\theta/\delta) \sim \alpha x/U_\infty \tag{5.52}$$

or, in view of Eq. (5.36),

$$\delta_\theta \sim (\nu x/U_\infty)^{1/6}(\alpha x/U_\infty)^{1/3} \tag{5.53}$$

or

$$\frac{x}{\delta_\theta} \sim Re_x^{1/2} Pr^{1/3}, \tag{5.54}$$

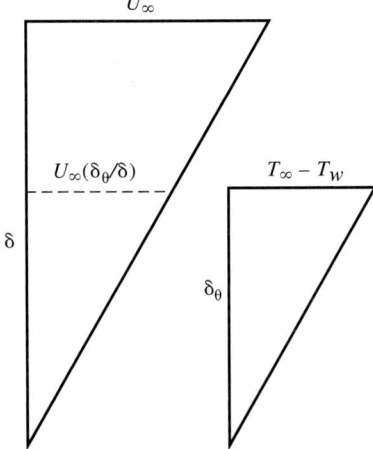

Figure 5.11 Similar velocity and temperature profiles.

$Pr = \nu/\alpha = Pe_x/Re_x$ being the **Prandtl number**. Note from the ratio of Eqs. (5.37) and (5.54) that

$$\frac{\delta}{\delta_\theta} \sim Pr^{1/3}. \qquad (5.55)$$

That is, δ and δ_θ change along x, but their ratio is a fluid property and does not depend on x. This is an important result which needs more attention.

Another approach leading to Eq. (5.55) and based on the **analogy between momentum and heat** assumes[7]

$$\text{Change in} \left(\frac{\text{Axial momentum flow}}{\text{Transversal momentum flux}} = \frac{\text{Axial enthalpy flow}}{\text{Transversal heat flux}} \right),$$

which in terms of Fig. 5.11 yields

$$\frac{\rho U_\infty \dfrac{U_\infty}{x}}{\mu \dfrac{U_\infty}{\delta^2}} = \frac{\rho c_p \left(U_\infty \dfrac{\delta_\theta}{\delta}\right)\dfrac{T_\infty - T_w}{x}}{k \dfrac{T_\infty - T_w}{\delta_\theta^2}}, \qquad (5.56)$$

or

$$\boxed{\frac{\delta}{\delta_\theta} = Pr^{1/3}} \qquad (5.57)$$

Thus, under analogy between momentum and heat transfer, the unknown constant of Eq. (5.55) becomes unity.

The heat transfer coefficient in the present case (corresponding to $\delta > \delta_\theta$) is readily obtained by inserting Eq. (5.54) into Eq. (5.25). The result is

$$Nu_x \sim Re_x^{1/2} Pr^{1/3}. \qquad (5.58)$$

Dividing this result by $Re_x Pr$, introducing the local **Stanton number**,

$$St_x = \frac{Nu_x}{Pe_x} = \frac{Nu_x}{Re_x Pr}, \qquad (5.59)$$

and employing the definition of the friction coefficient, an alternative form of Eq. (5.58) is found to be

$$St_x Pr^{2/3} \sim \frac{1}{2} f_x. \qquad (5.60)$$

---

[7] When two different fields have similar governing equations and boundary conditions they are called analogous. A majority of momentum and thermal boundary layers, however, are not analogous. For example, a pressure gradient in the momentum boundary layer or an energy generation in the thermal boundary layer, or an incompatibility between momentum and thermal boundary conditions, eliminates this analogy.

Actually, with the analogy between momentum and heat, we have the equality

$$\frac{q_w}{\tau_w} = \frac{k(T_\infty - T_w)/\delta_\theta}{\mu U_\infty/\delta}, \tag{5.61}$$

because the proportionality constant involved with q_w is identical to that involved with τ_w. In terms of the heat transfer coefficient and the friction coefficient, Eq. (5.61) may be rearranged as

$$\frac{h/k}{U_\infty/\nu} = \frac{1}{2}\left(\frac{\delta}{\delta_\theta}\right) f_x, \tag{5.62}$$

or, in view of Eq. (5.57), as

$$\boxed{St_x Pr^{2/3} = \frac{1}{2} f_x,} \tag{5.63}$$

which implies a proportionality constant of unity in Eq. (5.60). This result (of analogy) coupled with computations (or measurements) only on the skin friction (available from isothermal studies) provides indirect information about heat transfer in forced laminar flow over a horizontal plate. This is an important result, because isothermal flows are easier to study analytically or experimentally than thermal flows.

For the constant involved with Eq. (5.58), actual velocity and temperature distributions are needed. In what follows is a first-order approximation based on linear profiles.

EXAMPLE 5.3

Reconsider the governing equation given by Eq. (5.28).

In terms of assumed linear profiles,

$$u = \frac{U_\infty y}{\delta} \tag{5.64}$$

and

$$\frac{T - T_\infty}{T_w - T_\infty} = 1 - \frac{y}{\delta_\theta}, \tag{5.65}$$

Eq. (5.28) becomes

$$U_\infty \int_0^{\delta_\theta} \frac{y}{\delta}\left(1 - \frac{y}{\delta_\theta}\right) dy = \alpha \int_0^x \frac{dx}{\delta_\theta}, \tag{5.66}$$

or, after the integration of the left-hand side with respect to y,

$$\frac{U_\infty \delta_\theta^2}{6\delta} = \alpha \int_0^x \frac{dx}{\delta_\theta}. \tag{5.67}$$

Now, assuming δ/δ_θ to be constant for similar velocity and temperature profiles, taking the derivative on both sides of Eq. (5.67) with respect to x leads to

$$\delta_\theta \frac{d\delta_\theta}{dx} = 6\left(\frac{\alpha}{U_\infty}\right)\frac{\delta}{\delta_\theta}, \tag{5.68}$$

which can be rearranged as

$$\frac{d\delta_\theta^2}{dx} = 12\left(\frac{\alpha}{U_\infty}\right)\frac{\delta}{\delta_\theta}. \tag{5.69}$$

Integration of this result readily gives

$$\delta_\theta = 3.464 \left(\frac{\alpha x}{U_\infty}\right)^{1/2} \left(\frac{\delta}{\delta_\theta}\right)^{1/2}, \tag{5.70}$$

and the ratio of Eq. (5.49) and (5.70),

$$\frac{\delta}{\delta_\theta} = Pr^{1/2} \left(\frac{\delta_\theta}{\delta}\right)^{1/2}, \tag{5.71}$$

which is

$$\frac{\delta}{\delta_\theta} = Pr^{1/3}. \tag{5.57}$$

Then, inserting Eq. (5.57) into Eq. (5.70) and inverting the result relative to x yields

$$\frac{x}{\delta_\theta} = 0.289 Re_x^{1/2} Pr^{1/3}. \tag{5.72}$$

In terms of the assumed linear temperature profile, the heat transfer is

$$Nu_x = \frac{x}{\delta_\theta} = \left(\frac{x}{\delta}\right)\frac{\delta}{\delta_\theta} \tag{5.73}$$

or, in view of Eq. (5.72) or the product of Eqs. (5.50) and (5.57),

$$Nu_x = 0.289 Re_x^{1/2} Pr^{1/3}. \tag{5.74}$$

Table 5.3 shows the numerical constants resulting from the use of higher order profiles. ◆

Table 5.3 Constants from various profiles.

Constants	Linear	Parabolic	Cubic	Exact
$Nu_x Re_x^{-1/2} Pr^{-1/3}$	0.289	0.182	0.323	0.332

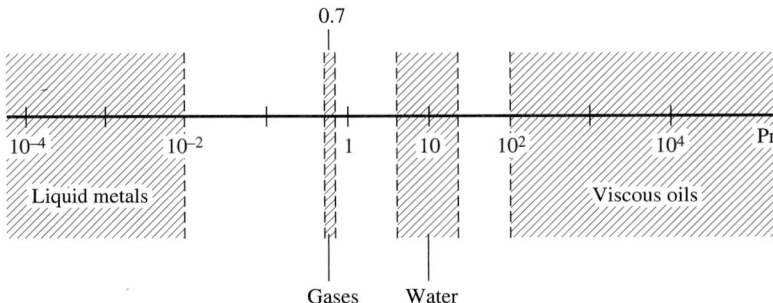

Figure 5.12 Prandtl number ranges for various material.

So far we have considered the case $\delta > \delta_\theta$. Equation (5.57) indicates that $\delta/\delta_\theta = Pr^{1/3}$, and δ can only be greater than δ_θ if $Pr > 1$. However, as shown in Fig. 5.12, the Prandtl number for various boundary layers changes over the wide range

$$0 < Pr < \infty$$

and corresponds to $\delta/\delta_\theta \ll 1$ for liquid metals, $\delta/\delta_\theta \sim 1$ for gases, $\delta/\delta_\theta > 1$ for water, and $\delta/\delta_\theta \gg 1$ for viscous oils. Accordingly, Eqs. (5.57) and (5.63) apply to viscous oils, water, and approximately to gases; Eq. (5.26) applies to liquid metals.

Here we terminate laminar forced convection and proceed to laminar natural convection.

5.2 LAMINAR NATURAL CONVECTION ○

Consider a heated vertical plate in a quiescent fluid. The plate heats the fluid in its neighborhood, which then becomes lighter and moves upward. The force resulting from the product of gravity and density difference and causing this upward motion is called **buoyancy**. The fluid moving under the effect of buoyancy develops a vertical boundary layer about the plate. Within the boundary layer the temperature decreases from the plate temperature to the fluid temperature, while the velocity vanishes on the plate walls and beyond the boundary layer and has a maximum in between (Fig. 5.13). Actually, in a manner similar to forced convection, the momentum boundary layer of natural convection is expected to be thicker for larger Prandtl numbers than the thermal boundary layer. However, the characteristic velocity for the enthalpy flow across δ_θ should be scaled relative to δ_θ rather than δ.

In the preceding study on forced convection we neglected the buoyancy force relative to the inertial force. Here, we neglect the inertial force relative to the buoyancy force. The momentum balance for the control volume involving a fluid of height x and thickness δ (Fig. 5.14) gives then

$$x\delta g \Delta \rho \sim \tau_w x. \tag{5.75}$$

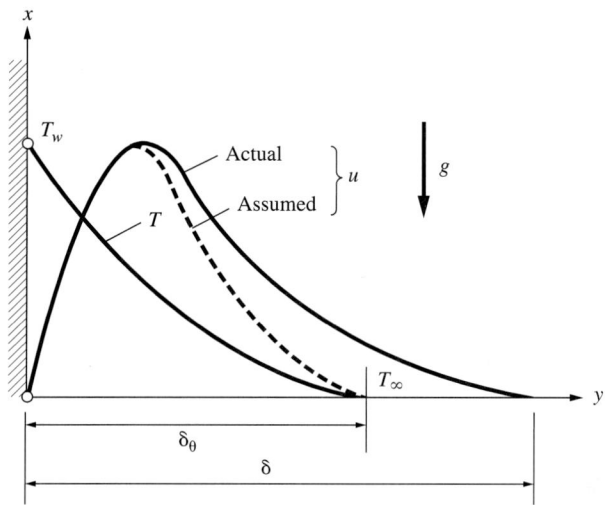

Figure 5.13 Velocity and temperature of natural convection about a vertical plate.

For an estimate on τ_w in a manner similar to the estimate on q_w [recall Eq. 5.19)], we may assume, in terms of a mean velocity U for the upward motion,

$$\tau_w \sim \mu \frac{U}{\delta}. \tag{5.76}$$

Inserting Eq. (5.76) into Eq. (5.75) yields

$$U \sim \frac{g}{\nu}\left(\frac{\Delta\rho}{\rho}\right)\delta^2. \tag{5.77}$$

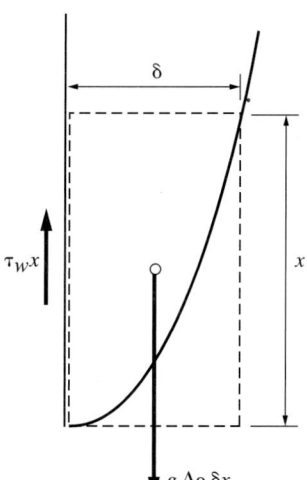

Figure 5.14 Momentum balance for natural convection.

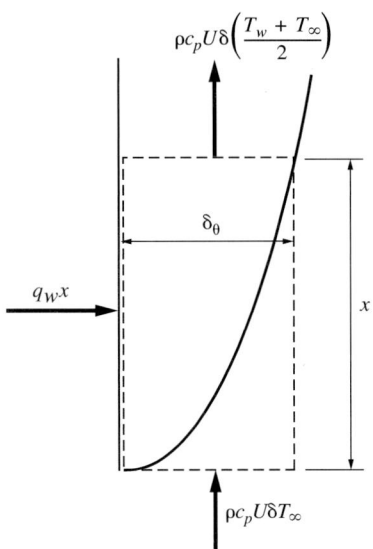

Figure 5.15 Thermal energy balance for natural convection.

For the thermal energy, consider the balance between the axial enthalpy flow and the transversal convection for the control volume shown in Fig. 5.15. Thus,

$$\frac{1}{2}\rho c_p U \delta_\theta \frac{T_w - T_\infty}{x} \sim k\frac{T_w - T_\infty}{\delta_\theta}, \tag{5.78}$$

from which we obtain

$$U \sim \frac{\alpha x}{\delta_\theta^2}. \tag{5.79}$$

Elimination of U between Eqs. (5.77) and (5.79), with the assumption[8] that the difference between δ and δ_θ is of secondary importance for heat transfer, yields

$$\frac{\alpha x}{\delta_\theta^2} \sim \frac{g}{\nu}\left(\frac{\Delta\rho}{\rho}\right)\delta_\theta^2$$

or

$$\frac{x}{\delta_\theta^4} \sim \frac{g}{\nu\alpha}\left(\frac{\Delta\rho}{\rho}\right)$$

or

$$\frac{x}{\delta_\theta} \sim Ra_x^{1/4}, \tag{5.80}$$

---

[8] Introduced by Squire [14].

Sec. 5.2 Laminar Natural Convection

where

$$Ra_x = \frac{g}{\nu\alpha}\left(\frac{\Delta\rho}{\rho}\right)x^3 \tag{5.81}$$

is the local **Rayleigh number**.

A measure of the heat transfer from the vertical plate [recall Eq. 5.25)] is

$$Nu_x \sim \frac{x}{\delta_\theta}, \tag{5.82}$$

which gives, in terms of Eq. (5.80),

$$Nu_x \sim Ra_x^{1/4}. \tag{5.83}$$

Next, we neglect the viscous force rather than the inertial force. The momentum balance for the control volume shown in Fig. 5.15 now gives [recall Eq. (5.75)]

$$x\delta g\Delta\rho \sim \frac{1}{4}\rho U^2 \delta \tag{5.84}$$

or

$$U^2 \sim g\left(\frac{\Delta\rho}{\rho}\right)x. \tag{5.85}$$

Equations (5.78) and (5.79) continue to hold for the thermal energy. Then, elimination of U between Eqs. (5.85) and (5.79) yields

$$g\left(\frac{\Delta\rho}{\rho}\right)x \sim \left(\frac{\alpha x}{\delta_\theta^2}\right)^2 \tag{5.86}$$

or

$$\frac{x}{\delta_\theta^4} \sim \frac{g}{\alpha^2}\left(\frac{\Delta\rho}{\rho}\right). \tag{5.87}$$

After multiplying and dividing the right-hand side of Eq. (5.87) by U and recalling the definition of Prandtl number ($Pr = \nu/\alpha$),

$$\frac{x}{\delta_\theta} \sim (Pr\, Ra_x)^{1/4} \tag{5.88}$$

and[9]

$$Nu_x \sim (Pr\, Ra_x)^{1/4}. \tag{5.89}$$

Now the implicit heat transfer relation,

$$Nu = f(\text{Motion}), \tag{1.62}$$

introduced in Chapter 1 becomes, in terms of Eq. (5.83) or (5.89),

$$Nu = f(\text{Buoyancy}).$$

---

[9] $Pr\, Ra_x$ is sometimes called the Boussinesq number, Bo_x. In terms of this number, Eq. (5.89) becomes $Nu_x \sim Bo_x^{1/4}$.

For the constant involved with Eqs. (5.83) and (5.89), actual velocity and temperature distributions are needed. In what follows is a first-order approximation based on linear profiles.

EXAMPLE 5.4

Consider the control volume shown in Fig. 5.16.

The momentum balance given by Eq.(5.44), noting $U_\infty \equiv 0$ and including the buoyancy term, now becomes

$$-\int_0^x \int_0^{\delta_\theta} g\left(\frac{\Delta\rho}{\rho}\right) dy\, dx = \int_0^{\delta_\theta} u^2\, dy + \nu \int_0^x \left(\frac{\partial u}{\partial y}\right)_{wall} dx. \qquad (5.90)$$

For small temperature differences, assume

$$\Delta\rho = \rho - \rho_\infty = \left(\frac{d\rho}{dT}\right)(T - T_\infty) = \left(\frac{d\rho}{dT}\right)\Delta T, \qquad (5.91)$$

which may be rearranged in terms of the coefficient of thermal expansion, $\beta = -(1/\rho)(d\rho/dT)$, as

$$\Delta\rho = -\beta\rho\Delta T. \qquad (5.92)$$

Therefore, Eq. (5.90) becomes

$$\int_0^x \int_0^{\delta_\theta} g\beta(T - T_\infty)\, dy\, dx = \int_0^{\delta_\theta} u^2\, dy + \nu \int_0^x \left(\frac{\partial u}{\partial y}\right)_{wall} dx. \qquad (5.93)$$

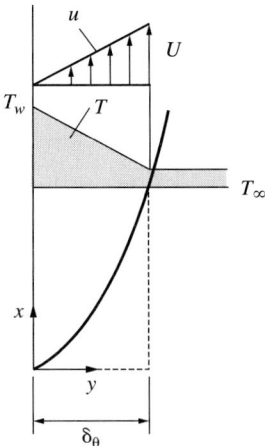

Figure 5.16
Schematic of the control volume for $Pr \geq 1$.

For $Pr \geq 1$ (which includes water, viscous oils, and approximately gases) the inertial effect is neglected. Then, after eliminating the common integrals in x, Eq. (5.93) is reduced to

$$\int_0^{\delta_\theta} g\beta(T - T_\infty)\, dy = \nu \left(\frac{\partial u}{\partial y}\right)_{wall}. \tag{5.94}$$

The first law applied to this control volume leading to Eq. (5.28) continues to hold in terms of δ_θ:

$$\int_0^{\delta_\theta} u(T - T_\infty)\, dy = -\alpha \int_0^x \left(\frac{\partial T}{\partial y}\right)_w dx. \tag{5.95}$$

With assumed linear profiles as shown in Fig. 5.16,

$$u = U\frac{y}{\delta_\theta} \tag{5.96}$$

and

$$\frac{T - T_\infty}{T_w - T_\infty} = 1 - \frac{y}{\delta_\theta}. \tag{5.97}$$

Eqs. (5.94) and (5.95) yield, in terms of δ_θ and U,

$$\delta_\theta^2 = \frac{2\nu U}{g\beta \Delta T}, \tag{5.98}$$

and

$$\frac{U\delta_\theta}{6} = \alpha \int_0^x \frac{dx}{\delta_\theta}. \tag{5.99}$$

Elimination of U between Eqs. (5.98) and (5.99) gives

$$\frac{g\beta \Delta T \delta_\theta^3}{12\alpha\nu} = \int_0^x \frac{dx}{\delta_\theta}, \tag{5.100}$$

or, after differentiating with respect to x,

$$\delta_\theta^3 \frac{d\delta_\theta}{dx} = 4\frac{\alpha\nu}{g\beta \Delta T} \tag{5.101}$$

which can be rearranged as

$$\frac{d\delta_\theta^4}{dx} = 16\frac{\alpha\nu}{g\beta \Delta T}. \tag{5.102}$$

Integration of this result readily gives

$$\delta_\theta = 2\left(\frac{\alpha\nu x}{g\beta \Delta T}\right)^{1/4}. \tag{5.103}$$

Therefore, the local Nusselt number is

$$Nu_x = \frac{x}{\delta_\theta} = 0.5 Ra_x^{1/4}. \tag{5.104}$$

264 Chap. 5 Foundations of Convection

For $Pr \ll 1$ (which includes liquid metals), the viscous force is neglected. Then, Eq. (5.93) is reduced to

$$\int_0^{\delta_\theta} \int_0^x g\beta(T - T_\infty)\, dx\, dy = \int_0^{\delta_\theta} u^2\, dy. \tag{5.105}$$

Also, the first law applied to this control volume leading to Eq. (5.28) continues to hold in terms of δ_θ:

$$\int_0^{\delta_\theta} u(T - T_\infty)\, dy = -\alpha \int_0^x \left(\frac{\partial T}{\partial y}\right)_w dx. \tag{5.106}$$

Because of the neglected viscosity, the velocity is no longer zero on the wall. Then, with a uniform velocity

$$u = U \tag{5.107}$$

and a linear temperature profile, as shown in Fig. 5.17,

$$\frac{T - T_\infty}{T_w - T_\infty} = 1 - \frac{y}{\delta_\theta}. \tag{5.108}$$

In terms of these profiles Eqs. (5.105) and (5.106) yield

$$\int_0^x \delta_\theta\, dx = \frac{2U^2 \delta_\theta}{g\beta \Delta T} \tag{5.109}$$

and

$$\frac{U\delta_\theta}{2} = \alpha \int_0^x \frac{dx}{\delta_\theta}, \tag{5.110}$$

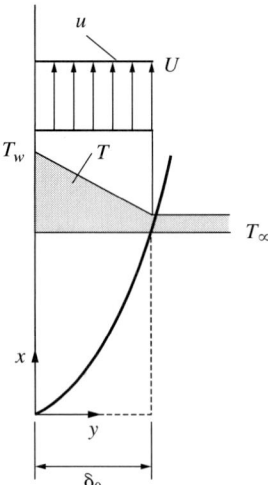

Figure 5.17 Schematic of the control volume for $Pr \ll 1$.

which are **nonlinear integrodifferential** equations and in general are difficult to solve. However, on dimensional grounds, Eq. (5.109) leads to

$$U \sim x^{1/2} \tag{5.111}$$

and Eq. (5.110) leads to

$$U\delta_\theta^2 \sim x \tag{5.112}$$

or, their combination, to

$$\delta_\theta \sim x^{1/4}. \tag{5.113}$$

Then, in terms of

$$U = Cx^{1/2} \text{ and } \delta_\theta = Dx^{1/4}, \tag{5.114}$$

Equations (5.109) and (5.110) yield, respectively,

$$C = \left(\frac{2}{5}g\beta\Delta T\right)^{1/2} \tag{5.115}$$

$$CD^2 = \frac{8}{3}\alpha, \tag{5.116}$$

and their ratio

$$D = \left(\frac{160\alpha^2}{9g\beta\Delta T}\right)^{1/4}. \tag{5.117}$$

Consequently,

$$\delta_\theta = 2.053\left(\frac{\alpha}{\nu}\frac{\alpha\nu x}{g\beta\Delta T}\right)^{1/4}, \tag{5.118}$$

and the local Nusselt number is

$$Nu_x = \frac{x}{\delta_\theta} = 0.487(Pr\,Ra_x)^{1/4}. \tag{5.119}$$

Table 5.4 shows the numerical constants resulting from the use of higher-order profiles. ◆

Table 5.4 Constants from various profiles

Constants	Linear	Squire's	Exact
$Nu_x\,Ra_x^{-1/4}$	0.500	0.508	0.503
$Nu_x(Pr\,Ra_x)^{-1/4}$	0.487	0.514	0.600

The concept of analogy between momentum and heat does not apply to natural convection. In forced convection, momentum is independent of thermal energy, and the temperature distribution may or may not be similar to the velocity distribution. In natural convection, momentum and thermal energy are coupled; although the velocity and temperature distributions are determined simultaneously, they are not similar.

So far we have learned the evaluation of heat transfer by analytical means and by the analogy between heat and momentum transfer. When an analytical solution is beyond our reach, or when there exists no analogy between momentum and heat, we rely on experimental measurements. Dimensional analysis provides an effective way of organizing experimental data. The next section is devoted to a review of the methods of dimensional analysis, arranged in a manner particularly suitable to heat transfer studies.

5.3 DIMENSIONAL ANALYSIS ○

When we have a complete understanding of the physics of a problem and have no difficulty with the formulation but are mathematically stuck on the solution, we refer to dimensional analysis for a functional (implicit) form of the solution. Three distinct methods exist for dimensional analysis:

1) Formulation (nondimensionalized): Whenever a formulation is readily available, a term-by-term nondimensionalization of this formulation leads directly to the related dimensionless numbers. The procedure is not suitable to problems which cannot be readily formulated.

2) Π-Theorem:[10] If a formulation is not readily accessible but *all* physical and geometric quantities which characterize a physical situation are *clearly* known, we write an implicit relation among these quantities,

$$f(Q_1, Q_2, \ldots, Q_n) = 0. \tag{5.120}$$

Expressing these quantities in terms of appropriate fundamental units, and making Eq. (5.120) independent of these fundamental units by an appropriate combination of Q's, yields the dimensionless numbers.

3) Physical similitude: Ratios established from the individual terms of the appropriate general principles (force balance, energy balance) give the physically relevant dimensionless numbers. The great convenience of this method is that there is no need to worry about an explicit formulation (required for the first method), except for a clear understanding of the terms comprising a general principle. Also, there is no need to go through the nondimensionalization process (required for the second method), since a ratio between any two terms of a general principle is automatically dimensionless.

Let us illustrate the application of the foregoing methods in terms of an illustrative example based on a simple oscillating pendulum of length ℓ and mass m (in a vacuum). Let ϕ_0 be the initial angle displacement. We wish to determine the period of this pendulum by dimensional analysis.

---

[10] Dimensionless numbers obtained by this method are usually called Π's.

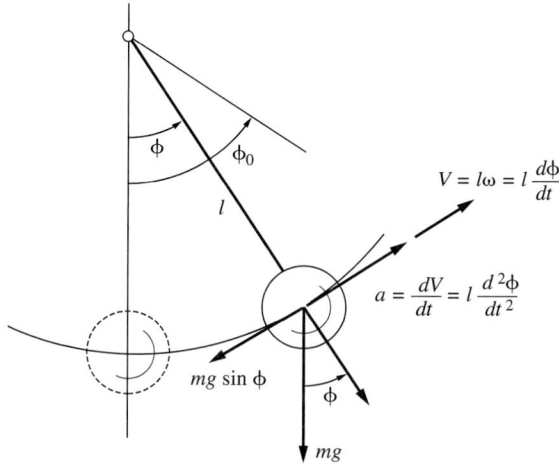

Figure 5.18 Simple pendulum.

Newton's law states that $F = ma$, where F is the weight of the pendulum and a its acceleration. The projections of F and a in the directions along and normal to ℓ are shown in Fig. 5.18. From the normal component of Newton's law of motion, we have the governing equation,

$$\frac{d^2\phi}{dt^2} + \frac{g}{\ell}\sin\phi = 0. \tag{5.121}$$

For the **first method**, we have, from the nondimensionalization of Eq. (5.121) in terms of period T,

$$\frac{d^2\phi}{d(t/T)^2} + \left(T^2\frac{g}{\ell}\right)\sin\phi = 0,$$

which suggests the functional (implicit) relationship

$$\phi = f\left(\frac{t}{T}, T^2\frac{g}{\ell}\right). \tag{5.122}$$

However, we are not interested in the instantaneous position ϕ (of the pendulum) but rather its extremum ϕ_0, for which t/T assumes integer values, $1, 2, 3, \ldots$. Consequently,

$$\phi_0 = f\left(T^2\frac{g}{\ell}\right).$$

Inverting this functional relationship, and expressing the result in terms of the period rather than its square, we have

$$T\sqrt{\frac{g}{\ell}} = f(\phi_0). \tag{5.123}$$

For the **second method** (the Π-theorem) we recall that the tangential momentum is balanced by the tangential component of the gravitational body force, and from the inspection of this balance we conclude that

$$T = f(m, g, \ell, \phi_0), \tag{5.124}$$

where m, g, ℓ, and ϕ_0 all are independent quantities. In terms of three fundamental units of mechanics [M], [L], [T],[11] Eq. (5.124) may be expressed as

$$[T] \equiv f\left[M, \frac{L}{T^2}, L, 0\right]. \tag{5.125}$$

Now we begin rearranging Eq. (5.124) in such a way that, with each arrangement, it becomes independent of one fundamental unit. First of all, the dimensional homogeneity in [M] suggests

$$T = f_1(g, \ell, \phi_0)$$

or, in terms of the fundamental units,

$$[T] \equiv f_1\left[\frac{L}{T^2}, L, 0\right].$$

Eliminating [L], for example, by the ratio g/ℓ yields

$$T = f_2\left(\frac{g}{\ell}, \phi_0\right)$$

or, in terms of the fundamental units,

$$[T] \equiv f_2\left[\frac{1}{T^2}, 0\right].$$

Finally, eliminating [T] by the product $T\sqrt{\frac{g}{\ell}}$ gives the dimensionless relation

$$T\sqrt{\frac{g}{\ell}} = f_3(\phi_0),$$

which is identical to Eq. (5.123). Note that the number of steps in the foregoing nondimensionalization procedure is equal to the number of fundamental units. Consequently, the number of dimensionless numbers is equal to the difference between the number of dimensional quantities in the original statement of a problem and the number of fundamental units. That is, Eq. (5.124) is in terms of 5 quantities, and, since there are 3 fundamental units, the result involves $5 - 3 = 2$ dimensionless numbers.

---

[11] Or [F], [L], [T].

For the **third method** (physical similitude) consider the tangential balance between the inertial and gravitational forces, $F_I \sim F_g$, or the ratio

$$F_I/F_g \sim 1. \tag{5.126}$$

From Fig. 5.18 the normal component of the gravitational force is $mg \sin \phi$, which, on dimensional grounds, indicates that $F_g \sim mg f_0(\phi_0)$, where $f_0(\phi_0)$ shows the angle dependence. The normal component of acceleration is $dV/dt = \ell(d^2\phi/dt^2)$. On dimensional grounds, the inertial force becomes $F_I = ma \sim m\ell(\phi_0/T^2)$. Then

$$\frac{F_I}{F_g} \sim \frac{\ell \phi_0/T^2}{g f_0(\phi_0)} \sim 1,$$

which leads to

$$T\sqrt{\frac{g}{\ell}} = f(\phi_0), \tag{5.123}$$

$f(\phi_0)$ replacing $[\phi_0/f_0(\phi_0)]^{1/2}$ for notational convenience. Thus, by three distinct methods we are able to show that the dimensionless period of a simple pendulum in a vacuum depends only on its initial displacement. Now, combining Eq. (5.123) with a simple experiment to be performed by one pendulum with a number of ϕ_0's (Fig. 5.19), we can determine the explicit form of Eq. (5.123).

As is well known, for small displacements, ϕ_0, from equilibrium, assuming $\sin \phi \cong \phi$, Eq. (5.121) is reduced to

$$\frac{d^2\phi}{dt^2} + \frac{g}{\ell}\phi = 0, \tag{5.127}$$

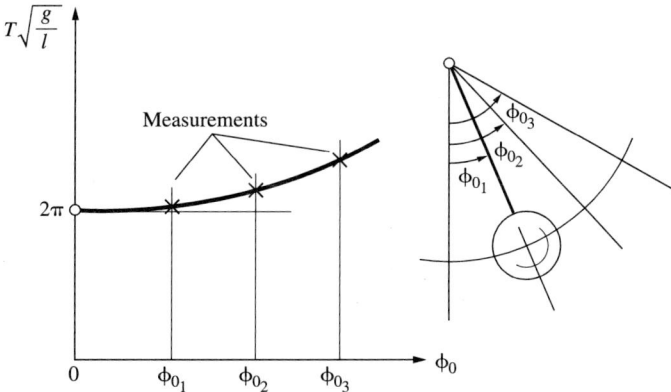

Figure 5.19 Experiments with a simple pendulum.

which characterizes a harmonic motion with angular frequency $\omega = \sqrt{g/\ell}$. Consequently, for small displacements,

$$T = \frac{2\pi}{\omega} = 2\pi\sqrt{\frac{\ell}{g}}, \tag{5.128}$$

which turns out to be independent of ϕ_0 because of the assumed small oscillations.

As we have seen with this pendulum example, dimensional analysis, when done properly, yields the correct dimensionless parameters that characterize a problem. Most technologically significant problems are much more complex than a simple pendulum problem and thus cannot be solved using simple mathematics. For such problems, dimensional analysis can yield the appropriate dimensionless parameters. Experiments can then be conducted to relate these parameters for a wide range of values and thus develop useful correlations that can be used in engineering problems. Our ultimate goal in the following chapter will be to determine values of the heat transfer coefficient h. With that in mind, we now proceed to examples incrementally more relevant to our convection studies, starting with an isothermal flow problem which will be useful for the enthalpy terms of our convection problems.

5.4 A FORCED FLOW ○

Let a solid sphere of diameter D be immersed and held stationary in an incompressible fluid streaming by steadily with a uniform velocity V (Fig. 5.20). The density and viscosity of the fluid are ρ and μ, respectively. The sphere is restrained from moving in any direction. We wish to determine the drag force F on the sphere.

Since the differential formulation of a viscous flow near a sphere is beyond the scope of this text, we proceed with the Π-theorem. In view of the fact that the drag force is balanced by the inertial and viscous forces, we assume

$$F = f(V, D, \rho, \mu), \tag{5.129}$$

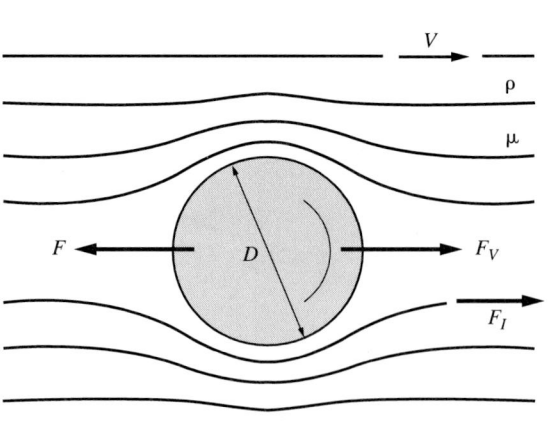

Figure 5.20 Forced flow over a sphere.

which may be expressed in terms of the fundamental units as

$$\left[\frac{ML}{T^2}\right] \equiv f\left[\frac{L}{T}, L, \frac{M}{L^3}, \frac{M}{LT}\right].$$

Now we begin rearranging Eq. (5.129) by making it independent of one fundamental unit at a time. Since the mass dependence is the simplest one, we begin with mass. To eliminate [M], we pick any one of the mass dependent terms on the right-hand side, ρ or μ. Let us pick μ, for example (later we comment on what would happen if we would have picked ρ instead), and combine it with F and ρ in such a way that the mass dependence disappears. Thus

$$\frac{F}{\mu} = f_1\left(V, D, \frac{\rho}{\mu}\right), \qquad (5.130)$$

which may be expressed in terms of the remaining fundamental units as

$$\left[\frac{L^2}{T}\right] \equiv f_1\left[\frac{L}{T}, L, \frac{T}{L^2}\right].$$

Clearly, the time dependence of Eq. (5.130) is simpler than its length dependence. To eliminate [T], we pick any one of the time dependent terms on the righthand side, V or ρ/μ. Since the final dimensionless numbers will ultimately involve all quantities describing Eq. (5.129), and since we have already manipulated with μ, let us pick V this time and combine it with F/μ and ρ/μ in such a way that the time dependence disappears. Thus

$$\frac{F}{\mu V} = f_2\left(D, \frac{\rho V}{\mu}\right), \qquad (5.131)$$

which, in terms of the length unit, may be expressed as

$$[L] \equiv f_2\left[L, \frac{1}{L}\right].$$

Finally, eliminating [L] by combining D with the other terms of Eq. (5.131), we get

$$\frac{F}{\mu V D} = f_3\left(\frac{\rho V D}{\mu}\right) = f_3(Re), \qquad (5.132)$$

where $Re = \rho V D/\mu$ is the **Reynolds number**.

Now, let us go back to Eq. (5.129) and this time make this equation independent of [M] by manipulating the mass-dependent terms with ρ rather than μ. This leads to

$$\frac{F}{\rho V^2 D^2} = f_4\left(\frac{\mu}{\rho V D}\right) = f_4(Re), \qquad (5.133)$$

Since dimensional analysis can provide only a functional (implicit) relationship between dimensionless numbers, Eqs. (5.132) and (5.133) are synonymous dimensionless results. That is, by suitable transformations, a dimensionless result can be made identical to another dimensionless result. However, Eq. (5.132), representing the ratio of drag force and the viscous force, would be most useful in flows dominated by viscous forces, while Eq. (5.133), representing the ratio of drag force and inertial force, would be most useful in flows dominated by inertial forces.

Dimensional analysis offers no clue as to which one of Eqs. (5.132) and (5.133) may be most convenient. Aside from the obvious fact that the dependent variable F should be included in only one dimensionless number, it is necessary to rely on past experience and physical insight in the selection of one of these relations.

Next, we proceed to the dimensional analysis of the same problem by the method of physical similitude. Since the force F on the sphere exerted by the moving fluid is balanced by the inertial and viscous forces,

$$F = f(F_I, F_V),$$

from which we may establish ratios

$$\frac{F}{F_V} \sim \frac{F}{D^2 \mu (V/D)} = \frac{F}{\mu V D}, \tag{5.134}$$

$$\frac{F_I}{F_V} \sim \frac{\rho D^3 (V^2/D)}{D^2 \mu (V/D)} = \frac{\rho V D}{\mu} = Re, \tag{5.135}$$

and

$$\frac{F}{F_I} \sim \frac{F}{\rho D^3 (V^2/D)} = \frac{F}{\rho V^2 D^2}. \tag{5.136}$$

Clearly, from $F_I = m(dV/dt)$ we have $F_I \sim mV/t$ and (in view of $m \sim \rho D^3$ and $t \sim D/V$) get $F_I \sim \rho D^3 (V^2/D) = \rho V^2 D^2$. Also, for a Newtonian fluid, from $\tau = \mu(dU/dy)$ we have $F_V \sim \tau A \sim D^2 \mu(V/D) = \mu V D$. Equations (5.134) and (5.135) are the dimensionless numbers associated with Eq. (5.132), and Eqs. (5.135) and (5.136) are those associated with Eq. (5.133).

One advantage of using physical similitude is that, working with the forces involved in a problem, one does not have to worry about a long list of relevant properties. In developing the list for the Π-theorem, one could easily overlook and exclude certain properties which prevent a successful completion of dimensional arguments. The physical similitude becomes also quite useful for experiments to be conducted with scaled models rather than the actual prototype. Physical similitude is said to exist between two systems if the corresponding dimensionless numbers have the same value. Geometric similitude is a prerequisite for physical similitude. Further elaborations on similitude, however, belong to texts on fluid mechanics.

Having learned the dimensionless numbers associated with forced flows, we proceed next to the dimensionless numbers associated with buoyancy driven flows.

5.5 A FREE FALL ○

Consider the solid sphere of the preceding example. Let the sphere now fall, under the effect of gravity, in a fluid of density ρ and viscosity μ (Fig. 5.21). The difference between the density of the sphere and that of the fluid is $\Delta\rho$. We wish to determine the terminal velocity of the sphere.

In a manner similar to the preceding problem, we begin with the Π-theorem. First, replacing F of Eq. (5.129) with the buoyant force per unit volume $g\Delta\rho$, we write

$$g\Delta\rho = f(V, D, \rho, \mu),$$

and, because V now is the dependent variable, rearrange for V,

$$V = f(g\Delta\rho, D, \rho, \mu). \tag{5.137}$$

In terms of the fundamental units, Eq. (5.147) is equivalent to

$$\left[\frac{L}{T}\right] \equiv f\left[\frac{M}{L^2T^2}, L, \frac{M}{L^3}, \frac{M}{LT}\right].$$

Next, we begin rearranging Eq. (5.137) so that it becomes suitable to the elimination of one fundamental unit at a time. To eliminate [M], we pick ρ, for example, and combine it with $g\Delta\rho$ and μ in such a way that the mass dependence disappears. Thus

$$V = f_1\left(g\frac{\Delta\rho}{\rho}, D, \frac{\mu}{\rho}\right), \tag{5.138}$$

which, in terms of the remaining fundamental units, is equivalent to

$$\left[\frac{L}{T}\right] \equiv f_1\left[\frac{L}{T^2}, L, \frac{L^2}{T}\right].$$

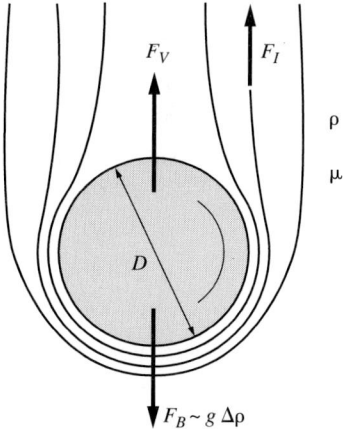

Figure 5.21 Free fall of a sphere.

To eliminate [T] we pick $\mu/\rho = \nu$, for example, and combine it with V and $g(\Delta\rho/\rho)$ in such a way that the time dependence disappears. Thus

$$\frac{\rho V}{\mu} = f_2\left(\frac{g}{\nu^2}(\frac{\Delta\rho}{\rho}), D\right), \qquad (5.139)$$

which, in terms of the length unit, is equivalent to

$$\left[\frac{1}{L}\right] \equiv f_2\left[\frac{1}{L^3}, L\right].$$

Finally, eliminating [L] by combining D with the other terms of Eq. (5.139), we get

$$\frac{\rho V D}{\mu} = f_3\left(\frac{g}{\nu^2}(\frac{\Delta\rho}{\rho})D^3\right), \qquad (5.140)$$

where

$$\frac{g}{\nu^2}\left(\frac{\Delta\rho}{\rho}\right)D^3 = Gr \qquad (5.141)$$

is the **Grashof number**. Thus, the terminal velocity of a buoyancy-driven body is found to be governed by the dimensionless relation

$$Re = f(Gr). \qquad (5.142)$$

Next, we proceed to the dimensional analysis of the same problem by the method of physical similitude. For the sphere, the buoyant force is balanced by the viscous force,[12] and for the fluid, the viscous force is balanced by the inertial force,

$$\begin{array}{c|c} \text{Sphere} & \text{Fluid} \\ F_B \sim F_V & \sim F_I \\ | & \end{array}$$

which lead to the following force ratios:

$$\frac{F_I}{F_V} \sim \frac{\rho V D}{\mu} = Re, \qquad (5.135)$$

$$\frac{F_B}{F_V} \sim \frac{g\Delta\rho D^3}{\mu(V/D)D^2} = \frac{g\Delta\rho D^2}{\mu V}. \qquad (5.143)$$

Now, we refer to the general fact that the physics of any problem may be described by one dimensionless number for the unknown quantity depending on other dimensionless number(s) composed only of independent quantities. For the present problem, velocity is unknown and

$$\text{Velocity} = f(\text{Buoyancy}).$$

---

[12] At the terminal velocity, the acceleration of the sphere is zero.

In the final form of this relation we have only one V to appear in one dimensionless number. Velocity is nondimensionalized with Re, and with the following combination, obtained from the product of Eqs. (5.135) and (5.143), buoyancy is nondimensionalized with

$$\frac{F_B}{F_V} \times \frac{F_I}{F_V} \sim \frac{g \Delta \rho D^2}{\mu V} \times \frac{\rho V D}{\mu} = \frac{g}{\nu^2} \left(\frac{\Delta \rho}{\rho}\right) D^3 = Gr,$$

which is independent of velocity. Thus the physical similitude leads us to a relation among force ratios

$$\frac{F_I}{F_V} = f \left(\frac{F_B}{F_V} \times \frac{F_I}{F_V} \right), \tag{5.144}$$

which is identical to Eq. (5.142), the result already obtained by employing the Π-theorem.

Now we are ready for the dimensional analysis of convection problems. We begin with forced convection because of its relative simplicity.

5.6 FORCED CONVECTION

In this chapter we have already learned in terms of Fig. 5.1 that the Nusselt number is the dimensionless wall gradient of fluid temperature. Ignoring the method of nondimensionalized governing equations because of its complexity, we proceed with the Π-theorem.

Consider, for example, a steady two-dimensional problem with temperature distribution

$$\frac{T_w - T}{T_w - T_\infty} = f(x, y, \underbrace{\dot{Q}_H}_{c_p(V, \rho, \mu)}, \underbrace{\dot{Q}_K}_{k}),$$

$\dot{Q}_H$ being the enthalpy flow and $\dot{Q}_K$ the conduction. Then

$$\frac{T_w - T}{T_w - T_\infty} = f(x, y, V, \rho, \mu, c_p, k).$$

The wall gradient of this temperature gives the **local heat transfer coefficient**,

$$h_x = f(x, V, \rho, \mu, c_p, k),$$

whose average over distance D (or ℓ) gives the (**average**) **heat transfer coefficient**,

$$h = f(D, V, \rho, \mu, c_p, k), \tag{5.145}$$

where the righthand side is composed only of independent quantities. For thermal problems, a fourth fundamental unit is needed in addition to the three fundamental units of mechanics. This unit is usually assumed to be temperature $[\theta]$.[13] In terms of the four fundamental units, Eq. (5.145) may be expressed as

$$\left[\frac{M}{T^3\theta}\right] \equiv f\left[L, \frac{L}{T}, \frac{M}{L^3}, \frac{M}{LT}, \frac{L^2}{T^2\theta}, \frac{ML}{T^3\theta}\right].$$

(For the units of c_p, recall the definition of stagnation enthalpy,[14] $\hat{h}^0 = \hat{h} + V^2/2 + gz$, and use $\hat{h} \sim c_p T \sim V^2/2$, which gives $c_p[\theta] \sim [L^2/T^2]$; for the units of h, use $q \sim$ Power/Area = Force × Velocity/Area, which yields $h[\theta] \sim [ML/T^2][L/T]/[L^2]$; and for the units of k, note that $h[L] \sim k$.) We proceed now to the dimensionless numbers associated with forced convection by successively eliminating the fundamental units from Eq. (5.145).

Again we begin the elimination process by mass. Combining ρ, for example, with other mass-dependent quantities, we have from Eq. (5.145)

$$\frac{h}{\rho} = f_1\left(D, V, \nu, c_p, \frac{k}{\rho}\right), \qquad (5.146)$$

where $\nu = \mu/\rho$ is the kinematic viscosity (or momentum diffusivity). In terms of the fundamental units, Eq. (5.146) is equivalent to

$$\left[\frac{L^3}{T^3\theta}\right] \equiv f_1\left[L, \frac{L}{T}, \frac{L^2}{T}, \frac{L^2}{T^2\theta}, \frac{L^4}{T^3\theta}\right].$$

Now the temperature dependence appears to be the simplest one. Elimination of $[\theta]$ from Eq. (5.146) by combining c_p, for example, with other temperature-dependent quantities yields

$$\frac{h}{\rho c_p} = f_2(D, V, \nu, \alpha), \qquad (5.147)$$

where $\alpha = k/\rho c_p$ is the thermal diffusivity. Equation (5.147) is equivalent to

$$\left[\frac{L}{T}\right] \equiv f_2\left[L, \frac{L}{T}, \frac{L^2}{T}, \frac{L^2}{T}\right].$$

Time dependence of Eq. (5.147) is somewhat easier than its length dependence. Eliminating [T] by combining V, for example, with other time-dependent quantities[15] gives

$$\frac{h}{\rho c_p V} = f_3\left(D, \frac{V}{\nu}, \frac{V}{\alpha}\right), \qquad (5.148)$$

---

[13] Or heat flux $[\dot{Q}]$.

[14] $\hat{h}$, used for enthalpy, should not be confused with heat transfer coefficient h.

[15] Also [T] may be eliminated with either α or ν.

which is equivalent to

$$[0] \equiv f_3\left[L, \frac{1}{L}, \frac{1}{L}\right],$$

where 0 corresponds to a dimensionless quantity. Finally, eliminating [L], we have

$$\frac{h}{\rho c_p V} = f_4\left(\frac{VD}{\nu}, \frac{VD}{\alpha}\right)$$

or

$$St = f_4(Re, Pe), \tag{5.149}$$

where $St = h/\rho c_p V$ is the **Stanton number** and $Pe = VD/\alpha$ is the **Peclet number** (or thermal Reynolds number). Noting that

$$Pe = Re Pr,$$

$$St = Nu/Pe = Nu/Re Pr,$$

where $Pr = \nu/\alpha$ is the Prandtl number, and $Nu = hD/k$ is the Nusselt number, Eq. (5.149) may be written alternatively as

$$\boxed{Nu = f(Re, Pr)} \tag{5.150}$$

where Re is the only dimensionless number involving the effect of velocity. This is the form used most frequently. The following arguments will clarify the fundamental reason for using Pr rather than Pe.

Next, we proceed to dimensional analysis of the same problem by the method of physical similitude. From the definition of

$$\text{Convection} = \text{Conduction in} \underbrace{\underbrace{\text{moving media}}_{\text{Inertial, Viscous forces}}}_{\dot{Q}_H},$$

we have

$$\dot{Q}_C = f(F_I, F_V, \ldots \dot{Q}_H, \ldots \dot{Q}_K), \tag{5.151}$$

where $\dot{Q}_C$ denotes convection, $\dot{Q}_H$ enthalpy flow, $\dot{Q}_K$ conduction, F_I inertial force, and F_V viscous force. Next, we establish the ratios

$$\frac{\dot{Q}_C}{\dot{Q}_K} \sim \frac{hD^2\theta}{kD^2(\theta/D)} = \frac{hD}{k} = Nu, \tag{5.152}$$

$$\frac{F_I}{F_V} \sim \frac{VD}{\nu} = Re, \tag{5.136}$$

$$\frac{\dot{Q}_H}{\dot{Q}_K} \sim \frac{\rho c_p VD^2\theta}{kD^2(\theta/D)} = \frac{VD}{\alpha} = Pe, \tag{5.153}$$

where Nu is the heat transfer coefficient nondimensionalized relative to conduction and Re and Pe are obtained from the nondimensionalization of momentum and thermal energy, respectively. For an incompressible flow, momentum is decoupled from energy and is characterized by Re. However, energy is coupled to momentum through enthalpy flow, and the Pe number is silent to this coupling. Eliminating the velocity between Eqs. (5.135) and (5.153), we obtain

$$\frac{\dot{Q}_H}{\dot{Q}_K} \times \frac{F_V}{F_I} \sim \frac{\nu}{\alpha} = Pr, \qquad (5.154)$$

which describes the coupling of energy to momentum. Also, Pr characterizes the diffusion of momentum relative to that of heat and is the only dimensionless number in terms of physical properties.

Experimental data associated with gases, water, and viscous oils may be correlated with Eq. (5.150) as shown in Fig. 5.22(a), where Re_c denotes the critical Reynolds number at which the laminar flow is unstable. Beyond Re_c the forced convection eventually becomes turbulent. Equation (5.150) does not correlate the liquid metal data. For liquid metals, viscous forces are small, the momentum equation degenerates to a limit of uniform velocity, and the importance of the Reynolds number diminishes. Consequently, as shown in Fig. 5.22(b),

$$\boxed{Nu = f(Pe)} \qquad (5.155)$$

correlates the data on liquid metals [recall Eq. 5.26)].

Having learned the dimensionless numbers associated with forced convection we now proceed to those for natural convection.

5.7 NATURAL CONVECTION

So far we have learned that the correlation of forced convection, begins with

$$h = f(D, V, \rho, \mu, c_p, k), \qquad (5.145)$$

where the righthand terms are made up only of independent quantities. Since the velocity of natural convection depends on buoyancy and is not a "given" parameter, we may utilize Eq. (5.145) for natural convection after replacing V with a buoyancy term.

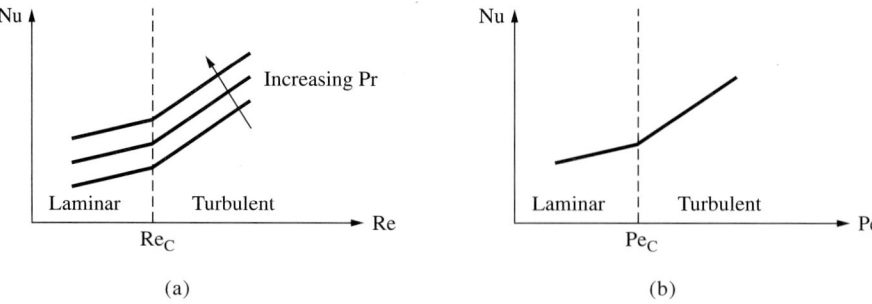

Figure 5.22 Correlation of forced convection data on (a) gases, water, and viscous oils, (b) liquid metals.

The buoyancy force/volume is

$$g\Delta\rho.$$

For small temperature differences, assume

$$\Delta\rho = \left(\frac{d\rho}{dT}\right)\Delta T,$$

which may be rearranged in terms of the coefficient of thermal expansion, $\beta = -(1/\rho)(d\rho/dT)$, as

$$\Delta\rho \sim \beta\rho\Delta T.$$

Consequently, the buoyancy force/volume becomes

$$g\beta\rho\Delta T.$$

Noting that Eq. (5.145) already involves ρ because of the inertial force, and now replacing V of this equation with $g\beta\Delta T$, we have

$$h = f(D, g\beta\Delta T, \rho, \mu, c_p, k). \tag{5.156}$$

Consider first the Π-theorem and, accordingly, express Eq. (5.156) in terms of the four fundamental units as

$$\left[\frac{M}{T^3\theta}\right] \equiv f\left[L, \frac{L}{T^2}, \frac{M}{L^3}, \frac{M}{LT}, \frac{L^2}{T^2\theta}, \frac{ML}{T^3\theta}\right]$$

and begin the elimination process with the mass-dependent terms. We have already used μ (in connection with flow around a sphere) and ρ (in connection with forced convection) for this elimination. Let us see what happens when we eliminate [M] by combining k with other mass-dependent quantities:

$$\frac{h}{k} = f_1\left(D, g\beta\Delta T, \frac{\rho}{k}, \frac{\mu}{k}, c_p\right), \tag{5.157}$$

which, in terms of fundamental units, is equivalent to

$$\left[\frac{1}{L}\right] \equiv f_1\left[L, \frac{L}{T^2}, \frac{T^3\theta}{L^4}, \frac{T^2\theta}{L^2}, \frac{L^2}{T^2\theta}\right].$$

At this stage the temperature dependence appears to be the simplest one. Elimination of $[\theta]$ from Eq. (5.157) by combining c_p, for example, with other temperature-dependent quantities yields

$$\frac{h}{k} = f_2(D, g\beta\Delta T, \alpha, Pr), \qquad (5.158)$$

where $\alpha = k/\rho c_p$ is the thermal diffusivity and $Pr = \mu c_p/k = \nu/\alpha$ is the Prandtl number. Equation (5.158) is equivalent to

$$\left[\frac{1}{L}\right] \equiv f_2\left[L, \frac{L}{T^2}, \frac{L^2}{T}, 0\right].$$

Eliminating the time dependence from Eq. (5.158) by combining, for example, α with $g\beta\Delta T$, we get

$$\frac{h}{k} = f_3\left(D, \frac{g\beta\Delta T}{\alpha^2}, Pr\right), \qquad (5.159)$$

which is equivalent to

$$\left[\frac{1}{L}\right] \equiv f_3\left[L, \frac{1}{L^3}, 0\right].$$

Finally, eliminating $[L]$ yields

$$\frac{hD}{k} = f_4\left(\frac{g\beta\Delta T D^3}{\alpha^2}, Pr\right)$$

or

$$Nu = f(\Pi, Pr), \qquad (5.160)$$

where $\Pi = g\beta\Delta T D^3/\alpha^2$ is a dimensionless number. Noting that

$$\frac{\Pi}{Pr} = \frac{g\beta\Delta T D^3}{\nu\alpha} = Ra, \qquad (5.161)$$

Eq. (5.160) may be rearranged in terms of Ra as

$$\boxed{Nu = f(Ra, Pr)} \qquad (5.162)$$

where Ra is the **Rayleigh number**. The following arguments will clarify the fundamental reason for using Ra rather than Π.

Next, we proceed to the dimensional analysis of the same problem by the method of physical similitude. The fluid motion now involves inertial, viscous, and buoyant forces (F_I, F_V, F_B). From the definition of

$$\text{Natural Convection} = \text{Conduction in } \underbrace{\text{moving media}}_{\text{Buoyant, Inertial, Viscous forces}}$$

$$\underbrace{\phantom{\text{Natural Convection} = \text{Conduction in moving media}}}_{\dot{Q}_H}$$

we have
$$\dot{Q}_C = f(F_B, F_I, F_V, \dot{Q}_H, \dot{Q}_K).$$

Next we establish the following ratios:

$$\frac{\dot{Q}_C}{\dot{Q}_K} \sim \frac{hD}{k} = Nu, \tag{5.152}$$

$$\frac{F_I}{F_V} \sim \frac{VD}{\nu} = Re, \tag{5.135}$$

$$\frac{F_B}{F_I} \sim \frac{mg\beta\Delta T}{mV^2/D} = \frac{g\beta\Delta T D}{V^2}, \tag{5.163}$$

$$\frac{\dot{Q}_H}{\dot{Q}_K} \sim \frac{VD}{\alpha} = Pe. \tag{5.153}$$

Here, we recall that dimensionless numbers are composed only of independent quantities, and note that V is *not* an independent quantity for natural convection. Consequently, Eqs. (5.135), (5.153), and (5.163) may describe natural convection after they are combined and made independent of V. For example, the following combination of Eqs. (5.153) and (5.135),

$$\frac{F_B}{F_V} \times \frac{\dot{Q}_H}{\dot{Q}_K} \sim \frac{(g\beta\Delta T)D^2}{\nu V} \times \frac{VD}{\alpha} = \frac{g}{\nu\alpha}(\beta\Delta T)D^3 = Ra, \tag{5.164}$$

which is independent of velocity, shows how the momentum is coupled to thermal energy through its buoyancy. Also, the following combination of Eqs. (5.135) and (5.153),

$$\frac{\dot{Q}_H}{\dot{Q}_K} \times \frac{F_V}{F_I} \sim \frac{VD}{\alpha} \times \frac{\nu}{VD} = \frac{\nu}{\alpha} = Pr, \tag{5.154}$$

which also is independent of velocity, shows how the thermal energy is coupled to momentum through its enthalpy flow.

For $Pr > 1$, the inertial effect is negligible and Eq. (5.162) is reduced to

$$Nu = f(Ra). \tag{5.165}$$

Accordingly, the experimental data on gases, water, and viscous oils are correlated with Eq. (5.165) as shown in Fig. 5.23(a).

For $Pr \ll 1$ the viscous effect is negligible, and Eq. (5.160) is reduced to

$$Nu = f(\Pi), \quad \Pi = Ra Pr. \tag{5.166}$$

Experimental data on liquid metals are correlated with Eq. (5.166) as shown in Fig. 5.23(b).

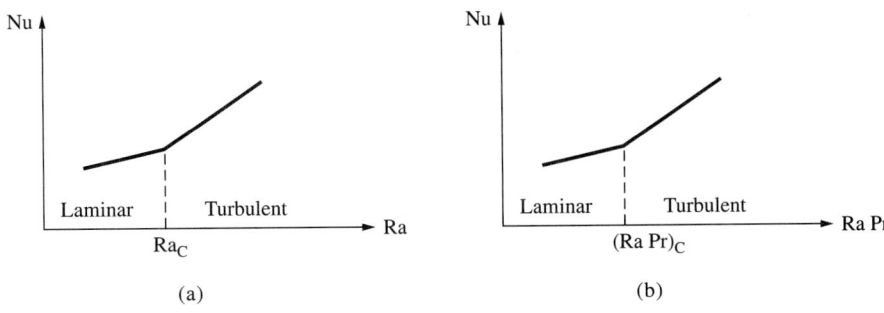

Figure 5.23 Correlation of natural convection data on (a) gases, water, and viscous oils, (b) liquid metals.

It is important to note here the conceptual difference between the Reynolds number of forced convection and the Rayleigh number of natural convection. Re results from the nondimensionalized momentum (of forced convection) which is uncoupled from thermal energy of incompressible (and constant property) fluids. On the other hand, Ra characterizes the coupling (through buoyancy) of momentum to energy.[16]

Having learned the functional (implicit) relation among the dimensionless numbers of forced convection and of natural convection, we proceed to Chapter 6 for explicit relations among these numbers.

■ REFERENCES

5.1 V. S. Arpacı, *Conduction Heat Transfer*. Addison-Wesley, Reading, MA, 1966.

5.2 V. S. Arpacı, and P. S. Larsen, *Convection Heat Transfer*. Prentice-Hall, New York, 1984.

5.3 V. S. Arpacı, *Microscales of Turbulence-Heat and Mass Transfer Correlations*. Gordon and Breach Science Publishers, Amsterdam, The Netherlands, 1997.

5.4 V. S. Arpacı, and S.-H. Kao, "Thermocapillary driven turbulent heat transfer," *ASME J. Heat Transfer*, 120, 214–219, 1998.

5.5 V. S. Arpacı, and S.-H. Kao, "Microscales of rotating turbulent flows," *Int. J. Heat Mass Transfer*, 40, 3819–3826, 1997.

5.6 V. S. Arpacı, "Buoyant turbulent flow driven by internal energy generation," *Int. J. Heat Mass Transfer*, 38, 2761–2770, 1995.

5.7 V. S. Arpacı, "Microscales of turbulence and heat transfer correlations," *Int. J. Heat Mass Transfer*, 29, 1071, 1986.

5.8 A. Bejan, *Convection Heat Transfer*. Wiley, New York, 1984.

5.9 A. Bejan, *Heat Transfer*. Wiley, New York, 1993.

5.10 J. C. Hunsaker and B. G. Rightmire, 1947, *Engineering Applications of Fluid Mechanics*, McGraw-Hill, New York, 1947.

5.11 F. Kreith, and M. S. Bohn, *Principles of Heat Transfer*, 4th ed. Harper & Row, New York, 1986.

---

[16] Grashof number, $Gr = g\beta \Delta T D^3/\nu^2$, results from the nondimensionalized momentum only; it alone cannot characterize and should not be used for natural convection resulting from a temperature difference.

5.12 H. L. Langhaar, *Dimensional Analysis and Theory of Models*, Wiley, New York, 1951.

5.13 L. I. Sedov, *Similarity and Dimensional Methods in Mechanics*. Academic Press, New York, 1959.

5.14 H. B. Squire, "Free convection from a heated vertical plate," *Modern Developments in Fluid Mechanics*, ed. Goldstein, vol. 2, p. 638. Oxford, 1939.

■ EXERCISES

5.1 For forced convection, consider the hybrid (differential in x and integral in y) control volume shown in Fig. 5P–1. Write the conservation of mass, the balance of momentum in terms of the momentum boundary-layer thickness δ and the conservation of thermal energy in terms of the thermal boundary-layer thickness δ_θ. Compare the results with Eqs. (5.44) and (5.28).

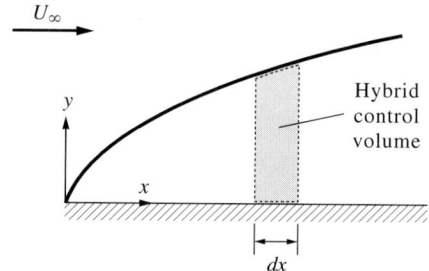

Figure 5P–1

5.2 For forced convection of liquid metals ($Pr \ll 1$) over a horizontal flat plate subject to a uniform heat flux q_w, evaluate the Nusselt number based on a uniform velocity and linear temperature profiles (Fig. 5P–2).

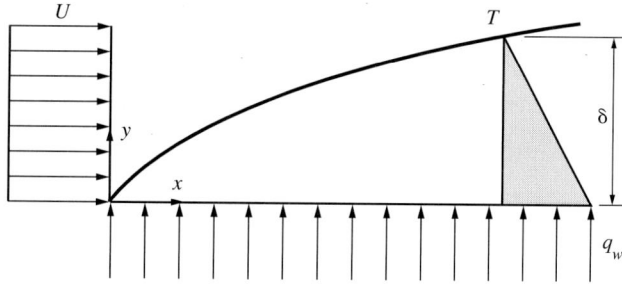

Figure 5P–2

5.3 For forced convection of any fluid having $Pr \geq 1$ over a horizontal flat plate subject to a uniform heat flux q_w, evaluate the Nusselt number based on linear velocity and temperature profiles (Fig. 5P–3).

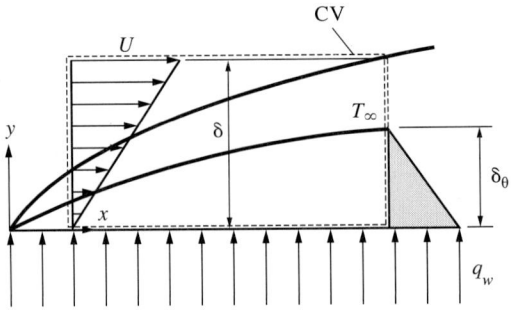

Figure 5P–3

5.4 For natural convection, consider the hybrid (differential in x and integral in y) control volume shown in Fig. 5P–4. Write the conservation of mass, the balance of momentum in terms of δ, and the conservation of thermal energy in terms of δ_θ. Compare the results with Eqs. (5.90) and (5.95).

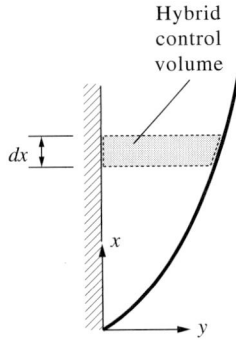

Figure 5P–4

5.5 In terms of a linear velocity profile, we have already obtained the constant involved with Eq. (5.40) to be $C = 0.289$ (see Table 5.2). Also, the exact relationship is known to be

$$\frac{1}{2}f = \frac{0.332}{Re_x^{1/2}}.$$

For both cases, evaluate the mean friction coefficient by averaging the local coefficient over a length ℓ.

5.6 In terms of a linear temperature profile, we have already obtained the constant involved with Eq. (5.58) to be $C = 0.289$ (see Table 5.3). Also, the exact relation is known to be

$$Nu_x = 0.332 Re_x^{1/2} Pr^{1/3}.$$

For both cases, evaluate the mean Nusselt number by averaging the local Nusselt number over a length ℓ.

5.7 Find the local Nusselt number for free convection of a vertical plate subjected to a constant wall heat flux by using linear profiles for cases $Pr > 1$ (Fig. 5P–5). Compare the result with Eq. (5.104).

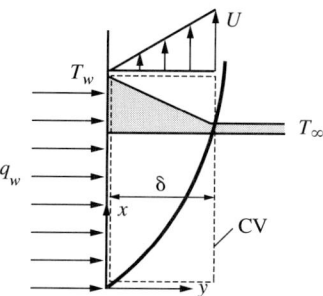

Figure 5P–5

5.8 Find the local Nusselt number for free convection of a vertical plate subjected to a constant wall heat flux by using linear profiles for cases $Pr < 1$ (Fig. 5P–6). Compare the result with Eq. (5.119).

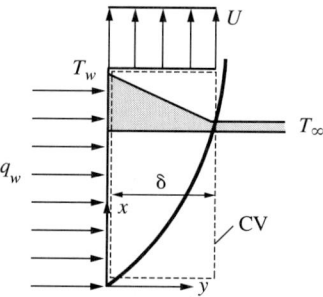

Figure 5P–6

5.9 Consider the natural convection for $Pr > 1$ from a vertical plate at a temperature T_w in an ambient at temperature T_∞. Evaluate the local heat transfer for the following three cases. **(a)** linear temperature, parabolic velocity profiles, **(b)** parabolic temperature, linear velocity profiles, **(c)** parabolic temperature and velocity profiles. Compare the results with Eq. (5.104).

5.10 Assuming a parabolic temperature and cubic velocity for natural convection from a vertical plate at temperature T_w in an ambient air temperature T_∞, show that for any Prandtl number

$$Nu_x = 0.508 \left(\frac{Pr}{0.952 + Pr} \right)^{1/4} Ra_x^{1/4}$$

or

$$Nu_x = 0.508 \Pi_x^{1/4}, \quad \Pi_x = \frac{Ra_x}{1 + 0.952/Pr}.$$

5.11 Repeat Prob. 5.10 for the limit $Pr \to 0$ and $Pr \to \infty$.

5.12 Determine the numerical constant obtained from averaging the local friction and heat transfer coefficients over a longitudinal length, say ℓ. Use only the exact solution given by Tables 5.1, 5.2, 5.3, and 5.4.

5.13 From $F_B \sim F_I + F_V$, $Q_H \sim Q_K$. Show that

$$\Pi_N = \frac{Ra}{1 + 1/Pr}, \quad Nu = f(\Pi_N).$$

5.14 Repeat Prob. 5.13 for the limit $Pr \to 0$ and $Pr \to \infty$.

5.15 Reconsider the problem stated in Section 5.4. Including the effect of gravity g and surface tension σ (force/length), now assume

$$F = f(V, D, \rho, \mu, g, \sigma).$$

Show by the Π-theorem and by physical similitude that the drag force may be nondimensionalized to give

$$\frac{F}{\rho V^2 D^2} = f(Re, Fr, We),$$

where $Fr = V/\sqrt{gD}$ is the **Froude number** and $We = \rho V^2 D/\sigma$ is the **Weber number**. Clearly explain the physics characterized by these numbers.

5.16 Adding the effect of surface tension σ to Eq. (5.137), extend the problem stated in Section 5.5 to a gas bubble rising in a liquid. For the terminal velocity of the bubble, assume

$$V = f(g\Delta\rho, D, \rho, \mu, \sigma).$$

Show by the Π-theorem and by physical similitude that the velocity may be nondimensionalized to give

$$Re = f(Gr, Bo),$$

where $Bo = g\Delta\rho D^2/\sigma$ is the **Bond number**.

5.17 Express in terms of the appropriate dimensionless numbers the diameter of droplets formed by a liquid discharging with a specified velocity (or under the effect of a pressure gradient) from a horizontal tube.

5.18 Express in terms of the appropriate dimensionless numbers the diameter of droplets formed by a liquid discharging under the effect of gravity from a vertical tube.

5.19 Consider the flow between two coaxial cylinders in relative rotation (Fig. 5P-7). Write a dimensionless relation between torque and angular frequency.

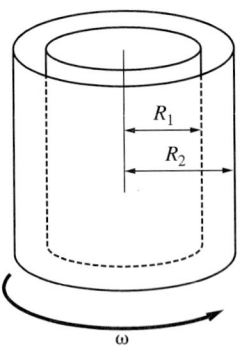

Figure 5P–7

5.20 A wooden sphere is held by a string in a water stream (Fig. 5P–8). Determine the string force by means of a dimensional analysis.

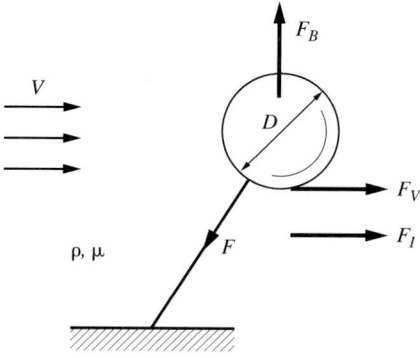

Figure 5P–8

5.21 What are the dimensionless numbers of combined forced-natural convection?

5.22 Consider the natural convection from a horizontal cylinder rotating with an angular frequency ω (Fig. 5P–9). The peripheral surface temperature of the cylinder is T_w and the ambient temperature is T_∞. The diameter of the cylinder is D. Assuming that the natural convection resulting from rotation and that from gravity can be superimposed, express the Nusselt number in terms of the appropriate dimensionless numbers.

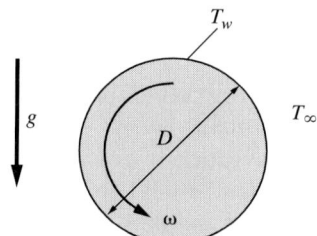

Figure 5P–9

5.23 Discuss the physics of a hot air balloon in terms of the appropriate dimensionless numbers.

CHAPTER 6

CORRELATIONS FOR CONVECTION

In Chapter 5, we learned the foundations of convection. Integrating the governing equations for laminar boundary layers, we obtained expressions for the heat transfer associated with forced convection over a horizontal plate and natural convection about a vertical plate. We also found analytically, as well as by the analogy between heat and momentum, that the thermal and momentum characteristics of laminar flow over a flat plate are related by

$$\boxed{St\, Pr^{2/3} = \frac{1}{2} f}, \tag{5.63}$$

provided that any pressure gradient in momentum and dissipation in energy are negligible. Actually, Eq. (5.63) is a fundamental relation independent of flow conditions, holding for turbulent as well as laminar flow. However, its validity for turbulent flow cannot be shown within the scope of this text. Equation (5.63) is instrumental in providing indirect information on heat transfer from (an analytical or experimental) knowledge on friction. Consequently, the first section of this chapter is devoted to the friction involved with two classes of frequently encountered problems: flow through pipes and flow over submerged bodies. Pipe flow is usually characterized by pressure drop or its dimensionless form, the **friction coefficient**, while external flow is usually characterized by drag force or its dimensionless form, the **drag coefficient**. The subsequent sections are devoted to heat transfer correlations for forced and natural convection.

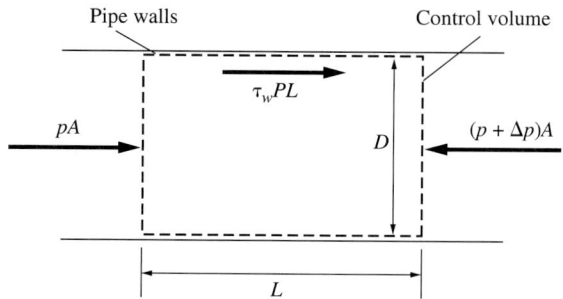

Figure 6.1 Steady, fully developed momentum for lumped control volume.

6.1 FRICTION FACTOR, DRAG COEFFICIENT ○

First consider steady, fully developed pipe flow. The conservation of momentum for the control volume shown in Fig. 6.1, noting that momentum flow $\rho A V^2$ does not change in fully developed flow, gives

$$-\Delta p A + \tau_w P L = 0, \qquad (6.1)$$

where Δp is the pressure drop over a length L of the pipe, τ_w is the wall shear stress, $A = \pi D^2/4$, and $P = \pi D$, D being the diameter of pipe. The **friction coefficient**, defined as

$$f = \frac{-\tau_w}{\frac{1}{2}\rho V^2}, \qquad (6.2)$$

may be rearranged in terms of Eq. (6.1), as

$$f = \frac{\Delta p}{4(L/D)(\frac{1}{2}\rho V^2)}. \qquad (6.3)$$

Equation (6.2) is suitable for the evaluation of the friction coefficient by analytical means, while Equation (6.3) is most useful when experimental data on pressure drop is available. For the fully developed pipe flow under consideration, the differential control volume shown in Fig. 6.2 yields

$$-\frac{d}{dr}(r\tau) + r\frac{dp}{dx} = 0. \qquad (6.4)$$

To proceed further, we need to know the flow conditions. Since an analytical approach to turbulent flow is beyond the scope of this text, let the flow be laminar for the time being. What we learn from laminar flow will be helpful for the interpretation of the experimental data on turbulent flows. For laminar flow of incompressible viscous fluids,

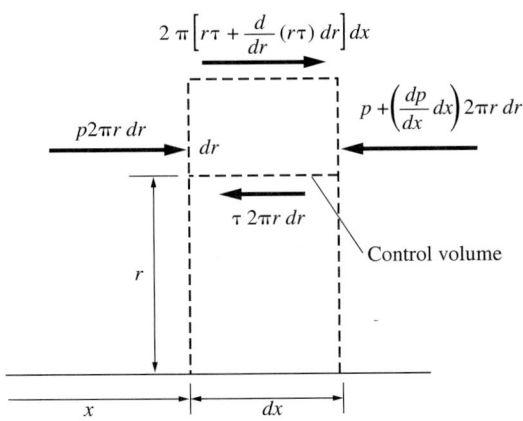

Figure 6.2 Steady, fully developed momentum for differential control volume.

Newton's law[1] states that

$$\tau = \mu \frac{du}{dr}. \tag{6.5}$$

Inserting Eq. (6.5) into Eq. (6.4), assuming the viscosity to be constant, and rearranging gives

$$\frac{d}{dr}\left(r\frac{du}{dr}\right) = \frac{1}{\mu}\left(\frac{dp}{dx}\right)r. \tag{6.6}$$

Integrating Eq. (6.6) twice relative to r, and evaluating the resulting constants by the boundary conditions,

$$u(0) = \text{finite}, \quad u(R) = 0, \tag{6.7}$$

yields[2]

$$u = \frac{1}{4\mu}\left(-\frac{dp}{dx}\right)(R^2 - r^2). \tag{6.8}$$

This result may be rearranged in terms of a **bulk velocity**

$$V = \frac{1}{\pi R^2} \int_0^R u(r) 2\pi r \, dr. \tag{6.9}$$

Inserting Eq. (6.8) into Eq. (6.9) yields

$$V = \frac{1}{8\mu}\left(-\frac{dp}{dx}\right)R^2. \tag{6.10}$$

---

[1] This law on the diffusion of momentum and the Fourier law of conduction (on the diffusion of heat) are special cases of diffusion phenomena.

[2] Note that the fully developed laminar flow in a pipe which leads to Eq. (6.8) is identical in form to the one-dimensional cylindrical conduction with energy generation governed by Eq. (2.71) for $h \to \infty$.

Eliminating the pressure gradient between Eqs. (6.8) and (6.10) gives

$$u = 2V\left(1 - \frac{r^2}{R^2}\right). \tag{6.11}$$

Inserting wall shear stress,

$$\tau_w = \mu \left.\frac{du}{dr}\right|_w = -\frac{8\mu V}{D},$$

into Eq. (6.2) results in

$$f = \frac{16}{Re}, \tag{6.12}$$

where $Re = \rho V D/\mu$ is the Reynolds number based on the bulk velocity. Equation (6.12) gives the exact expression for the friction coefficient in steady, fully developed laminar flow in a smooth pipe.

Most pipe flows are turbulent and involve rough pipes. For fully developed laminar or turbulent flow in a pipe, noting Eq. (6.12), and introducing a characteristic length e as a measure for the surface roughness of the pipe, the friction factor may be functionally expressed as

$$f = f(Re, e/D). \tag{6.13}$$

Figure 6.3 shows the explicit form of Eq. (6.13), obtained from experimental data on fully developed pipe flow. Data in the laminar region agree with Eq. (6.12) and are independent of surface roughness. Two linear approximations for data on turbulent flows, drawn by dotted lines in Fig. 6.3, are

$$f \cong \frac{0.079}{Re^{1/4}}, \quad 3 \times 10^3 \leq Re \leq 10^5, \tag{6.14}$$

and

$$\boxed{f \cong \frac{0.046}{Re^{1/5}}, \quad Re \geq 2 \times 10^4}. \tag{6.15}$$

We have so far discussed friction in pipe flow. Actually, the simplest case of turbulent friction is that of flow over a flat plate with negligible pressure gradient. This case applies to many important technological problems, such as drag on ships, airplanes, turbines, compressors, and propellers. An approximate analytical approach, which is beyond the scope of this text, leads to

$$f \cong \frac{0.072}{Re_L^{1/5}} \tag{6.16}$$

for turbulent flow past a horizontal plate, where $Re_L = V_\infty L/\nu$ is the Reynolds number based on the free-stream velocity (far away from the plate) and the length of the plate.

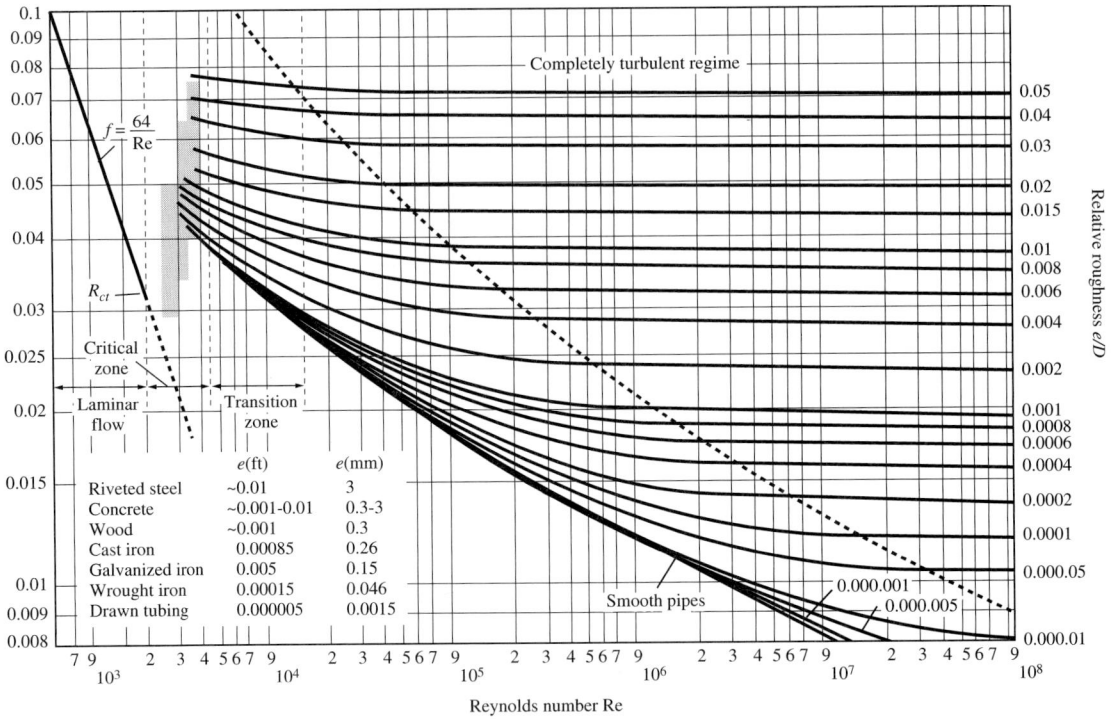

Figure 6.3 Friction factor for flow in circular pipes (from Moody [41]).

Equation (6.16) agrees well with experimental data for $Re \leq 10^7$ (Fig. 6.4). When Re_L exceeds this value, the deviation of Eq. (6.16) from experimental data becomes significant. A more elaborate analysis leads to a logarithmic relation shown in Fig. 6.4. Also, included in the same figure is the laminar friction evaluated analytically in Chapter 5.

We now proceed to the **drag coefficient**, which provides the drag force on a body moving steadily through an infinite fluid or, equivalently, the drag force on a stationary body in an infinite fluid streaming with uniform velocity V_∞. The drag coefficient is defined as

$$C_D = \frac{F_D/A}{\frac{1}{2}\rho V_\infty^2}, \qquad (6.17)$$

where F_D is the total drag force on the body. It results from the difference between the pressure forces which act on the front and back projected area of the body, plus the friction force over the surface of the body. For blunt objects such as cylinders and spheres, A is customarily assumed to be the projected area normal to the flow.

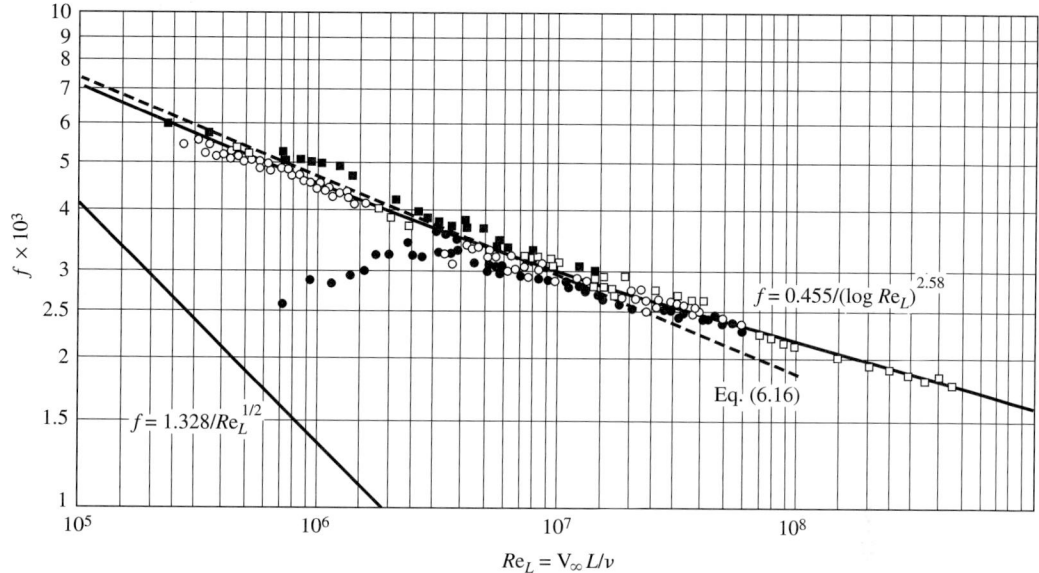

Figure 6.4 Friction coefficient for smooth flat plate (from Rohsenow & Choi [14]).

Consequently, F_D/A has the dimensions of stress but is not an average of the true stress over the body. It can be shown, in a manner similar to the discussion leading to Eq. (6.13), that

$$C_D = f(Re, e/D).$$

Figure 6.5 shows C_D for a few common shapes with polished (smooth) surfaces (i.e., with negligible ϵ/D). Variation of the drag coefficient as a function of the Reynolds number depends on boundary-layer separation and, relative to this separation, the transition to turbulence. These aspects are beyond the scope of this text.

At very low Reynolds numbers no flow separation occurs. In Chapter 5 the Reynolds number was shown to represent a ratio of inertial and viscous forces. At very low Reynolds numbers the inertial forces are small, and the inertial terms of the momentum equation become negligibly small compared to the viscous terms. Under these conditions the drag force on a sphere of diameter D is found to be

$$F_D = 3\pi D \mu V_\infty, \tag{6.18}$$

which may be combined with Eq. (6.17) to yield

$$C_D = \frac{24}{Re_D}. \tag{6.19}$$

Details of the development leading to Eq. (6.19) are again beyond the scope of this text. As shown in Fig. 6.5, Eq. (6.19) agrees well with experimental data when $Re_D \leq 1$.

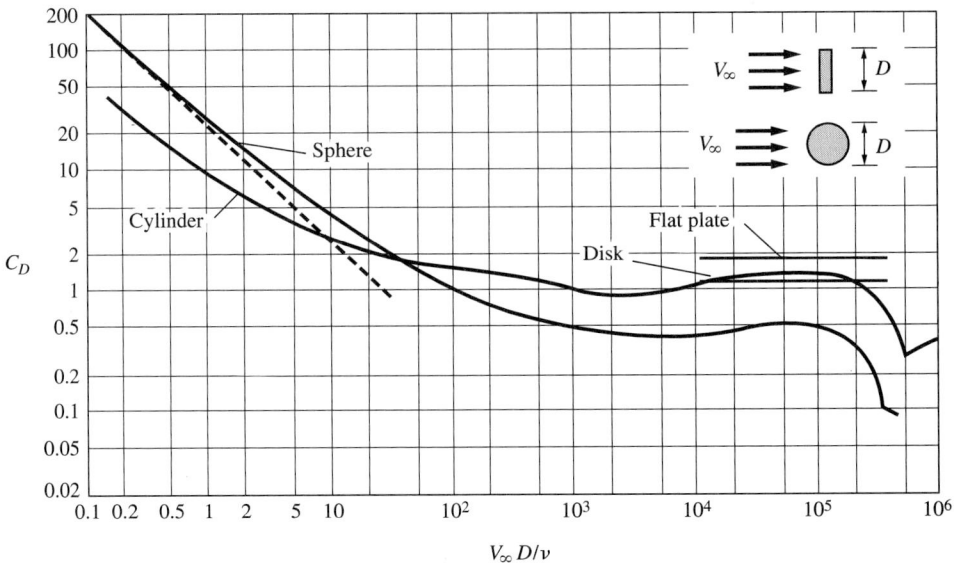

Figure 6.5 Drag coefficient for some common shapes (from Eisner [40]).

So far we have studied the friction factor and the drag coefficient associated with a number of common cases. We may now utilize the analogy between heat and momentum transfer, obtaining the heat transfer indirectly from the friction associated with these cases. Combining Eq. (6.15) with Eq. (5.63) yields a relation for the heat transfer in fully developed turbulent pipe flow,

$$Nu_D = 0.023 Re_D^{4/5} Pr^{1/3}. \tag{6.20}$$

The analogy between heat and momentum does not account for the pressure drop in pipes and is not, strictly speaking, valid for pipe flow. However, the effect of pressure drop on this analogy appears to remain within the uncertainty of the available experimental data and is usually ignored. Next, introducing Eq. (6.16) into Eq. (5.63) gives the heat transfer in fully developed turbulent flow over a flat plate,

$$Nu_L = 0.036 Re_L^{4/5} Pr^{1/3}, \tag{6.21}$$

where Nu_L and Re_L respectively are the Nusselt and Reynolds numbers based on the plate length. Since Eq. (6.20) holds exactly in this case, Eq. (6.21) agrees very well with experimental data. It is, however, important to keep in mind the fact that the analogy between heat and momentum transfer is not universal and applies only to a few (including the foregoing two) special cases. In general, flows in most real cases are much more complicated, and heat transfer expressions cannot be obtained analytically. Instead, correlations are developed by relating experimental data with the help of the appropriate dimensionless numbers introduced in Chapter 5. The remainder of this chapter is devoted to heat transfer correlations for a number of different flow conditions.

6.2 FORCED CONVECTION

In Section 5.3 of the preceding chapter we learned from dimensional considerations that forced-convection heat transfer is described by

$$Nu = f(Re, Pr). \tag{5.150}$$

Also, from Section 5.1 of the same chapter dealing analytically with laminar forced convection, and from the analogy between heat and momentum discussed in Section 6.1, we found that Eq. (5.150) explicitly becomes

$$Nu = C Re^n Pr^m, \tag{6.22}$$

where the unknown parameter and exponents depend on the flow conditions. What remain to be specified are the specific values of C, n, and m for a number of common cases classified as **internal flow** through a pipe or duct and **external flow** over an object.

6.2.1 Internal Flow

The prevalence of pipe flows in engineering (heating, cooling, power plants, water transport, etc) makes pipe flow the most important application of internal flows. Because of this importance, there exist a number of correlations of experimental data on pipe flow. Before listing these correlations, however, let us recall Eq. (6.20), obtained from the analogy between heat and momentum transfer. All of the physical properties associated with the dimensionless numbers of this equation depend on the fluid temperature. Therefore a reference temperature is needed for the evaluation of the properties. A commonly used temperature for this purpose is the bulk temperature T_b associated with the enthalpy flow in the first law (recall $\sum \dot{m}_i h_i^\circ$ of Eq. 1.10),

$$\rho c_p A V T_b = \int_A \rho c_p u(A) T(A) \, dA, \tag{6.23}$$

which, for a constant-property fluid, reduces to

$$A V T_b = \int_A u(A) T(A) \, dA, \tag{6.24}$$

where A is the cross-sectional area of the pipe, and V is the bulk velocity defined by Eq. (6.9).

Although the bulk temperature is a measure for the fluid temperature averaged over the cross section of a pipe, it does not distinguish between the cases $T_w < T_b$ or $T_w > T_b$, where T_w is the temperature of the pipe walls. Note that, because of the dependence of viscosity on temperature, cold or warm walls relative to the bulk temperature make viscosity near the walls heavy or light, which affects the heat transfer. The film temperature T_f, defined as

$$T_f = \frac{1}{2}(T_b + T_w), \tag{6.25}$$

not only takes into account the effect of $T_b \lessgtr T_w$ but also represents a more meaningful temperature for the physical properties. We now proceed to heat transfer correlations for pipe flow in terms of T_b and/or T_f.

Actually, the first relation is the interpretation, say with T_f, of the analogy between heat and momentum transfer [recall Eq. (6.20)],

$$Nu_D = 0.023 Re_D^{0.8} Pr^{1/3}, \quad L/D \geq 60. \tag{6.26}$$

Two improved versions of Eq. (6.26), which incorporate the effect of variable properties by considering the wall temperature relative to the fluid bulk temperature, are given in Table 6.1, L being the length of the pipe. All properties for these correlations are evaluated at T_b. The Prandtl-number exponent of the first correlation and the (μ/μ_w) ratio of the second correlation take into account the temperature dependence of viscosity. Note that, because of condition $L/D \geq 60$, Eq. (6.26) applies only to fully developed flow. For shorter pipes the entrance effect associated with the development of the flow needs to be taken into account. One correlation including this effect, and correlated with T_f, is

$$Nu_D = 0.036 Re_D^{0.8} Pr^{1/3} (D/L)^{0.055}, \tag{6.27}$$

valid for

$$10 < L/D < 400.$$

Note that these correlations evaluate an average heat transfer coefficient. The heat transfer coefficient actually varies along the length of the pipe. In the next section, a procedure for the evaluation of the heat transfer coefficient is given in terms of five computational steps suitable to one of the correlations of Table 6.1.

Table 6.1 Correlation of internal forced convection

Forced Convection (Internal)	
(diagram: V_b, T_b, D, T_w, L)	$T_w =$ Const
	$0.7 \leq Pr \leq 160$
	$Re_D \geq 10^5$
	$L/D \geq 60$
	T_b for properties
$Nu = 0.023 Re_D^{0.8} Pr^n$	$T_w > T_b, \quad n = 0.4$
	$T_w < T_b, \quad n = 0.3$
Property Corrected	$0.7 \leq Pr \leq 16.700$
	$Re_D \geq 10^5$
$Nu = 0.027 Re_D^{0.8} Pr^{1/3} \left(\dfrac{\mu}{\mu_w}\right)^{0.14}$	$L/D \geq 60$
	T_b except T_w for μ_w

6.2.2 Computation of the Heat Transfer Coefficient of Internal Flow

Key Problem.

Given: Bulk temperature T_b and velocity V of fluid flowing through a pipe with specified wall temperature T_w.

Required: Heat transfer coefficient.

Computational Steps.

1. Adjust the bulk temperature to the closest temperature of the related property table.

2. Read the following five properties from Appendix B–2 or B–3 for the fluid at the bulk temperature T_b:

$$\nu \left[\frac{m^2}{s}\right], \quad Pr \text{ [Dimensionless]}, \quad k \left[\frac{W}{m \cdot K}\right], \quad \rho \left[\frac{kg}{m^3}\right], \quad c_p \left[\frac{kJ}{kg \cdot K}\right]. \tag{6.28}$$

For a first trial, pick the numerical values corresponding to the listed temperature closest to the available T_b. Note that, to evaluate the heat transfer coefficient, one needs only the first three properties. The last two are used in the enthalpy flow terms of an energy balance, if they are needed.

3. Using V, ν, and D, compute the Reynolds number

$$Re_D = \frac{VD}{\nu}. \tag{6.29}$$

4. From Table 6.1 pick the power of Pr, $n = 0.3$ for $T_w < T_b$, and $n = 0.4$ for $T_w > T_b$, and compute Nu from

$$Nu = 0.023 Re_D^{0.8} Pr^n. \tag{6.30}$$

5. Employing Nu, k, and D, compute the heat transfer coefficient from

$$h \left[\frac{W}{m^2 \cdot K}\right] = Nu \text{ [Dimensionless]} \frac{k \left[\frac{W}{m \cdot K}\right]}{L \text{ [m]}}. \tag{6.31}$$

EXAMPLE 6.1[3]

Consider water at a bulk temperature $T_b = 40\,°C$ flowing with a velocity $V = 3$ m/s through a pipe of diameter $D = 2$ cm in a condensing steam bath at a temperature $T_\infty = 120\,°C$ (Fig. 6.6). We wish to determine (a) the rate of heat transfer from the steam bath to the water flow, (b) the temperature rise in the water flow per unit length of pipe.

---

[3] The FORTRAN program EX6–1.F is listed in the appendix of this chapter.

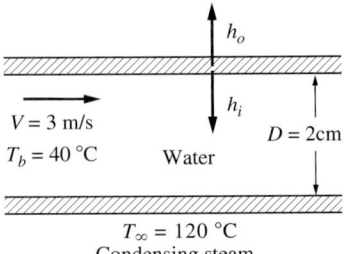

Figure 6.6 Water flow through a pipe.

a) Neglecting the effect of pipe wall curvature, the total heat transfer coefficient between the steam bath and water flow is given by

$$\frac{1}{U} = \frac{1}{h_0} + \frac{\delta}{k_w} + \frac{1}{h_i},$$

h_0 being the outside (or steam side) and h_i the inside (or water side) heat transfer coefficient, δ the thickness of pipe, and k_w the pipe conductivity. Note from Table 1.2 that

$$h_0 \gg h_i$$

(h_0 for condensation, h_i for forced convection in water) and the effect of h_0 on U is thus negligible. Since there is no information about the thickness of the pipe walls, we also neglect δ/k_w, which usually turns out to be small. Thus $U \cong h_i$ and $T_w \cong T_\infty$, and we then need only compute water-side coefficient h_i to determine the heat transfer between the steam bath and water flow.

Computation of heat transfer coefficient. Following the five computational steps:

1. Adjust $T_b = 40 + 273 = 313$ to 315 K (closest list temperature).
2. From Appendix B–3, for water (saturated liquid), read the following five thermomechanical properties,

Water at $T_b = 315$ K	$\nu = 0.637 \times 10^{-6}\,\text{m}^2/\text{s}$
	$Pr = 4.16$
	$k = 0.634\,\text{W/m·K}$
	$\rho = 991\,\text{kg/m}^3$
	$c_p = 4{,}179\,\text{J/kg·K}$

3. Compute the Reynolds number,

$$Re_D = \frac{VD}{\nu} = \frac{3\,\text{m/s} \times 0.02\,\text{m}}{0.637 \times 10^{-6}\,\text{m}^2/\text{s}} = 0.94 \times 10^5.$$

4. From Table 6.1, pick $n = 0.4$ for $T_w \cong T_\infty > T_b$ and compute Nu,

$$Nu_D = 0.023(94{,}000)^{0.8}(4.16)^{0.4} = 387.$$

5. Compute h,

$$h_i = \frac{k Nu_D}{D} = \frac{0.634 \text{ W/m·K} \times 387}{0.02 \text{ m}},$$

$$\boxed{h_i = 12{,}268 \text{ W/m}^2\text{·K}},$$

which agrees with the upper bound expected for forced convection in a water flow (see Table 1.2). A posteriori, we learn from this numerical result that the present convection problem is highly turbulent.

The heat transfer from the steam bath to the water flow, assuming T_b = Constant over ℓ, is

$$\dot{Q} = h_i P \ell (T_\infty - T_b)$$

or, per unit length,

$$\dot{Q}/\ell = h_i \pi D (T_\infty - T_b) = 12{,}268 \text{ W/m}^2\text{·K} \times \pi \times 0.02 \text{ m} \times (120 - 40)\text{K},$$

$$\boxed{\dot{Q}/\ell \cong 61.7 \text{ kW/m}}.$$

b) Neglecting the effect of axial conduction, we balance the increase in the water enthalpy flow with the heat received by the water,[4]

$$\rho c_p V A \Delta T = h_i P \ell (T_\infty - T_b),$$

which gives

$$\frac{\Delta T}{\ell} = \frac{4 h_i (T_\infty - T_b)}{\rho c_p V D}$$

or

$$\frac{\Delta T}{\ell} = \frac{4 \times 12{,}268 \text{ W/m}^2\text{·K} \times (120 - 40)\text{K}}{991 \text{ kg/m}^3 \times 4{,}179 \text{ J/kg·K} \times 3 \text{ m/s} \times 0.02 \text{ m}}$$

$$\cong 15.8 \text{ K (or °C)/m}.$$

Note the appreciable temperature rise over 1 m along the pipe. In calculating h_i, we evaluated the properties at T_b, thus neglecting any temperature variation of the fluid along the length of the pipe. By the computer program provided, the interested reader may evaluate $\Delta T/\ell$ for a number of fluids.

Let us see how this rise in temperature affects the heat transfer. The bulk exit temperature is

$$T_{be} = 40 + 15.8 \cong 56 \text{ °C} = 329 \text{ K} \cong 330 \text{ K}.$$

We wish now to recompute the heat transfer coefficient based on this exit temperature. Again, from Appendix B–3,

Water at	$\nu = 0.497 \times 10^{-6}$ m^2/s
$T_{be} = 330$ K	$Pr = 3.15$
	$k = 0.650$ W/m·K

---

[4] We assume the reader follows the five steps of formulation.

Then, for

$$Re_D = \frac{VD}{\nu} = \frac{3 \text{ m/s} \times 0.02 \text{ m}}{0.497 \times 10^{-6} \text{ m}^2/\text{s}} \cong 1.2 \times 10^5,$$

we have

$$Nu_D = 0.023(120{,}000)^{0.8}(3.15)^{0.4} \cong 422,$$

and from

$$h_i = \frac{k Nu_D}{D} = \frac{0.650 \text{ W/m·K} \times 422}{0.02 \text{ m}},$$

$$\boxed{h_i = 13{,}715 \text{ W/m}^2\text{·K}}.$$

For the longitudinally averaged heat transfer coefficient

$$h_i = \frac{12{,}268 + 13{,}715}{2} = 12{,}992 \text{ W/m}^2\text{·K}$$

and the longitudinally averaged bulk temperature

$$\overline{T_b} = \frac{40 + 56}{2} = 48\,°\text{C},$$

we have

$$\dot{Q}/\ell = 12{,}992 \times \pi \times 0.02 \times (120 - 48)$$

or

$$\boxed{\dot{Q}/\ell = 58.8 \text{ kW/m}}.$$

As a rule of thumb, the uncertainty involved with a heat transfer coefficient is usually assumed to be about ±30%. In view of this uncertainty, the difference between the first and second estimates of $\dot{Q}/\ell$ is negligible, and there is no need to carry out the computations based on the averaged values. Had the answers differed by more than 30%, the heat transfer coefficient, heat transfer, and exit temperature would have had to be reevaluated based on the averaged T_b. ◆

For flows through noncircular cross sections and ducts, the heat transfer correlations developed for pipes can be used based on a **hydraulic diameter**,

$$D_h = 4 \frac{\text{Flow cross sectional area}}{\text{Wetted perimeter}}. \tag{6.32}$$

For example, for a rectangular duct of side dimensions a and b, the hydraulic diameter is

$$D_h = \frac{4ab}{2(a+b)} = \frac{2ab}{a+b}.$$

The correlations that we have studied so far are valid only for $Pr \geq 0.7$, which exclude liquid metals (recall Fig. 5.12). As we learned in Chapter 5, the forced-convection heat transfer associated with liquid metals is described by the relation

$$Nu = f(Pe), \tag{5.155}$$

where $Pe = RePr$ is the Peclet number. In general, liquid metals have small Peclet numbers, resulting from their high conductivity. This allows for the removal of larger heat fluxes than those in other liquids and gases. Since high levels of energy, relative to conventional systems, are generated in the core of nuclear reactors, liquid metals are suitable for the cooling of such systems.

For fully developed turbulent flow in tubes subject to a uniform heat flux, the experimental data are correlated with

$$Nu_D = 0.625 Pe_D^{0.4}, \qquad (6.33)$$

under conditions

$$100 < Pe_D < 10{,}000, \; L/D > 60,$$

provided all properties are evaluated at a bulk temperature T_b. No correlation is available in the literature for shorter pipes with $L/D < 60$. The correction factor

$$\left(1 + 2\frac{D}{L}\ln\frac{L}{D}\right), \qquad (6.34)$$

obtained from the averaging of the local Nusselt number may be utilized until a more accurate relation becomes available. Having learned the correlations for internal flows through pipes and ducts, we now proceed to correlations for external flows.

6.2.3 External Flow

We learned in Chapter 1 (recall Fig. 1.1) that the heat transfer involved with thermal systems is composed of a convective internal resistance, conductive resistance(s), and a convective external resistance,

$$\frac{1}{U} = \frac{1}{h_i} + \frac{\delta}{k} + \frac{1}{h_0},$$

or, for a significant curvature effect,

$$\frac{1}{U} = \frac{1}{h_i}\left(\frac{A_0}{A_i}\right) + \frac{\delta}{k}\left(\frac{A_0}{\overline{A}}\right) + \frac{1}{h_0},$$

where U, h_i, and h_0, are the total, inside, and outside coefficients of heat transfer, δ and k are the wall thickness and conductivity, and A_i, A_0 and $\overline{A}$ are the inner, outer and mean surface area of the walls of the thermal system. Consequently, the convection heat transfer from the outer surfaces of cylinders, spheres, tube banks, nozzles, combustors, engines, etc. is also important in technological problems.

Since the flow conditions strongly influence the heat transfer (recall $Nu \sim Re^{1/2}$ for laminar flows and $Nu \sim Re^{0.8}$ for turbulent flows over a flat plate not involving liquid metals), let us recall from fluid mechanics the results of experimental observations on flow around a cylinder, which are sketched in Fig. 6.7. We learn from this figure that the flow around a cylinder may assume different forms, depending on the Reynolds number. The separation, beginning with the second sketch from the top, is associated

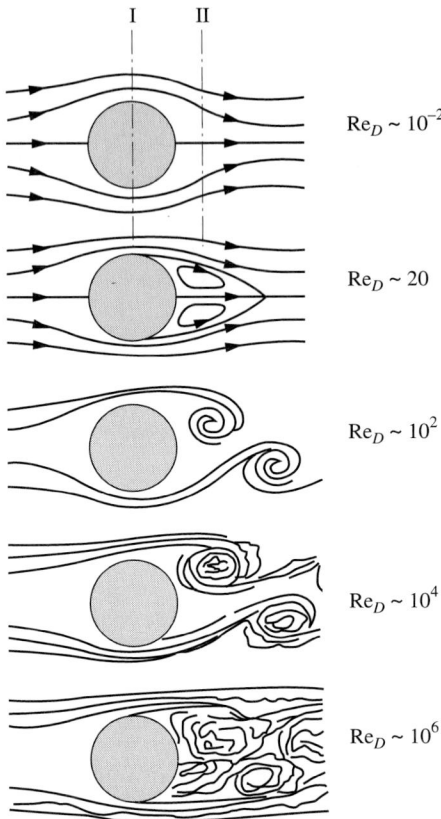

Figure 6.7 Flow regimes around a cylinder.

with an adverse pressure gradient. This gradient results from the fact that the upstream velocity (at location I) is greater than the downstream velocity (at location II) because of the conservation of mass, while the pressure at I is smaller than the pressure at II because of the Bernoulli theorem. Under the influence of the pressure gradient, the transition from laminar flow to turbulent flow occurs apparently in several steps, each corresponding to a different flow regime.

For each regime involving separation, the characteristics of the downstream flow separated from the cylinder are quite different from the characteristics of the upstream flow that is attached to the cylinder. This difference is also reflected in the upstream and downstream heat transfers. Figure 6.8 shows, for free-stream Reynolds numbers between 70,000 and 220,000, the change in the local Nusselt number as a function of angular distance from the stagnation point. Note the difference between the maximum and minimum Nusselt number, as well as the location of maximum heat transfer.

For most problems of technological importance, however, we usually need the average heat transfer coefficient. For example, the experimental data for air flowing

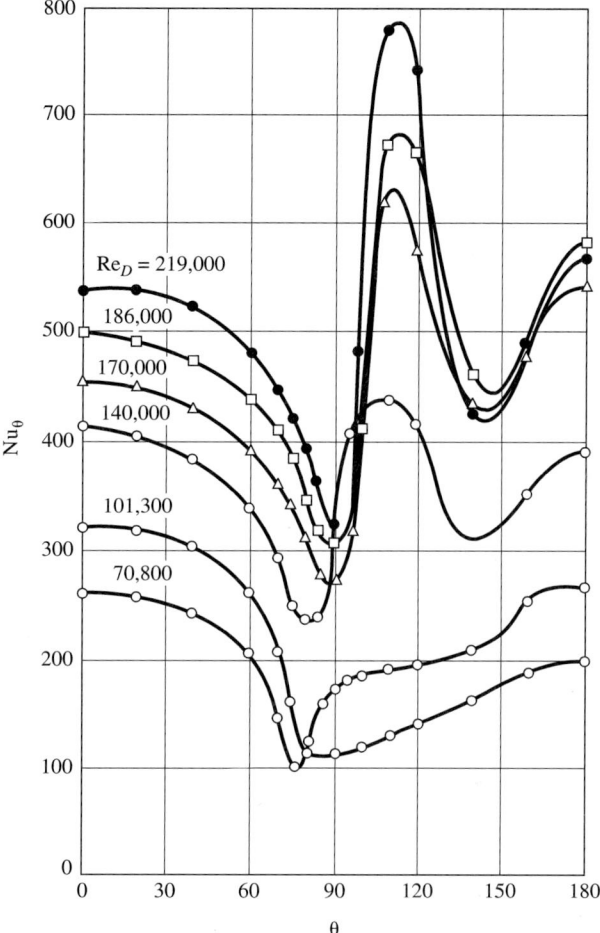

Figure 6.8 Circumferential variation of h for cross flow of air over a cylinder (from Giedt [42]).

normal to a single cylinder yields the average Nusselt number versus the Reynolds number, as shown in Fig. 6.9. Because of the aforementioned different flow regimes, these data plotted on a log-log scale fall on a curve rather than on a straight line, indicating different power laws (i.e., Reynolds-number exponents). However, approximating this curve by piecewise straight lines, the data are correlated by the usual form given by Eq. (6.22). For each straight line and $m = 1/3$, C and n assume a different set of values: $Nu = C Re_D^n Pr^{1/3}$. In terms of the foregoing arguments, the experimental data for ordinary fluids as well as gases including air are correlated by the relation given in Table 6.2. In this table, $Re_D = V_\infty D/\nu_f$ is the Reynolds number based on the free-stream velocity, and all properties are evaluated at the film temperature $T_f = (T_w + T_\infty)/2$. Equation (6.22) continues to apply for flow over a noncircular cylinder, provided the values of C and n are now taken from Table 6.3 and the Reynolds number is evaluated based on the appropriate D_e.

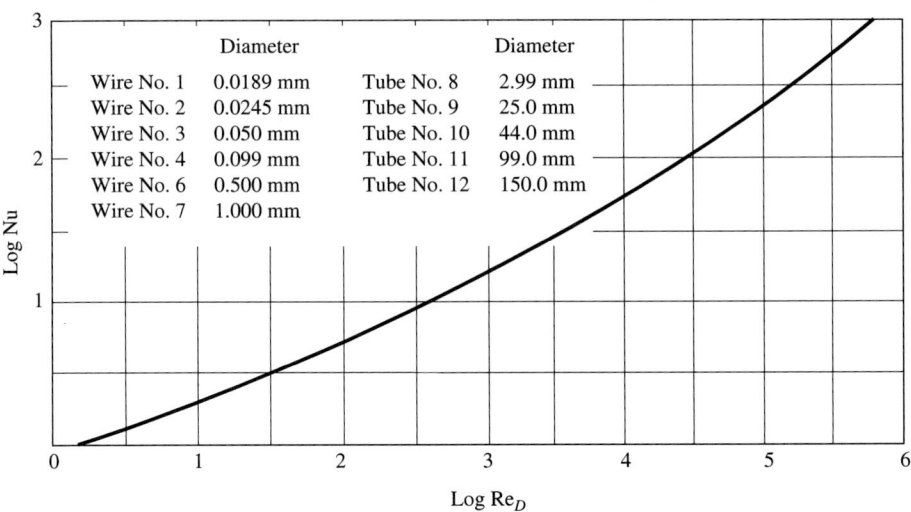

Figure 6.9 Average heat transfer coefficient versus Reynolds number for a circular cylinder in cross flow with air (from McAdams [25]).

Table 6.2 C and n for circular tubes

$Nu_D = C Re_D^n Pr^{1/3}$		T_f properties
$Re_D = VD/\nu_f$	C	n
0.4–4	0.989	0.330
4–40	0.911	0.385
40–4,000	0.683	0.466
4,000–40,000	0.193	0.618
40,000–400,000	0.0266	0.805

For flow over spheres, experimental data is correlated by

$$Nu_D = 2 + \left(0.4 Re_D^{1/2} + 0.06 Re_D^{2/3}\right) Pr^{0.4} \left(\frac{\mu}{\mu_w}\right)^{1/4} \tag{6.35}$$

under conditions

$$3.5 < Re_D < 8 \times 10^4, \quad 0.7 < Pr < 380,$$

where all properties, except for μ_w at T_w, are evaluated at the free-stream temperature T_∞.

Table 6.3 C and n for flow over noncircular tubes (from Jakob [43]).

Geometry	$Re = V_\infty D_e/\nu$	n	C
$V_\infty \rightarrow$ ◇ D_e	5,000–100,000 2,500–7,500	0.588 0.624	0.222 0.261
$V_\infty \rightarrow$ ⬭ D_e	2,500–15,000	0.612	0.224
$V_\infty \rightarrow$ ⬣ D_e	5,000–100,000	0.638	0.138
$V_\infty \rightarrow$ ⬡ D_e	5,000–19,500 19,500–100,000	0.638 0.782	0.144 0.035
$V_\infty \rightarrow$ ◻ D_e	2,500–8,000	0.699	0.160
$V_\infty \rightarrow$ | D_e	4,000–15,000	0.731	0.205
$V_\infty \rightarrow$ ⬭ D_e	3,000–15,000	0.804	0.085

In the next section a procedure for the evaluation of the heat transfer coefficient is given in terms of five computational steps suitable to one of the correlations of Table 6.2.

6.2.4 Computation of the Heat Transfer Coefficient for External Flow

Key Problem.

Given: Free-stream velocity V_∞ and temperature T_∞, and temperature T_w of the object.

Required: Heat loss q_w from or energy generation u''' within the object.

Computational Steps.

1. Compute the film temperature

$$T_f = \frac{1}{2}(T_w + T_\infty). \tag{6.36}$$

2. From Appendix B–2 for gases or B–3 for liquids, read the following three properties for ambient fluid at the film temperature T_f:

$$\nu \left[\frac{m^2}{s}\right], \quad Pr \text{ [Dimensionless]}, \quad k \left[\frac{W}{m \cdot K}\right]. \tag{6.37}$$

Do not use any interpolation; pick available numerical values closest to T_f.

3. Using V_∞, ν, and a characteristic length D, compute the Reynolds number

$$Re_D = \frac{V_\infty D}{\nu}, \tag{6.38}$$

where D is the length of the object in the direction of the flow (circular cylinder diameter or appropriate D_e from Table 6.3 for noncircular cylinder).

4. Depending on the Re_D range that the numerical value of Re_D falls within, pick appropriate C and n from Table 6.2 or 6.3 and compute Nu.

5. Employing Nu, k, and D (or D_e), compute the heat transfer coefficient h from

$$h\left[\frac{W}{m^2 \cdot K}\right] = Nu \text{ [Dimensionless]} \frac{k\left[\frac{W}{m \cdot K}\right]}{D\,[m]}. \tag{6.39}$$

EXAMPLE 6.2[5]

Air at 1 atm and $T_\infty = 25\,°C$ flowing with a velocity $V = 50$ m/s crosses an industrial heater made of a long solid rod of diameter $D = 2$ cm. The surface temperature of the heater is to be kept no higher than $T_w = 425\,°C$. We wish to determine the allowable electrical power density (u''' W/m^3) within the heater.

Computation of heat transfer coefficient. Following the procedure described in Section 6.2.4:

1. Compute the film temperature

$$T_f = \frac{T_w + T_\infty}{2} = \frac{425 + 25}{2} = 225\,°C \cong 500\,K.$$

---

[5] The FORTRAN program EX6–2.F is listed in the appendix of this chapter.

2. The three thermomechanical air properties at this temperature are

Air at $T_b = 500$ K	$\nu = 38.79 \times 10^{-6}\,\text{m}^2/\text{s}$
	$Pr = 0.684$
	$k = 40.7 \times 10^{-3}\,\text{W/m}\cdot\text{K}$

3. The Reynolds number

$$Re_D = \frac{VD}{\nu} = \frac{50 \text{ m/s} \times 0.02 \text{ m}}{38.79 \times 10^{-6}\,\text{m}^2/\text{s}} \cong 25{,}780$$

is within

$$4 \times 10^3 < Re_D < 4 \times 10^4,$$

corresponding from Table 6.2 to

$$C = 0.193 \text{ and } n = 0.618.$$

4. From Table 6.2,

$$Nu = 0.193 Re_D^{0.618} Pr^{1/3}$$

or

$$Nu = 0.193(25{,}780)^{0.618}(0.684)^{1/3} = 90.5.$$

5. Then

$$h = \frac{kNu}{D} = \frac{40.7 \times 10^{-3} \text{ W/m}\cdot\text{K} \times 90.5}{0.02 \text{ m}},$$

$$\boxed{h \cong 184 \text{ W/m}^2\cdot\text{K}}.$$

Note that the foregoing $Re_D = 25{,}780$ corresponds approximately to the flow regime for $Re \sim 10^4$ sketched in Fig. 6.7, which is not fully turbulent. Accordingly, the above heat transfer coefficient, being less than the upper bound of 300 for forced air flow (recall Table 1.2), is justified. The allowable electrical power density within the heater is

$$u''' = 4h/D(T_w - T_\infty) = 4 \times 184 \text{ W/m}^2\cdot\text{K} \times (425 - 25) \text{ K}/0.02 \text{ m}$$

$$= 1.47 \times 10^4 \text{ kW/m}^3$$

By the computer program provided, the interested ready may parametrically study the effect of fluid velocity on allowable energy generation. ◆

EXAMPLE 6.3[6]

Hot air at 1 atm $T_b = 325\,°\text{C}$ flows with a velocity $V = 20$ m/s through a pipe of diameter $D = 5$ cm while cooling ambient air at 1 atm and $T_\infty = 25\,°\text{C}$, flowing with a stream velocity $V_\infty = 40$ m/s across the pipe (Fig. 6.10). We wish to determine the heat loss per unit length of pipe.

---

[6] The FORTRAN program EX6–3.F is listed in the Appendix.

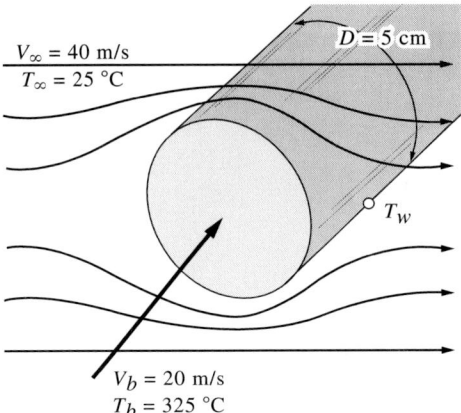

Figure 6.10 System configuration.

Computation of inside heat transfer coefficient. The inside heat transfer coefficient involves internal forced convection in a circular pipe, and thus the use of Table 6.1. Following the five computational steps:

1. Adjust $T_b = 325 + 273 = 598$ to 600 K.
2. From Appendix B–2 for air, read the following three thermomechanical properties:

Air at	$\nu = 52.69 \times 10^{-6}$ m^2/s
$T_b = 600$ K	$Pr = 0.685$
	$k = 46.9 \times 10^{-3}$ W/m·K

3. Compute the Reynolds number,

$$Re_D = \frac{V_b D}{\nu} = \frac{20 \text{ m/s} \times 0.05 \text{ m}}{52.69 \times 10^{-6} \text{ m}^2/\text{s}} \cong 1.9 \times 10^4.$$

4. From Table 6.1, pick $n = 0.3$ for for $T_w < T_b$ and compute Nu,

$$Nu = 0.023(19,000)^{0.8}(0.685)^{0.3} = 54.4.$$

5. Compute h_i,

$$h_i = \frac{kNu}{D} = \frac{46.9 \times 10^{-3} \text{ W/m·K} \times 54.4}{0.05 \text{ m}},$$

for the inside heat transfer coefficient,

$$\boxed{h_i \cong 51 \text{ W/m}^2 \cdot \text{K}},$$

which remains well within the range (10 − 300) for forced gas flow given in Table 1.2. However, for a turbulent flow, this result is considerably smaller than the expected value, which should be closer to the upper rather than the lower bound. We will return this issue in the next example.

Computation of outside heat transfer coefficient. For the outside heat transfer coefficient (flow over circular cylinder, Table 6.2), we need the wall-temperature, which is unknown. As a first trial, letting $h \sim V$ (rather than $V^{0.8}$) and noting that the outside flow velocity is twice the inside flow velocity, we may assume $h_0 \sim 2h_i$ and

$$h_i(T_b - T_w) = h_0(T_w - T_\infty)$$

and get an estimate for

$$T_w \cong \frac{T_b + 2T_\infty}{3} = \frac{325 + 50}{3} = 125\,°C.$$

Note that this type of reasoning applies only because both flows involve the same type of liquid. Had one of the fluids been water, for example, we would need to use Table 1.2 as a guide to estimate the order of magnitude of h_0.

Following the procedure described in Section 6.3.1:

1. Compute the film temperature

$$T_f = \frac{T_w + T_\infty}{2} = \frac{125 + 25}{2} = 75\,°C \cong 350\,K.$$

2. Read from Appendix B–2 for air:

Air at	$\nu = 20.92 \times 10^{-6}$ m^2/s
$T_f = 350$ K	$Pr = 0.7$
	$k = 30 \times 10^{-3}$ W/m·K

3. Then, for

$$Re_D = \frac{V_\infty D}{\nu} = \frac{40 \text{ m/s} \times 0.05 \text{ m}}{20.92 \times 10^{-6} \text{ m}^2/\text{s}} = 0.956 \times 10^5,$$

Table 6.2 corresponding to

$$4 \times 10^4 < Re_D < 4 \times 10^5$$

gives

$$C = 0.0266 \quad \text{and} \quad n = 0.805.$$

Note that, for a specified pipe wall thickness, the inside and outside Reynolds numbers should respectively be based on D_i and D_0.

4. Compute Nusselt number from

$$Nu = 0.0266\, Re_D^{0.805}\, Pr^{1/3},$$

which gives

$$Nu = 0.0266(95{,}600)^{0.805}(0.7)^{1/3} \cong 241.$$

5. Then, from

$$h_0 = \frac{kNu}{D} = \frac{30 \times 10^{-3} \text{ W/m·K} \times 241}{0.05 \text{ m}},$$

$$\boxed{h_0 \cong 145 \text{ W/m}^2\text{·K}}.$$

As a second trial, in view of $h_0/h_i = 2.84$, one may assume $h_0 \sim 3h_i$ and get for the wall temperature

$$T_w = \frac{T_b + 3T_\infty}{4} = \frac{325 + 75}{4} = 100 \text{ °C}$$

rather than $T_w = 125\,°\text{C}$ obtained as a first trial. However, a wall temperature difference of $\Delta T_w = 25\,°\text{C}$ does not significantly affect h_0 and will not be taken into account.

In terms of the total heat transfer coefficient,

$$\frac{1}{U} = \frac{1}{h_i} + \frac{1}{h_0} = \frac{1}{51} + \frac{1}{145} \cong \frac{1}{38},$$

$$\boxed{U = 38 \text{ W/m}^2\text{·K}},$$

we have

$$\dot{Q}/\ell = \pi D U (T_b - T_\infty)$$

or

$$\dot{Q}/\ell = \pi \times 0.05 \text{ m} \times 38 \text{ W/m}^2\text{·K}(325 - 25)\text{K}$$

or

$$\boxed{\dot{Q}/\ell = 1{,}790 \text{ W/m}}.$$

◆

EXAMPLE 6.4

Reconsider Ex. 6.3 for the hot air flow now pressurized to $p_b = 10$ atm and the cold air flow to $p_\infty = 3$ atm.

For a compressible fluid at constant temperature,

$$\frac{p}{\rho} = \text{Constant},$$

and we now have a tenfold increase in the density of the hot air flow compared to the preceding example. Accordingly, the Reynolds number becomes

$$Re_D = 1.9 \times 10^5,$$

the Nusselt number for this Re_D gives

$$Nu = 0.023(190{,}000)^{0.8}(0.685)^{0.3} \cong 343,$$

and the heat transfer coefficient becomes

$$h_i = \frac{kNu}{D} = \frac{46.9 \times 10^{-3} \text{ W/m·K} \times 343}{0.05 \text{ m}},$$

$$\boxed{h_i \cong 322 \text{ W/m}^2\text{·K}},$$

which now corresponds to the upper bound for a gas flow given in Table 1.2.

In the preceding example we already had a lower heat transfer coefficient for the inside flow than for the outside flow. With an increased pressure, the inside coefficient will be considerably higher. Accordingly, as a first trial, assume $h_0 \sim h_i$. Then

$$T_w \sim \frac{T_b + T_\infty}{2} = \frac{325 + 25}{2} = 175 \,°\text{C}.$$

Now, in terms of the five steps of computation:

1. Compute the film temperature

$$T_f = \frac{T_w + T_\infty}{2} = \frac{175 + 25}{2} = 100\,°\text{C} \cong 375 \text{ K}.$$

2. Read from Appendix B–2 for air:

Air at	$\nu = 23.62 \times 10^{-6}$ m^2/s
$T_f = 375$ K	$Pr = 0.695$
$P_f = 1$ atm	$k = 31.9 \times 10^{-3}$ W/m·K

3. Then, for

$$Re_D = \frac{V_\infty D}{\nu} = \frac{3 \times 40 \text{ m/s} \times 0.05 \text{ m}}{23.62 \times 10^{-6} \text{ m}^2/\text{s}} = 2.54 \times 10^5,$$

Table 6.2 corresponding to

$$4 \times 10^4 < Re_D < 4 \times 10^5$$

gives

$$C = 0.0266 \text{ and } n = 0.805.$$

4. Compute Nusselt number from

$$Nu = 0.0266 Re_D^{0.805} Pr^{1/3},$$

which gives

$$Nu = 0.0266(254,000)^{0.805}(0.695)^{1/3} = 528.6.$$

5. Then, from

$$h_0 = \frac{kNu}{D} = \frac{31.9 \times 10^{-3} \text{ W/m·K} \times 528.6}{0.05 \text{ m}},$$

$$\boxed{h_0 \cong 337 \text{ W/m}^2\text{·K}},$$

312   Chap. 6   Correlations For Convection

which turns out to be quite close to our initial assumption. Then, with

$$\frac{1}{U} = \frac{1}{h_i} + \frac{1}{h_0} = \frac{1}{321} + \frac{1}{337} \cong \frac{1}{164},$$

$$\boxed{U = 164 \text{ W/m}^2\cdot\text{K}},$$

we have

$$\dot{Q}/\ell = \pi D U (T_b - T_\infty)$$

or

$$\dot{Q}/\ell = \pi \times 0.05 \text{ m} \times 164 \text{ W/m}^2\cdot\text{K} \times (325 - 25) \text{ K}$$

or

$$\boxed{\dot{Q}/\ell = 7{,}728 \text{ W/m}}.$$

The comparison of this result for $\dot{Q}/\ell$ with that obtained in the preceding example shows the importance of pressure for gas flows.

◆

6.3   NATURAL CONVECTION

In Chapter 5, following some dimensional arguments, we learned that the independent dimensionless numbers characterizing buoyancy driven flows are the Rayleigh number and the Prandtl number (Ra, Pr) and the heat transfer in (Nusselt number Nu for) natural convection is governed by

$$\boxed{Nu = f(Ra, Pr)}. \tag{5.162}$$

A recently proposed dimensionless number (see Arpacı and co-worker references at the end of the chapter) describes these flows by a combination of Ra and Pr and needs to be considered here.

Let a buoyancy-driven momentum balance be

$$F_B \sim F_I + F_V, \tag{6.40}$$

where F_B, F_I, and F_V respectively denote buoyant, inertial, and viscous forces. Also, let the associated thermal energy balance be

$$\dot{Q}_H \sim \dot{Q}_K, \tag{6.41}$$

where $\dot{Q}_H$ and $\dot{Q}_K$ respectively denote enthalpy flow and conduction. Then, the following ratios from Eq. (6.40),

$$\frac{F_B}{F_I + F_V} = \frac{F_B/F_V}{F_I/F_V + 1}, \tag{6.42}$$

and, from Eq. (6.41),

$$\dot{Q}_H/\dot{Q}_K, \tag{6.43}$$

can be formed, the numeral 1 in Eq. (6.42) implying order of magnitude. The actual value of this numeral depends on the geometry and the flow conditions of the problem. Although the force ratios of Eq. (6.42) and the energy ratio of Eq. (6.43) are dimensionless, they are usually expressed in terms of velocity, which is a dependent variable in buoyancy-driven flows:

$$\frac{F_B}{F_V} \sim \frac{g(\Delta \rho)\ell^2}{\mu V}, \quad \frac{F_I}{F_V} \sim \frac{\rho V \ell}{\mu}, \quad \frac{\dot{Q}_H}{\dot{Q}_K} \sim \frac{\rho c V \ell}{k}, \quad (6.44)$$

where ℓ is a characteristic length, the rest of the notation being conventional. The only combination of Eq. (6.42) and (6.43) independent of velocity is

$$\Pi_N \sim \frac{(F_B/F_V)\dot{Q}_H/\dot{Q}_K}{(F_I/F_V)\dot{Q}_K/\dot{Q}_H + 1}, \quad (6.45)$$

or,

$$\Pi_N \sim \frac{Ra}{1 + Pr^{-1}} = \frac{Pr\,Ra}{1 + Pr}, \quad (6.46)$$

which is the appropriate dimensionless number for natural convection in any fluid (recall Prob. 5.13). Two limits of Π_N are

$$\lim_{Pr \to \infty} \Pi_N \to Ra \quad (6.47)$$

and

$$\lim_{Pr \to 0} \Pi_N \to Ra\,Pr. \quad (6.48)$$

Now, Eq. (5.162) can be written more explicitly in terms of Π_N as

$$Nu = f(\Pi_N) \quad (6.49)$$

and two of its limits as

$$\lim_{Pr \to \infty} Nu = f(Ra), \quad (5.165)$$

which correlates data on most fluids (gas, light and heavy liquids) except for liquid metals, and

$$\lim_{Pr \to 0} Nu = f(Ra\,Pr), \quad (5.166)$$

which correlates data on liquid metals.

The rest of the chapter introduces a new correlation based on Eq. (6.49) and outlined in Table 6.4, and reviews an existing correlation related to Eq. (5.165) and outlined in Table 6.7. We demonstrate the use of these tables for the computation of the heat transfer coefficient in terms of two key problems.

6.3.1 Computation of the Heat Transfer Coefficient for Given T_w

First Key Problem.

Given: Temperature T_w of an object and ambient temperature T_∞.

Required: Heat loss q_w from or energy generation u''' within object.

Computational Steps.

1. Compute the film temperature

$$T_f = \frac{1}{2}(T_w + T_\infty). \tag{6.50}$$

   Do not use any interpolation. For a first trial, pick available numerical values closest to T_f.

2. Read from Appendix B–2 or B–3 the following three properties for ambient fluid at film temperature T_f:

$$\frac{g\beta}{\nu\alpha}\left[\frac{1}{\text{m}^3\cdot\text{K}}\right], \quad Pr\text{ [Dimensionless]}, \quad k\left[\frac{\text{W}}{\text{m}\cdot\text{K}}\right]. \tag{6.51}$$

3. Using $g\beta/\nu\alpha$, $(T_w - T_\infty)$, and a characteristic length L, compute the Rayleigh number

$$Ra = \frac{g\beta}{\nu\alpha}(T_w - T_\infty)L^3, \tag{6.52}$$

   where L is the **height of the object in the direction of gravity** unless otherwise specified.

4. Depending on the geometry and orientation of the object, pick the numerical value of C_0 from Table 6.4 and compute the fundamental dimensionless number for natural convection,

$$\Pi_N = \frac{Ra}{1 + \dfrac{C_0}{Pr}}, \tag{6.53}$$

   and, using the range that the numerical value of Π_N falls within, pick C and n of

$$Nu = C\Pi_N^n \tag{6.54}$$

   from Table 6.4 and compute Nu.

5. Employing Nu, k, and L, compute the heat transfer coefficient from

$$h\left[\frac{\text{W}}{\text{m}^2\cdot\text{K}}\right] = Nu\text{ [Dimensionless]}\,\frac{k\left[\frac{\text{W}}{\text{m}\cdot\text{K}}\right]}{L\,[\text{m}]}. \tag{6.55}$$

Table 6.4 Natural convection correlations.

NATURAL CONVECTION		$T_w =$ Const.
$Nu = C\Pi_N^n$, $\quad \Pi_N = \dfrac{Ra}{1 + C_0/Pr}$		$0 < Pr < \infty$
		T_f

Geometry	Range	Correlation		
		C_0	C	n
(vertical plate/cylinder, L)	$10^4 \leq \Pi_N \leq 10^9$	0.492	0.670	1/4
	$10^9 < \Pi_N \leq 10^{12}$		0.150	1/3
(horizontal cylinder, D)	$10^4 \leq \Pi_N \leq 10^9$	0.559	0.518	1/4
	$10^9 < \Pi_N \leq 10^{12}$		0.150	1/3
(sphere, D)	$10^6 \leq \Pi_N \leq 10^9$	0.469	0.589	1/4
	$10^9 < \Pi_N \leq 10^{12}$		0.150	1/3

EXAMPLE 6.5[7]

An electrically heated, square (0.4 m × 0.4 m × 0.005 m), vertical flat plate is to be kept at $T_w = 95\ °C$ in an ambient at $T_\infty = 25\ °C$. We wish to determine the power supply to the plate in ambient (a) air, (b) water, oil, or mercury.

Assuming the plate temperature to be uniform, consider a system for the entire plate. The first law of thermodynamics for this system is

$$0 = -2\dot{Q}_C + \dot{W}_e, \tag{6.56}$$

$\dot{Q}_C$ being the rate of convective heat loss,

$$\dot{Q}_C = hA(T_w - T_\infty), \tag{6.57}$$

---

[7] The FORTRAN program EX6–5.F is listed in the appendix of this chapter.

from one side of the plate and $\dot{W}_e$ the electric power supply. Insert Eq. (6.57) into Eq. (6.56) for the governing equation,

$$\dot{W}_e = 2hA(T_w - T_\infty). \tag{6.58}$$

Because of the assumed lumped system, the governing equation is algebraic rather than differential. No boundary conditions are needed, and Eq. (6.58) is also the solution of the problem.

Computation of heat transfer coefficient.

(a) AIR: Following the procedure described in Section 6.3.1:

1. Film temperature is

$$T_f = \frac{1}{2}(95 + 25) = 60\,°\text{C}(333\,\text{K}).$$

The temperature closest to T_f in the air table of Appendix B2 is

$$T_f = 350\,\text{K}.$$

2. Three thermomechanical air properties at this temperature are

Air at	$\dfrac{g\beta}{\nu\alpha} = 44.8 \times 10^6\,[\text{m}^3\cdot\text{K}]^{-1}$
$T_f = 350\,\text{K}$	$Pr = 0.7$
	$k = 30 \times 10^{-3}\,\text{W/m}\cdot\text{K}$

3. For a vertical plate, the appropriate length to be used is L. The Rayleigh number is

$$Ra_L = \frac{g\beta}{\nu\alpha}(T_w - T_\infty)L^3 = 44.8 \times 10^6 (95 - 25)(0.4)^3 = 2 \times 10^8.$$

4. For a vertical plate, $C_0 = 0.492$ from Table 6.4 and

$$\Pi_N = \frac{2 \times 10^8}{1 + \dfrac{0.492}{0.7}} = \frac{2 \times 10^8}{1.70} \cong 1.17 \times 10^8 < 10^9.$$

The flow is laminar. From Table 6.4, for buoyancy-driven laminar flow next to a vertical plate,

$$C = 0.670,\ n = \frac{1}{4},$$

and

$$Nu = 0.670\Pi_N^{1/4} = 0.670(1.17 \times 10^8)^{1/4} = 69.7.$$

5. The heat transfer coefficient is then

$$h = Nu\left(\frac{k}{L}\right) = 69.7\frac{30 \times 10^{-3}}{0.4} \cong 5.2\,\text{W/m}^2\cdot\text{K}$$

Solution: From Eq. (6.58),

$$\dot{W}_e = 2 \times 5.2(0.4)^2(95 - 25) \cong 116 \text{ W}$$

or

$$\frac{\dot{W}_e}{2A} = 5.2(95 - 25) \cong 364 \text{ W/m}^2$$

is the power need to keep the plate at $T_w = 95\,°\text{C}$ in air at $T_\infty = 25\,°\text{C}$.

Repeat the foregoing calculations for $T_f = 300$ K and 325 K. What is your conclusion? By the computer program provided, the interested reader may parametrically study the effect of T_w on the power supply.

(b) The problem may also be repeated by considering water, oil, and mercury as the ambient fluid. Table 6.5 summarizes the results obtained for these cases. Note the several orders of magnitude difference between the power need of each case. Note also the agreement between the computed h values given in Table 6.5 with the orders of magnitude, suggested in Table 1.2. This comparison assures the reader about the soundness of the computed values. ◆

Table 6.5

	Air	Water	Oil	Mercury
T_f K	350	335	330	350
Table	Appendix B–2	Appendix B–3	Appendix B–3	Appendix B–3
$\dfrac{g\beta}{v\alpha}\left[\dfrac{1}{\text{m}^3\cdot\text{K}}\right]$	44.8×10^6	71.38×10^9	0.888×10^9	3.66×10^9
Pr	0.7	2.88	1,205	0.0196
$k\left[\dfrac{\text{W}}{\text{m}\cdot\text{K}}\right]$	30×10^{-3}	656×10^{-3}	141×10^{-3}	$9{,}180 \times 10^{-3}$
Ra	1.97×10^8	3.19×10^{11}	3.98×10^9	1.64×10^{10}
C_0	0.492	0.492	0.492	0.492
Π_N	1.17×10^8	2.73×10^{11}	3.98×10^9	6.28×10^8
Flow	Laminar	Turbulent	Turbulent	Laminar
C	0.670	0.150	0.150	0.670
n	$\frac{1}{4}$	$\frac{1}{3}$	$\frac{1}{3}$	$\frac{1}{4}$
Nu	69.7	973	238	106
$h\left[\dfrac{\text{W}}{\text{m}^2\cdot\text{K}}\right]$	5.2	1,600	84	2,430
$\dot{W}_e$ [kW]	0.116	35.8	1.88	54.5
$\dfrac{\dot{W}_e}{2A}\left[\dfrac{\text{kW}}{\text{m}^2}\right]$	0.364	112	5.88	170

EXAMPLE 6.6

Reconsider Ex. 6.5 for an infinitely long horizontal cylinder of diameter $D = 0.4$ m to be kept at $T_w = 95\,°\mathrm{C}$ in an ambient $T_\infty = 25\,°\mathrm{C}$.

Computation of heat transfer coefficient.

(a) Air: The first three steps of the computation are identical to those of the preceding example, leading to
$$Ra_D = 2 \times 10^8.$$

4. For a horizontal cylinder, $C_0 = 0.559$ from Table 6.4 and
$$\Pi_N = \frac{2 \times 10^8}{1 + \dfrac{0.559}{0.7}} = \frac{2 \times 10^8}{1.80} \cong 1.11 \times 10^8 < 10^9.$$

The flow is laminar. From Table 6.4, for a horizontal cylinder,
$$C = 0.518, \quad n = \frac{1}{4},$$

and
$$Nu = 0.518\Pi_N^{1/4} = 0.518(1.11 \times 10^8)^{1/4} = 53.2.$$

5. Then,
$$h = Nu\left(\frac{k}{D}\right) = 53.2\frac{30 \times 10^{-3}}{0.4} = 3.99 \text{ W/m}^2\cdot\text{K}.$$

Solution:
For the power need per $L = 0.4$ m length of the cylinder,
$$\dot{W}_e = hA(T_w - T_\infty), \quad A = \pi DL,$$
$$\dot{W}_e = 3.99 \times \pi \times 0.4 \times 0.4(95 - 25) \cong 140 \text{ W},$$

we have
$$\frac{\dot{W}_e}{A} = 3.99(95 - 25) \cong 279 \text{ W/m}^2,$$

which keeps the cylinder at $T_w = 95\,°\mathrm{C}$ in air at $T_\infty = 25\,°\mathrm{C}$.

Note that the total power need for the cylinder is greater than that for the flat plate while the power/area need for the cylinder is smaller than that for the flat plate. (Why?) The problem may also be repeated by considering water, oil, and mercury as the ambient fluid. Table 6.6 summarizes the results obtained for these cases. Note the several orders of magnitude difference between the power need of each case. ◆

Table 6.6

	Air	Water	Oil	Mercury
T_f K	350	335	330	350
Pr	0.7	2.88	1,205	0.0196
$k \left[\dfrac{W}{m \cdot K} \right]$	30×10^{-3}	656×10^{-3}	141×10^{-3}	$9{,}180 \times 10^{-3}$
Ra	1.97×10^{8}	3.19×10^{11}	3.98×10^{9}	1.64×10^{10}
C_0	0.559	0.559	0.559	0.559
Π_N	1.09×10^{8}	2.67×10^{11}	3.98×10^{9}	5.56×10^{8}
Flow	Laminar	Turbulent	Turbulent	Laminar
C	0.518	0.150	0.150	0.518
n	$\dfrac{1}{4}$	$\dfrac{1}{3}$	$\dfrac{1}{3}$	$\dfrac{1}{4}$
Nu	52.9	966	238	79.5
$h \left[\dfrac{W}{m^2 \cdot K} \right]$	3.99	1,580	84	1,830
$\dot{W}_e$ [kW]	0.140	55.6	2.96	64.4
$\dfrac{\dot{W}_e}{A} \left[\dfrac{W}{m^2} \right]$	0.279	111	5.88	128

EXAMPLE 6.7

Reconsider Ex. 2.13, in which the heat transfer coefficients are given. Here, we wish to compute the outer heat transfer coefficient.

Computation of heat transfer coefficient.

Following the procedure described in Section 6.3.1:

1. **Film temperature:**
   Since $h_i \gg h_0$ (forced convection in water $\gg$ natural convection in air), the internal convective resistance and the radial temperature change are negligible, and
   $$T_w \cong T_i, \quad U \cong h_0.$$

We also learned in Ex. 2.13 that the longitudinal change in the water temperature is small. Accordingly,

$$T_f \cong \frac{1}{2}(T_i + T_\infty) = \frac{1}{2}(97 + 27) = 62\,°C\,(335\,K)$$

can safely be used in the computation of h_0. The temperature closest to T_f in the air table of Appendix B–2 is $T_f = 350$ K.

2. We have already used this temperature in Exs. 6.5 and 6.6 for the thermomechanical properties of air:

Air at	$\dfrac{g\beta}{\nu\alpha} = 44.8 \times 10^6 \, [\text{m}^3 \cdot \text{K}]^{-1}$
$T_f = 350 \text{ K}$	$Pr = 0.7$
	$k = 30 \times 10^{-3} \text{ W/m·K}$

3. Here, the Rayleigh number

$$Ra_D = \frac{g\beta}{\nu\alpha}(T_w - T_\infty)D^3 = 44.8 \times 10^6 (97 - 27)(0.0334)^3 = 1.17 \times 10^5$$

is different from those in Exs. 6.5 and 6.6 because of the geometry difference.

4. For a horizontal pipe (cylinder), $C_0 = 0.559$ from Table 6.4 and

$$\Pi_N = \frac{1.17 \times 10^5}{1 + \dfrac{0.559}{0.7}} = \frac{1.17 \times 10^5}{1.80} \cong 0.651 \times 10^5 < 10^9$$

The flow is laminar. From Table 6.4, for buoyancy-driven laminar flow around a horizontal cylinder,

$$C = 0.518, \quad n = \frac{1}{4},$$

and

$$Nu = 0.518 \Pi_N^{1/4} = 0.518(0.651 \times 10^5)^{1/4} = 8.3.$$

5. Then, from

$$h_0 = Nu \left(\frac{k}{D}\right) = 8.3 \frac{30 \times 10^{-3}}{0.0334},$$

we have

$$\boxed{h_0 \cong 7.5 \text{ W/m}^2 \cdot \text{K}},$$

which is somewhat less than the assumed value of $h_0 = 10 \text{ W/m}^2 \cdot \text{K}$ in Ex. 2.13. With the new value of h_0, the exit temperature becomes slightly higher, with the computation (using the result from the first key problem of Section 2.5) leading to $T_e \cong 94.5 \, °\text{C}$.

6.3.2 Computation of the Heat Transfer Coefficient for Given q_w

Second Key Problem.

Given: Power generation $\dot{W}_e$ within an object and ambient temperature T_∞.

Required: Temperature T_w of the object.

Computational Steps.

1. Assume flow to be turbulent, which is the case for most practical situations. The laminar cases of Exs. 6.6 and 6.7 are exceptions, provided for illustration. The computation of h requires knowledge of T_w, which is the question. For an initial estimate of T_w, pick an h from Table 1.2 corresponding to the upper limit of natural convection in the ambient fluid.
2. Make an initial estimate of T_w from

$$\dot{W}_e = hA(T_w - T_\infty), \qquad (6.59)$$

   A being the total heat transfer area ($2L^2$ for square plate).
3. Follow the five steps of the first key problem to obtain an h.
4. Iterate the problem until the difference between the assumed and found h becomes negligible.
5. Evaluate T_w with the final value of h. ◆

EXAMPLE 6.8[8]

Electrical power $\dot{W}_e = 2$ kW is steadily applied to a square (0.4 m × 0.4 m × 0.005 m) vertical plate kept in an engine oil ambient at 25 °C. We wish to determine the surface temperature T_w of the plate.

Following the computational steps above:

1. From the Table 1.2, assume $h = 100$ W/m²·K, which is close to the upper limit for turbulent natural convection in an oil.
2. From

$$\dot{W}_e = hA(T_w - T_\infty),$$

   $$2 \text{ kW} = 0.1 \text{ kW/m}^2\cdot\text{K} \times (2 \times 0.4 \times 0.4)\text{m}^2 \times (T_w - 25) \text{ K}.$$

   make an initial estimate of

   $$T_w \cong 88 \,°\text{C}.$$

3. Now, following the five steps of the first key problem:

   (a)
   $$T_f = \frac{1}{2}(88 + 25) \cong 57\,°\text{C}(330 \text{ K}).$$

   (b) From Appendix B–3, the three thermomechanical properties of engine oil are:

Engine oil at	$\dfrac{g\beta}{\nu\alpha} = 0.888 \times 10^9 \,[\text{m}^3\cdot\text{K}]^{-1}$
$T_f = 330$ K	$Pr = 1{,}205$
	$k = 141 \times 10^{-3}$ W/m·K

---

[8] The FORTRAN program EX6–8.F is listed in the appendix of this chapter.

(c) The Rayleigh number is

$$Ra_L = \frac{g\beta}{\nu\alpha}(T_w - T_\infty)L^3 = 0.888 \times 10^9 (88 - 25)(0.4)^3 = 3.58 \times 10^9.$$

(d) For a vertical plate, $C_0 = 0.492$ from Table 6.4 and

$$\Pi_N = \frac{3.58 \times 10^9}{1 + \dfrac{0.492}{1205}} \cong 3.58 \times 10^9 > 10^9.$$

The flow is turbulent. From Table 6.4, for buoyancy-driven turbulent flow next to a vertical plate,

$$C = 0.150, \quad n = \frac{1}{3},$$

and

$$Nu = 0.150\Pi_N^{1/3} = 0.150(3.58 \times 10^9)^{1/3} \cong 229.$$

(e) From

$$h = Nu\left(\frac{k}{L}\right) = 229\frac{141 \times 10^{-3}}{0.4} \cong 81 \text{ W/m}^2 \cdot \text{K}$$

we assumed $h = 100$ W/m²·K and found $h = 81$ W/m²·K. Averaging the assumed and found values of h,

$$h = \frac{1}{2}(100 + 81) \cong 90 \text{ W/m}^2 \cdot \text{K},$$

recompute the value of T_w,

$$2 \text{ kW} = 0.09 \text{ kW/m}^2 \cdot \text{K} \times (2 \times 0.4 \times 0.4) \text{ m}^2 \times (T_w - 25) \text{ K},$$

$$T_w = 94 \, °\text{C}.$$

4. Since $T_w = 88\,°\text{C}$ obtained by the assumed h is quite close to $T_w = 94\,°\text{C}$ obtained by the computed h, no iteration is needed. ◆

EXAMPLE 6.9[9]

Ethylene glycol is pumped through a pipeline of diameter $D = 0.4$ m which runs across a lake $\ell = 200$ m wide (Fig. 6.11). The velocity and inlet temperature of the oil are 2.5 m/s and $T_i = 50\,°\text{C}$. The temperature of the lake water is $T_\infty = 5\,°\text{C}$. Evaluate the exit temperature of the oil.

---

[9] The FORTRAN program EX6–9.F is listed in the Appendix.

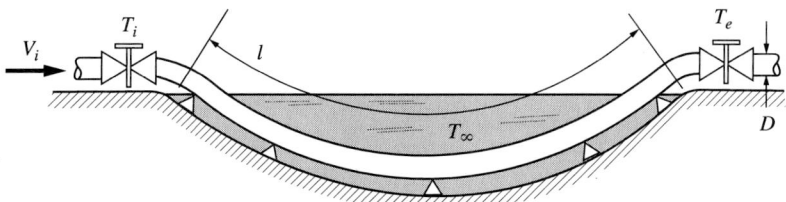

Figure 6.11 System configuration.

For an initial estimate, use Table 1.2 to approximate the values of heat transfer coefficient:

$$h_0 \cong 1{,}000 \text{ W/m}^2\cdot\text{K} \quad \text{for forced oil,}$$

$$h_w \cong 1{,}000 \text{ W/m}^2\cdot\text{K} \quad \text{for stagnant water,}$$

and use the inlet oil temperature for properties from Appendix B–2:

Ethylene glycol at	$\rho = 1{,}096 \text{ kg/m}^3$
$T_i = 50\,°\text{C} \cong 320 \text{ K}$	$c_p = 2{,}505 \text{ J/kg·K}$

Then, the exit oil temperature may be evaluated from Eq. (2.160):

$$\frac{T_e - T_\infty}{T_i - T_\infty} = \exp\left[-\frac{4U\ell}{\rho c_p V D}\right],$$

where

$$U = 500 \text{ W/m}^2\cdot\text{K},$$

$$\frac{1}{U} = \frac{1}{h_0} + \frac{1}{h_w} = \frac{1}{1{,}000} + \frac{1}{1{,}000},$$

and

$$\frac{4U\ell}{\rho c_p V D} = \frac{4 \times 500 \text{ W/m}^2\cdot\text{K} \times 200 \text{ m}}{1{,}096 \text{ kg/m}^3 \times 2{,}505 \text{ J/kg·K} \times 2.5 \text{ m/s} \times 0.4 \text{ m}} \cong 0.147.$$

Then, Eq. (2.160) gives

$$\frac{T_e - 5}{50 - 5} = e^{-0.147} \cong 0.864$$

or

$$T_e = 5 + 45 \times 0.864 \cong 44\,°\text{C}$$

and

$$T_i - T_e = 50 - 44 = 6\,°\text{C}.$$

324 Chap. 6 Correlations For Convection

This small temperature drop along the length of the pipe does not significantly affect the property values of oil and water. Accordingly, we evaluate h_o and h_w at the inlet conditions.

Forced convection in oil.

1. Adjust $T_b \cong T_i = 50\,°C$ to 320 K.
2. From Appendix B–2, read

Ethylene glycol at	$\nu = 6.91 \times 10^{-6}\,\mathrm{m^2/s}$
$T_b = 320\,K$	$Pr = 73.5$
	$k = 0.258\,\mathrm{W/m\cdot K}$

3. Compute the Reynolds number:

   $$Re = \frac{VD}{\nu_b} = \frac{2\,\mathrm{m/s} \times 0.4\,\mathrm{m}}{6.91 \times 10^{-6}\,\mathrm{m^2/s}} = 1.16 \times 10^5.$$

   The flow is turbulent.

4. From Table 6.1 for $T_b > T_w$, pick $n = 0.3$,

   $$Nu = 0.023\,Re_D^{0.8}\,Pr^{0.3} = 0.023(116{,}000)^{0.8}(73.5)^{0.3} = 940.$$

5. Then,

   $$h_o = Nu\left(\frac{k}{D}\right) = 940\,\frac{0.258}{0.4},$$

   $$\boxed{h_o = 606\,\mathrm{W/m^2\cdot K}}$$

Natural convection in water.

1. Because we assumed $h_w = h_o$,

   $$T_w \cong \frac{1}{2}(T_i + T_\infty) = \frac{1}{2}(50 + 5) \cong 28\,°C$$

   and

   $$T_f = \frac{1}{2}(T_w + T_\infty) = \frac{1}{2}(28 + 5) \cong 17\,°C(290\,K).$$

2. From Appendix B–2, read

Water at	$\dfrac{g\beta}{\nu\alpha} = 11.03 \times 10^9\,[\mathrm{m^3\cdot K}]^{-1}$
$T_f = 290\,K$	$Pr = 7.56$
	$k = 0.598\,\mathrm{W/m\cdot K}$

3. The Rayleigh number is

$$Ra_D = \frac{g\beta}{\nu\alpha}(T_w - T_\infty)D^3 = 11.03 \times 10^9(28 - 5)(0.4)^3 = 16.24 \times 10^9.$$

4. For a horizontal pipe, $C_0 = 0.559$ from Table 6.4 and

$$\Pi_N = \frac{16.24 \times 10^9}{1 + \dfrac{0.559}{7.56}} = \frac{16.24 \times 10^9}{1.07} = 15.18 \times 10^9 > 10^9.$$

The flow is turbulent.

5. From Table 6.4, for buoyancy-driven turbulent flow around a horizontal pipe,

$$C = 0.150, \quad n = \frac{1}{3},$$

and

$$Nu = 0.150\Pi_N^{1/3} = 0.150(15.18 \times 10^9)^{1/3} \cong 371.$$

6. Then, from

$$h_w = Nu\left(\frac{k}{D}\right) = 371\frac{0.598}{0.4}$$

we have

$$\boxed{h_w = 555 \text{ W/m}^2 \cdot \text{K}}$$

Our initial assumption of $h_w \cong h_0$ turns out to be holding approximately, and no iterations are needed for better values. However, the calculated heat transfer coefficients are about one-half of the assumed coefficients, so the exit temperature must be recalculated. For these new values,

$$\frac{1}{U} = \frac{1}{h_0} + \frac{1}{h_w} = \frac{1}{606} + \frac{1}{555},$$

$$U = 290 \text{ W/m}^2 \cdot \text{K},$$

and

$$\frac{4U\ell}{\rho c_p V D} = 0.147 \times \frac{290}{500} \cong 0.085.$$

Then, Eq. (<2-180>) gives

$$\frac{T_e - 5}{50 - 5} = e^{-0.085} \cong 0.918$$

or

$$T_e = 5 + 45 \times 0.918 \cong 46\,°\text{C},$$

which is quite close to the initial estimate of $T_e = 44\,°\text{C}$. By the computer program provided, the interested reader may parametrically study the exit temperature for a number of fluids under different conditions.

◆

Table 6.7 Natural convection correlations for horizontal plates

Natural Convection			T_w = Const.	
$\bar{N}_u = C\,Ra^n$			$Pr \geq 0.7$ T_f	
Geometry	Range	C	n
Hot ($T_w > T_\infty$) horizontal surface facing up, or,	$10^5 \leq Ra \leq 10^7$	0.54	1/4
Cold ($T_w < T_\infty$) horizontal surface facing down	$10^7 \leq Ra \leq 10^{10}$	0.15	1/3
Hot ($T_w > T_\infty$) horizontal surface facing down, or, Cold ($T_w < T_\infty$) horizontal surface facing up	$10^5 \leq Ra \leq 10^{10}$	0.27	1/4

FOR HORIZONTAL SURFACES:

The characteristic length is one side of a square plate,

the mean of the two sides of a rectangular plate,

0.9 diameter of a circular disk, and

4 area/perimeter of an odd surface.

So far, for computation of heat transfer from vertical flat plates and cylinders, horizontal cylinders and spheres, we have used Table 6.4 correlations in the form of

$$Nu = C\Pi_N^n, \tag{6.54}$$

which applies for any Prandtl number. For computation of heat transfer from horizontal plates, the only correlations available are in the form of

$$Nu = C\,Ra^n, \tag{6.60}$$

which applies for $Pr > 1$. These correlations are collected in Table 6.7. Due to different buoyancy effects, the top and bottom heat transfer coefficients of a horizontal plate are different. Before we illustrate the use of this table, let us demonstrate the shortcomings of correlations based on the Rayleigh number by rearranging Eq. (6.46) as

$$\frac{\Pi_N}{Ra} = \frac{1}{1 + Pr^{-1}}. \tag{6.61}$$

The definition of Prandtl number,

$$Pr = \frac{\dot{Q}_H}{\dot{Q}_K} \times \frac{F_I}{F_V} \tag{5.154}$$

shows that Pr^{-1} is a measure of the inertial effect on natural convection. For viscous oils, $Pr^{-1} \ll 1$ and the inertial effect is negligible. For liquid metals, $Pr^{-1} \gg 1$ and the inertial effect dominates (recall Fig. 5.12). A correlation based on Eq. (6.60) applies strictly to viscous oils, water and somewhat approximately to gases but not to liquid metals. Now, we proceed to the use of Table 6.7 correlations based on Eq. (6.60) for the two key problems associated with horizontal plates. Since these problems are special cases of the previous key problems, and require only the computation of Ra rather than Π_N, they are illustrated in terms of the following two examples.

EXAMPLE 6.10

We wish to reconsider Ex. 6.5 for a horizontal plate in ambient air. Eq. (6.58) continues to be valid for the present case provided the upward and downward coefficients are distinguished. Thus

$$\dot{W}_e = (h_1 + h_2)A(T_w - T_\infty). \quad (6.62)$$

Computation of heat transfer coefficient.

1. Film temperature closest to T_f in the air table of Appendix B2 is

$$T_f = 350 \text{ K}.$$

2. Three thermomechanical air properties, already used in Ex. 6.6, are

Air at	$\dfrac{g\beta}{\nu\alpha} = 44.8 \times 10^6 \, [\text{m}^3 \cdot \text{K}]^{-1}$
$T_f = 350$ K	$Pr = 0.7$
	$k = 30 \times 10^{-3}$ W/m·K

3. The Rayleigh number is

$$Ra = 2 \times 10^8.$$

4. The plate temperature T_w is higher than the ambient temperature T_∞. Table 6.7 gives for the upward heat transfer coefficient,

$$C = 0.15, \; n = \frac{1}{3},$$

indicating turbulent conditions, and gives for the downward heat transfer coefficient,

$$C = 0.27, \; n = \frac{1}{4},$$

indicating laminar conditions. Then we have for the upward heat transfer

$$Nu_1 = 0.15 Ra^{1/3} = 0.15(1.97 \times 10^8)^{1/3} = 87.3$$

and for the downward heat transfer

$$Nu_2 = 0.27 Ra^{1/4} = 0.27(1.97 \times 10^8)^{1/4} = 32.$$

5. The heat transfer coefficients are

$$h_1 = Nu_1 \left(\frac{k}{L}\right) = 87.3 \frac{30 \times 10^{-3}}{0.4} \cong 6.5 \text{ W/m}^2\cdot\text{K}$$

and

$$h_2 = Nu_2 \left(\frac{k}{L}\right) = 32 \frac{30 \times 10^{-3}}{0.4} \cong 2.4 \text{ W/m}^2\cdot\text{K}$$

As a rule of thumb, keep in mind that the **downward heat transfer coefficient** usually turns out to be **20–30% of the upward heat transfer coefficient**.

Solution: From Eq. (6.62),

$$\dot{W}_e = (6.5 + 2.4)(0.4)^2(95 - 25) \cong 100 \text{ W}$$

is the power needed to keep the horizontal plate at $T_w = 95\,°\text{C}$ in air at $T_\infty = 25\,°\text{C}$. ◆

EXAMPLE 6.11

Example 6.8 is reconsidered for a horizontal flat plate. We wish to determine the wall temperature of the plate.

Computational Steps.

1. Equation (6.62),

$$\dot{W}_e = (h_1 + h_2)A(T_w - T_\infty), \tag{6.62}$$

obtained from the first four steps of formulation, continues to apply. To determine T_w, we need to know h_1 and h_2, which in turn depend on T_w. We use a trial-and-error procedure as follows.

First, based on the insight gained in the preceding example, assume the downward heat transfer coefficient to be about 30% of the upward coefficient,

$$h_2 = 0.3\, h_1,$$

and, as an estimate for the upward coefficient, pick from Table 1.2

$$h_1 = 120 \text{ kW/m}^2\cdot\text{K},$$

which is the upper bound for buoyant oil flows.

2. From

$$\dot{W}_e = 1.3\, h_1 A(T_w - T_\infty)$$

or

$$2{,}000 \text{ W} = 1.3 \times 120 \text{ W/m}^2\cdot\text{K} \times (0.4 \times 0.4) \text{ m}^2 \times (T_w - 25) \text{ K}$$

we get an estimate

$$T_w = 105\,°\text{C}.$$

3. The problem is reduced, in terms of the assumed h_1 and h_2, to the first key problem to be carried out for $T_w = 105\,°\text{C}$.

(a) The film temperature is

$$T_f = \frac{T_w + T_\infty}{2} = \frac{1}{2}(105 + 25) \cong 65\,°C \cong 340\,K.$$

(b) The thermomechanical properties are

Engine oil at	$\dfrac{g\beta}{\nu\alpha} = 1.428 \times 10^9\,[m^3 \cdot K]^{-1}$
$T_f = 340\,K$	$k = 141 \times 10^{-3}\,W/m \cdot K$

For the correlations in Table 6.7, no Prandtl number is needed because of the neglected inertial effect.

(c) The Rayleigh number is

$$Ra_L = \frac{g\beta}{\nu\alpha}(T_w - T_\infty)L^3 = 1.428 \times 10^9 (105 - 25)(0.4)^3 = 7.31 \times 10^9.$$

(d) Table 6.7 suggests, for this Rayleigh number, turbulent upward heating,

$$Nu_1 = 0.15 Ra_L^{1/3} = 0.15(7.31 \times 10^9)^{1/3} = 291,$$

and downward laminar heating,

$$Nu_2 = 0.27 Ra_L^{1/4} = 0.27(7.31 \times 10^9)^{1/4} = 79.$$

(e) We can then get

$$h_1 = Nu_1 \left(\frac{k}{L}\right) = 291 \frac{141 \times 10^{-3}}{0.4} = 102\,W/m^2 \cdot K,$$

$$h_2 = Nu_2 \left(\frac{k}{L}\right) = 79 \frac{141 \times 10^{-3}}{0.4} = 28\,W/m^2 \cdot K.$$

We assumed $h_1 + h_2 = 156\,W/m^2 \cdot K$, and with a temperature $T_w = 105\,°C$ resulting from this assumption, obtained the foregoing $h_1 + h_2 = 130\,W/m^2 \cdot K$. Actual coefficients are expected to be in between the assumed and computed values, say

$$h_1 + h_2 = \frac{156 + 130}{2} \cong 143\,W/m^2 \cdot K,$$

which yields

$$T_w = 112\,°C.$$

4. This temperature is close enough to the assumed temperature $T_w = 105\,°C$, and no further iteration is needed. ◆

In this chapter we learned, in terms of a few selected correlations, how to use a given correlation rather than exploring the entire literature on correlations. Appendix A contains a list of all correlations available in the literature. The reader may refer to any one of these correlations particularly suited to a problem under consideration.

■ REFERENCES

6.1 V. S. Arpacı, *Microscales of Turbulence-Heat and Mass Transfer Correlations*, Gordon and Breach Science Publishers, Amsterdam, The Netherlands, 1997.

6.2 V. S. Arpacı, and S.-H. Kao, "Thermocapillary driven turbulent heat transfer," *ASME J. Heat Transfer*, 120, 214–219, 1998.

6.3 V. S. Arpacı, and S.-H. Kao, "Microscales of rotating turbulent flows," *Int. J. Heat Mass Transfer*, 40, 3819–3826, 1997.

6.4 V. S. Arpacı, "Microscales of turbulence and heat transfer correlations," *Int. J. Heat Mass Transfer*, 29, p. 1071, 1986a.

6.5 V. S. Arpacı, "Two thermal microscales for natural convection and heat transfer correlations," *Significant Questions in Buoyancy Affected Enclosure or Cavity Flows*, ASME HTD–60, 117, 1986b.

6.6 V. S. Arpacı, "Microscales of turbulence and heat transfer correlations," *Annual Review of Heat Transfer*, vol. 3, p. 195. Hemisphere, New York, 1990.

6.7 V. S. Arpacı, (Keynote Lecture), "Microscales of turbulence, mass transfer correlations," *International Symposium on Turbulence, Heat and Mass Transfer*, Lisbon, Portugal, 1994a.

6.8 V. S. Arpacı, (Keynote Lecture), "Microscales of turbulence, heat transfer correlations," *Tenth International Heat Transfer Conference*, Brighton, United Kingdom, 1994b.

6.9 V. S. Arpacı, "Buoyant turbulent flow driven by internal energy generation," *Int. J. Heat Mass Transfer*, 38, pp. 2761–2770, 1995a.

6.10 V. S. Arpacı, "Microscales of turbulent combustion," *Prog. Energy Combust. Sci.*, 21, 153–171, 1995b.

6.11 V. S. Arpacı, and J. E. Dec, "A theory for buoyancy-driven turbulent flows," *The 24th ASME National Heat Transfer Conference*, ASME 87-HT, vol. 5, p. 1, Pittsburgh, PA, 1987.

6.12 V. S. Arpacı, and S.-H. Kao, "Microscales of turbulent two-phase film," in *ANS Proceedings 1996 National Heat Transfer Conference*, HTC 9, 417–424. Houston, TX, 1996.

6.13 V. S. Arpacı, and A. Selamet, "Buoyancy driven turbulent diffusion flames," *Combust. Flame*, 86, 203, 1991.

6.14 W. M. Rohsenow and H. Y. Choi, *Heat, Mass and Momentum Transfer*. Prentice Hall, Englewood Cliffs, NJ, 1961, p. 59.

6.15 V. S. Arpacı, and P. S. Larsen, *Convection Heat Transfer*. Prentice Hall, Englewood Cliffs, NJ, 1984, p. 405.

6.16 E. N. Sieder and C. E. Tate, "Heat transfer and pressure drop of liquids in tubes," *Ind. Eng. Chem.*, 28, 1936. 1429.

6.17 A. P. Colburn, "A method of correlating forced convection heat transfer data and a comparison with fluid friction," *Trans. AIChE*, 29, 174–210, 1933.

6.18 B. S. Petukhov, "Heat transfer and friction in turbulent pipe flow with variable physical properties," *Advances in Heat Transfer* (J.P. Hartnett and T.F. Irvine, eds.), Academic Press. New York, 1970, pp. 501–564.

6.19 B. Lubarsky and S. J. Kaufman, *Review of Experimental Investigations of Liquid-Metal Heat Transfer*, NACA TN3336, 1956.

6.20 W. M. Kays and H. C. Perkins, *Handbook of Heat Transfer*, chap. 7 (W.M. Rohsenow and J.P. Hartnett, eds.). McGraw-Hill, New York, 1972.

6.21 S. Whitaker, "Forced convection heat-transfer correlations for flow in pipes, past flat plates, single cylinders, single spheres, and flow in packed beds and tube bundles," *AIChE J.*, 18, 361–371, 1972.

6.22 A. Zhukauskas, "Heat transfer from tubes in cross flow," *Advances in Heat Transfer*, 8 (J.P. Hartnett and T.F. Irvine, eds.). Academic Press, New York, 1972, pp. 93–160.

6.23 S. W. Churchill and M. Berstein, "A correlating equation for forced convection from gases and liquids to a circular cylinder in crossflow," *J. Heat Transfer*, 99, 300–306, 1977.

6.24 S. Nakai and T. Okazaki, "Heat transfer from a horizontal circular wire at small Re and Gr numbers–1: Pure convection," *Int. J. Heat Mass Transfer*, 18, 387–396, 1975.

6.25 W. H. McAdams, *Heat Transmission*, 3d ed. McGraw-Hill, New York, 1954, p. 265.

6.26 E. Ackenbach, "Heat transfer from spheres up to $Re = 6$ times 10^6," *Proc. Sixth Int. Heat Trans. Conf.*, vol. 5. Hemisphere, Washington, DC, 1978, 341–346.

6.27 E. D. Grimson, "Correlation and utilization of new data on flow resistance and heat transfer for cross flow of gases over tube banks," *Trans. ASME*, 59, 583–594, 1937.

6.28 C. Y. Warner and V. S. Arpacı, "An experimental investigation of turbulent natural convection in air at low pressure for a vertical heated flat plate," *Int. J. Heat Mass Transfer*, 11, 397–406, 1968.

6.29 S. W. Churchill and, H. H. S. Chu, "Correlating equations for laminar and turbulent free convection from a vertical plate," *Int. J. Heat Mass Transfer*, 18, 1323–1328, 1975.

6.30 T. Fujii and H. Imura, "Natural convection heat transfer from a plate with arbitrary inclination," *Int. J. Heat Mass Transfer*, 15, 755–767, 1972.

6.31 V. T. Morgan, "The overall convective heat transfer from smooth circular cylinders," *Advances in Heat Transfer* (T.F. Irvine and J.P. Hartnett, eds.), 11. Academic Press, New York, 1975, pp. 199–264.

6.32 S. W. Churchill and H. H. S. Chu, "Correlating equations for laminar and turbulent free convection from a horizontal cylinder," *Int. J. Heat Mass Transfer*, 18, 1049–1053, 1975.

6.33 S. Globe and D. Dropkin, "Natural convection heat transfer in liquids confined between two horizontal plates," *J. Heat Transfer*, C81, 24–29, 1959.

6.34 R. K. McGregor and A. P. Emery, "Free convection through vertical plane layers: Moderate and high Pr number fluids," *J. Heat Transfer*, 91, 391–403, 1969.

6.35 I. Catton, "Natural convection in enclosures," *Proc. 6th Int. Heat Transfer Conf.*, 6, 13 Toronto, Canada, 1978.

6.36 K. G. T. Hollands, S. E. Unny, G. D. Raithby and L. Konicek, "Free convective heat transfer across inclined air layers," *J. Heat Transfer*, 98, 189–193, 1976.

6.37 P. S. Ayyaswamy and I. Catton, "The boundary layer regime for natural convection in a differentially heated, tilted rectangular cavity," *J. Heat Transfer*, 95, 543–545, 1973.

6.38 A. F. Mills, Heat Transfer. 2d ed., Prentice-Hall, Upper Saddle River, NJ, 1999.

332 Chap. 6 Correlations For Convection

6.39 F. P. Incropera and D. P. DeWitt, *Introduction to Heat Transfer*. Wiley, New York, 1990.

6.40 F. Eisner, *3rd Int. Cong. Appl. Mech.*, Stockholm, 1930.

6.41 L. F. Moody, "Friction factor for pipe flow," *Trans. ASME*, 66, 671, 1944.

6.42 W. H. Giedt, *"Trans. ASME*, 71, 375, 1949.

6.43 M. Jakob, *Heat Transfer*, vol. 1, New York: Wiley, 1949.

6.4 APPENDIX

```
C----------------------------------
C    EX6-1.F (START)
C----------------------------------
      PROGRAM MAIN
      IMPLICIT REAL*8 (A-H,K-Z)
      PI=4*ATAN(1.)
      WRITE(*,*) 'EXAMPLE 6.1....'
C----------------------------------------------------
C    INPUT DATA
C----------------------------------------------------
      WRITE(*,*) 'INPUT THE FOLLOWING DATA...'
      WRITE(*,*) 'T_B: C'
      READ(*,*) TB
      WRITE(*,*) 'V: m/s'
      READ(*,*) V
      WRITE(*,*) 'D: cm'
      READ(*,*) D
      WRITE(*,*) 'T_INFTY: C'
      READ(*,*) TINFTY
C----------------------------------------------------
C    UNIT CONVERSION
C----------------------------------------------------
      D=0.01*D
C----------------------------------------------------
C    CHART READINGS
C----------------------------------------------------
      WRITE(*,*) 'ADJUST TB=',TB+273,' K'
      WRITE(*,*) 'TO THE CLOSET LISTED TEMPERATURE T_C ON Appendix B-3'
      WRITE(*,*) 'READ FROM Appendix B-3 FOR WATER PROPERTIES AT T_C'
      WRITE(*,*) 'INPUT NU: m^2/s'
      READ(*,*) NU
      WRITE(*,*) 'INPUT PR'
      READ(*,*) PR
          WRITE(*,*) 'INPUT K: W/m.K'
      READ(*,*) K
      WRITE(*,*) 'INPUT RHO: kg/m^3'
      READ(*,*) RHO
          WRITE(*,*) 'INPUT CP: J/kg.K'
      READ(*,*) CP
C----------------------------------------------------
C    CALCULATION
C----------------------------------------------------
      RED=V*D/NU
C----------------------------------------------------
C    CHART READINGS
C----------------------------------------------------
      WRITE(*,*) 'PICK N FROM TABLE 6.1'
      WRITE(*,*) 'INPUT N'
      READ(*,*) N
C----------------------------------------------------
C    CALCULATION
```

```
C-------------------------------------------------
      NU=0.023*RED**0.8*PR**N
      H=K*NU/D
      DOTQ=H*PI*D*(TINFTY-TB)
      DT=4*H*(TINFTY-TB)/(RHO*CP*V*D)
      TE=TB+DT
C-------------------------------------------------
C     ANSWER
C-------------------------------------------------
      WRITE(*,*) 'HEAT TRANSFER FROM STEAM TO WATER IS'
      WRITE(*,*) DOTQ/1000,' kW/m^2 PER UNIT LENGTH'
      WRITE(*,*) 'TEMPERATURE RISE IN THE WATER IS'
      WRITE(*,*) DT,' K/m PER UNIT LENGTH'
      WRITE(*,*) 'EXIT TEMPERATURE OF WATER FOR 1 M LONG TUBE IS'
      WRITE(*,*) TE+273,' K'
C-------------------------------------------------
C     CHART READINGS
C-------------------------------------------------
      WRITE(*,*) 'ADJUST TBE=',TE+273,' K'
      WRITE(*,*) 'TO THE CLOSET LISTED TEMPERATURE T_C ON Appendix B-3'
      WRITE(*,*) 'READ FROM Appendix B-3 FOR WATER PROPERTIES AT T_C'
      WRITE(*,*) 'INPUT NU: m^2/s'
      READ(*,*) NU
      WRITE(*,*) 'INPUT PR'
      READ(*,*) PR
          WRITE(*,*) 'INPUT K: W/m.K'
      READ(*,*) K
C-------------------------------------------------
C     CALCULATION
C-------------------------------------------------
      RED=V*D/NU
      NU=0.023*RED**0.8*PR**0.4
      HI=K*NU/D
      HAVE=(H+HI)/2
      TBAVE=(TB+TE)/2
      DOTQ=HAVE*PI*D*(TINFTY-TBAVE)
C-------------------------------------------------
C     ANSWER
C-------------------------------------------------
      WRITE(*,*) 'HEAT TRANSFER FROM STEAM TO WATER IS'
      WRITE(*,*) DOTQ/1000,' kW/m^2 PER UNIT LENGTH'
      STOP
      END
C----------------------------------------
C     EX6-1.F (END)
C----------------------------------------
C----------------------------------------
C     EX6-2.F (START)
C----------------------------------------
      PROGRAM MAIN
      IMPLICIT REAL*8 (A-H,K-Z)
      PI=4*ATAN(1.)
      WRITE(*,*) 'EXAMPLE 6.2....'
C-------------------------------------------------
C     INPUT DATA
```

```fortran
C-------------------------------------------------
      WRITE(*,*) 'INPUT THE FOLLOWING DATA...'
      WRITE(*,*) 'T_INFTY: C'
      READ(*,*) TINFTY
      WRITE(*,*) 'V: m/s'
      READ(*,*) V
      WRITE(*,*) 'D: cm'
      READ(*,*) D
      WRITE(*,*) 'T_W: C'
      READ(*,*) TW
C-------------------------------------------------
C     UNIT CONVERSION
C-------------------------------------------------
      D=0.01*D
C-------------------------------------------------
C     CALCULATION
C-------------------------------------------------
      TF=(TW+TINFTY)/2
C-------------------------------------------------
C     CHART READINGS
C-------------------------------------------------
      WRITE(*,*) 'ADJUST TF=',TF+273,' K'
      WRITE(*,*) 'TO THE CLOSET LISTED TEMPERATURE T_C ON Appendix B-2'
      WRITE(*,*) 'READ FROM Appendix B-2 FOR AIR PROPERTIES AT T_C'
      WRITE(*,*) 'INPUT NU: m^2/s'
      READ(*,*) NU
      WRITE(*,*) 'INPUT PR'
      READ(*,*) PR
      WRITE(*,*) 'INPUT K: W/m.K'
      READ(*,*) K
C-------------------------------------------------
C     CALCULATION
C-------------------------------------------------
      RED=V*D/NU
C-------------------------------------------------
C     CHART READINGS
C-------------------------------------------------
      WRITE(*,*) 'READ FROM TABLE 6.2 FOR C AND N WITH'
      WRITE(*,*) 'REYNOLDS NUMBER=',RED
      WRITE(*,*) 'INPUT C'
      READ(*,*) C
      WRITE(*,*) 'INPUT N'
      READ(*,*) N
C-------------------------------------------------
C     CALCULATION
C-------------------------------------------------
      NU=C*RED**N*PR**(1./3.)
      H=K*NU/D
      U=H*PI*D*(TW-TINFTY)
C-------------------------------------------------
C     ANSWER
C-------------------------------------------------
      WRITE(*,*) 'ALLOWABLE ELECTRICAL POWER DENSITY IS'
      WRITE(*,*) U/1000,' kW/m^3'
      STOP
      END
C---------------------------------------
C     EX6-2.F (END)
C---------------------------------------
C---------------------------------------
C     EX6-3.F (START)
```

```fortran
C----------------------------------------
      PROGRAM MAIN
      IMPLICIT REAL*8 (A-H,K-Z)
      PI=4*ATAN(1.)
      WRITE(*,*) 'EXAMPLE 6.3....'
C--------------------------------------------------
C     INPUT DATA
C--------------------------------------------------
      WRITE(*,*) 'INPUT THE FOLLOWING DATA...'
      WRITE(*,*) 'T_B: C'
      READ(*,*) TB
      WRITE(*,*) 'V: m/s'
      READ(*,*) V
      WRITE(*,*) 'D: cm'
      READ(*,*) D
      WRITE(*,*) 'T_INFTY: C'
      READ(*,*) TINFTY
      WRITE(*,*) 'VINFTY: m/s'
      READ(*,*) VINFTY
C--------------------------------------------------
C     UNIT CONVERSION
C--------------------------------------------------
      D=0.01*D
C--------------------------------------------------
C     CHART READINGS
C--------------------------------------------------
      WRITE(*,*) 'ADJUST TB=',TB+273,' K'
      WRITE(*,*) 'TO THE CLOSET LISTED TEMPERATURE T_C ON Appendix B-2'
      WRITE(*,*) 'READ FROM Appendix B-2 FOR AIR PROPERTIES AT T_C'
      WRITE(*,*) 'INPUT NU: m^2/s'
      READ(*,*) NU
      WRITE(*,*) 'INPUT PR'
      READ(*,*) PR
         WRITE(*,*) 'INPUT K: W/m.K'
      READ(*,*) K
C--------------------------------------------------
C     CALCULATION
C--------------------------------------------------
      RED=V*D/NU
C--------------------------------------------------
C     CHART READINGS
C--------------------------------------------------
      WRITE(*,*) 'PICK N FROM TABLE 6.1'
      WRITE(*,*) 'INPUT N'
      READ(*,*) N
C--------------------------------------------------
C     CALCULATION
C--------------------------------------------------
      NU=0.023*RED**0.8*PR**N
      HI=K*NU/D
C--------------------------------------------------
C     TW ESTIMATE
C--------------------------------------------------
      WRITE(*,*) 'ESTIMATE HO=N*HI'
      WRITE(*,*) 'INPUT N'
      READ(*,*) N
      TW=(TB+N*TINFTY)/(N+1)
      TF=(TW+TINFTY)/2
C--------------------------------------------------
C     CHART READINGS
```

```
C-----------------------------------------------
      WRITE(*,*) 'ADJUST TF=',TF+273,' K'
      WRITE(*,*) 'TO THE CLOSET LISTED TEMPERATURE T_C ON Appendix B-2'
      WRITE(*,*) 'READ FROM Appendix B-2 FOR AIR PROPERTIES AT T_C'
      WRITE(*,*) 'INPUT NU: m^2/s'
      READ(*,*) NU
      WRITE(*,*) 'INPUT PR'
      READ(*,*) PR
           WRITE(*,*) 'INPUT K: W/m.K'
      READ(*,*) K
C-----------------------------------------------
C     CALCULATION
C-----------------------------------------------
      RED=VINFTY*D/NU
C-----------------------------------------------
C     CHART READINGS
C-----------------------------------------------
      WRITE(*,*) 'READ FROM TABLE 6.2 FOR C AND N WITH'
      WRITE(*,*) 'REYNOLDS NUMBER=',RED
      WRITE(*,*) 'INPUT C'
      READ(*,*) C
      WRITE(*,*) 'INPUT N'
      READ(*,*) N
      NU=C*RED**N*PR**(1./3.)
      HO=K*NU/D
C-----------------------------------------------
C     CALCULATION
C-----------------------------------------------
      U=1/(1/HI+1/HO)
      DOTQ=PI*D*U*(TB-TINFTY)
C-----------------------------------------------
C     ANSWER
C-----------------------------------------------
      WRITE(*,*) 'HEAT LOSS OF THE PIPE IS'
      WRITE(*,*) DOTQ,' W/m PER UNIT LENGTH'
      STOP
      END
C---------------------------------------
C     EX6-3.F (END)
C---------------------------------------
C---------------------------------------
C     EX6-5.F (START)
C---------------------------------------
      PROGRAM MAIN
      IMPLICIT REAL*8 (A-H,K-Z)
      PI=4*ATAN(1.)
      WRITE(*,*) 'EXAMPLE 6.5....'
C-----------------------------------------------
C     INPUT DATA
C-----------------------------------------------
      WRITE(*,*) 'INPUT THE FOLLOWING DATA...'
      WRITE(*,*) 'VERTICAL HEIGHT L: m'
      READ(*,*) L
      WRITE(*,*) 'T_W: C'
      READ(*,*) TW
      WRITE(*,*) 'T_INFTY: C'
      READ(*,*) TINFTY
C-----------------------------------------------
C     CALCULATION
C-----------------------------------------------
      TF=(TW+TINFTY)/2
C-----------------------------------------------
C     CHART READINGS
```

```
C-------------------------------------------------
      WRITE(*,*) 'ADJUST TF=',TF+273,' K'
      WRITE(*,*)'TO THE CLOSET LISTED TEMPERATURE T_C ON Appendix B-2,3'
      WRITE(*,*) 'READ FROM Appendix B-2,3 FOR PROPERTIES AT T_C'
      WRITE(*,*) 'INPUT (G*BETA/NU*ALPHA): 1/m^3.K'
      READ(*,*) GBETA
      WRITE(*,*) 'INPUT PR'
      READ(*,*) PR
          WRITE(*,*) 'INPUT K: W/m.K'
      READ(*,*) K
C-------------------------------------------------
C     CALCULATION
C-------------------------------------------------
      RA=GBETA*(TW-TINFTY)*L**3
C-------------------------------------------------
C     CHART READINGS
C-------------------------------------------------
      WRITE(*,*) 'PICK C0 FROM TABLE 6.4 FOR A VERTICAL PLATE'
      WRITE(*,*) 'INPUT C0'
      READ(*,*) C0
C-------------------------------------------------
C     CALCULATION
C-------------------------------------------------
      PIN=RA/(1+C0/PR)
C-------------------------------------------------
C     CHART READINGS
C-------------------------------------------------
      WRITE(*,*) 'PICK C AND N FROM TABLE 6.4 FOR A VERTICAL PLATE'
      WRITE(*,*) 'ACCORDING TO PIN=', PIN
      WRITE(*,*) 'INPUT C'
      READ(*,*) C
      WRITE(*,*) 'INPUT N'
      READ(*,*) N
C-------------------------------------------------
C     CALCULATION
C-------------------------------------------------
      NU=C*PIN**N
      H=K*NU/L
      DOTWE=H*(TW-TINFTY)
C-------------------------------------------------
C     ANSWER
C-------------------------------------------------
      WRITE(*,*) 'POWER NEED IS'
      WRITE(*,*) DOTWE,' W/m^2 PER UNIT AREA'
      STOP
      END
C-------------------------------------------------
C     EX6-5.F (END)
C-------------------------------------------------
C-------------------------------------------------
C     EX6-8.F (START)
C-------------------------------------------------
      PROGRAM MAIN
      IMPLICIT REAL*8 (A-H,K-Z)
      PI=4*ATAN(1.)
      WRITE(*,*) 'EXAMPLE 6.8....'
C-------------------------------------------------
C     INPUT DATA
```

```
C------------------------------------------------
      WRITE(*,*) 'INPUT THE FOLLOWING DATA...'
      WRITE(*,*) 'DOTWE: kW'
      READ(*,*) DOTWE
      WRITE(*,*) 'VERTICAL HEIGHT L: m'
      READ(*,*) L
      WRITE(*,*) 'WIDTH W: m'
      READ(*,*) W
      WRITE(*,*) 'T_INFTY: C'
      READ(*,*) TINFTY
C------------------------------------------------
C     UNIT CONVERSION
C------------------------------------------------
      DOTWE=1000*DOTWE
C------------------------------------------------
C     ESTIMATE H
C------------------------------------------------
      WRITE(*,*) 'INPUT H (W/m^2.K) ACCORDING TO TABLE 1.2'
      READ(*,*) H
C------------------------------------------------
C     CALCULATION
C------------------------------------------------
      A=2*L*W
      TW=TINFTY+DOTWE/(A*H)
      TF=(TW+TINFTY)/2
C------------------------------------------------
C     CHART READINGS
C------------------------------------------------
      WRITE(*,*) 'ADJUST TF=',TF+273,' K'
      WRITE(*,*)'TO THE CLOSET LISTED TEMPERATURE T_C ON Appendix B-3'
      WRITE(*,*) 'READ FROM Appendix B-3 FOR OIL PROPERTIES AT T_C'
      WRITE(*,*) 'INPUT (G*BETA/NU*ALPHA): 1/m^3.K'
      READ(*,*) GBETA
      WRITE(*,*) 'INPUT PR'
      READ(*,*) PR
            WRITE(*,*) 'INPUT K: W/m.K'
      READ(*,*) K
C------------------------------------------------
C     CALCULATION
C------------------------------------------------
      RA=GBETA*(TW-TINFTY)*L**3
C------------------------------------------------
C     CHART READINGS
C------------------------------------------------
      WRITE(*,*) 'PICK C0 FROM TABLE 6.4 FOR A VERTICAL PLATE'
      WRITE(*,*) 'INPUT C0'
      READ(*,*) C0
C------------------------------------------------
C     CALCULATION
C------------------------------------------------
      PIN=RA/(1+C0/PR)
C------------------------------------------------
C     CHART READINGS
C------------------------------------------------
      WRITE(*,*) 'PICK C AND N FROM TABLE 6.4 FOR A VERTICAL PLATE'
      WRITE(*,*) 'ACCORDING TO PIN=', PIN
      WRITE(*,*) 'INPUT C'
      READ(*,*) C
      WRITE(*,*) 'INPUT N'
      READ(*,*) N
C------------------------------------------------
C     CALCULATION
```

```
C-------------------------------------------------
      NU=C*PIN**N
      H1=K*NU/L
      HAVE=(H+H1)/2
      TW=TINFTY+DOTWE/(A*HAVE)
C-------------------------------------------------
C     ANSWER
C-------------------------------------------------
      WRITE(*,*) 'SURFACE TEMPERATURE OF THE PLATE IS'
      WRITE(*,*) TW,' C'
      STOP
      END
C-----------------------------------
C     EX6-8.F (END)
C-----------------------------------
C-----------------------------------
C     EX6-9.F (START)
C-----------------------------------
      PROGRAM MAIN
      IMPLICIT REAL*8 (A-H,K-Z)
      PI=4*ATAN(1.)
      WRITE(*,*) 'EXAMPLE 6.9....'
C-------------------------------------------------
C     INPUT DATA
C-------------------------------------------------
      WRITE(*,*) 'D: m'
      READ(*,*) D
      WRITE(*,*) 'L: m'
      READ(*,*) L
      WRITE(*,*) 'V: m/s'
      READ(*,*) V
      WRITE(*,*) 'T_I: C'
      READ(*,*) TI
      WRITE(*,*) 'T_INFTY: C'
      READ(*,*) TINFTY
C-------------------------------------------------
C     ESTIMATE H
C-------------------------------------------------
      WRITE(*,*) 'INPUT HO (W/m^2.K) ACCORDING TO TABLE 1.2'
      READ(*,*) HO
      WRITE(*,*) 'INPUT HW (W/m^2.K) ACCORDING TO TABLE 1.2'
      READ(*,*) HW
C-------------------------------------------------
C     CHART READINGS
C-------------------------------------------------
      WRITE(*,*) 'ADJUST TI=',TI+273,' K'
      WRITE(*,*)'TO THE CLOSET LISTED TEMPERATURE T_C ON Appendix B-2'
      WRITE(*,*) 'READ FROM Appendix B-2'
      WRITE(*,*) 'FOR ETHYLENE GLYCOL PROPERTIES AT T_C'
      WRITE(*,*) 'INPUT RHO: kg/m^3'
      READ(*,*) RHO
            WRITE(*,*) 'INPUT CP: J/kg.K'
      READ(*,*) CP
C-------------------------------------------------
C     CALCULATION
C-------------------------------------------------
      U=1/(1/HO+1/HW)
      TE=TINFTY+(TI-TINFTY)*EXP(-4*U*L/(RHO*CP*V*D))
C-------------------------------------------------
C     CHART READINGS
```

```
C------------------------------------------------
  WRITE(*,*) 'ADJUST TB=',TI+273,' K'
  WRITE(*,*)'TO THE CLOSET LISTED TEMPERATURE T_C ON Appendix B-2'
  WRITE(*,*) 'READ FROM Appendix B-2'
  WRITE(*,*) 'FOR ETHYLENE GLYCOL PROPERTIES AT T_C'
  WRITE(*,*) 'INPUT NU: m^2/s'
  READ(*,*) NU
  WRITE(*,*) 'INPUT PR'
  READ(*,*) PR
       WRITE(*,*) 'INPUT K: W/m.K'
  READ(*,*) K
C------------------------------------------------
C     CALCULATION
C------------------------------------------------
  RED=V*D/NU
C------------------------------------------------
C     CHART READINGS
C------------------------------------------------
  WRITE(*,*) 'PICK N FROM TABLE 6.1'
  WRITE(*,*) 'INPUT N'
  READ(*,*) N
C------------------------------------------------
C     CALCULATION
C------------------------------------------------
  NU=0.023*RED**0.8*PR**N
  HO=K*NU/D
  TW=(TI+TINFTY)/2
  TF=(TW+TINFTY)/2
C------------------------------------------------
C     CHART READINGS
C------------------------------------------------
  WRITE(*,*) 'ADJUST TF=',TF+273,' K'
  WRITE(*,*)'TO THE CLOSET LISTED TEMPERATURE T_C ON Appendix B-2'
  WRITE(*,*) 'READ FROM Appendix B-2 FOR WATER PROPERTIES AT T_C'
  WRITE(*,*) 'INPUT (G*BETA/NU*ALPHA): 1/m^3.K'
  READ(*,*) GBETA
  WRITE(*,*) 'INPUT PR'
  READ(*,*) PR
       WRITE(*,*) 'INPUT K: W/m.K'
  READ(*,*) K
C------------------------------------------------
C     CALCULATION
C------------------------------------------------
  RA=GBETA*(TW-TINFTY)*D**3
C------------------------------------------------
C     CHART READINGS
C------------------------------------------------
  WRITE(*,*) 'PICK C0 FROM TABLE 6.4 FOR A HORIZONTAL PIPE'
  WRITE(*,*) 'INPUT C0'
  READ(*,*) C0
C------------------------------------------------
C     CALCULATION
C------------------------------------------------
  PIN=RA/(1+C0/PR)
C------------------------------------------------
C     CHART READINGS
C------------------------------------------------
  WRITE(*,*) 'PICK C AND N FROM TABLE 6.4 FOR A HORIZONTAL PIPE'
  WRITE(*,*) 'ACCORDING TO PIN=', PIN
  WRITE(*,*) 'INPUT C'
  READ(*,*) C
  WRITE(*,*) 'INPUT N'
  READ(*,*) N
C------------------------------------------------
```

```
      C     CALCULATION
      C-------------------------------------------------
            NU=C*PIN**N
            HW=K*NU/D
            U=1/(1/HO+1/HW)
            TE=TINFTY+(TI-TINFTY)*EXP(-4*U*L/(RHO*CP*V*D))
      C-------------------------------------------------
      C     ANSWER
      C-------------------------------------------------
            WRITE(*,*) 'EXIT TEMPERATURE OF THE OIL IS'
            WRITE(*,*) TE,' C'
            STOP
            END
      C-------------------------------------------------
      C     EX6-9.F (END)
      C-------------------------------------------------
```

■ EXERCISES

6.1 Steam at atmospheric pressure condenses on the outside of an $\ell = 5$ m long tube of diameter $D = 0.3$ m. Water at temperature $T_i = 20\,°\text{C}$ flows, with a mass flow rate of 0.2 kg/s, inside the tube. Find the mass flow rate of the condensing steam. For steam: $p = 1$ atm, $T_s = 100\,°\text{C}$, $h_{fg} = 2.257$ kJ/kg.

6.2 Air at $T_\infty = 20\,°\text{C}$ with velocity $V_\infty = 80$ m/s crosses over a pipe ($D = 5$ cm) filled with condensing steam at $p = 1$ atm. Determine the heat transfer from the steam to the air.

6.3 Each element of an air heater is made of a solid cylinder ($k = 400$ W/m·K) of diameter $D = 2$ cm and length $L = 1$ m. Air at temperature $T_\infty = 27\,°\text{C}$ crosses over this cylinder with free-stream velocity $V_\infty = 50$ m/s. The maximum allowable temperature for heating elements is $T_0 = 827\,°\text{C}$. Find the rate of energy generation for each element.

6.4 A solid spherical nuclear ($D = 0.4$ m) is kept at $95\,°\text{C}$ in an ambient at $25\,°\text{C}$. We wish to determine the nuclear power generation within the sphere in ambient (a) air, (b) water, (c) oil or (d) mercury.

6.5 Consider a horizontal flat plate (1 m × 1 m) in stagnant air at a temperature $T_\infty = 20\,°\text{C}$. The upper surface of the plate is subjected to a solar flux $q'' = 600$ W/m². Determine the steady temperature of the plate.

6.6 High-voltage electric power transmitted to a city is dropped to a low voltage in a transformer before use. To keep the transformer at a steady temperature, Joulean power dissipation (recall $EI = I^2 R$) needs to be transferred to an ambient. Assume the transformer to be a vertical cylinder ($D = 0.4$ m, $L = 1$ m) whose walls are kept at $90\,°\text{C}$ in a stagnant water bath at $15\,°\text{C}$. Determine the maximum allowable power dissipation.

6.7 Repeat Prob. 6.6 for a stagnant oil bath and compare the results. Why is oil rather than water used in commercial transformers?

6.8 Repeat Prob. 6.6 for a horizontal cylinder and compare the results. What is your conclusion?

6.9 Hydrogen at 500 kPa (gage) and $T_b = 40°C$ flows with $V = 60$ m/s in an insulated tube of square cross section (Fig 6P–1). Electrical energy $u''' = 0.5$ MW/m^3 is dissipated within the tube[10]. The thermal conductivity of the tube walls is $k = 350$ W/m·K. Compute **(a)** the increase of the bulk hydrogen temperature per unit length of pipe, **(b)** the inner surface temperature of the tube walls, and **(c)** the outer surface temperature of the tube walls.

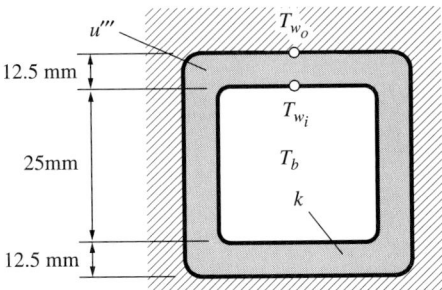

Figure 6P–1

6.10 Water at a temperature $T_b = 77 °C$ flows with a velocity $V = 6$ m/s through a horizontal tube. Air at temperature $T_\infty = 27 °C$ crosses the tube with a velocity $V_\infty = 30$ m/s (Fig. 6P–2). Find the exit temperature of the water at $L = 1$ m.

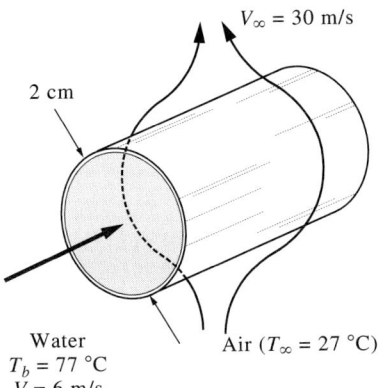

Figure 6P–2

6.11 Bismuth at a temperature $T_b = 400 °C$ flows with velocity $V = 2$ m/s through a horizontal tube of diameter $D = 2.5$ cm and wall thickness $\delta = 1.5$ mm. Water at temperature $T_\infty = 40°C$ crosses the tube with velocity $V_\infty = 2$ m/s. The system is sufficiently pressurized to prevent boiling. Calculate the rate of heat transfer between the bismuth and the water.

6.12 Assume that an industrial burner made of a tube of diameter $D = 40$ cm is internally exposed to a uniform heat flux $q'' = 24$ kW/m^2 resulting from combustion in the tube (Fig. 6P–2). The heat flux is to be transferred to the surrounding stagnant air at temperature $T_\infty = 25 °C$ by convective means.

---

[10] This problem is related to the electric-generator conduits of a modern power plant. Hollow conduits are cooled with a gas flow.

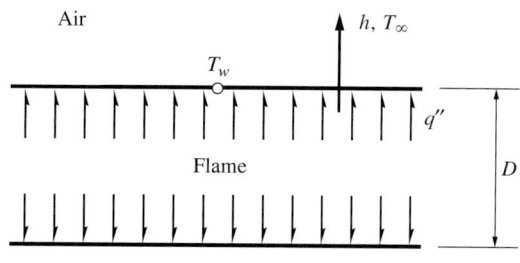

Figure 6P–3

(a) Determine the outside heat transfer coefficient h_o and wall temperature T_w.

(b) Due to material restrictions, T_w cannot be allowed to exceed 1000 °C. Keeping the geometry and fluid the same, what do you propose to reduce T_w? How do you go about determining T_w this time?

6.13 For a calorimeter experiment, an insulated cylindrical shell of diameter $D_0 = 102$ cm is placed concentrically around a $2\ell = 2$ m long furnace of diameter $D_i = 100$ cm (Fig. 6P–4). The distribution of flame heat flux $q''(x)$ [W/m²] acting axisymetrically on the inner surface of the furnace wall is linear with a maximum of $q_0'' = 210,000$ W/m² in the middle of the furnace. Heat is removed by water flowing coaxially between the furnace and the shell. The water inlet temperature and velocity are $T_i = 285$ K and $V = 1$ m/s, respectively. *Neglecting* the furnace-wall thickness, determine (a) the heat transfer coefficient on the outer surface of the furnace, (b) the longitudinal temperature of the furnace wall.

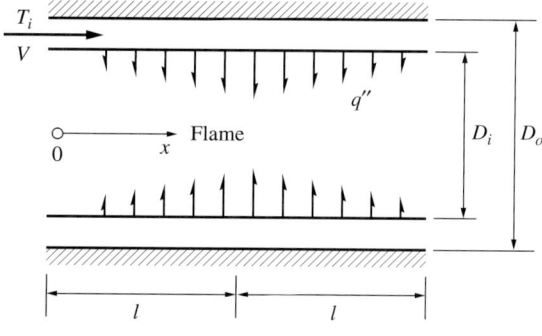

Figure 6P–4

6.14 Natural gas pressurized to 400 kPa at 500 K steadily flows with $V = 60$ m/s through a pipe of 0.3 m diameter lying in a lake (stagnant water) at 300 K. Determine the drop in the bulk temperature of the gas per meter length of the pipe. Use the air table for the gas properties. *Hint:* $\rho_{\text{pressurized}} = 4\rho_{\text{Table}}$, $\nu_{\text{pressurized}} = \dfrac{1}{4}\nu_{\text{Table}}$.

6.15 Consider the exhaust manifold of an internal combustion engine running at 5500 rpm. The exhaust gases enter the cylindrical exhaust duct of diameter 5 cm and length 2 m with a velocity of 100 m/s and an average inlet temperature of 1,100 K. Use air properties for the exhaust gases. The stagnant air is at 27 °C. Neglecting the radiation effect, determine the duct-wall temperature distribution.

344 Chap. 6 Correlations For Convection

6.16 In a certain segment of an Alaska pipeline the pipes are located above the ground. Pumping stations are to be installed at intervals, where the oil is also heated to reduce pumping power requirements. Estimate the distance between the pumping stations and the corresponding pressure drop Δp. *Data:* A carbon steel pipe 1 m OD by 1 cm wall thickness is used, insulated on the outside with a glass fiber blanket of density 32 kg/m³ and thickness 1.3 cm. The oil at $T_{b1} = 95\,°C$ leaves a pumping station with $V = 1.5$ m/s and arrives at the next station at $T_{b2} = 40\,°C$. For properties of crude oil, use the properties of engine oil. The design condition is for wind at $-20\,°C$ blowing cross country at 30 km/hr.

6.17 A tube of diameter $D = 1.25$ cm is placed at the focus of a parabolic solar collector. The solar energy received by the tube is $q'' = 12$ kW/m² (Fig. 6P–5). Water is heated while it flows with velocity $V = 1$ m/s through the tube. The ambient air temperature and the inlet water temperature are the same, $T_i = T_\infty = 25\,°C$. Determine the length of the tube (and the collector) at which the water temperature reaches $T_e = 50\,°C$.

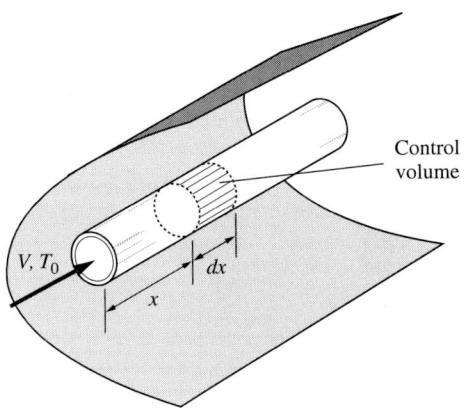

Figure 6P–5

6.18 A 40-liter tank full of water at $15\,°C$ is to be heated to $50\,°C$ by means of a 1-cm-OD copper steam coil having 10 turns of 30 cm diameter. The steam is at atmospheric pressure and its thermal resistance is negligibly small. Neglecting the heat losses from the tank, estimate the heating time required. (*Hint:* Assume the coil as a horizontal cylinder.)

6.19 Consider a spherical fuel element of diameter $D = 5$ cm in pressurized ($p = 145$ kPa, $T_s \cong 110\,°C$) stagnant water at temperature $T_\infty = 20\,°C$.

    (a) Compute the maximum possible (and radially averaged) energy generation u''' without boiling the water.

    (b) Repeat part (a) for a coolant velocity $V_\infty = 1$ m/s.

    (c) Assuming $k_{\text{fuel}} = 20$ W/m·K, determine the temperature of the center of the fuel for (a) and (b).

6.20 Steam having a quality of 96% at a pressure of 175 kPa is flowing at 10 m/s through a steel pipe with inner and outer diameters of 20 mm and 26 mm, respectively. The thermal conductivity of the pipe is 40 W/m·K, and the temperature of the ambient air is $20\,°C$. Estimate the change of steam quality per 10 m length of pipe for (a) stagnant ambient air, (b) ambient air at 10 m/s. (c) repeat (a) and (b) for oils. (d) Repeat (a) and (b) for water.

6.21 Uniform internal energy $u''' = 6$ MW/m^3 is generated in a long, vertical, cylindrical fuel element ($D = 5$ cm, $k = 20$ W/m·K) immersed in a stagnant water pool at $T_\infty = 40\,°$C. Assuming the water is pressurized (no boiling), determine (a) the surface temperature of the fuel element and (b) the center temperature of the fuel element. (c) Repeat (a) and (b) for a water flow with $V_\infty = 5$ m/s.

6.22 A water heater is to be constructed by an electrically heated copper rod ($D = 1$ cm, $\ell = 1$ m) surrounded by an insulated cylindrical shell ($D = 3$ cm) as shown in Fig. 6P–6. The water inlet temperature is $T_i = 25\,°$C. Hot water at temperature $T_0 = 90\,°$C with steady mass flow rate $\dot{m} = 1$ kg/s is required. Determine the electrical power need.

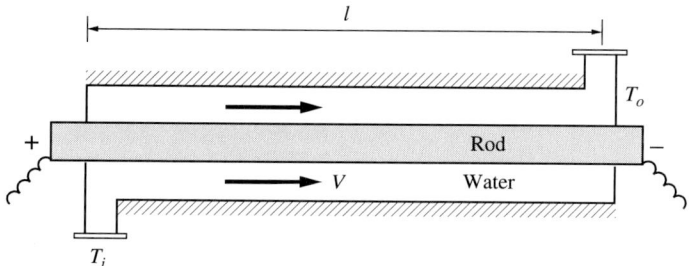

Figure 6P–6

6.23 A reactor core element is simulated by a coaxial sodium flow over a solid fuel rod. In terms of the following data, $u''' = 35$ MW/m^3, $D_i = 25$ mm, $D_0 = 37.5$ mm, $\ell = 6$ m, $T_i = 90\,°$C, $V = 1.2$ m/s, and $k_{\text{fuel rod}} = 9$ W/m·K, evaluate (a) the exit temperature of sodium, (b) the surface temperature of the fuel rod, and (c) the centerline temperature of the fuel rod.

Chapter 7

HEAT EXCHANGERS

A heat exchanger is a device in which heat is transferred from a fluid at a high temperature to a fluid at a low temperature. The usual objective of this transfer is to control the temperature of one of the fluids for a technological purpose. For example, the coolant (water or antifreeze) used in a car engine is cooled in the radiator (heat exchanger) by air flowing over the radiator. From the standpoint of a radiator designer, the coolant temperature drop through the radiator is of critical importance. Additional mechanisms, such as a fan and a thermostat, help to control the coolant temperature.

Conceptually speaking, heat transfer from one fluid to another can be accomplished by mixing the fluids directly or, if mixing is undesirable, through a partition between the fluids. A customary example for direct mixing is cream poured into a cup of coffee. Of course, cream is added to coffee for palatal reasons rather than for heat transfer. However, we are quite conscious of the associated heat transfer process which lowers the temperature of the coffee. An example for heat transfer through a partition is the aforementioned car radiator which keeps the coolant liquid separated from the ambient air. Almost all technological heat exchange problems require a partition because of the need to keep one fluid separated from the other. The heat transfer between the two fluids takes the form of convection on the fluid side and conduction through the partition walls.

In practice, we encounter three types of heat exchangers, classified according to the flow of one fluid relative to that of the other. In a **parallel-flow** heat exchanger both fluids flow in the same direction; in a **counter-flow** heat exchanger the fluids flow in opposite directions; and in a **cross-flow** heat exchanger the fluids flow at right angles

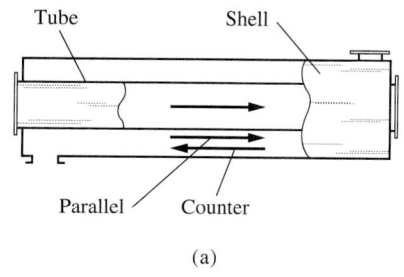

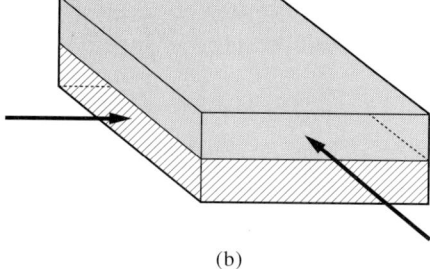

Figure 7.1 Heat exchanger concepts. (a) Parallel-and counter-flow concepts, (b) cross-flow concept.

to each other (Fig. 7.1). Furthermore, as a subclassification, the case in which both fluids traverse the exchanger only once is called a **single-pass** heat exchanger. The two exchangers sketched in Fig. 7.1 are of this type. Figure 7.2(a) shows an exchanger with **two passes** on the tube side. One-half of this exchanger operates under parallel-flow conditions and the other half under counter-flow conditions. Baffles (partitions) improve the mixing (turbulence) and even out the heat transfer between the two fluids. Cross-flow heat exchangers may also be classified as **mixed** and **unmixed**. In some flow

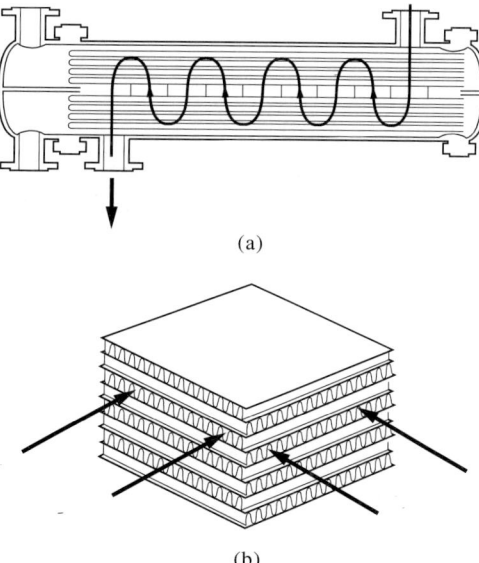

Figure 7.2 Heat exchanger types. (a) Arrangement of one shell pass and multiple of two tube passes, (b) arrangement of cross flow with unmixed fluids.

arrangements [such as the one illustrated in Fig. 7.2(b)] a fluid is partitioned and forced to flow through individual channels, thus making the fluid stream unmixed. In other flow configurations, a fluid is not partitioned and is said to be mixed. We will comment later on the relative merits of the different types of heat exchangers. Among their many applications we may cite heaters (in addition to the already mentioned car radiator), air conditioners, refrigerators, and conventional and nuclear power plants (Fig. 7.3).

Having learned their purpose, types, and applications, we now proceed to the design of heat exchangers, which follows three stages: thermal design, mechanical design, and manufacturing design. Thermal design involves the selection of the type of exchanger and the evaluation of the required heat transfer area between the two fluids. Mechanical design is associated with the pressure drop in and the corrosive properties of both fluids, with provisions being made for thermal expansion and for thermal stresses

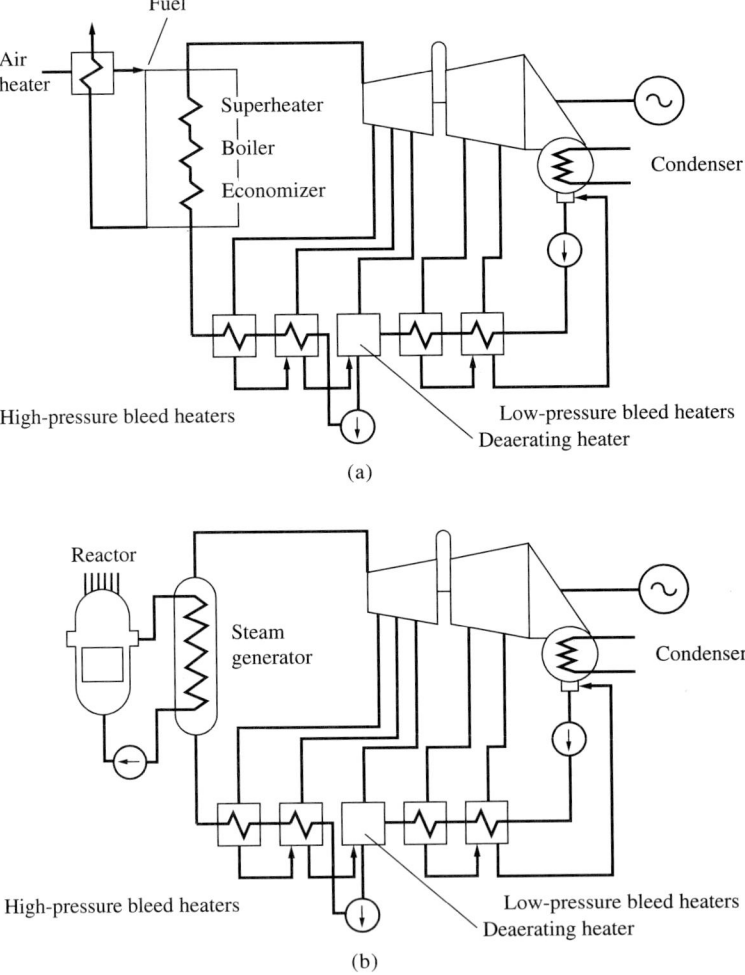

Figure 7.3 Types of power plants. (a) Conventional power plant, (b) nuclear power plant.

resulting from unavoidable constrictions. Manufacturing design is based on satisfying the thermal and mechanical design requirements at the lowest cost. Also, to reduce the costs further, one may select a standard (commercially available) heat exchanger which fulfills the thermal and mechanical requirements of a particular situation.

In this text we are mainly concerned with the thermal design of heat exchangers, a topic treated in the next section. An important related problem, the performance of an available heat exchanger under different conditions, is treated in Section 7.2.

7.1 THERMAL DESIGN. LMTD METHOD

In a typical heat-exchanger design problem, the desired temperature change of one of the fluids (say the hot fluid) is given as well as the inlet temperature of the other fluid. The mass flow rate of each fluid is also specified. With this information in hand, we want to evaluate the heat transfer area, A, required to achieve the necessary temperature change. As in the previous chapters, we will make use of the steps of formulation to relate the known parameters and heat transfer area. For simplicity, we first consider the shell-and-tube heat exchanger shown in Fig. 7.1(a). Following the formulation method set forth in Section 1.7 and used throughout this text, we refer next to the **five steps of formulation**. Since bulk temperatures are adequate for heat-exchanger problems, consider a radially lumped and axially differential control volume for both the hot and cold fluids [Fig. 7.4(a)]. Let us assume that the hot fluid flows through the pipe and the cold fluid through the shell,[1] and the direction of the cold flow is changeable. We will also carry the analysis for both parallel flows and counterflows simultaneously. Furthermore, we will assume that the heat exchanger loses no heat to the ambient, i.e., that there is heat exchange only between the two fluids.

After neglecting the effect of axial conduction, the first law of thermodynamics applied to the control volumes shown in Fig. 7.4(b) gives for the hot fluid

$$\dot{m}h^o \big|_h - \dot{m}\left(h^o + dh^o\right)\big|_h - q_c P dx = 0,$$

which may be rewritten for constant cross section, and thus constant velocity, as

$$\dot{m}c_p T_h - \dot{m}c_p (T + dT)_h - q_c P dx = 0, \tag{7.1}$$

assuming that there is no phase change. Here subscript h refers to the hot fluid, and the other nomenclature is as usual. For convenience, introducing the **heat-capacity flow rate**[2]

$$\dot{m}c_p = C,$$

Equation (7.1) may be rearranged as

$$-C_h dT_h - d\dot{Q} = 0, \tag{7.2}$$

---

[1] In order to minimize heat losses to the ambient, this is typically the arrangement used in actual heat exchangers.

[2] Note that C alone does not have any physical significance.

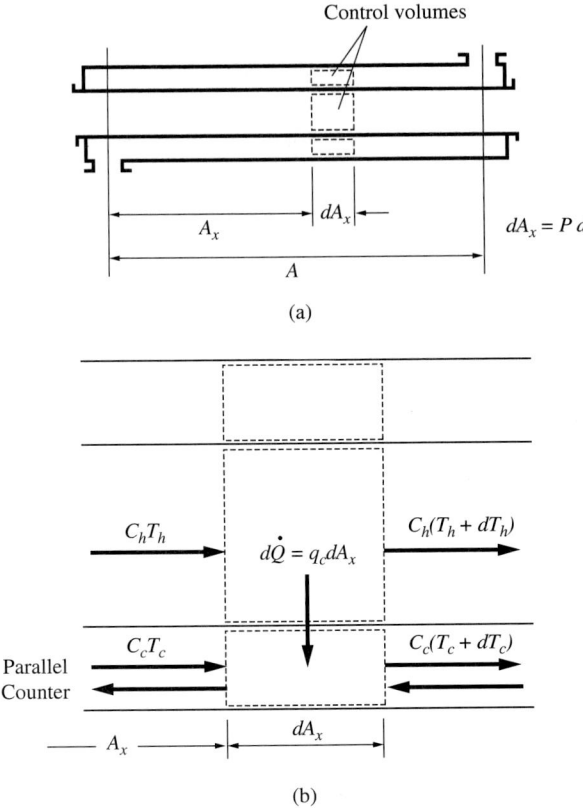

Figure 7.4 (a) Radially lumped axially differential control volumes for both hot and cold fluids, (b) the first law of thermodynamics for radially lumped axially differential hot and cold fluids.

$d\dot{Q}$ denoting the heat transfer from the hot fluid to the cold fluid. Following the same steps, we have for the case of the cold fluid flowing parallel to the hot fluid,

$$-C_c dT_c + d\dot{Q} = 0, \qquad (7.3)$$

where the subscript c indicates the cold fluid. For the case of cold fluid[3] flowing in the opposite direction (counter-flow configuration),

$$+C_c dT_c + d\dot{Q} = 0. \qquad (7.4)$$

Rearranging Eqs. (7.2)-(7.4), we obtain

$$d\dot{Q} = -C_h dT_h = \pm C_c dT_c, \qquad (7.5)$$

where the $\pm$ signs associated with the cold fluid correspond respectively to parallel and counter flows. Equation (7.5) summarizes the (differential) first law of thermodynamics

---

[3] Note that $T_c + dT_c$ corresponds algebraically to $A_x + dA_x$.

for parallel-and counter-flow heat exchangers. For convenience, rearrange the first equality of Eq. (7.5) as

$$dT_h = -\frac{1}{C_h} d\dot{Q}, \qquad (7.6)$$

the second equality of Eq. (7.5) as

$$dT_c = \pm \frac{1}{C_c} d\dot{Q}, \qquad (7.7)$$

and the difference between Eq (7.6) and (7.7) as

$$d(T_h - T_c) = -\left(\frac{1}{C_h} \pm \frac{1}{C_c}\right) d\dot{Q}, \qquad (7.8)$$

where the $\pm$ signs in parenthesis refer respectively to parallel and counter flows. To proceed further, introduce the particular law

$$q_c = U(T_h - T_c), \quad d\dot{Q} = q_c dA_x, \qquad (7.9)$$

U being the total heat transfer coefficient and $dA_x = Pdx$ the differential heat transfer area. Recalling Chapter 2, when the effect of pipe curvature is negligible, say $r_o/r_i < 2$, the total heat transfer coefficient based on the mean of the inner and outer surface areas of the pipe wall is given by

$$\frac{1}{U} = \frac{1}{h_h} + \frac{\delta}{k} + \frac{1}{h_c},$$

and when the effect of curvature is appreciable, say $r_0/r_i > 2$, the total heat transfer coefficient based on the pipe outer surface area is given by

$$\frac{1}{U} = \frac{A_c/A_h}{h_h} + \frac{\delta A_c}{k\bar{A}} + \frac{1}{h_c}$$

or, explicitly,

$$\frac{1}{U} = \frac{r_0/r_i}{h_h} + \frac{r_0 \ln(r_0/r_i)}{k} + \frac{1}{h_c},$$

where r_i and r_0 are the inner and outer radii of the pipe walls and $r_0 - r_i = \delta$ is the pipe wall thickness [recall the criterion given by Eq. (2.17) and the development associated with Eq. (2.31)]. Inserting Eq. (7.9), into Eq. (7.8), and introducing $\Delta T = T_h - T_c$, we obtain

$$d\Delta T = -\left(\frac{1}{C_h} \pm \frac{1}{C_c}\right) U \Delta T dA_x, \qquad (7.10)$$

which is the (differential) equation governing parallel- and counter-flow heat exchangers. Equation (7.10) subject to the specified ΔT at both ends of the parallel- and counter-flow heat exchangers completes our formulation. We now proceed to its integration. For the time being, let U remain uniform[4] over the entire length of the heat

---

[4] See Section 7.5 for some remarks on a variable U.

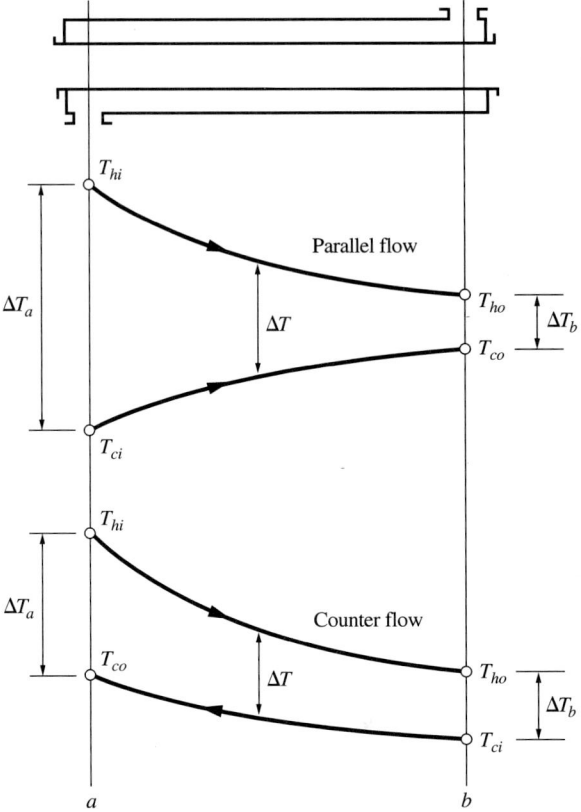

Figure 7.5 Cold and hot fluid temperatures for parallel and counter flows.

exchanger and integrate Eq. (7.10) from one end of the heat exchanger (say a) to the other (say b). This gives, in terms of Fig. 7.5,

$$\int_a^b \frac{d\Delta T}{\Delta T} = -\left(\frac{1}{C_h} \pm \frac{1}{C_c}\right) U \int_a^b dA_x,$$

or

$$\boxed{\ln \frac{\Delta T_b}{\Delta T_a} = -\left(\frac{1}{C_h} \pm \frac{1}{C_c}\right) UA}, \qquad (7.11)$$

where $\int_a^b dA_x = A$ denotes the total heat transfer area between the hot and cold fluids. Equation (7.11) is **one** of the **two important relations** needed for the thermal design of heat exchangers (developed here only for parallel and counter flows). We now proceed to the development of the second relation.

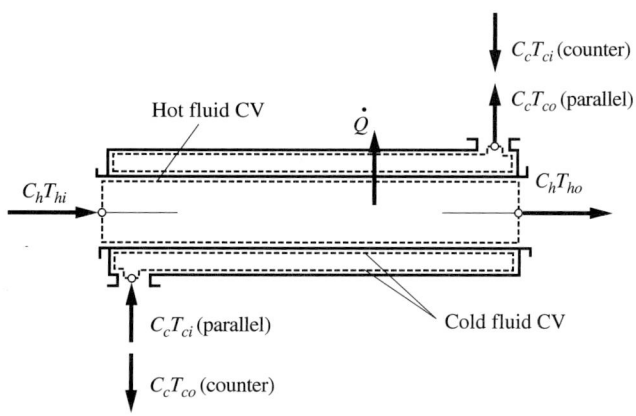

Figure 7.6 Control volumes for the entire hot and cold fluids.

For the hot and cold fluids, consider again a radially lumped control volume, extending this time over the entire length of the heat exchanger (Fig. 7.6). The first law of thermodynamics applied to these control volumes gives for the hot fluid

$$C_h(T_{hi} - T_{ho}) - \dot{Q} = 0$$

and for the cold fluid

$$C_c(T_{ci} - T_{co}) + \dot{Q} = 0,$$

where $\dot{Q}$ is the (total) heat transfer from the hot to the cold fluid, and subscripts i and o refer to inlet and outlet conditions. Then

$$\dot{Q} = C_h(T_{hi} - T_{ho}) = C_c(T_{co} - T_{ci}), \qquad (7.12)$$

which is the **second relation** needed. Note that while Eq. (7.11) depends on the type of heat exchanger, Eq. (7.12) is general and independent of the type used.

From Eq. (7.12), solving for $1/C_h$ and $1/C_c$ in terms of $\dot{Q}$ and inserting these into Eq. (7.11) yields

$$\ln \frac{\Delta T_b}{\Delta T_a} = -[(T_{hi} - T_{ho}) \pm (T_{co} - T_{ci})] \frac{UA}{\dot{Q}}. \qquad (7.13)$$

Recalling Fig. 7.5, the bracketed terms in Eq. (7.13) may be rearranged for parallel flow (+ sign between parentheses) as

$$(T_{hi} - T_{ho}) + (T_{co} - T_{ci}) = (T_{hi} - T_{ci}) - (T_{ho} - T_{co}) = \Delta T_a - \Delta T_b,$$

and for counter flow (− sign between parentheses) as

$$(T_{hi} - T_{ho}) - (T_{co} - T_{ci}) = (T_{hi} - T_{co}) - (T_{ho} - T_{ci}) = \Delta T_a - \Delta T_b.$$

Accordingly, Eq. (7.13) becomes for both parallel and counter flows

$$\ln \frac{\Delta T_b}{\Delta T_a} = -(\Delta T_a - \Delta T_b) \frac{UA}{\dot{Q}},$$

or
$$\dot{Q} = UA\Delta T_{\text{LM}},\tag{7.14}$$

where
$$\Delta T_{\text{LM}} = \frac{\Delta T_a - \Delta T_b}{\ln \dfrac{\Delta T_a}{\Delta T_b}} = \frac{\Delta T_{\max} - \Delta T_{\min}}{\ln \dfrac{\Delta T_{\max}}{\Delta T_{\min}}} \tag{7.15}$$

is the **logarithmic mean temperature difference**. Because of this logarithmic difference, the foregoing development is called the LMTD method. Here $\Delta T_{\max}$ and $\Delta T_{\min}$ respectively refer to the larger and smaller value of ΔT_a and ΔT_b. When $\Delta T_{\max}/\Delta T_{\min} \leq 2$, ΔT_{LM} may be replaced by the **arithmetic mean temperature difference**

$$\Delta T_{\text{AM}} = \frac{1}{2}(\Delta T_{\min} + \Delta T_{\max}). \tag{7.16}$$

The error involved with this approximation is $\leq 4\%$ [recall Eq. (2.21)]. In the case of a counter-flow heat exchanger, equal values of ΔT_a and ΔT_b result from $C_h = C_c$, which leads to $\Delta T_{\text{LM}} = \Delta T_a = \Delta T_b$.[5]

We are now ready to solve problems related to heat-exchanger design with the foregoing background. By solving for the heat transfer between the two fluids and the one unknown temperature using Eq. (7.12), we can then obtain the heat transfer area A using Eq. (7.14). The overall heat transfer coefficient U may be known or can be calculated using the appropriate correlations given in Chapter 6. Note that Eq. (7.14) is valid only for single-pass shell-and-tube heat exchangers. The use of the LMTD method for other types of heat exchangers is the focus of the next section. Having learned the LMTD method, we now proceed to an example.

EXAMPLE 7.1[6]

In a heat exchanger, oil with a mass flow rate $\dot{m}_h = 20$ kg/s is to be cooled from $T_{hi} = 120\,°\text{C}$[7] to $T_{ho} = 60\,°\text{C}$ by the help of water with a flow rate $\dot{m}_c = 15$ kg/s available at $T_{ci} = 10\,°\text{C}$. The specific heats of oil and water are $c_{ph} = 2$ kJ/kg·K and $c_{pc} = 4$ kJ/kg·K, respectively. The total heat transfer coefficient is approximately $U = 1{,}100$ W/m²·K.[8] We wish to determine the heat transfer area of this exchanger for (a) parallel-flow and counter-flow arrangements, and (b) re-solve the problem for $\dot{m}_h$ increased to 30 kg/s and T_{hi} decreased to $100\,°\text{C}$. Assume that U remains the same as in part (a).

---

[5] If $C_h = C_c$, Eq. (7.12) can be rearranged as $T_{hi} - T_{co} = T_{ho} - T_{ci}$ or $\Delta T_a = \Delta T_b$ for a counter-flow heat exchanger.

[6] The FORTRAN program EX7–1.F is listed in the appendix of this chapter.

[7] The boiling temperature of viscous oils at atmospheric pressure may be as high as 150–200 °C.

[8] Note from Table 1.2 that $h_h \cong 1{,}200$ W/m²·K and $h_c \cong 12{,}000$ W/m²·K are typical values for forced convection in oil and water.

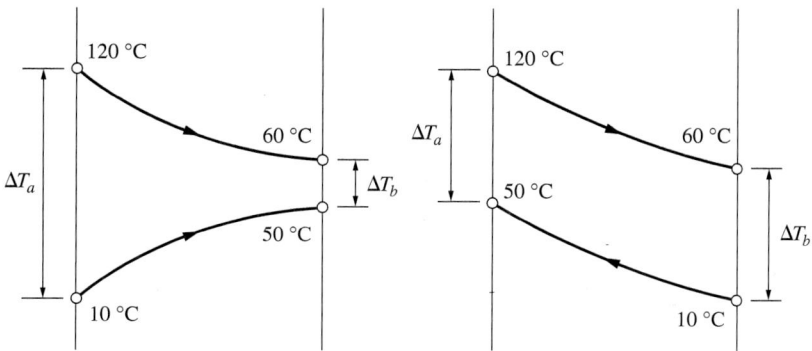

Figure 7.7 Parallel and counter flow temperatures.

(a) The heat-capacity flow rates are

$$C_h = \dot{m}_h c_{ph} = 20 \text{ kg/s} \times 2 \text{ kJ/kg·K} = 40 \text{ kW/K},$$

$$C_c = \dot{m}_c c_{pc} = 15 \text{ kg/s} \times 4 \text{ kJ/kg·K} = 60 \text{ kW/K}.$$

From Eq. (7.12), $\dot{Q} = C_h(T_{hi} - T_{ho}) = C(T_{co} - T_{ci})$,

$$\dot{Q} = 40 \text{ kW/K} \times (120 - 60) \text{ K} = 60 \text{ kW/K} \times (T_{co} - 10) \text{ K} = 2{,}400 \text{ kW}$$

which gives

$$T_{co} = 50\,°\text{C}.$$

Sketched in Fig. 7.7 are the temperature distributions corresponding to parallel-and counter-flow arrangements.

For parallel flow,

$$\Delta T_a = T_{hi} - T_{ci} = 120 - 10 = 110\,°\text{C},$$

$$\Delta T_b = T_{ho} - T_{co} = 60 - 50 = 10\,°\text{C},$$

and the LMTD is

$$\Delta T_{LM} = \frac{\Delta T_a - \Delta T_b}{\ln \dfrac{\Delta T_a}{\Delta T_b}} = \frac{110 - 10}{\ln \dfrac{110}{10}} = 41.7\,°\text{C}.$$

For counter flow,

$$\Delta T_a = T_{hi} - T_{co} = 120 - 50 = 70\,°\text{C},$$

$$\Delta T_b = T_{ho} - T_{ci} = 60 - 10 = 50\,°\text{C},$$

and the LMTD is

$$\Delta T_{LM} = \frac{\Delta T_a - \Delta T_b}{\ln \dfrac{\Delta T_a}{\Delta T_b}} = \frac{70 - 50}{\ln \dfrac{70}{50}} = 59.4\,°\text{C}.$$

Accordingly, the respective areas follow from Eq. (7.14) as

$$A = \frac{\dot{Q}}{U \Delta T_{LM}} = \frac{2,400,000 \text{ W}}{1,100 \text{ W/m}^2 \cdot \text{K} \times (41.7; 59.4) \,°\text{C}}$$

or

$$A = \begin{cases} 52.3 \text{ m}^2 & \text{for parallel flow,} \\ 36.7 \text{ m}^2 & \text{for counter flow.} \end{cases}$$

As a general rule, a **counter-flow heat exchanger requires 20–30% less heat transfer area than a parallel-flow heat exchanger.** For the present case, the actual percentage difference turns out to be

$$\frac{52.3 - 36.7}{52.3} \times 100 \cong 30\%.$$

That is why the parallel-flow arrangement is **not** usually considered in industry.

(b) The new heat-capacity flow rate of the hot fluid is

$$C_h = \dot{m}_h c_{ph} = 30 \text{ kg/s} \times 2 \text{ kJ/kg} \cdot \text{K} = 60 \text{ kW/K},$$

and the heat capacity flow rate of the cold fluid from part 1 is

$$C_c = 60 \text{ kW/K}.$$

Employing Eq. (7.12),

$$\dot{Q} = 60 \text{ kW/K} \times (100 - 60) \,°\text{C} = 60 \text{ kW/K} \times (T_{co} - 10) \,°\text{C} = 2,400 \text{ kW},$$

which gives

$$T_{co} = 50 \,°\text{C}.$$

The LMTD for parallel flow is

$$\Delta T_{LM} = \frac{\Delta T_a - \Delta T_b}{\ln \frac{\Delta T_a}{\Delta T_b}} = \frac{(100 - 10) - (60 - 50)}{\ln \frac{100 - 10}{60 - 50}} = \frac{90 - 10}{\ln \frac{90}{10}} = 36.4 \,°\text{C}.$$

For the counter-flow heat exchanger, ΔT remains constant throughout the exchanger[9] when $C_h = C_c$ [a fact readily shown by the rearrangement of Eq. (7.12)]. Then

$$\Delta T_{LM} = \Delta T_a = \Delta T_b = \Delta T.$$

Here, $\Delta T = 100 - 50 = 60 - 10 = 50 \,°\text{C}.$

Accordingly, the respective areas follow from Eq. (7.14) as

$$A = \frac{\dot{Q}}{U \Delta T_{LM}} = \frac{2,400,000 \text{ W}}{1,100 \text{ W/m}^2 \cdot \text{K} \times (36.4; 50) \,°\text{C}}$$

or

$$A = \begin{cases} 59.8 \text{ m}^2 \text{ for parallel flow,} \\ 43.6 \text{ m}^2 \text{ for counter flow.} \end{cases}$$

---

[9] Note that any average of two identical numbers is equal to the numbers themselves.

The comparison of the new areas reveals

$$\frac{59.9 - 43.6}{59.9} \times 100 \cong 27\%,$$

which is consistent with our general rule. By the computer program provided, the interested reader may parametrically study the heat transfer area depending on the total heat transfer coefficient. ◆

So far we have considered parallel-and counter-flow heat exchangers. We now proceed to heat exchangers with more complex flow arrangements.

EXAMPLE 7.2

Consider a cross-flow heat exchanger with both fluids mixed [Fig. 7.8.(a)]. Let the velocity and the inlet temperature of the hot and cold fluids be (V_1, T_0) and $(V_2, 0)$, respectively. We wish to determine the temperature distribution in the hot and cold fluids.

In terms of the differential control volumes shown in Fig. 7.8(b), the first law of thermodynamics rearranged with Newton's cooling law yields for the hot fluid

$$0 = -\rho_1 c_1 \delta_1 V_1 \frac{\partial T_1}{\partial x} - U(T_1 - T_2) \tag{7.17}$$

and for the cold fluid

$$0 = -\rho_2 c_2 \delta_2 V_2 \frac{\partial T_2}{\partial y} + U(T_1 - T_2) \tag{7.18}$$

subject to the inlet (boundary) conditions

$$T_1(0, y) = T_0 \quad \text{and} \quad T_2(x, 0) = 0. \tag{7.19}$$

Introducing

$$b_1 = U/\rho_1 c_1 \delta_1 V_1 \quad \text{and} \quad b_2 = U/\rho_2 c_2 \delta_2 V_2,$$

Eqs. (7.17) and (7.18) may be respectively rearranged as

$$0 = \frac{\partial T_1}{\partial x} + b_1(T_1 - T_2) \tag{7.20}$$

and

$$0 = -\frac{\partial T_2}{\partial y} + b_2(T_1 - T_2) \tag{7.21}$$

or, in terms of

$$\theta = \frac{T}{T_0}, \quad \xi = b_1 x, \quad \text{and} \quad \eta = b_2 y,$$

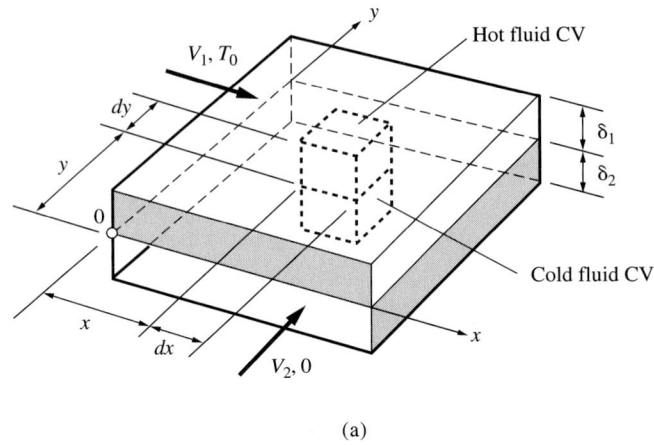

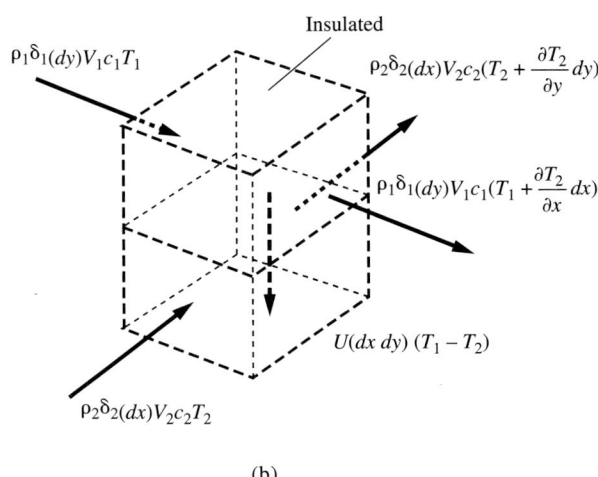

Figure 7.8 System configuration.

as

$$0 = \frac{\partial \theta_1}{\partial \xi} + (\theta_1 - \theta_2) \tag{7.22}$$

and

$$0 = -\frac{\partial \theta_2}{\partial \eta} + (\theta_1 - \theta_2) \tag{7.23}$$

subject to

$$\theta_1(0, \eta) = 1, \quad \theta_2(\xi, 0) = 0. \tag{7.24}$$

We had no difficulty in formulating the problem. However, the solution process is considerably involved and remains beyond the scope of this text. For example, the use of Laplace transforms (see Chapter 7 of *Conduction Heat Transfer* by Arpacı) conveniently leads to

$$\theta_1(\xi, \eta) = e^{-(\xi+\eta)} I_0 \left[2(\xi\eta)^{1/2} \right] + \theta_2(\xi, \eta) \tag{7.25}$$

and

$$\theta_2(\xi, \eta) = e^{-\xi} \int_0^\eta e^{-\eta^*} I_0 \left[2(\xi\eta^*)^{1/2} \right] d\eta^*, \tag{7.26}$$

where I_0 is the modified Bessel function of the first kind. Also, the (dimensionless) local heat transfer between the hot and cold fluids is

$$q/UT_0 = e^{-(\xi+\eta)} I_0 \left[2(\xi\eta)^{1/2} \right]. \tag{7.27}$$

Furthermore, the cross sectional average of the θ_1 and θ_2 outlet temperatures and the average of q over the heat transfer area are needed. All these manipulations show that the temperature distributions and heat transfer in cross-flow heat exchanger are mathematically more complicated than those in parallel-and counter-flow heat exchangers. Because of this fact, practical problems associated with cross flow and other heat exchangers of similar complexity are handled by an approximation of mathematical results such as Eqs. (7.26) and (7.27) or by simpler results obtained from approximate formulations. These results are usually expressed in terms of a correction factor relative to the counter-flow (which is the most efficient) heat exchanger. ◆

7.2 CORRECTION FACTOR

In its simplest form, a shell-and-tube heat exchanger operates under mixed-flow conditions as sketched in Fig. 7.9(a),(b). Consequently, a part of the shell fluid recirculates in the heat exchanger, resulting in uneven and poor heat transfer. To improve and even-out the heat transfer, transverse baffles are placed in the heat exchanger as shown in Fig. 7.10. These baffles provide alternatively cross-flow and counter-flow conditions;

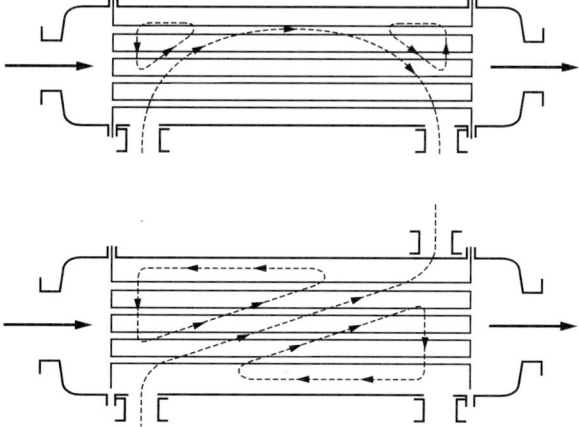

Figure 7.9 Shell-and-tube heat exchangers without baffles.

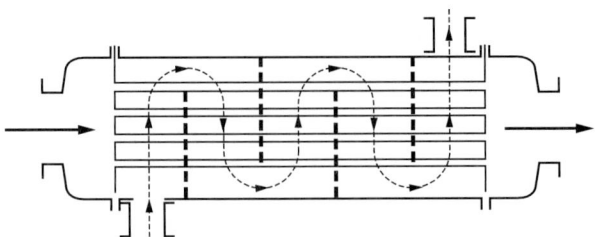

Figure 7.10 Shell-and-tube heat exchanger with baffles.

they reduce the average cross section of the shell flow and increase its velocity, thus improving the shell-side coefficient of heat transfer.[10] Accordingly, the cross flow produced by baffles may be a more efficient arrangement than the counter flow without baffles.

Conceptually, the procedure developed in Section 7.1 for the thermal design (heat transfer area) of parallel-and counter-flow heat exchangers is general and applies also to more complex heat exchangers as shown in Ex. 7.2. However, the algebra gets rather involved and is beyond the scope of this text. Therefore, we are forced to follow an alternative approach for the design of complex exchangers. Recalling Eq. (7.14), we write for any heat exchanger

$$\dot{Q} = UA\Delta T_{TM}, \tag{7.28}$$

where ΔT_{TM} denotes the **true mean temperature difference**. Normalizing ΔT_{TM} relative to ΔT_{LM} of a counter-flow heat exchanger, we may introduce the definition of a **correction factor**

$$F = \Delta T_{TM}/\Delta T_{LM}, \tag{7.29}$$

which is a measure of the degree of departure from counter-flow conditions. The involved algebra we just talked about actually leads to ΔT_{TM} or F as a function of parameters

$$P = \frac{T_{to} - T_{ti}}{T_{si} - T_{ti}} \quad \text{and} \quad R = \frac{T_{si} - T_{so}}{T_{to} - T_{ti}} = \frac{C_T}{C_S}. \tag{7.30}$$

Here P indicates the effectiveness of the heat exchanger (to be elaborated in Section 7.4) and R (from its definition) is the ratio of the heat-capacity flow rates. Note the change in nomenclature from subscripts h and c to t and s, the latter two referring to tube and shell, respectively. An important fact is that whether the hot (or cold) fluid is flowing in the shell side or in the tubes has no effect on F as long as the heat transfer to the ambient is negligible. Otherwise, the cold fluid should be in the shell side to reduce heat losses. Combination of Eqs. (7.28) and (7.29) gives

$$\boxed{\dot{Q} = UA\Delta T_{LM}F(P, R)}. \tag{7.31}$$

References 4 and 5 contain $F(P, R)$ charts for heat exchangers encountered in practice. Four of these charts, corresponding to the most common types used, are reproduced in

---

[10] From Chapter 6 recall increase of h with V.

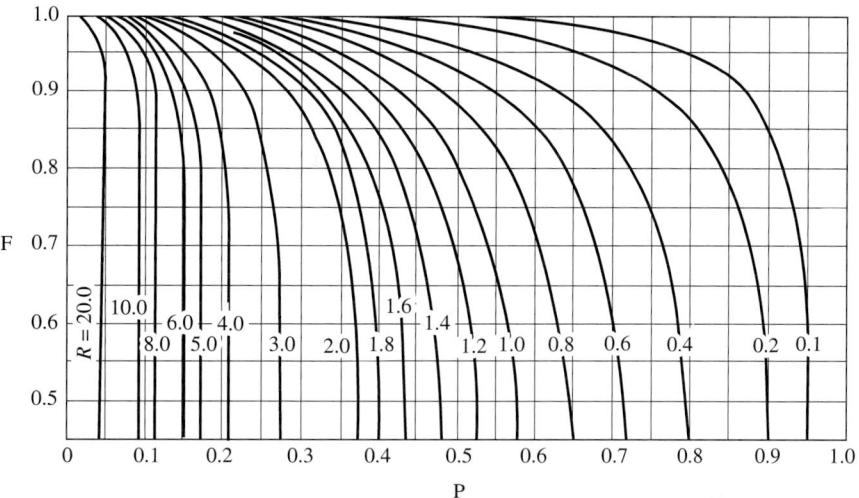

Figure 7.11 Correction-factor plot for exchanger with one shell pass and two, four, or any multiple of two tube passes

Figs. 7.11–7.14.[11] The first two charts are associated with shell-and-tube heat exchangers and the last two charts with cross-flow heat exchangers. Among these, the shell-and-tube types are inherently heavy and are considered for stationary applications, while the cross-flow types are inherently light and are considered for mobile applications. Here we illustrate the use of the correction factor in terms of an example.

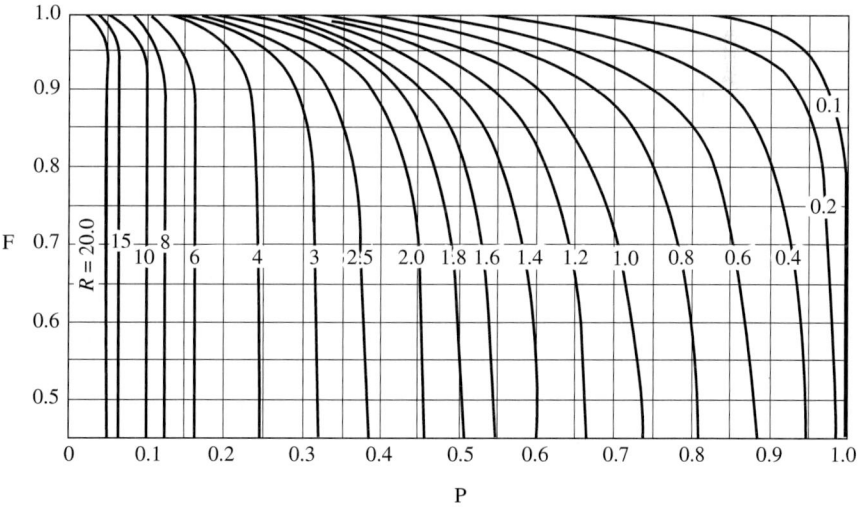

Figure 7.12 Correction-factor plot for exchanger with two shell passes and two, four, or any multiple of two tube passes

---

[11] Figures 7.11 through 7.14 are from Jakob[3].

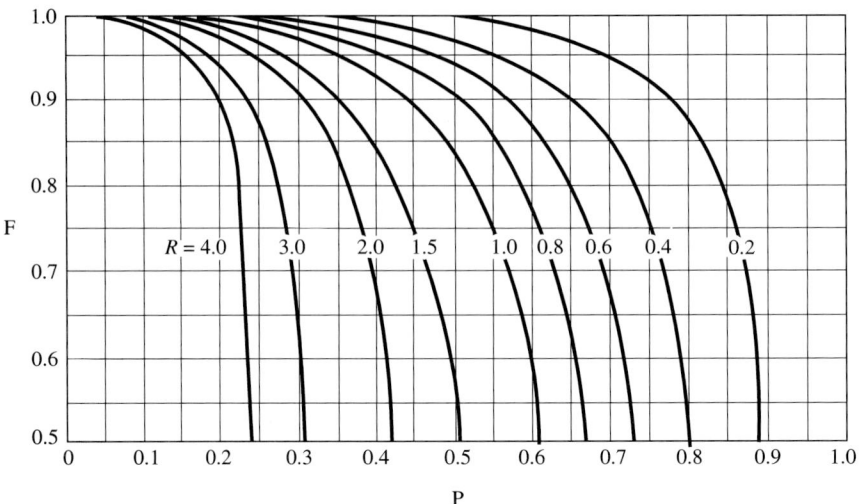

Figure 7.13 Correction-factor plot for single-pass cross-flow exchanger one fluid mixed, the other unmixed

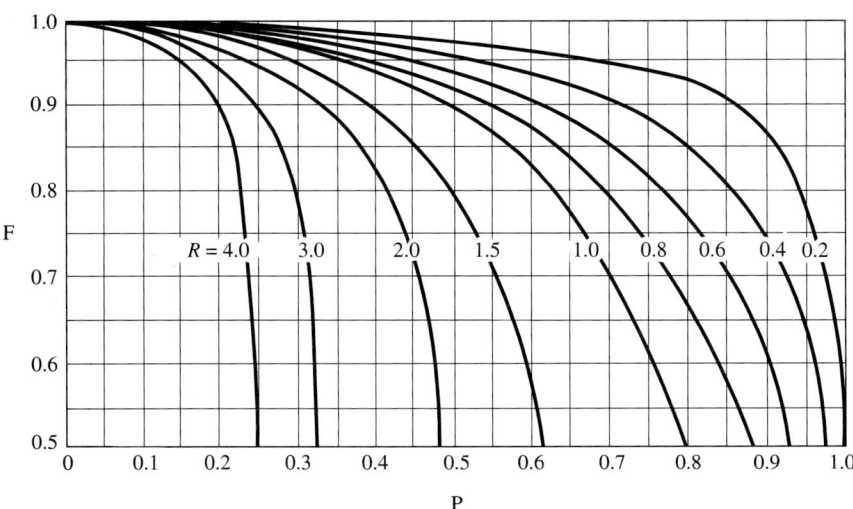

Figure 7.14 Correction-factor plot for single-pass cross-flow exchanger both fluids unmixed

Example 7.3

We wish to evaluate the heat transfer area of a (a) a one-shell pass and multiple of two-tube passes heat exchanger, (b) a two-shell pass and multiple of four-tube passes heat exchanger, which would satisfy the conditions of Ex. 7.1(a).

Since the inlet and outlet temperatures and the flow rates remain the same, from Eqs. (7.14) and (7.31)

$$\dot{Q} = U_1 A_1 \Delta T_{\text{LM}} = U_2 A_2 \Delta T_{\text{LM}} F(P, R),$$

where subscript 1 indicates the counter-flow exchanger from the first part of Ex. 7.1 and subscript 2 the shell-and-tube exchanger. Consequently,

$$A_2 = A_1(U_1/U_2)/F(P, R). \qquad (7.32)$$

For cold fluid (water) flowing in the shell side,

$$P = \frac{T_{to} - T_{ti}}{T_{si} - T_{ti}} = \frac{T_{ho} - T_{hi}}{T_{ci} - T_{hi}} = \frac{60 - 120}{10 - 120} = \frac{60}{110} \cong 0.53$$

and

$$R = \frac{C_t}{C_s} = \frac{C_h}{C_c} = \frac{40}{60} = 0.67.$$

(a) Employing these parameters with Fig. 7.11 gives $F \cong 0.87$.[12] Now, from Eq. (7.32) we have

$$A_2 = 1.15 A_1(U_1/U_2).$$

If both the counter-flow and the shell-and-tube heat exchangers have the same total heat transfer coefficient, the area of the shell-and-tube type turns out to be 15% more than that of the counter-flow type.

(b) Figure 7.12 with the foregoing P and R values gives $F \cong 0.98$. Then, from Eq. (7.32)

$$A_2 = 1.02 A_1(U_1/U_2),$$

which is 2% more than the area of the counter-flow type if U is the same. ◆

Example 7.4[13]

The efficiency of a gas turbine can be improved by increasing the intake air temperature to $210\,°C$. A counter-flow heat exchanger is to be designed using the exhaust gases of this turbine as the hot fluid to heat the air to the desired temperature. The flow rates are $\dot{m}_h = \dot{m}_c = 10$ kg/s, the heat transfer coefficients are $h_h = h_c = 150$ W/m²·K, the specific heats are $c_{ph} = c_{pc} = 1$ kJ/kg·K, and the three given temperatures are $T_{hi} = 425\,°C$, $T_{ci} = 25\,°C$, and $T_{co} = 210\,°C$. We wish to determine the area of this heat exchanger.

Noting that both flow rates and specific heats of the hot and cold gases are identical, the heat-capacity flow rates are found to be

$$C_h = C_c = \dot{m} c_p = 10 \text{ kg/s} \times 1 \text{ kJ/kg·K} = 10 \text{ kW/K}.$$

From Eq. (7.12)

$$\dot{Q} = 10 \text{ kW/K} \times (425 - T_{ho})\,°C = 10 \text{ kW/K} \times (210 - 25)\,°C = 1{,}850 \text{ kW},$$

which gives

$$T_{ho} = 240\,°C.$$

---

[12] For the case of hot fluid (oil) flowing in the shell side we obtain the same result for F with $P = \dfrac{T_{to} - T_{ti}}{T_{si} - T_{ti}} = \dfrac{T_{co} - T_{ci}}{T_{hi} - T_{ci}} = \dfrac{50 - 10}{120 - 10} = \dfrac{40}{110} \cong 0.37$ and $R = \dfrac{C_t}{C_s} = \dfrac{C_c}{C_h} = \dfrac{60}{40} \cong 1.5.$

[13] The FORTRAN program EX7–4.F is listed in the Appendix.

The overall heat transfer coefficient is given by

$$\frac{1}{U} = \frac{1}{h_h} + \frac{1}{h_c} = \frac{1}{150} + \frac{1}{150}$$

or

$$U = 75 \text{ W/m}^2 \cdot \text{K}.$$

Because of the fact that $C_h = C_c$, and $\Delta T_a = \Delta T_b$, recalling Example 7.1(b), the LMTD is

$$\Delta T_{\text{LM}} = 425 - 210 = 240 - 25 = 215\,°\text{C}.$$

Combining these into Eq. (7.14), we get

$$A = \frac{\dot{Q}}{U\Delta T_{\text{LM}}} = \frac{1{,}850{,}000 \text{ W}}{75 \text{ W/m}^2 \cdot \text{K} \times 215\,°\text{C}} \cong 115 \text{ m}^2.$$

◆

EXAMPLE 7.5

Reconsider Ex. 7.4. We wish to determine the area of a cross-flow heat exchanger with (a) one fluid unmixed and (b) both fluids unmixed, which would satisfy the conditions of Ex. 7.4.

We have the values for $\dot{Q}$, U, and ΔT_{LM} from Ex. 7.4. Here, we need to evaluate the correction factor F for each case.

(a) For hot gas flowing through the tubes, Eq. (7.30) gives

$$P = \frac{T_{to} - T_{ti}}{T_{si} - T_{ti}} = \frac{T_{ho} - T_{hi}}{T_{ci} - T_{hi}} = \frac{240 - 425}{25 - 425} = \frac{185}{400} \cong 0.46$$

and

$$R = \frac{C_t}{C_s} = \frac{C_h}{C_c} = 1,$$

and we have from Fig. 7.13,

$$F(P, R) = F(0.46, 1) \cong 0.88.$$

From Eqs. (7.28) and (7.29), the area of any heat exchanger (excluding parallel flow) relative to that of the counter flow is

$$A = A_{\text{counter}}/F(P, R).$$

Accordingly,

$$A = 115 \text{ m}^2/0.88 \cong 131 \text{ m}^2$$

for one fluid mixed.

(b) From Fig. 7.14,

$$F(P, R) = F(0.46, 1) \cong 0.92$$

and

$$A = 115 \text{ m}^2/0.92 \cong 125 \text{ m}^2$$

for both fluids unmixed.

◆

When the temperature of either fluid remains constant during the heat exchange resulting from a change of phase (condensation or evaporation), the LMTD method continues to apply, but it may be given a more convenient form. This is done in the next section.

7.3 CONDENSER. EVAPORATOR (BOILER) ◯

In a condenser, vapor condenses at a constant temperature (and pressure) while heating a cold fluid [Fig. 7.15(a)] and exits as a liquid. In an evaporator (or boiler), liquid evaporates at a constant temperature while cooling a hot fluid [Fig. 7.15(b)] and exits as a vapor. Two important features of condensers and evaporators and related to the fact that the temperature remains constant in the two-phase (condensing or evaporating) side:

1. The change of enthalpy flow in the two-phase side can no longer be expressed in terms of $C\Delta T$, because $\Delta T = 0$. In this case the change of enthalpy flow is $\dot{m}\Delta h$, $\dot{m}$ being the mass flow rate and Δh the difference between inlet and outlet

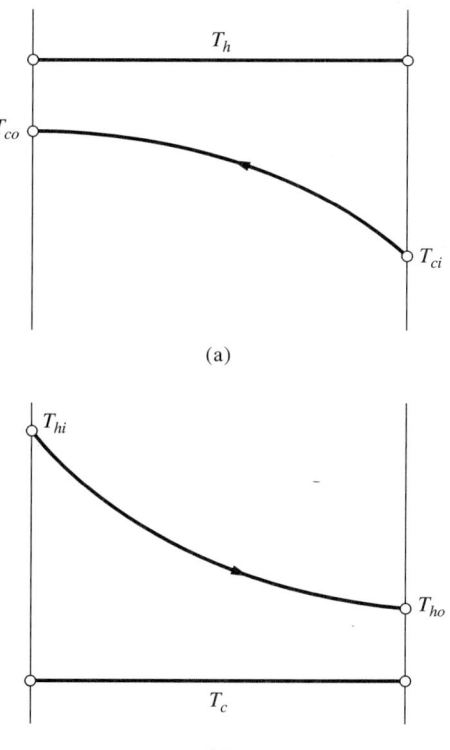

Figure 7.15 Cold and hot fluid temperatures for condenser and evaporator (boiler). (a) Condenser, (b) evaporator (boiler).

enthalpies. The enthalpy for a two-phase fluid is given by

$$h = xh_g + (1-x)h_f, \tag{7.33}$$

h_g being the enthalpy of saturated vapor, h_f the enthalpy of saturated liquid and x the quality.

2. The change of enthalpy is much larger in the two-phase side. Consequently, the mass flow rate is much smaller in the two-phase side. Assuming the two-phase side to remain approximately stagnant, we may eliminate the effect of the type of heat exchanger and set $F \cong 1$.

In terms of a condenser, for example, we have from Eqs. (7.12) and (7.14)

$$\dot{m}_h(h_i - h_0) = (\dot{m}c_p)_c(T_{co} - T_{ci}) = UA\Delta T_{LM}, \tag{7.34}$$

where h_i and h_o are the inlet and outlet enthalpies of the hot (condensing) fluid, and

$$\Delta T_{LM} = \frac{(T_{co} - T_{ci})}{\ln\left(\dfrac{T_h - T_{ci}}{T_h - T_{co}}\right)}. \tag{7.35}$$

In view of Eq. (7.33), $h_i = x_i h_g + (1-x_i)h_f$ and $h_0 = h_f$ (no vapor at the outlet!), and

$$h_i - h_0 = x_i h_{fg}, \tag{7.36}$$

where $h_{fg} = h_g - h_f$ is the latent heat. Inserting Eq. (7.36) into the first equality of Eq. (7.34), we obtain the ratio of mass flow rates in a condenser

$$\frac{\dot{m}_h}{\dot{m}_c} = \frac{c_{pc}(T_{co} - T_{ci})}{x_i h_{fg}}. \tag{7.37}$$

Introducing Eq. (7.35) into the second equality of Eq. (7.34), we obtain for the heat transfer area of a condenser

$$A = \frac{C_c}{U}\ln\left(\frac{T_h - T_{ci}}{T_h - T_{co}}\right), \tag{7.38}$$

or, in terms of Eq. (7.36), alternatively,

$$A = \frac{x_i \dot{m}_h h_{fg}}{U\Delta T_{LM}}. \tag{7.39}$$

We may wish to utilize an already built heat exchanger as a condenser. In this case the heat transfer area is known and the exit temperature of the cold fluid becomes unknown. Rearranging Eq. (7.38),

$$\frac{T_h - T_{co}}{T_h - T_{ci}} = \exp(-UA/C_c) \tag{7.40}$$

and, subtracting each side of Eq. (7.40) from 1, we get the unknown exit temperature,

$$\frac{T_{co} - T_{ci}}{T_h - T_{ci}} = 1 - \exp(-UA/C_c). \tag{7.41}$$

We will return to this result in Section 7.4. Let us now illustrate the foregoing considerations in terms of two examples.

EXAMPLE 7.6

A condenser for a small, conventional power plant is to be designed. The flow rate of steam through this condenser is going to be $\dot{m}_h = 10$ kg/s. The conditions of the steam at the inlet of the condenser are given in Fig. 7.16. The heat transfer coefficient on the cold-water side is $h_c = 12{,}000$ W/m²·K. Steam will be condensed by water to be taken from a nearby river. The temperature of the river is $T_{ci} = 10\,°C$. Ecological considerations suggest that the river temperature be raised no more than 10 °C. We wish to determine (a) the heat transfer area of the condenser, (b) the flow rate of the water to be taken from the river.

(a) First, from Eq. (7.35)

$$\Delta T_{LM} = \frac{20 - 10}{\ln \dfrac{30}{20}} \cong 24.66\,°C.$$

From Table 1.2, the heat transfer coefficient for condensation $h_h \cong 120{,}000$ W/m²·K is found to be an order of magnitude greater than h_c. Thus, neglecting the effect of curvature, the conductive resistance of the pipe walls, and the convective resistance of the steam, the overall heat transfer coefficient is $U \cong h_c = 12{,}000$ W/m²·K. Then from Eq. (7.39)

$$A = \frac{x_i \dot{m}_h h_{fg}}{U \Delta T_{LM}} = \frac{0.96 \times 10 \times 2{,}406{,}700}{12{,}000 \times 24.66} \cong 78\,\text{m}^2.$$

Since the type of heat exchanger has negligible effect on A, the least expensive design should be selected.

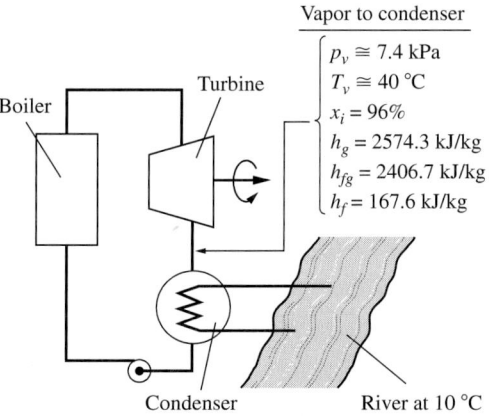

Figure 7.16 Condenser design.

(b) From Eq. (7.37), after letting $c_{pc} \cong 4.19$ kJ/kg·K,

$$\frac{\dot{m}_h}{\dot{m}_c} = \frac{c_{pc}(T_{co} - T_{ci})}{x_i h_{fg}} = \frac{4.19 \times (20 - 10)}{0.96 \times 2406.7} \cong 0.018.$$

That is, the mass flow rate of the steam is about 1.8% of the mass flow rate of the cooled water, which is consistent with what was assumed previously. In view of $\dot{m}_h = 10$ kg/s,

$$\dot{m}_c \cong 555 \text{ kg/s}.$$

If the neighboring river (or lake) cannot supply this flow rate, a supplementary cooling-tower system becomes necessary to achieve the required amount of cooling. ◆

EXAMPLE 7.7

Saturated water at 120 °C with a quality of $x_i = 0.2$ and a mass flow rate $\dot{m}_h = 10$ kg/s is to be cooled to 60 °C with a water flow of $\dot{m}_c = 40$ kg/s at 20 °C. The heat transfer coefficient for liquid water is $h_w = 8{,}000$ W/m²·K and for condensing water is $h_s = 24{,}000$ W/m²·K. We wish to determine the required heat transfer area of a counter-flow heat exchanger.

The saturated (hot) water will first condense at constant temperature ($T_h = 120$ °C) until there is no more vapor left. Then, by losing more heat to the cold water, its temperature will proceed to drop down to the desired 60 °C. The solution thus needs to be treated in two parts, as illustrated in Fig. 7.17. The first and second parts correspond to a counter-flow heat exchanger and to a condenser, respectively.

Part I: Heat exchanger

We determine A_1 from Eq. (7.14), which, in turn, requires the knowledge of $\dot{Q}_1$, U, and ΔT_{LM} depending on T_{co}. For the heat-capacity flow rates, assuming $c_{ph} \cong c_{pc} \cong 4{,}200$ J/kg·K, we get

$$C_h = \dot{m}_h c_p = 10 \text{ kg/s} \times 4{,}200 \text{ J/kg·K} = 42{,}000 \text{ W/K},$$

$$C_c = \dot{m}_c c_p = 40 \text{ kg/s} \times 4{,}200 \text{ J/kg·K} = 168{,}000 \text{ W/K}.$$

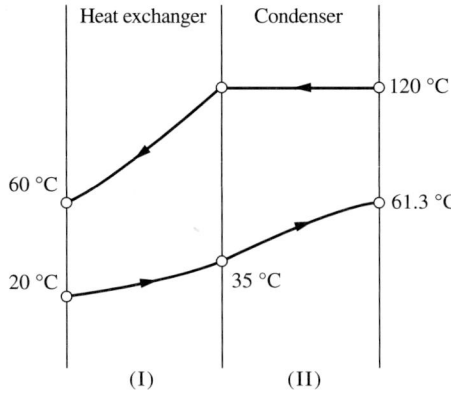

Figure 7.17 (I) Counter-flow exchanger, (II) condenser.

In view of Eq. (7.12),

$$\dot{Q}_1 = 42{,}000 \text{ W/K} \times (120 - 60)\,°\text{C} = 168{,}000 \text{ W/K} \times (T_{co} - 20)\,°\text{C} = 2.52 \times 10^6 \text{ W},$$

which also gives

$$T_{co} = 35\,°\text{C}.$$

Both flows here involve liquid water. Thus $h_h \cong h_c \cong h_w = 8{,}000 \text{ W/m}^2\cdot\text{K}$. For the overall heat transfer coefficient

$$\frac{1}{U_1} = \frac{1}{h_h} + \frac{1}{h_c} = \frac{1}{8{,}000} + \frac{1}{8{,}000}$$

or

$$U_1 = 4{,}000 \text{ W/m}^2\cdot\text{K},$$

and from Eq. (7.15) we have for the LMTD of the heat exchanger

$$\Delta T_{\text{LM1}} = \frac{\Delta T_a - \Delta T_b}{\ln \dfrac{\Delta T_a}{\Delta T_b}} = \frac{(120 - 35) - (60 - 20)}{\ln \dfrac{120 - 35}{60 - 20}} = \frac{85 - 40}{\ln \dfrac{85}{40}} = 59.7\,°\text{C}.$$

In terms of this average, Eq. (7.14) yields, for the heat-exchanger area,

$$A_1 = \frac{\dot{Q}_1}{U_1 \Delta T_{\text{LM1}}} = \frac{2{,}520{,}000 \text{ W}}{4{,}000 \text{ W/m}^2\cdot\text{K} \times 59.7\,°\text{C}} = 10.6 \text{ m}^2.$$

Part II: Condenser

Having determined the area of and the cold-fluid outlet temperature from the heat exchanger (which is also the inlet temperature for the condenser), we proceed to the area of the condenser. Employing Eq. (7.34) and Eq. (7.36),

$$\dot{Q}_2 = \dot{m}_h x_i h_{fg} = (\dot{m}c_p)_c (T_{co} - T_{ci}) = U_2 A_2 \Delta T_{\text{LM2}}$$

with $h_{fg} = 2{,}202.3 \text{ kJ/kg}$ (at $120\,°\text{C}$), and we get

$$\dot{Q}_2 = 10 \text{ kg/s} \times 0.2 \times 2{,}202.3 \text{ kJ/kg} = 168 \text{ kW/K} \times (T_{co} - 35)\,°\text{C} = 4.4 \times 10^6 \text{ W},$$

which gives

$$T_{co} = 61.3\,°\text{C}.$$

For ΔT_{LM2} of the condenser, we have from Eq. (7.35)

$$\Delta T_{\text{LM2}} = \frac{(T_{co} - T_{ci})}{\ln\left(\dfrac{T_h - T_{ci}}{T_h - T_{co}}\right)} = \frac{61.3 - 35}{\ln \dfrac{120 - 35}{120 - 61.3}} \cong 71\,°\text{C},$$

and for the overall heat transfer coefficient, noting $h_h = h_s = 24{,}000 \text{ W/m}^2\cdot\text{K}$,

$$\frac{1}{U_2} = \frac{1}{h_h} + \frac{1}{h_c} = \frac{1}{24{,}000} + \frac{1}{8{,}000}$$

or

$$U_2 = 6{,}000 \text{ W/m}^2\cdot\text{K}.$$

Finally, employing Eq. (7.39),

$$A_2 = \frac{x_i \dot{m}_h h_{fg}}{U_2 \Delta T_{LM2}} = \frac{0.2 \times 10 \text{ kg/s} \times 2{,}202.3 \times 10^3 \text{ J/kg}}{6{,}000 \text{ W/m}^2 \cdot \text{K} \times 71 \, ^\circ\text{C}},$$

which yields, for the condenser area,

$$A_2 = 10.34 \text{ m}^2.$$

Then, the total heat-exchanger area A is

$$A = A_1 + A_2 = 10.6 + 10.34 \cong 21 \text{ m}^2.$$

Note that while the areas in both parts are about the same, the heat transfer in the condenser is about 75% higher than that in the counter-flow exchanger. ◆

Having learned the thermal design of heat exchangers, we now proceed to their performance under different conditions.

7.4 PERFORMANCE. NTU METHOD

The thermal design of heat exchangers discussed in the preceding sections rests usually on the following first-prototype problem:

Given: Three temperatures (T_{hi}, T_{ci}, and one of the outlet temperatures), C_h and C_c, and U (to be estimated if not given).

Determine: Heat transfer area A.

The usual procedure, based on the LMTD method, is to get the remaining outlet temperature from Eq. (7.12), estimate U if not given, and determine the heat transfer area from Eq. (7.31). The **performance** of an existing heat exchanger **under different operating conditions**, or the utilization of a heat exchanger for a purpose different than its original design objective, is the second-prototype problem. In this case:

Given: A, inlet temperatures T_{hi} and T_{ci}, C_h and C_c, and U (to be estimated if not given).

Determine: Outlet temperatures T_{ho} and T_{co}.

The outlet temperatures can be determined by trial and error using the LMTD method. If one of the outlet temperatures is guessed, then the problem is reduced to the first prototype. However, because this would not be the actual exit temperature, the heat transfer values given by Eqs. (7.12) and (7.31) will not match. As the assumed value of the outlet temperature approaches its actual value (through successive guesses), the difference between Eqs. (7.12) and (7.31) diminishes. This is a tedious trial-and-error procedure, and a new method different from the LMTD method is needed. The so-called NTU method[14] eliminates the foregoing trial-and-error and is the topic of this section.

---

[14] The definition of NTU will become clear with Eq. (7.45).

Return to Eq. (7.12) and let this equation be equal to a product $\epsilon C \Delta T$, where the proportionality constant ϵ is the so-called **heat-exchanger effectiveness** and ΔT is a temperature difference. Since the inlet temperatures are given, let $\Delta T = T_{hi} - T_{ci}$, which is the largest temperature difference in a heat exchanger. C_h or C_c are two possibilities for C. Consider the smaller of the two and designate it as $C_{\min}$. Then, for any heat exchanger, we have

$$\boxed{\epsilon = \frac{C_h(T_{hi} - T_{ho})}{C_{\min}(T_{hi} - T_{ci})} = \frac{C_c(T_{co} - T_{ci})}{C_{\min}(T_{hi} - T_{ci})}.} \qquad (7.42)$$

This definition accepts an interpretation in terms of a counter-flow heat exchanger as

$$\epsilon = \frac{\text{Actual HT rate in any heat exchanger}}{\text{Maximum possible HT rate in a counter-flow heat exchanger}}$$

To demonstrate that $C_{\min}(T_{hi} - T_{ci})$ is the maximum attainable heat transfer with a counter-flow heat exchanger, consider Eq. (7.11) for a counter-flow heat exchanger,

$$\ln \frac{\Delta T_b}{\Delta T_a} = -\left(\frac{1}{C_h} - \frac{1}{C_c}\right) UA.$$

For $C_h > C_c$ or $1/C_h < 1/C_c$, $\Delta T_a \to 0$ as $A \to \infty$. In terms of Fig. 7.18(a), $\Delta T_a \to 0$ implies $T_{co} \to T_{hi}$, and $C_c(T_{co} - T_{ci}) \to C_{\min}(T_{hi} - T_{ci})$. Since $T_{co} - T_{ci}$ is the largest temperature difference but C_c is the smaller of two heat capacities, $C_c(T_{hi} - T_{ci})$ is the maximum realizable heat transfer with a counter-flow heat exchanger.

Also, for $C_h < C_c$ or $1/C_h > 1/C_c$, $\Delta T_b \to 0$ as $A \to \infty$. In terms of Fig. 7.18(b), $\Delta T_b \to 0$ implies $T_{ho} \to T_{ci}$, and $C_h(T_{hi} - T_{ho}) \to C_{\min}(T_{hi} - T_{ci})$. Now $T_{hi} - T_{ho}$ is the largest temperature difference and C_h is the smaller of two heat-capacity flow rates. For $C_h = C_c$, the foregoing interpretation does not hold.

Now, we proceed to the evaluation of ϵ depending on the type of heat exchanger. Consider first a parallel-flow heat exchanger. From Eq. (7.11)

$$\ln \frac{\Delta T_b}{\Delta T_a} = -\left(\frac{1}{C_h} + \frac{1}{C_c}\right) UA,$$

or, in terms of Fig. 7.5,

$$\ln \frac{T_{ho} - T_{co}}{T_{hi} - T_{ci}} = -\left(\frac{1}{C_h} + \frac{1}{C_c}\right) UA. \qquad (7.43)$$

Also, from the first equality of Eq. (7.42)

$$T_{ho} = T_{hi} - \epsilon \frac{C_{\min}}{C_h}(T_{hi} - T_{ci}),$$

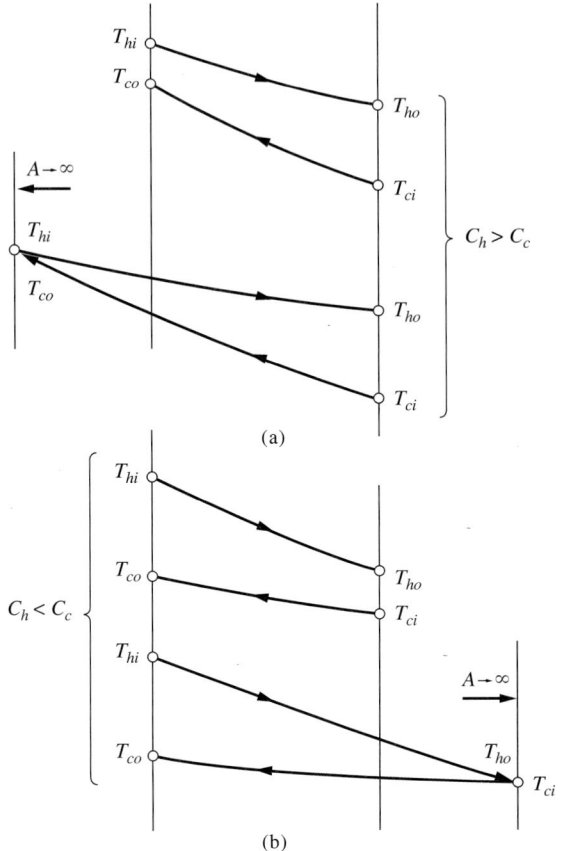

Figure 7.18 Counter flow as $A \to \infty$.
(a) $C_h > C_c$, (b) $C_h < C_c$.

from the second equality of the same equation

$$T_{co} = T_{ci} + \epsilon \frac{C_{min}}{C_c}(T_{hi} - T_{ci}),$$

and from their difference, after dividing by $T_{hi} - T_{ci}$,

$$\frac{T_{ho} - T_{co}}{T_{hi} - T_{ci}} = 1 - \epsilon \left(\frac{C_{min}}{C_h} + \frac{C_{min}}{C_c} \right). \quad (7.44)$$

Inserting Eq. (7.44) into Eq. (7.43)

$$\ln \left[1 - \epsilon \left(\frac{C_{min}}{C_h} + \frac{C_{min}}{C_c} \right) \right] = -\left(\frac{1}{C_h} + \frac{1}{C_c} \right) UA$$

or

$$1 - \epsilon \left(\frac{C_{min}}{C_h} + \frac{C_{min}}{C_c} \right) = \exp\left[-\left(\frac{1}{C_h} + \frac{1}{C_c} \right) UA \right]$$

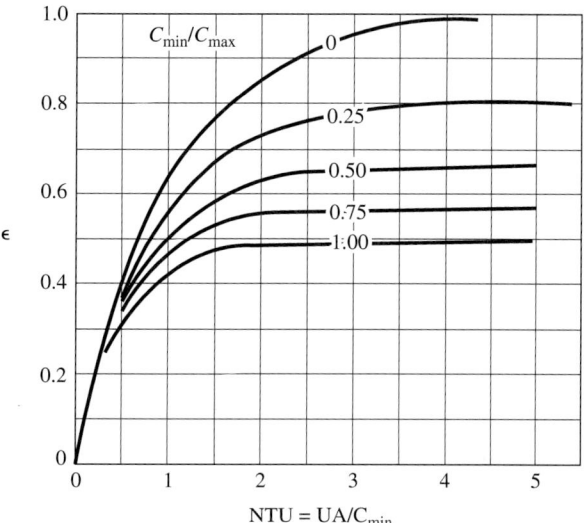

Figure 7.19 Effectiveness for parallel-flow heat exchanger.

or

$$\epsilon = \frac{1 - \exp\left[-\left(\frac{1}{C_h} + \frac{1}{C_c}\right) UA\right]}{\frac{C_{\min}}{C_h} + \frac{C_{\min}}{C_c}}.$$

Finally, noting that either C_h or C_c is the smaller of the two heat-capacity flow rates, redefining it as $C_{\min}$, and the other as $C_{\max}$, we have for a parallel-flow heat exchanger,

$$\epsilon = \frac{1 - \exp\left[-(1 + C_{\min}/C_{\max}) UA/C_{\min}\right]}{1 + C_{\min}/C_{\max}}, \qquad (7.45)$$

where $UA/C_{\min}$ is usually referred to as the number of transfer units, or NTU in short.[15] For convenience in heat-exchanger calculations, Eq. (7.45) is plotted in Fig. 7.19.

Following the same steps, we have for a counter-flow heat exchanger

$$\epsilon = \frac{1 - \exp\left[-(1 - C_{\min}/C_{\max}) UA/C_{\min}\right]}{1 - (C_{\min}/C_{\max}) \exp\left[-(1 - C_{\min}/C_{\max}) UA/C_{\min}\right]}, \qquad (7.46)$$

which is plotted in Fig. 7.20. The algebra gets quite involved for more complex heat exchangers and is beyond the scope of this text.[16] The effectivenesses of four of these heat exchangers are reproduced in Figs. 7.21 through 7.25.

---

[15] Actually $UA/C_{\min}$ is a Stanton number based on the total heat transfer coefficient.

[16] Figs. 7.19 through 7.25 are from Kays and London[5].

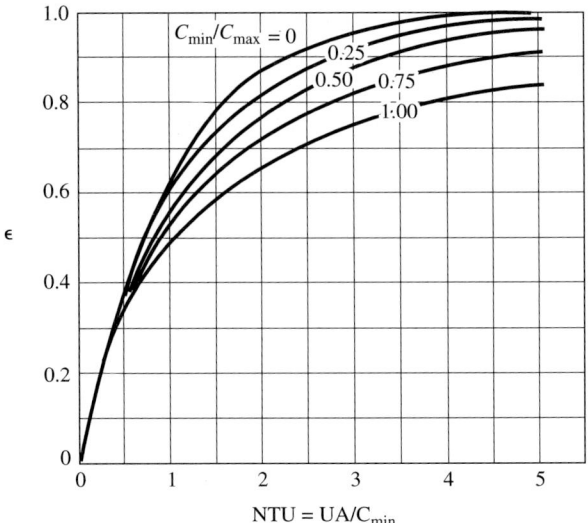

Figure 7.20 Effectiveness for counter-flow heat exchanger.

For a condenser or evaporator, the flow rate $\dot{m}$ of the two-phase side is small but finite. However, since the temperature remains constant on this side, $C = C_{max} \to \infty$ for finite $\dot{Q} = C\Delta T$ (also from $C = \dot{m}c_p$ we may conclude that $c_p \to \infty$ and the definition of c_p is degenerate). Accordingly, $C_{min}/C_{max} = 0$ for a condenser or evaporator. Since the type of heat exchanger is irrelevant because of negligibly small $\dot{m}$, from Eq. (7.45) or Eq. (7.46) for $C_{min}/C_{max} = 0$, we have the efficiency of condensers

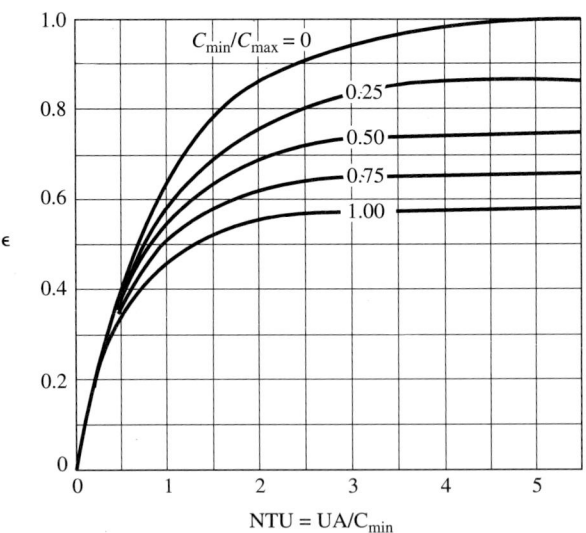

Figure 7.21 Effectiveness of 1–2 parallel-counter-flow heat exchanger.

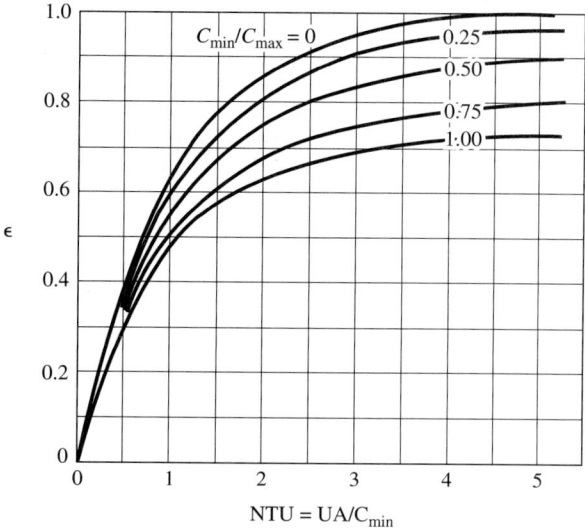

Figure 7.22 Effectiveness of 2–4 multipass heat exchanger.

and evaporators,

$$\epsilon = 1 - \exp(-UA/C_{\min}). \tag{7.47}$$

For example, from Figs. 7.21 through 7.25, for $C_{\min}/C_{\max} = 0$ and for a fixed NTU, say 1.2, we read the same $\epsilon \cong 0.70$.

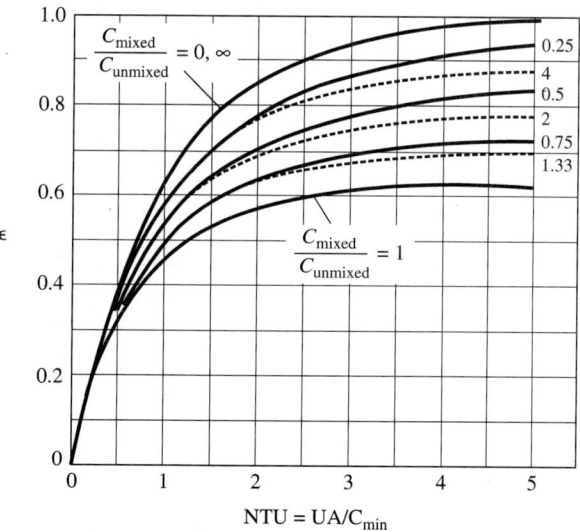

Figure 7.23 Effectiveness of one-fluid-mixed cross-flow heat exchanger.

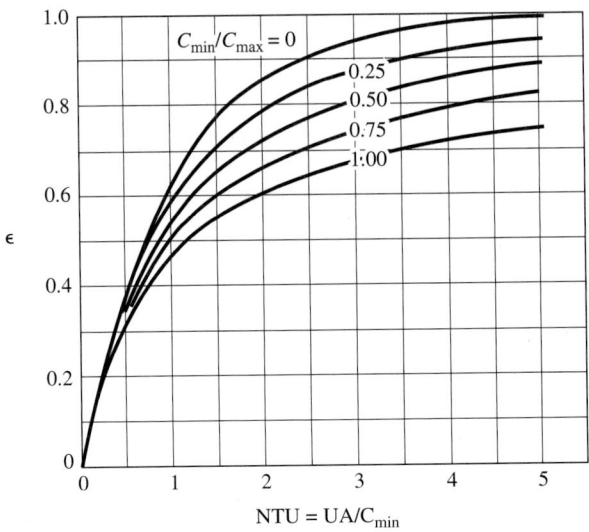

Figure 7.24 Effectiveness of cross-flow heat exchanger with unmixed fluids.

In a typical gas-to-gas heat exchanger, the C_{min}/C_{max} Ratio is approximately equal to unity. From Eq. (7.45), the efficiency of a parallel-flow heat exchanger for $C_{min}/C_{max} = 1$ is

$$\epsilon = \frac{1}{2}[1 - \exp(-2UA/C_{min})], \tag{7.48}$$

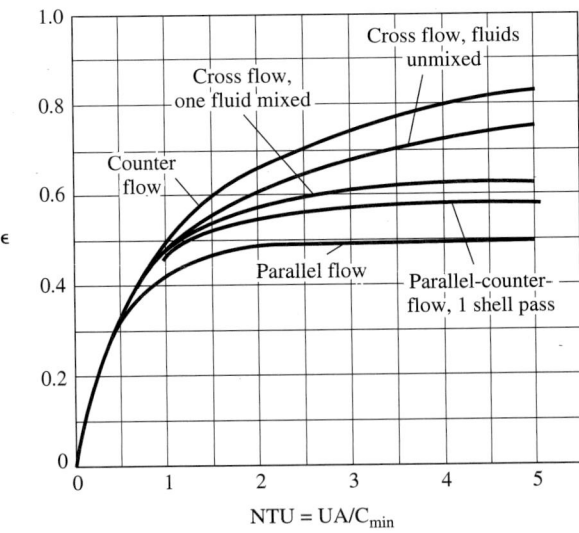

Figure 7.25 Effectiveness of heat exchangers.

and from the limit of Eq. (7.46) as $C_{min}/C_{max} \to 1$, the efficiency of a counter-flow heat exchanger for $C_{min}/C_{max} \cong 1$ is

$$\epsilon = \frac{UA/C_{min}}{1 + UA/C_{min}}. \tag{7.49}$$

As $UA/C_{min} \to \infty$, the efficiency of a parallel-flow heat exchanger approaches 1/2 while that of a counter-flow heat exchanger approaches 1. The efficiency of more complex heat exchangers is known to remain in between.

It is important to note that the NTU method, although devised for performance under different conditions, equally applies to the design of heat exchangers. Therefore, it is a more general method than the LMTD method. Here we recapitulate both methods for a ready reference in heat-exchanger calculations:

Independent of type	Dependent on type	Method
$C_h(T_{hi} - T_{ho}) = C_c(T_{co} - T_{ci}) =$	$UA\Delta T_{LM}F(P, R)$	LMTD
	$C_{min}(T_{hi} - T_{ci})\epsilon(\gamma, S)$	NTU

(7.50)

where $\gamma = C_{min}/C_{max}$ and $S = \text{NTU} = UA/C_{min}$ are introduced for notational convenience. We proceed now to a couple of examples which illustrate the use of the NTU method in heat-exchanger operation and/or design.

EXAMPLE 7.8[17]

A one-shell-pass and multiple-of-two-tube passes heat exchanger is to be used to cool 12 kg/s of oil with an equal flow rate of water. The heat transfer area of the heat exchanger is 40 m². The inlet temperatures of the oil and water are $T_{hi} = 100\,°\text{C}$ and $T_{ci} = 20\,°\text{C}$, respectively. The specific heats of the oil and water are 2.09 kJ/kg·K and 4.18 kJ/kg·K, respectively. The total heat transfer coefficient is controlled by the oil side and is 1,000 W/m²·K. We wish to determine the exit temperature of the oil and water.

The heat-capacity flow rates are

$$C_h = 2.09 \times 12 \cong 25 \text{ kW/K} = C_{min},$$

$$C_c = 4.18 \times 12 \cong 50 \text{ kW/K} = C_{max},$$

and

$$C_{min}/C_{max} = 1/2.$$

The number of transfer units is

$$\text{NTU} = \frac{UA}{C_{min}} = \frac{1,000 \times 40}{25,000} = 1.6.$$

Then, from Fig. 7.21,

$$\epsilon \cong 0.65.$$

---

[17] The FORTRAN program EX7–8.F is listed in the appendix of this chapter.

378 Chap. 7 Heat Exchangers

Now, employing Eq. (7.42), we have

$$C_h(T_{hi} - T_{ho}) = C_c(T_{co} - T_{ci}) = \epsilon C_{\min}(T_{hi} - T_{ci})$$

or

$$25(100 - T_{ho}) = 50(T_{co} - 20) = 0.65 \times 25(100 - 20),$$

which gives

$$T_{ho} = 48\,°C \quad \text{and} \quad T_{co} = 46\,°C.$$

Using the computer program provided, the interested reader may parametrically study the effect of the total heat transfer coefficients on the exit temperatures.

If one of these exit temperatures were specified from the beginning and the heat transfer area were unknown, the foregoing problem would become one of design. In this case ϵ and $C_{\min}/C_{\max}$ would be known, and from Fig. 7.21 we would get NTU and A. ◆

EXAMPLE 7.9

A cross-flow recuperator (usual heat exchanger for gas turbines) with both fluids unmixed is to be designed under a set of conditions and to be operated under different conditions. The hot exhaust gases flow through tubes, and the cold intake air flows across these tubes. The wall thickness of the tubes is negligible. The heat exchanger is to be designed with mass flow rates $\dot{m}_h = \dot{m}_c = 10$ kg/s, heat transfer coefficients $h_h = h_c = 150$ W/m²·K, and the three end temperatures $T_{hi} = 425\,°C$, $T_{ci} = 25\,°C$, and $T_{co} = 210\,°C$. We wish to determine (a) the heat transfer area of this exchanger, and (b) the new outlet temperatures after doubling the flow of (1) the hot fluid, (2) the cold fluid, while maintaining the same inlet temperatures.

(a) The heat-capacity flow rates, assuming equal heat capacity $c_p \cong 1$ kJ/kg·K for both fluids and noting the equal flow rates, are

$$C_h = C_c = 10 \text{ kg/s} \times 1,000 \text{ J/kg·K} = 10^4 \text{ W/K}.$$

It follows from Eq. (7.12),

$$10^4 \text{ W/K} \times (425 - T_{ho})\,°C = 10^4 \text{ W/K} \times (210 - 25)\,°C,$$

which gives

$$T_{ho} = 240\,°C.$$

Recall that both the LMTD method and the NTU method are suitable to the evaluation of heat transfer area. From Eqs. (7.12) and (7.42)

$$\dot{Q} = C_h(T_{hi} - T_{ho}) = \epsilon C_{\min}(T_{hi} - T_{ci}),$$

which yields

$$10^4 \text{ W/K} \times 185\,°C = \epsilon \times 10^4 \text{ W/K} \times (425 - 25)\,°C$$

or

$$\epsilon = \frac{185}{400} \cong 0.46.$$

Also,

$$C_{\min}/C_{\max} = 1.$$

Then, from Fig. 7.24 we read

$$\text{NTU} = \frac{UA}{C_{\min}} \cong 0.95,$$

which gives

$$0.95 = \frac{75 \text{ W/m}^2\cdot\text{K} \times A}{10^4 \text{ W/K}}$$

or, solving for the area,

$$A \cong 127 \text{ m}^2.$$

For comparison, let us evaluate the same area by the LMTD method. With

$$P = \frac{T_{to} - T_{ti}}{T_{si} - T_{ti}} = \frac{240 - 425}{25 - 425} = \frac{185}{400} \cong 0.46$$

and

$$R = \frac{C_t}{C_s} = 1$$

we have from Fig. 7.14

$$F(P, R) = F(0.46, 1) \cong 0.93.$$

Then, from Eq. (7.31), in view of $\Delta T_{\text{LM}} = 215\ °\text{C}$, we get

$$10^4 \text{ W/K} \times 185\ °\text{C} = 75 \text{ W/m}^2\cdot\text{K} \times A \text{ m}^2 \times 215\ °\text{C} \times 0.93$$

or

$$A \cong 123 \text{m}^2,$$

which is slightly different from the area obtained by the NTU method. Clearly, this difference is due to the uncertainty involved with reading from Figs. 7.14 and 7.24.

(b)–(1) As we learned in Chapter 6, forced-convection heat transfer coefficients depend on the flow velocity and geometry. If the mass flow rate of the hot fluid is changed, then its velocity, and consequently h_h, will change also. The hot fluid flows through tubes. Since the effect of Prandtl number on heat transfer in gases is small, then, neglecting this effect, we have from Eq. (6.30)

$$h_h \sim V_h^{0.8},$$

where V_h is the velocity of the hot fluid, and for different velocities

$$(h_2/h_1)_h = (V_2/V_1)_h^{0.8}.$$

The new hot-side heat transfer coefficient, after doubling $\dot{m}_h$, is

$$h_{2h} = h_{1h} 2^{0.8} = 150 \text{ W/m}^2\cdot\text{K} \times 2^{0.8} \cong 261 \text{ W/m}^2\cdot\text{K}.$$

Accordingly, from

$$\frac{1}{U} = \frac{1}{h_h} + \frac{1}{h_c} = \frac{1}{261} + \frac{1}{150}$$

we have

$$U \cong 95 \text{ W/m}^2\cdot\text{K},$$

and the NTU is

$$\text{NTU} = \frac{UA}{C_{\min}} = \frac{95 \text{ W/m}^2\cdot\text{K} \times 127 \text{ m}^2}{10^4 \text{ W/K}} \cong 1.2.$$

Also, noting

$$\frac{C_{\min}}{C_{\max}} = 0.5,$$

we read from Fig. 7.24

$$\epsilon \cong 0.60.$$

Then, from Eq. (7.42),

$$C_h(T_{hi} - T_{ho}) = C_c(T_{co} - T_{ci}) = \epsilon C_{\min}(T_{hi} - T_{ci})$$

or

$$2(425 - T_{ho}) = (T_{co} - 25) = 0.60 \times (425 - 25),$$

which yields

$$T_{ho} = 425 - \frac{1}{2} \times 0.60 \times 400 \cong 305\,°\text{C},$$

$$T_{co} = 25 + 0.60 \times 400 \cong 265\,°\text{C}.$$

The changes in the outlet temperatures resulting from the doubled flow rate of the hot fluid are sketched in Fig. 7.26(a). Both outlet temperatures are higher because of the increased flow rate of the hot fluid.

(b)–(2) Approximating the cold-side heat transfer with the flow over a single cylinder (Fig. 7.27),[18] here we utilize Table 6.2. However, since the statement of the problem does not provide the velocity over a single tube, we assume $n \sim 0.6$ from Table 6.2 with the assumption that $4 \times 10^3 < \text{Re}_D < 4 \times 10^4$.

Again, neglecting the effect of Prandtl number, we have

$$(h_2/h_1)_c = (V_2/V_1)_c^{0.6},$$

and the new cold-side heat transfer coefficient is

$$h_{2c} = h_{1c} 2^{0.6} = 150 \times 2^{0.6} \cong 227 \text{ W/m}^2\cdot\text{K}.$$

---

[18] Actually the tubes are placed closely, and they affect the flow pattern and heat transfer. Yet 0.6 continues to represent closely the approximate value of n (see Appendix 2 for tube-bank correlations).

Accordingly, from

$$\frac{1}{U} = \frac{1}{h_h} + \frac{1}{h_c} = \frac{1}{150} + \frac{1}{227}$$

we have

$$U \cong 90 \text{ W/m}^2 \cdot \text{K}$$

and

$$\frac{UA}{C_{min}} = \frac{90 \text{ W/m}^2 \cdot \text{K} \times 127 \text{ m}^2}{10^4 \text{ W/K}} = 1.14.$$

Also, noting that

$$\frac{C_{min}}{C_{max}} = 0.5,$$

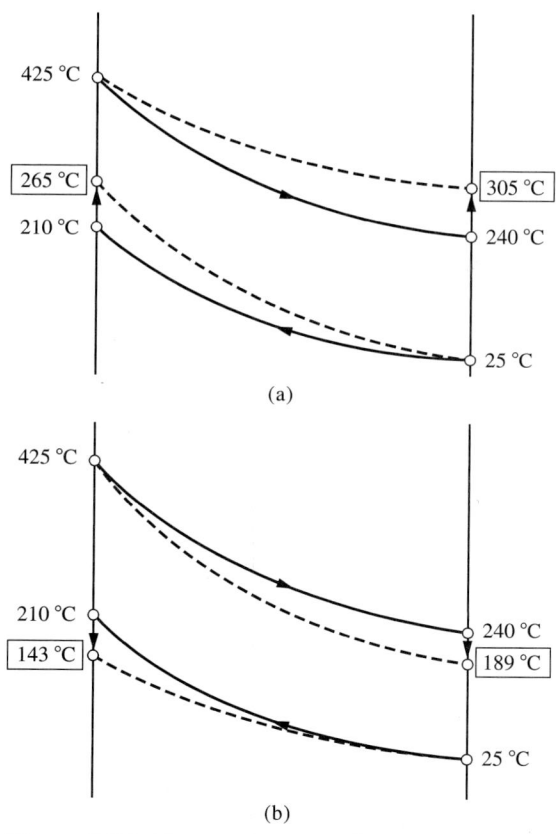

Figure 7.26 Recuperator performance
a) Effect of doubled flow of hot fluid b) Effect of doubled flow of cold fluid.

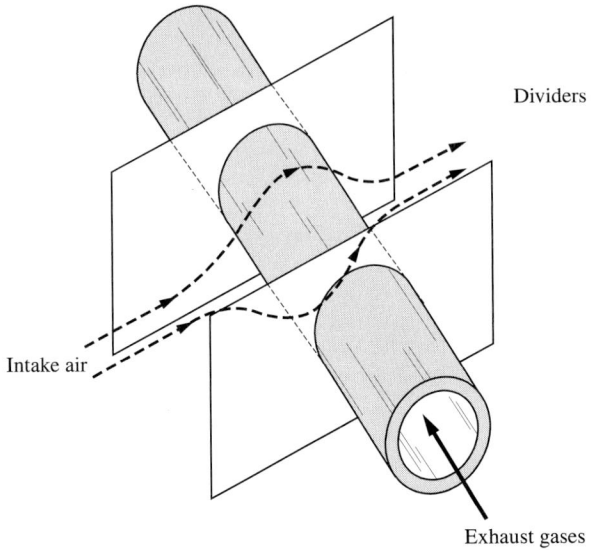

Figure 7.27 Recuperator idealized.

we read from Fig. 7.24

$$\epsilon \cong 0.59.$$

Then, from Eq. (7.42),

$$(425 - T_{ho}) = 2(T_{co} - 25) = 0.59 \times (425 - 25),$$

and we have

$$T_{ho} = 425 - 0.59 \times 400 \cong 189\,°C,$$

$$T_{co} = 25 + \frac{1}{2} \times 0.59 \times 400 \cong 143\,°C.$$

The changes in the outlet temperatures resulting from the doubled flow rate of the cold fluid are sketched in Fig. 7.26(b). Both outlet temperatures are lower because of the increased flow rate of the cold fluid.

It is worth noting that in a gas-turbine recuperator, the flow rate of the intake air and that of exhaust gases are interrelated and can only be changed simultaneously. However, to demonstrate the effect of each flow separately, here we increased either the flow of the hot fluid or that of the cold fluid. ◆

In terms of the foregoing examples, we have tried to improve and extend our understanding of the foundations of heat exchangers. In practice, however, the thermal design of a heat exchanger requires the evaluation of the heat transfer coefficients and decisions on the size and length of tubes to be used. The following example deals with these aspects of heat exchangers.

EXAMPLE 7.10

In terms of the cross-flow recuperator of the preceding example, we wish to evaluate the hot- and cold-side heat transfer coefficients from appropriate correlation formulas and select the diameter, length, number, and arrangement of the tubes.

An order of magnitude range for the gas velocity in tubes, pipes, conduits, and channels is $15 - 60$ m/s (refer to an introductory fluid mechanics text). Also, in heat-exchanger technology, the diameter of tubes used is usually in the range of $15 - 25$ mm. Assume, for example, a standard-size tube with 19 mm OD and 16.6 mm ID. Since the hot gases relative to the intake air have a lower density, let, for the time being, $V_h \cong 60$ m/s be the velocity corresponding to an averaged temperature of the hot gases. Accordingly, from Appendix B–2, we have, for air at $T_b = (T_{hi} + T_{ho})/2 = (425 + 240)/2 \cong 333°\text{C} = 606$ K,

Air at $T_b = 600$ K	$\rho = 0.580$ kg/m^3
	$\nu = 52.7 \times 10^{-6}$ m^2/s
	$Pr = 0.685$
	$k = 46.9 \times 10^{-3}$ W/m·K

and

$$\text{Re}_D = \frac{V_h D}{\nu} = \frac{60 \text{ m/s} \times 16.6 \times 10^{-3} \text{ m}}{52.7 \times 10^{-6} \text{ m}^2/\text{s}} = 18{,}900$$

Then, from Table 6.1,

$$\text{Nu}_{Dh} = 0.023(18{,}900)^{0.8}(0.685)^{0.3} \cong 54,$$

where subscript h indicates hot fluid, and

$$h_h = \frac{k}{D}\text{Nu}_{Dh} = \frac{46.9 \times 10^{-3} \text{ W/m·K}}{16.6 \times 10^{-3} \text{ m}} \times 54 \cong 153 \text{ W/m}^2\text{·K}.$$

For the cold side, let $V_c \cong 20$ m/s, because the intake air has higher density relative to the hot gases, and assume this velocity corresponding to inlet rather than to an averaged temperature of the air flow. Also, let $T_\infty = (T_{ci} + T_{co})/2 = (25 + 210)/2 \cong 118°\text{C}$, $T_w = (T_b + T_\infty)/2 = 225°\text{C}$, and $T_f = (T_w + T_\infty)/2 = 172°\text{C} = 445$ K. From Appendix B–2 we have

Air at $T_f = 450$ K	$\rho_f = 0.774$ kg/m^3
	$\nu_f = 52.7 \times 10^{-6}$ m^2/s
	$Pr_f = 0.686$
	$k_f = 37.3 \times 10^{-3}$ W/m·K

Since the outer cold fluid flows through a tube grid rather than over a single tube, we will employ tube-bank correlations for the external heat transfer coefficient. The use of these correlations requires that some quantities such as δ (the tube spacing), S_T, and S_L (see Appendix A–2) be known. For the tube size selected, the cross sectional area is $A_t = 2.16 \times 10^{-4}$ m^2. From the total flow rate of the hot gas inside the tubes we get, for the number of tubes,

$$n = \frac{\dot{m}_h}{\rho_h A_t V_h} = \frac{10 \text{ kg/s}}{0.580 \text{ kg/m}^3 \times 2.16 \times 10^{-4} \text{ m}^2 \times 60 \text{ m/s}} = 1{,}330$$

384 Chap. 7 Heat Exchangers

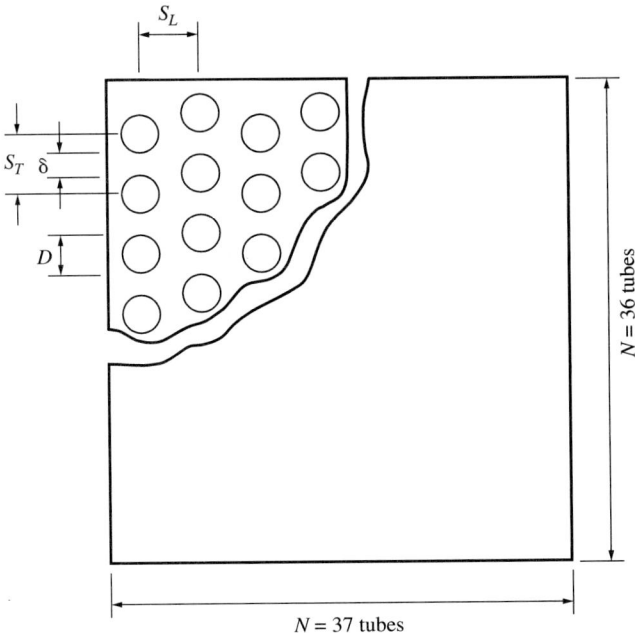

Figure 7.28 Size of the heat exchanger a) Effect of doubled flow of hot fluid b) Effect of doubled flow of cold fluid.

or

$$n = 1330 \sim 36 \times 37 = 1332 \quad \text{tubes approximately,}$$

which yields, in turn, $N \cong 37$, the number of zigzag distances shown in Fig. 7.28. The peripheral area of these tubes is $A_p = 0.0597 \text{ m}^2/\text{m}$ length. Assuming (for the time being) that the heat-exchanger area determined in Ex. 7.9 is approximately the area that we need, then the length of the tubes is

$$\ell = \frac{A}{nA_p} = \frac{127 \text{ m}^2}{1332 \times 0.0597 \text{ m}^2/\text{m}} \cong 1.6 \text{ m}.$$

Finally, we determine the spacing of tubes, δ, from the flow rate of cold air outside the tubes,

$$\delta = \frac{\dot{m}_c}{N\rho_c \ell V_c} = \frac{10 \text{ kg/s}}{37 \times 0.774 \text{ kg/m}^3 \times 1.6 \text{ m} \times 20 \text{ m/s}} \cong 1.1 \text{cm}.$$

The hot-fluid cross section of the heat exchanger may be now arranged as shown in Fig. 7.28. Accordingly, for S_T the spacing of tubes results in

$$S_T = D_0 + \delta = 1.9 + 1.1 = 3 \text{ cm},$$

which gives $S_T/D = 3/1.9 \cong 1.5$. Also, assuming $S_L/D \cong 1.5$ for staggered tube configuration, we get $C = 0.52$, and $m = 0.562$ from Appendix A–2, where

$$Nu_D = C Re_{D,\max}^m \Pr^{1/3}.$$

The Reynolds number,

$$\text{Re}_D = \frac{\rho_f V_c D}{\mu_f} = \frac{0.774 \text{ kg/m}^3 \times 20 \text{ m/s} \times 19 \times 10^{-3} \text{ m}}{250 \times 10^{-7} \text{ N·s/m}^2} = 11{,}765.$$

Then,[19] inserting C and m,

$$Nu_{Dc} = 0.52(11{,}765)^{0.562}(0.686)^{1/3} = 88.8$$

and

$$h_c = \frac{k}{D} Nu_{Dc} = \frac{37.3 \times 10^{-3} \text{ W/m·K}}{19 \times 10^{-3} \text{ m}} \times 88.8 = 174 \text{ W/m}^2\text{·K}.$$

Note that the substantially lower velocity assumed for the cold air does not yield a comparably low Reynolds number. Neglecting the conductive resistance of the tube walls, we have

$$\frac{1}{U} = \frac{1}{153} + \frac{1}{174}$$

or

$$U \cong 81 \text{ W/m}^2\text{·K},$$

which is quite close to the initially assumed $U = 75$ W/m²·K. ◆

The present example shows how to go about determining the size of a heat exchanger. No further attempt will be made here to refine the foregoing calculations. Clearly, like any problem of engineering design, this problem requires a number of trials until the assumed and found values of unknowns coincide. For example, we implicitly assumed atmospheric flow for both fluids. Otherwise the densities should be adjusted accordingly. Also, to determine the size of pumps, we need to know the pressure drop in the hot and cold fluids. The efficiency of our heat exchanger is reasonably low. It can be improved at the expense of increased heat transfer area, which in turn increases the pressure drop and the size of the pumps.

7.5 FOULING FACTOR. VARIABLE COEFFICIENT OF HEAT TRANSFER. CLOSURE ○

During the operation of a heat exchanger a film begins to form on both the hot and cold sides of the heat transfer surface. This film may be dirt, silt or another chemical deposit, rust or another oxide resulting from the interaction of the exchanger fluids with the solid material of the transfer surface. The effect of a film, called the **fouling factor**, is to increase the thermal resistance between the exchanger fluids. Table 7.1, taken from Ref. 4, gives typical values of the fouling resistance, R_f. The overall heat transfer coefficient then needs to be reduced by this resistance.

In Chapter 6 we have learned that heat transfer coefficient varies with the distance from the entrance of a pipe. However, so far in this chapter we have assumed a uniform coefficient throughout the exchanger. This assumption coupled with mean fluid properties usually gives satisfactory answers. Yet in situations involving chain polymers

---

[19] Since $2 \times 10^3 < 11{,}765 < 4 \times 10^4$, the tables in Appendix A2 are appropriate for this problem.

Table 7.1 Fouling resistances (From Ref. [4])

Fluid	Fouling Resistance, R_f $[W/m^2 \cdot K]^{-1}$
Fuel oil	0.005
Transformer oil	0.001
Vegetable oils	0.003
Light gas oil	0.002
Heavy gas oil	0.003
Asphalt	0.005
Gasoline	0.001
Kerosene	0.001
Caustic solutions	0.002
Refrigerant liquids	0.001
Hydraulic fluid	0.001
Molten salts	0.0005
Engine exhaust gas	0.01
Steam (non–oil bearing)	0.0005
Steam (oil-bearing)	0.001
Refrigerant vapors (oil-bearing)	0.002
Compressed air	0.002
Acid gas	0.001
Solvent vapors	0.001
Sea water	0.0005–0.001
Brackish water	0.001–0.003
Cooling tower water (treated)	0.001–0.002
Cooling tower water (untreated)	0.002–0.005
River water	0.001–0.004
Distilled or closed-cycle condensate water	0.0005
Treated boiler feedwater	0.0005–0.001

and viscous oils the strong temperature dependence of the fluid properties should be taken into account. Accordingly, the heat transfer area is divided into a finite number of area elements, each with an overall coefficient of heat transfer based on the local temperature of the hot and cold fluids.

In this chapter we studied the most commonly encountered heat exchangers, sometimes called recuperators. Two other types of heat exchangers, the **regenerator** and **cooling tower**, are also utilized, although much less frequently. In regenerators, the hot fluid and cold fluid successively occupy the same space, and the exchange of heat is

Table 7.2 Approximate overall coefficients for preliminary estimates (Ref. [7])

Heat Exchanger Duty	U, [W/m² · K]
Fuel oil	0.005
Gas to gas	10–30
Water to gas (e.g., gas cooler, gas boiler)	10–50
Condensing vapor-air (e.g., steam radiator, air heater)	5–50
Steam to heavy fuel oil	50–180
Water to water	800–2500
Water to other liquids	200–1000
Water to lubricating oil	100–350
Light organics to light organics	200–450
Heavy organics to heavy organics	50–200
Air-cooled condensors	50–200
Water-cooled steam condensors	1000–4000
Water-cooled ammonia condensors	800–1400
Water-cooled organic vapor condensors	300–1000
Steam boilers	10–40+radiation
Refrigerator evaporators	300–1000
Steam water evaporators	1500–6000
Steam-jacketed agitated vessels	150–1000
Heating coil in vessel, water to water	
Unstirred	50–250
Stirred	500–2000

unsteady. The theory associated with this unsteady process is beyond the scope of this text (for extensive treatments see, for example, Refs. 3 and 6). In cooling towers, both fluids (usually water and air) flow simultaneously through the same passage, and the heat transfer is coupled with mass transfer.

As we learned with the examples of the present chapter, we need to know the overall coefficient of heat transfer for the (thermal) design or performance of a heat exchanger. Since this coefficient usually depends on unknown exit temperatures, we are faced with a trial-and-error procedure. For a first trial, an order-of-magnitude value of the heat transfer coefficient is usually satisfactory. Table 7.2, taken from Ref. 7, gives the value of the overall coefficient for a number of frequently encountered cases. This table, which includes the effect of a fouling factor, is more conservative than Table 1.2.

Finally, we tried in this chapter to learn the most important and general aspects of heat exchangers. For more detailed studies the reader is referred to special handbooks and texts such as Refs. 5, 8 through 12, and the references cited therein.

388 Chap. 7 Heat Exchangers

■ REFERENCES

7.1 M. Jakob and G.A. Hawkins, *Elements of Heat Transfer*. Wiley, 1957.

7.2 W.M. Rohsenow and H. Choi, *Heat, Mass, and Momentum Transfer*. Prentice-Hall, Englewood Cliffs, NJ, 1961.

7.3 M. Jakob, *Heat Transfer*, vol. 2. Wiley, New York, 1957.

7.4 *Standards* TEMA. Tubular Exchanger Manufacturers Association, 1978.

7.5 W.M. Kays and A.L. London, *Compact Heat Exchangers*. McGraw-Hill, New York, 1964.

7.6 V.S. Arpacı, *Conduction Heat Transfer*. Addison-Wesley, Reading. MA, 1966.

7.7 A.C. Mueller, "Thermal design of shell-and-tube heat exchangers for liquid-to-liquid heat transfer," *Eng. Bull., Res. Ser.*, 121, Purdue University Eng. Exp. Sta., 1954.

7.8 W.J. Yang, "Industrial heat exchangers," *Univ. of Michigan Eng. Summer Conf.*, 1978.

7.9 N.H. Afgan and E.U. Schlunder, *Heat Exchanger: Design and Theory*. McGraw-Hill, New York, 1974.

7.10 A.P. Fraas and M.N. Özısık, *Heat Exchanger Design*. Wiley, New York, 1965.

7.11 D.Q. Kern, *Process Heat Transfer*. McGraw-Hill, New York, 1950.

7.12 R.H. Perry, D.W. Green, and J.O. Maloney, eds., *Perry's Chemical Engineers' Handbook*, 6th ed. McGraw-Hill, New York, 1984.

7.6 APPENDIX

```
C-----------------------------------
C     EX7-1.F (START)
C-----------------------------------
      PROGRAM MAIN
      IMPLICIT REAL*8 (A-H,K-Z)
      PI=4*ATAN(1.)
      WRITE(*,*) 'EXAMPLE 7.1....'
C-----------------------------------------------
C     INPUT DATA
C-----------------------------------------------
      WRITE(*,*) 'DOTMH: kg/s'
      READ(*,*) DOTMH
      WRITE(*,*) 'TH_I: C'
      READ(*,*) THI
      WRITE(*,*) 'TH_O: C'
      READ(*,*) THO
      WRITE(*,*) 'DOTMC: kg/s'
      READ(*,*) DOTMC
      WRITE(*,*) 'TC_I: C'
      READ(*,*) TCI
          WRITE(*,*) 'CPH: kJ/kg.K'
      READ(*,*) CPH
          WRITE(*,*) 'CPC: kJ/kg.K'
      READ(*,*) CPC
      WRITE(*,*) 'U: W/m^2.K'
      READ(*,*) U
C-----------------------------------------------
C     UNIT CONVERSION
C-----------------------------------------------
      CPH=1000*CPH
      CPC=1000*CPC
C-----------------------------------------------
C     CALCULATION
```

```fortran
C-------------------------------------------------
      CH=DOTMH*CPH
      CC=DOTMC*CPC
      DOTQ=CH*(THI-THO)
      TCO=TCI+DOTQ/CC
      DTA=THI-TCI
      DTB=THO-TCO
      IF(DTA.NE.DTB) THEN
      DTPARA=(DTA-DTB)/LOG(DTA/DTB)
      ELSE
      DTPARA=DTA
      ENDIF
      DTA=THI-TCO
      DTB=THO-TCI
      IF(DTA.NE.DTB) THEN
      DTCOUN=(DTA-DTB)/LOG(DTA/DTB)
      ELSE
      DTCOUN=DTA
      ENDIF
      APARA=DOTQ/(U*DTPARA)
      ACOUN=DOTQ/(U*DTCOUN)
C-------------------------------------------------
C     ANSWER
C-------------------------------------------------
      WRITE(*,*) 'PARALLEL HEAT TRANSFER AREA IS'
      WRITE(*,*) APARA,' m^2'
      WRITE(*,*) 'COUNTER HEAT TRANSFER AREA IS'
      WRITE(*,*) ACOUN,' m^2'
      STOP
      END
C-------------------------------------
C     EX7-1.F (END)
C-------------------------------------
C-------------------------------------
C     EX7-4.F (START)
C-------------------------------------
      PROGRAM MAIN
      IMPLICIT REAL*8 (A-H,K-Z)
      PI=4*ATAN(1.)
      WRITE(*,*) 'EXAMPLE 7.4....'
C-------------------------------------------------
C     INPUT DATA
C-------------------------------------------------
      WRITE(*,*) 'DOTMH: kg/s'
      READ(*,*) DOTMH
      WRITE(*,*) 'DOTMC: kg/s'
      READ(*,*) DOTMC
      WRITE(*,*) 'HH: W/m^2.K'
      READ(*,*) HH
      WRITE(*,*) 'HC: W/m^2.K'
      READ(*,*) HC
           WRITE(*,*) 'CPH: kJ/kg.K'
      READ(*,*) CPH
           WRITE(*,*) 'CPC: kJ/kg.K'
      READ(*,*) CPC
      WRITE(*,*) 'TH_I: C'
      READ(*,*) THI
      WRITE(*,*) 'TC_I: C'
      READ(*,*) TCI
      WRITE(*,*) 'TC_O: C'
      READ(*,*) TCO
C-------------------------------------------------
C     UNIT CONVERSION
```

```
C----------------------------------------------
  CPH=1000*CPH
  CPC=1000*CPC
C----------------------------------------------
C     CALCULATION
C----------------------------------------------
  CH=DOTMH*CPH
  CC=DOTMC*CPC
  DOTQ=CC*(TCO-TCI)
  THO=THI-DOTQ/CH
  DTA=THI-TCO
  DTB=THO-TCI
  IF(DTA.NE.DTB) THEN
  DTCOUN=(DTA-DTB)/LOG(DTA/DTB)
  ELSE
  DTCOUN=DTA
  ENDIF
  U=1/(1/HH+1/HC)
  ACOUN=DOTQ/(U*DTCOUN)
C----------------------------------------------
C     ANSWER
C----------------------------------------------
  WRITE(*,*) 'COUNTER HEAT TRANSFER AREA IS'
  WRITE(*,*) ACOUN,' m^2'
  STOP
  END
C----------------------------------
C     EX7-4.F (END)
C----------------------------------
C----------------------------------
C     EX7-8.F (START)
C----------------------------------
  PROGRAM MAIN
  IMPLICIT REAL*8 (A-H,K-Z)
  PI=4*ATAN(1.)
  WRITE(*,*) 'EXAMPLE 7.8....'
C----------------------------------------------
C     INPUT DATA
C----------------------------------------------
  WRITE(*,*) 'DOTMH: kg/s'
  READ(*,*) DOTMH
  WRITE(*,*) 'DOTMC: kg/s'
  READ(*,*) DOTMC
  WRITE(*,*) 'A: m^2'
  READ(*,*) A
  WRITE(*,*) 'TH_I: C'
  READ(*,*) THI
  WRITE(*,*) 'TC_I: C'
  READ(*,*) TCI
       WRITE(*,*) 'CPH: kJ/kg.K'
  READ(*,*) CPH
       WRITE(*,*) 'CPC: kJ/kg.K'
  READ(*,*) CPC
  WRITE(*,*) 'HH: W/m^2.K'
  READ(*,*) HH
C----------------------------------------------
C     UNIT CONVERSION
C----------------------------------------------
  CPH=1000*CPH
  CPC=1000*CPC
C----------------------------------------------
C     CALCULATION
```

```
C-------------------------------------------------
      CH=DOTMH*CPH
      CC=DOTMC*CPC
      IF(CH.GT.CC) THEN
      CMIN=CC
      CMAX=CH
      ELSE
      CMIN=CH
      CMAX=CC
      ENDIF
      CRATIO=CMIN/CMAX
      NTU=HH*A/CMIN
C-------------------------------------------------
C     CHART READINGS
C-------------------------------------------------
      WRITE(*,*) 'PICK EPSILON FROM FIG. 7.21 WITH'
      WRITE(*,*) 'CMIN/CMAX=',CRATIO,'  NTU=',NTU
      WRITE(*,*) 'INPUT EPSILON'
      READ(*,*) EPSILON
C-------------------------------------------------
C     CALCULATION
C-------------------------------------------------
      DOTQ=EPSILON*CMIN*(THI-TCI)
      THO=THI-DOTQ/CH
      TCO=TCI+DOTQ/CC
C-------------------------------------------------
C     ANSWER
C-------------------------------------------------
      WRITE(*,*) 'EXIT TEMPERATURE OF THE OIL IS'
      WRITE(*,*) THO,' C'
      WRITE(*,*) 'EXIT TEMPERATURE OF THE WATER IS'
      WRITE(*,*) TCO,' C'
      STOP
      END
C-------------------------------------------------
C     EX7-8.F (END)
C-------------------------------------------------
```

■ EXERCISES

7.1 A one-shell-pass and multiple-of-two-tube-passes heat exchanger is to be designed. Water with $\dot{m}_c = 20$ kg/s ($c_{pc} = 4$ kJ/kg·K) and $T_{ci} = 20\,°C$ flows in the tubes and cools hot shell oil with $\dot{m}_h = 80$ kg/s ($c_{ph} = 2$ kJ/kg·K) from $T_{hi} = 100\,°C$ to $T_{ho} = 80\,°C$. The total heat transfer coefficient is $U = 1.2$ kW/m²·K. (a) Evaluate the heat transfer area. (b) What type of heat exchanger will give the minimum heat transfer area? Evaluate this area. (c) What type of heat exchanger will give the maximum heat transfer area? Evaluate this area. (d) Comment on the given heat transfer coefficient.

7.2 A counter-flow heat exchanger is utilized to cool a 2.5 kg/s gas flow from 150 °C to 100 °C by means of a 3 kg/s air flow which enters the exchanger at 10 °C. We wish to increase the size of this exchanger and cool the gas to 90 °C rather than 100 °C. Assume that the other conditions remain the same, and determine the ratio of heat transfer areas for the two cases. What is your conclusion?

7.3 The exhaust gases of an industrial gas turbine are utilized to heat the intake (air). Consider a cross-flow heat exchanger with unmixed fluids under the following conditions: $C_h = C_c = 1{,}500$ W/K, $T_{hi} = 315\,°C$, $T_{ci} = 35\,°C$, $T_{co} = 175\,°C$, $h_h = h_c = 100$ W/m²·K. (a) Evaluate the heat transfer area. (b) One half of this exchanger is shut down for a periodic clean-up process. C_h and C_c are kept constant by increasing the pumping power. Find the new exit temperatures.

7.4 A counter-flow heat exchanger is to be designed. Hot oil ($\dot{m}_h = 600$ kg/s, $c_{ph} = 2$ kJ/kg·K) flows in tubes while heating shell water ($\dot{m}_c = 400$ kg/s, $c_{pc} = 4$ kJ/kg·K). The inlet and outlet temperatures of the oil and the inlet temperature of the water are $T_{hi} = 110$ °C, $T_{ho} = 70$ °C, and $T_{ci} = 20$ °C, respectively. Also, $h_h = 1{,}000$ W/m²·K and $h_h \sim V_h^{0.8}$. (a) Find the heat transfer area. (b) Because of faulty manufacturing, 1/3 of the pipes are sealed and the oil flow is reduced to $\dot{m}_h = 400$ kg/s. Find the new exit temperatures.

7.5 A one-shell pass and multiple-of-two-tube-passes heat exchanger will be designed. Water flows at the rate of 360,000 kg/hr ($V_c = 3$ m/s) in the tubes (2 cm ID, 2.5 cm OD, $k = 20$ W/m·K). The inlet and outlet temperature of the water are $T_{ci} = 10$ °C and $T_{co} = 65$ °C, respectively. Ethylene glycol flows ($V_h = 1.5$ m/s) through the shell at the rate of 720,000 kg/hr ($c_p = 2$ kJ/kg·K), and its temperature decreases from $T_{hi} = 180$ °C. Evaluate the shell-side heat transfer coefficient on the basis of cross flow. (a) Calculate the heat transfer area. (b) Recalculate the exit temperatures after reducing the water flow rate to 240,000 kg/hr.

7.6 A high-temperature-condensing water flow is cooled by a low-temperature-evaporating water flow in a counter-flow heat exchanger. (a) Calculate the exit quality of the evaporating flow. (b) Calculate the heat transfer area of the heat exchanger. *Data:* $T_c = 15$ °C, $h_{fc} = 62.98$ kJ/kg, $h_{gc} = 2528.9$ kJ/kg, $x_{ci} = 0$ %, $T_h = 140$ °C, $h_{fh} = 589.11$ kJ/kg, $h_{gh} = 2733.9$ kJ/kg, $x_{hi} = 100$%, $x_{h0} = 0$ %, $\dot{m}_h = \dot{m}_c = 600{,}000$ kg/hr, $c_{ph} = c_{pc} = 4.18$ kJ/kg·K, and $h_h = h_c = 500$ W/m²·K.

7.7 For gas-turbine recuperators and for most gas-to-gas heat exchangers, the $C_{\min}/C_{\max}$ ratio is approximately equal to unity. Show for this case that the heat-exchanger effectiveness for parallel flow is Eq. (7.48) and for counter flow is Eq. (7.49).

7.8 For condensers and evaporators the $C_{\min}/C_{\max}$ ratio is negligibly small. Show for this case that the effectiveness of any heat exchanger is given by Eq. (7.47).

7.9 The outlet vapor of a turbine is condensed by water flow in a one-shell-pass and multiple-passes (1–2 parallel-counter-flow type) heat exchanger. The water flows in tubes. (a) Find the water exit temperature. (b) Find the vapor flow rate. *Data:* $A = 3$ m², $U = 100$ W/m²·K, $T_{vapor} = 50$ °C ($h_f = 204$ kJ/kg, $h_g = 2{,}590$ kJ/kg), $T_{water, inlet} = 15$ °C, $\dot{m}_{water} = 3{,}600$ kg/hr, $c_{p,water} = 4.18$ kJ/kg·K, x_i (vapor quality) $= 0.8$, $x_0 = 0$.

7.10 A one-shell pass and multiple-of-two-tube-passes heat exchanger is to be designed. Water flows at the rate of 60 kg/s and with velocity 2.5 m/s in tubes of 25 mm OD and 20 mm ID. The thermal conductivity of the tubes is 20 W/m·K. The inlet and outlet temperatures of the water are 120 °C and 65 °C, respectively. Oil flows through the shell at the rate of 120 kg/s and its temperature increases from 10 °C. The specific heat of oil is 2,100 J/kg·K. Assume the shell-side heat transfer coefficient to be one-tenth of the tube-side heat transfer coefficient. Determine the heat transfer area.

7.11 To increase the overall efficiency of a steam power plant, part of the expanding steam is taken out from the turbine and used for heating (Fig. 7P–1). A heat exchanger in which low-pressure steam condenses at 120 °C while heating 20 kg/s of water flow from 65 °C to 90 °C will be designed for this purpose. The total heat transfer coefficient is 8,500 W/m²·K. (a) Determine the type of heat exchanger that would provide the minimum heat transfer area. (b) Evaluate this transfer area. (c) What flow rate of steam is to be drawn from the turbine to the heat exchanger? The steam quality at the inlet of the exchanger is $x_i = 1$.

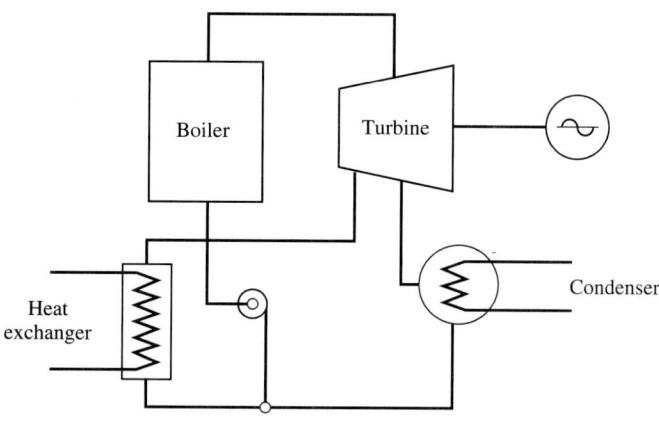

Figure 7P–1

7.12 We wish to determine the condenser size (heat transfer area) in the preceding problem. The vapor pressure and temperature at the inlet of the turbine are 12.5 MPa and 650 °C, respectively. The output of the turbine is 100 MW. The vapor pressure and quality at the inlet of the condenser are 50 kPa and 0.96, respectively. The temperature of the cooling water at the inlet of the condenser is 10 °C. The overall coefficient of heat transfer is 6,000 W/m²·K. (a) Assume a flow-rate ratio between the condensing vapor and the cooling water. (b) Evaluate the temperature of the cooling water at the outlet of the condenser. (c) Find the heat transfer area of the condenser.

7.13 Because of a slight leakage problem, the condenser pressure in the preceding problem rises from 50 to 100 kPa. Assume the quality of the steam to remain the same at the inlet of the condenser. (a) Find the reduction in the turbine output. (b) What are the means of retaining the turbine output at 100 MW? (c) Determine the heat transfer area of the condenser corresponding to 3 in. Hg vacuum pressure.

7.14 Low-pressure steam is condensed at the rate of $\dot{m} = 10$ kg/s by a pressurized and evaporating water flow in a multiple of-two-tube passes heat exchanger. From the saturated-steam tables, condensing steam at $p_s = 1.55$ MPa and $T_s = 200\,°C$, $h_f = 852$ kJ/kg, $h_{fg} = 1941$ kJ/kg, $h_g = 2{,}793$ kJ/kg, inlet quality 100%, outlet 0%; evaporating steam at $p_w = 0.2$ MPa and $T_w = 120\,°C$, $h_f = 503$ kJ/kg, $h_{fg} = 2203$ kJ/kg, $h_g = 2706$ kJ/kg, and and inlet quality 0%. (a) Sketch the temperature of the condensing steam and evaporating water. (b) Find the flow rate of water corresponding to 10% and 90% exit quality for the water. (c) Find the heat transfer area of the heat exchanger. (d) Which fluid should flow in tubes? (e) Which heat transfer coefficient do you need to compute?

7.15 Modify Eqs. (7.37), (7.39), and (7.41) for an evaporator.

7.16 Vapor at a saturation pressure p_s and superheat temperature T_h is condensed by a counter water flow in a heat exchanger of area A. The inlet temperature of the water is T_{wi}. The flow rates of the vapor and water are $\dot{m}_s$ and $\dot{m}_w$. Assume the total heat transfer coefficients to be U_1 and U_2 for parts 1 and 2 of the exchanger as shown in Fig. 7P–2. Log-mean averages may be replaced by arithmetic averages. Determine the exit temperature of the water and the exit quality of the vapor.

394 Chap. 7 Heat Exchangers

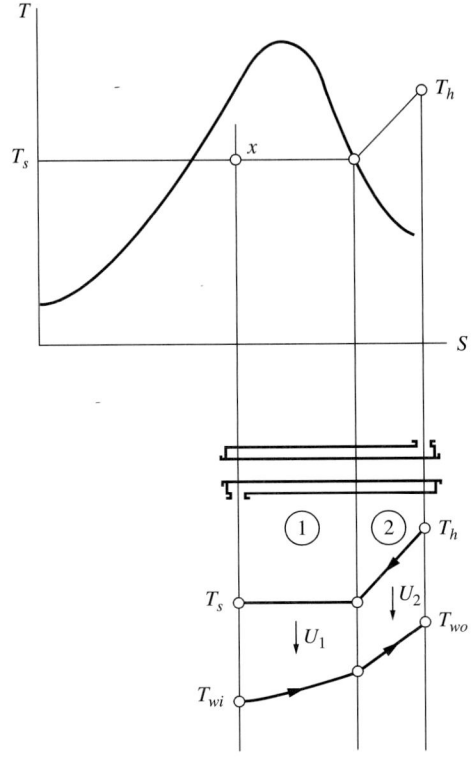

Figure 7P–2

7.17 Redesign the heat exchanger of Ex. 7.4 in terms of the design sketched in Fig. 7P–3.

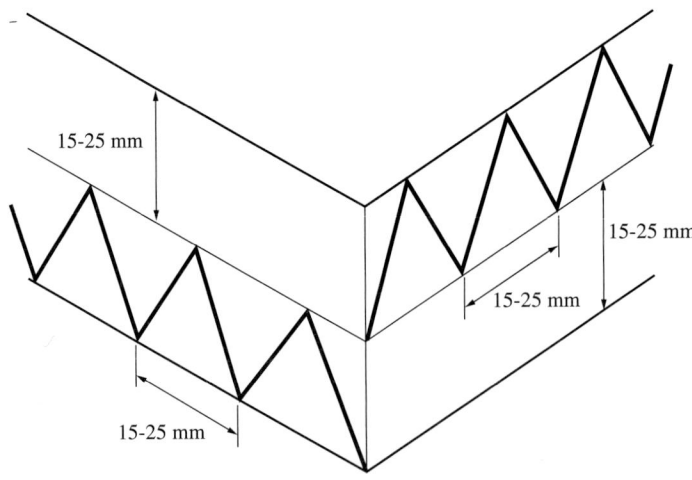

Figure 7P–3

7.18 A small space heater is to be designed consisting of steel tubes of 20 mm diameter arranged in equilateral, staggered triangles (Fig. 7P–4). Steam at 110 °C condenses inside the tubes. Air at the rate of 8 m³/s delivered by a fan is heated from 10 °C to 60 °C. (a) Evaluate the number and spacing of the tubes. (b) Estimate the pressure drop in the air flow.

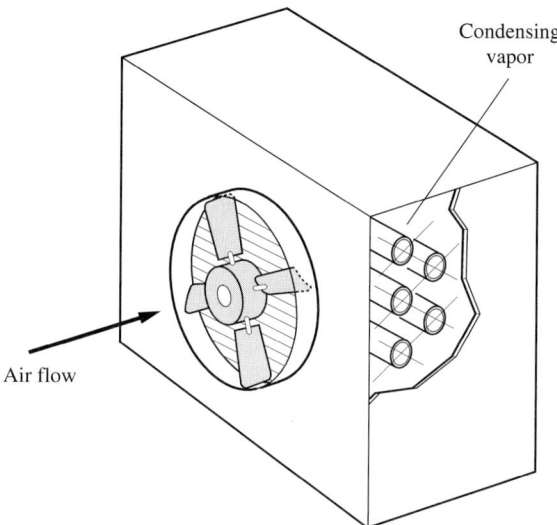

Figure 7P–4

7.19 The exhaust steam of a turbine with quality $x_i = 0.98$, saturation pressure $p_{sat} = 75$ kPa, and mass flow rate $\dot{m}_h = 1$ kg/s is fully condensed ($x_o = 0.0$) while transferring heat to river water which enters the condenser at $T_{ci} = 10$ °C with a mass flow rate of $\dot{m}_c = 8$ kg/s. A standard thermodynamic text yields a saturation temperature of $T_{sat} = 91.9$ °C and a latent heat $h_{fg} = 2278.6$ kJ/kg. Assuming a total heat transfer coefficient $U = 2{,}000$ W/m²·K for clean heat-exchanger surfaces, (a) determine the heat-exchanger area. (b) Due to residue in the river water, dirt is formed, creating an additional thermal resistance, the so-called fouling resistance. This additional thermal resistance is 0.0002 m²·K/W. For fixed $\dot{m}_h$, A, x_o, x_i, T_h, T_{ci}, determine the new exit temperature and the new mass flow rate of the cooling water.

7.20 A heat exchanger with one shell and two tube passes is to be designed. Hot oil flowing in tubes ($\dot{m}_h = 30$ kg/s, $c_{ph} = 2$ kJ/kg·K) heats water ($\dot{m}_c = 20$ kg/s, $c_{pc} = 4$ kJ/kg·K). The inlet and outlet temperatures of the water and the inlet temperature of the oil are $T_{ci} = 20$ °C, $T_{co} = 50$ °C, $T_{hi} = 100$ °C, respectively. The heat transfer coefficients are known to be $h_h = 1{,}000$ W/m²·K and $h_c = 9{,}000$ W/m²·K for oil and water, respectively. Find the heat transfer area of the heat exchanger.

CHAPTER 8

FOUNDATIONS OF RADIATION

We have so far studied the conduction and convection modes of heat transfer. We have learned that conduction depends on the properties of solids and fluids and that convection depends on the motion as well as the properties of fluids. In other words, heat by conduction or convection can be transferred only in matter, not in a vacuum. From our daily experience, however, we know that thermal energy can be transmitted between a source and a distant sink without any intervening carrier. For example, life on earth is supported by the energy radiated from sun. Feeling warm before an open fire is also associated with the same form of energy transfer. The mechanism by which energy is transmitted between a source and a distant sink is called **radiation**, a generic term covering thermal as well as other forms of radiation. Unlike to conduction and convection, radiation is hindered by matter and is at its best in vacuum. So the atmosphere of the earth and the air space around a fireplace hinder the propagation of radiation from sun and fireplace, respectively. That is something that we should actually be thankful for. Without the atmosphere, which acts as a radiation shield, too much radiation would reach the surface of the earth, and life as we know it would be impossible.

8.1 ORIGIN OF RADIATION. ELECTROMAGNETIC WAVES

The origin of radiation is electromagnetic and is based on three phenomenological facts: The **Ampere law**, the **Faraday law**, and the **Lorentz force** (Fig. 8.1) which are usually covered in a physics course. However, the usual form of Ampere's law, which gives

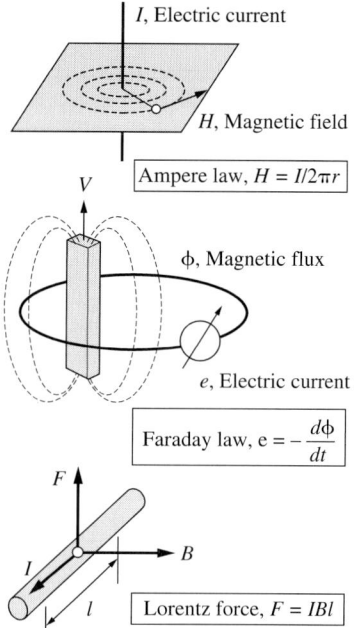

Figure 8.1 Foundations of electromagnetics.

the magnetic field generated by an electric current and is demonstrated by the simple experiment of Fig. 8.1(a), ignores the unsteady effect and is not complete. Including this effect to the Ampere law, and combining the result with the Faraday law, Maxwell analytically showed the existence of electromagnetic waves, which were later confirmed experimentally by Hertz. The velocity of electromagnetic waves in empty space turns out to be that of light,

$$c = 3 \times 10^8 \text{m/s}.$$

This fact originally led to the conviction that Maxwell's electromagnetic waves furnished a theory of light and that a variety of waves discovered following Hertz were of the same type. These waves appear in nature for wavelengths over an almost unlimited range (Fig. 8.2), and radiation takes on different names (optics, thermal, radio, x, and γ rays, etc.) depending on the wavelength. However, each form of radiation is characterized by electromagnetic waves within only a small part of this range. For example, the wavelength range of thermal radiation (which is type of radiation that we are interested in) is $0.1-100 \mu \text{m}$, approximately. Outside this range, radiation does not manifest itself in the form of heat (if this were not the case, human beings would be "cooked" by some of these waves). The thermal range is divided into the ultraviolet, the visible, and the infrared ranges. The radiation of one wavelength is called **monochromatic** radiation, while the dependence of radiation on wavelength is termed **spectral**. The spectral distribution of radiation depends on the temperature and the surface characteristics of the source. For example, with an effective surface temperature of about 6,000 K, the sun emits most of its energy at wavelengths below $3 \mu \text{m}$, while at a temperature of about 300 K the earth emits almost all of its energy at wavelengths above $3 \mu \text{m}$. This

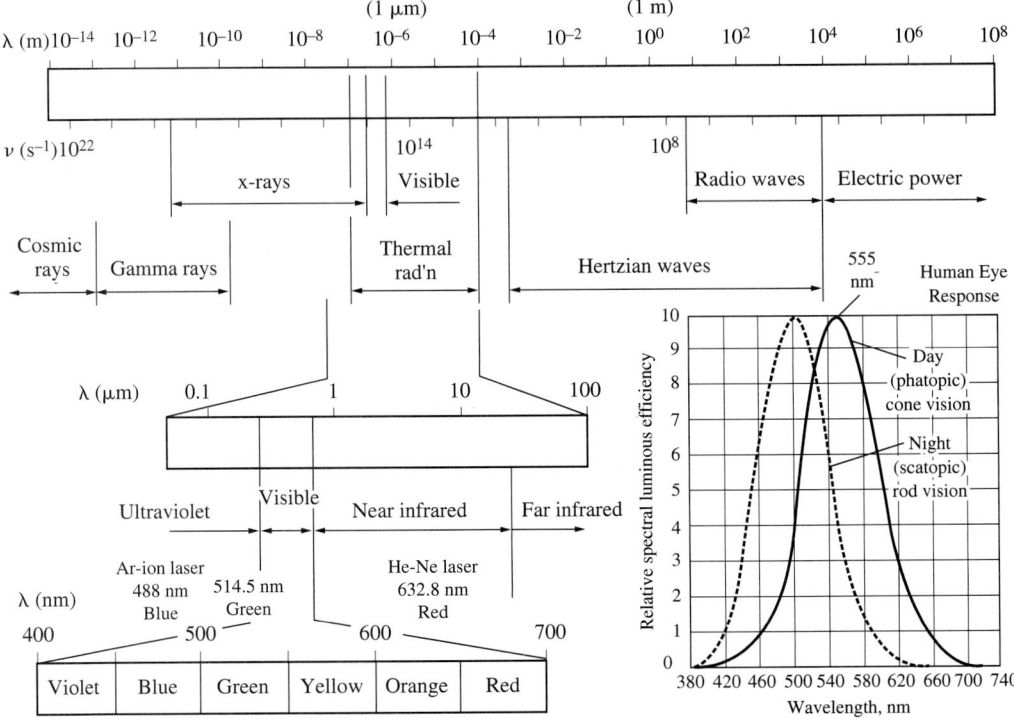

Figure 8.2 Electromagnetic spectrum with extended thermal and visible ranges.

difference between the spectral ranges results in the **greenhouse**[1] **effect**. The glass of a house permits radiation at wavelengths of the sun to pass but is almost opaque to radiation in wavelengths of the house interior. Thus, solar radiation enters but cannot leave the house, resulting in a warm inside even when the outside is cold.

The Lorentz force coupled with the Ampere law and the Faraday law leads to the balance of **electromagnetic momentum**[2] which involves the (divergence of) electromagnetic (Maxwell) stress. The **isotropic** limit of this stress is the electromagnetic pressure

$$p = \frac{1}{3}u, \tag{8.1}$$

where u is the electromagnetic **energy density**. Isotropic radiation is also called **blackbody** radiation or **ideal radiator**. An example is a cavity surrounded by isothermal walls. This cavity is filled with isotropic (equilibrium) radiation because of the thermal equilibrium between the radiation and the walls. Radiation escaping the cavity through a small hole is almost isotropic, because the hole disturbs the equilibrium only slightly.

---

[1] A greenhouse is a glass-enclosed structure used for cultivating plants with controlled temperature and humidity.

[2] The development of electromagnetic momentum is beyond the scope of this text.

To an observer the hole appears to be black. The radiation escaping the cavity is also black for reasons to be clarified in Section 8.4. Since the electromagnetic radiation is altered by but not attached to an intervening medium, its **energy** u defined per unit volume leads, for total internal energy, to

$$U = uV, \quad (8.2)$$

V being the volume.[3]

Now, assume the thermal radiation to be a gas occupying volume V, and consider a process involving an infinitesimal change in the volume of this gas. Then, rearranging the thermodynamic relation,

$$dU = TdS - pdV, \quad dS = \frac{dU}{T} + p\frac{dV}{T} \quad (8.3)$$

with Eqs. (8.1) and (8.2), we have

$$dS = \frac{V}{T}du + \frac{4}{3}\left(\frac{u}{T}\right)dV. \quad (8.4)$$

Furthermore, assuming the radiation to be an **ideal** gas, and recalling that $u = u(T)$ for an ideal gas, and $du = (du/dT)dT$, we may rearrange Eq. (8.4) as

$$dS = \frac{V}{T}\left(\frac{du}{dT}\right)dT + \frac{4}{3}\left(\frac{u}{T}\right)dV. \quad (8.5)$$

Since dS is an exact differential, from Eq. (8.5),

$$\frac{\partial}{\partial V}\left[\frac{V}{T}\left(\frac{du}{dT}\right)\right] = \frac{4}{3}\frac{\partial}{\partial T}\left(\frac{u}{T}\right),$$

which gives

$$\frac{du}{u} = 4\frac{dT}{T}$$

and, after integrating,

$$u \sim T^4 \quad (8.6)$$

or, with a proportionality constant a (to be elaborated later),

$$u = aT^4. \quad (8.7)$$

EXAMPLE 8.1

We wish to determine (a) the behavior of radiation as a gas, (b) the specific heat (at constant volume or pressure) of radiation, and (c) the relation governing the isentropic process of radiation.

(a) The enthalpy per volume V of radiation,

$$H = U + pV, \quad (8.8)$$

---

[3] Recall that the internal energy of matter is defined per unit mass rather than per unit volume.

in terms of Eqs. (8.1) and (8.2), becomes

$$H = \frac{4}{3}U \qquad (8.9)$$

or, per unit volume,

$$h = \frac{4}{3}u. \qquad (8.10)$$

Then, in view of Eq. (8.7), the internal energy and enthalpy of radiation depend only on its temperature. Consequently, the radiation is an ideal gas.

(b) The usual definitions of specific heats,

$$c_v = (\partial u/\partial T)_v, \quad c_p = (\partial h/\partial T)_p,$$

are reduced, for an ideal gas, to

$$c_v = du/dT, \quad c_p = dh/dT,$$

which give, in terms of Eqs. (8.7) and (8.10),

$$c_v = 4aT^3, \quad c_p = \frac{16}{3}aT^3. \qquad (8.11)$$

(c) The isentropic process of any ideal gas is governed by

$$pV^k = \text{Const.},$$

where $k = c_p/c_v$. The ratio of specific heats of radiation, obtained either from Eq. (8.11), or from the combination of

$$c_p/c_v = dh/du$$

with Eq. (8.10), is

$$k = c_p/c_v = 4/3.$$

Then, the isentropic process of radiation is governed by

$$pV^{4/3} = \text{Const.} \qquad (8.12)$$

(What matter gas satisfies the same relation?) ◆

For later convenience we may now introduce a simplification which closely approximates the thermal radiation.

8.2 APPROXIMATION OF RADIATION. OPTICAL RAYS

As we learned with Fig. 8.2, the major contribution of the electromagnetic spectrum to thermal radiation is provided by waves in the range of wavelengths from about 0.1 to 100 μm. Consequently, employing the basic assumption of geometric optics (that is, by letting $\lambda \to 0$), we may replace waves with **rays** and assume that **thermal radiation travels along straight lines**. Then we speak of the **color of a thermal ray**, characterized by the wavelength λ (or frequency ν) of the waves. However, in light of Eq. (8.2), a finite amount of energy can be attributed only to a volume, not to a single ray which has no volume. To account for this energy, let us draw a cone in an arbitrary direction,

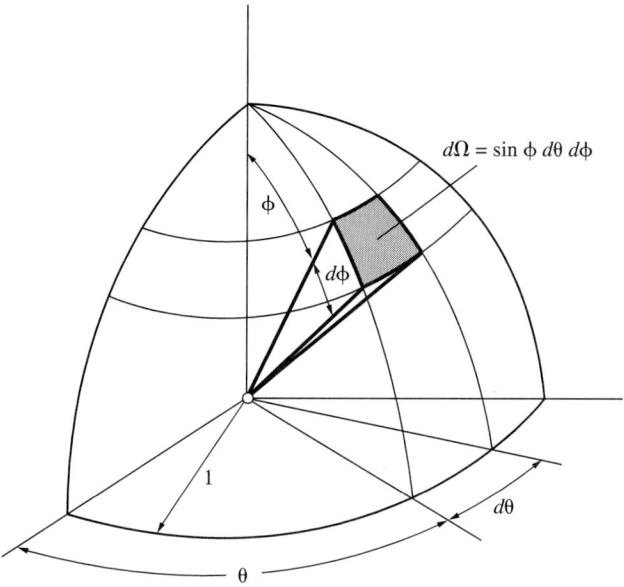

Figure 8.3 Solid angle (infinitesimal).

having any point of the radiation (gas) as the vertex, and describe around this same point a sphere of unit radius. This sphere intersects the cone in what is known as the **solid angle**. The **infinitesimal** solid angle $d\Omega$ (representing an infinitesimally small directional volume, as shown in Fig. 8.3) defines the radiation in a direction, the so-called **intensity** of radiation,

$$\text{Intensity} = \frac{\text{Radiation energy}}{\text{Solid angle} \times \text{Area} \times \text{Time}} = I.$$

To find the effect of radiation on thermal problems, however, we need to interpret the optical rays diverging in all directions in terms of thermal concepts such as energy density and heat flux. First, introduce the definition of the energy density of radiation,

$$\text{Energy density} = \frac{\text{Radiation energy}}{\text{Volume}} = u,$$

and express the flow of radiation energy from area dA_n over a time interval dt in terms of the flux of radiation energy (Fig. 8.4),

$$u\,dA_n(c\,dt) = q\,dA_n\,dt, \tag{8.13}$$

where q is the radiation flux (energy per unit area and time), c is the speed of light (recall that electromagnetic waves travel at the speed of light), and $c\,dt$ is the distance traveled over time interval dt. Since the intensity is the radiation flux per solid angle, area, and time, differentiating Eq. (8.13) with respect to Ω, we get

$$c\frac{du}{d\Omega} = \frac{dq}{d\Omega} = I, \tag{8.14}$$

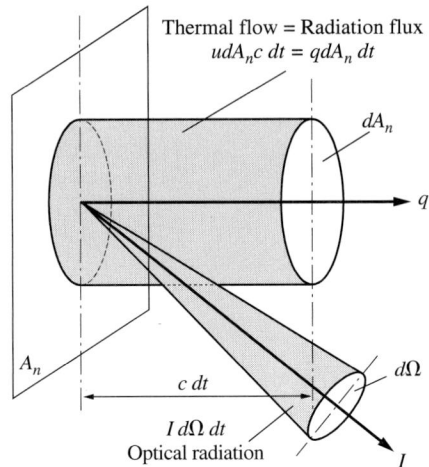

Figure 8.4 Optical and thermal interpretation of radiation energy.

or, after multiplying by $d\Omega$ and integrating over Ω,

$$u = \frac{1}{c} \int_\Omega I \, d\Omega. \tag{8.15}$$

For isotropic radiation, I is independent of Ω, and Eq. (8.15) for $\Omega = 4\pi$ (the total solid angle of a sphere) reduces to

$$\underset{\text{Thermal} \quad | \quad \text{Optical}}{u = \frac{4\pi}{c} I} \tag{8.16}$$

Now, introduce the **emissive power** E, or the radiation flux from an (actual) surface dA, as shown in Fig. 8.5. It follows that

$$E\,dA = q\,dA_n. \tag{8.17}$$

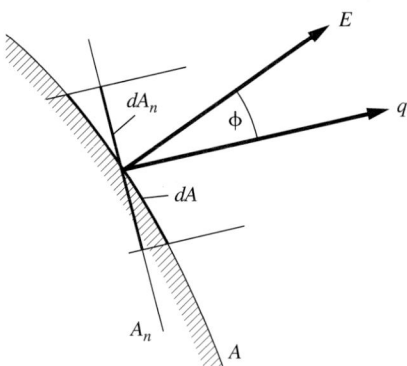

Figure 8.5 Emissive power and flux of radiation.

Differentiating Eq. (8.17) with respect to Ω, rearranging the result in terms of Eq. (8.14), and noting $dA_n = dA\cos\phi$, we obtain

$$\frac{dE}{d\Omega} = I\cos\phi,$$

and, after integrating over $\Omega (= 2\pi)$,[4]

$$E = \int_\Omega I\cos\phi\, d\Omega, \qquad (8.18)$$

where $d\Omega = \sin\phi\, d\phi\, d\theta$ (Fig. 8.3). For isotropic radiation, I is uniform, and Eq. (8.18) introduces the emissive power of a black body,

$$E_b = I\int_0^{2\pi} d\theta \int_0^{\pi/2} \sin\phi\cos\phi\, d\phi,$$

or, after the integrations, to

$$\begin{array}{c} E_b \;=\; \pi I \\ \text{Thermal} \;|\; \text{Optical} \end{array} \qquad (8.19)$$

Equation (8.19), together with Eq. (8.16), show how thermal concepts such as u and E_b are related to the optical concept of I.

Finally, inserting Eq. (8.7) into Eq. (8.16) and the result into Eq. (8.19) yields the **Stefan-Boltzmann law**,

$$E_b = \sigma T^4, \qquad (8.20)$$

where

$$\sigma = 5.67\times 10^{-8}\,\text{W/m}^2\cdot\text{K}^4$$

is the universal Stefan-Boltzmann constant. This constant is related to the constant in Eq. (8.7) by $a = 4\sigma/c$. For the time being, we assume that the numerical value of a or σ is determined experimentally and Eq. (8.20) is phenomenological. Later, it will theoretically be derived from quantum considerations.

EXAMPLE 8.2

Consider air at 300 K and 1 atm and the radiation gas at the same temperature. We wish to compare (a) internal energies, and (b) pressures of both gases.

(a) In view of $a = 4\sigma/c$, the internal energy of radiation follows from Eq. (8.7),

$$u_R = \frac{4\sigma}{c}T^4,$$

---

[4] Why are the integrations leading to Eqs. (8.16) and (8.19) over $\Omega = 4\pi$ and 2π, respectively? Note that u is local and E is directional.

which gives

$$u_R = \frac{4 \times 5.67\,\text{W/m}^2\cdot\text{K}^4 \times \left(\dfrac{300}{100}\right)^4 \text{K}^4}{3 \times 10^8\,\text{m/s}}$$

or

$$u_R = 6.12 \times 10^{-6}\,\text{J/m}^3.$$

For the internal energy of air per unit mass at 300 K and 1 atm (from air tables in a text on thermodynamics)

$$u_A^* = 214.09\,\text{kJ/kg}$$

or the internal energy per unit volume

$$u_A = \rho u_A^* = 1.1614\,\text{kg/m}^3 \times 214.09\,\text{kJ/kg},$$

which gives

$$u_A = 248.64\,\text{kJ/m}^3.$$

Then, the internal energy ratio

$$\frac{u_R}{u_A} = \frac{6.12 \times 10^{-5}\,\text{J/m}^3}{248.64 \times 10^3\,\text{J/m}^3} \cong 2.46 \times 10^{-11}.$$

(b) The radiation gas pressure from Eq. (8.1),

$$p_R = \frac{1}{3}u_R,$$

yields, when combined with part (a),

$$p_R = \frac{1}{3} \times 6.12 \times 10^{-6}\,\text{J/m}^3 \cong 2 \times 10^{-6}\,\text{Pa}.$$

Then the pressure ratio is

$$\frac{p_R}{p_A} = \frac{2 \times 10^{-6}\,\text{Pa}}{1 \times 1.013 \times 10^5\,\text{Pa}} \cong 2 \times 10^{-11},$$

which explains why the effect of the radiation pressure is ignored in the momentum equation. ◆

EXAMPLE 8.3

Assume that the sun at $T_S = 6000$ K, an incandescent bulb at $T_I = 3{,}000$ K, and the earth at $T_E = 300$ K are all black bodies. We wish to determine their emissive power.

Equation (8.20) yields

$$E_S = \sigma T_S^4 = 5.67\,\text{W/m}^2\cdot\text{K}^4 \times \left(\frac{6{,}000}{100}\right)^4 \text{K}^4 \cong 73{,}483\,\text{kW/m}^2,$$

$$E_I = \sigma T_I^4 = 5.67\,\text{W/m}^2\cdot\text{K}^4 \times \left(\frac{3{,}000}{100}\right)^4 \text{K}^4 \cong 4{,}593\,\text{kW/m}^2,$$

and
$$E_E = \sigma T_E^4 = 5.67 \text{W/m}^2 \cdot \text{K}^4 \times \left(\frac{300}{100}\right)^4 \text{K}^4 \cong 459 \text{W/m}^2$$

for the radiative fluxes from the sun, an incandescent bulb, and the earth, respectively. Then

$$\frac{E_S}{E_I} = \left(\frac{T_S}{T_I}\right)^4 = \left(\frac{6{,}000}{3{,}000}\right)^4 = 2^4 = 16$$

and

$$\frac{E_S}{E_E} = \left(\frac{T_S}{T_E}\right)^4 = \left(\frac{6{,}000}{300}\right)^4 = 20^4 = 16 \times 10^4.$$

♦

As we learned with Fig. 8.2, nature provides radiation over a wavelength spectrum. The energy of monochromatic radiation is the **monochromatic emissive power** E_λ, which we consider next.

8.3 MONOCHROMATIC RADIATION. QUANTUM MECHANICS ○

By definition, the relation between the monochromatic emissive power and the emissive power is

$$\int_0^\infty E_\lambda d\lambda = E. \tag{8.21}$$

For a black body, from Eq. (8.20),

$$\int_0^\infty E_{b\lambda} d\lambda = E_b = \sigma T^4. \tag{8.22}$$

This result (being temperature dependent after integration over the wavelength spectrum) proves that $E_{b\lambda} = E_{b\lambda}(\lambda, T)$. Here, we are interested in the explicit form of $E_{b\lambda}$. Equation (8.22) implies that

$$E_{b\lambda} d\lambda \sim dE_b \sim T^3 dT. \tag{8.23}$$

For an infinitesimal isentropic process of the radiation (gas), $dS = 0$, and Eq. (8.3) becomes

$$dU = -p dV, \tag{8.24}$$

which, in terms of Eqs. (8.1) and (8.2), may be rearranged to give

$$d(uV) = -\frac{1}{3} u dV$$

or

$$\frac{du}{u} = -\frac{4}{3} \frac{dV}{V}$$

or, after integrating, and recalling Eq. (8.6),

$$u \sim V^{-4/3} \sim T^4$$

or

$$T \sim V^{-1/3}. \tag{8.25}$$

Also, for the wavelength of isotropic waves we have (from the radius-volume relation of a sphere, for example)

$$\lambda \sim V^{1/3}. \tag{8.26}$$

Elimination of V between Eqs. (8.25) and (8.26) yields

$$\boxed{\lambda T = \text{Const.}}, \tag{8.27}$$

which is **Wien's law**. Differentiating Eq. (8.27), we get

$$\lambda dT + T d\lambda = 0,$$

which may be rearranged as

$$d\lambda \sim \frac{\lambda}{T} dT$$

or, in terms of Eq. (8.27), as

$$d\lambda \sim \frac{1}{T^2} dT. \tag{8.28}$$

Then, insertion of Eq. (8.28) into Eq. (8.23) results in

$$E_{b\lambda} \sim T^5$$

or, in view of Eq. (8.27),

$$\frac{E_{b\lambda}}{T^5} \sim \text{Const.} = \lambda T$$

or

$$\frac{E_{b\lambda}}{T^5} = f(\lambda T). \tag{8.29}$$

Classical electromagnetics and Boltzmann statistics, respectively, lead to explicit forms of Eq. (8.29) for $\lambda \to \infty$ and $\lambda \to 0$. However, both theories fail to provide the explicit form for an arbitrary λ. Extensive research for this explicit form eventually led Planck to the discovery of quantum mechanics, which explains radiation in terms of particles (photons) traveling with the speed of light. The energy and momentum associated with each photon, respectively, are

$$U = h\nu \quad \text{and} \quad U/c = h/\lambda, \tag{8.30}$$

where

$$h = 6.6262 \times 10^{-34} \text{J·s}$$

is the Planck constant, ν is the frequency, and $\lambda \nu = c$.

Sec. 8.3 Monochromatic Radiation. Quantum Mechanics ○ 407

For the constitution of a photon gas,[5] now reconsider Eq. (8.30) in terms of wavelength,

$$U = hc\lambda^{-1}, \tag{8.31}$$

which may be rearranged, in view of Eq. (8.26), as

$$U \sim V^{-1/3}. \tag{8.32}$$

An infinitesimal change in this energy is

$$dU \sim -\frac{1}{3}\frac{V^{-1/3}}{V}dV = -\frac{1}{3}u\,dV, \tag{8.33}$$

where $u = U/V$. For an infinitesimal isentropic process of any gas, Eq. (8.3) gives

$$dU = -p\,dV. \tag{8.24}$$

Then, from a comparison of Eqs. (8.33) with (8.24),

$$p = \frac{1}{3}u, \tag{8.1}$$

which is identical to the result already stated as the isotropic limit of electromagnetic stress.

It can be shown[6] by applying quantum (Bose-Einstein) statistics and quantum (Schrödinger) waves to photons that

$$E_{b\lambda} = \frac{C_1}{\lambda^5 \left(e^{C_2/\lambda T} - 1\right)}, \tag{8.34}$$

which is the **Planck law** of black-body emissive power. Here

$$C_1 = 2\pi hc^2 = 3.7415 \times 10^{-16}\,\text{W}\cdot\text{m}^2$$

and

$$C_2 = \frac{hc}{k} = 1.4388 \times 10^{-2}\,\text{m}\cdot\text{K}$$

are the first and second radiation constants, respectively, and

$$k = 1.3806 \times 10^{-23}\,\text{J/K} \tag{8.35}$$

---

[5] In place of the constitution of matter gas, $p = \rho RT$.

[6] Details are beyond the scope of the text.

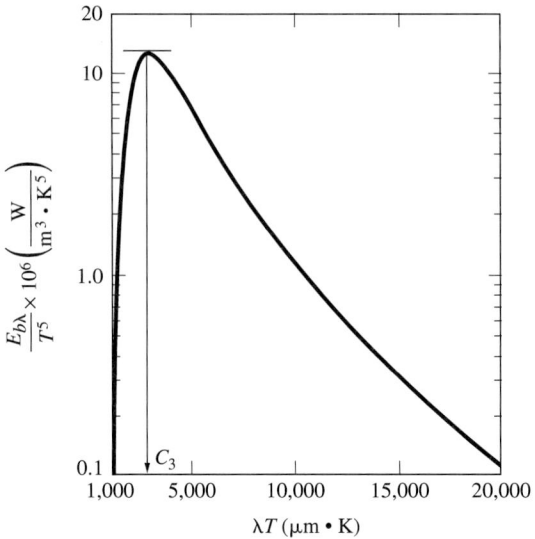

Figure 8.6 Black-body monochromatic emissive power.

is the Boltzmann constant. Equation (8.34), rearranged as

$$\frac{E_{b\lambda}}{T^5} = \frac{C_1}{(\lambda T)^5 \left(e^{C_2/\lambda T} - 1\right)}, \tag{8.36}$$

is in terms of λT only and provides the explicit form of Eq. (8.29). Inserting Eq. (8.34) into Eq. (8.22) and integrating, we get the Stefan-Boltzmann constant in terms of the radiation constants or the Planck and Boltzmann constants,

$$\sigma = \frac{\pi^4}{15} \frac{C_1}{C_2^4} = \frac{2\pi^5 k^4}{15 h^3 c^2} = 5.67 \times 10^{-8} \text{W/m}^2 \cdot \text{K}^4.$$

Figure 8.6 shows $E_{b\lambda}/T^5$ versus λT, and Fig. 8.7 gives $E_{b\lambda}$ versus λ, with T as a parameter. Note from Fig. 8.7 that, for higher temperatures, the maximum of $E_{b\lambda}$ is shifted to shorter wavelengths. Equating the derivative of Eq. (8.36) with respect to λT to zero gives the special value of the constant in Wien's law,

$$C_3 = \lambda_{max} T = \frac{C_2}{4.965} = 2{,}897.8 \mu\text{m} \cdot \text{K}, \tag{8.37}$$

where λ_{max} is the wavelength at which $E_{b\lambda}$ is a maximum and C_3 is the third radiation constant. The shift in this maximum explains the change in color of a body as it becomes heated. Since the visible band of wavelengths lies approximately between 0.35 and 0.75μm, only a very small part of the low-temperature radiation energy is detected by the human eye. When a solid body is heated, the maximum intensity of radiation is shifted to shorter wavelengths, and the first visible sign of increased temperature is a dark red color assumed by the body at about 700 °C. As the temperature is further

Sec. 8.3 Monochromatic Radiation. Quantum Mechanics 409

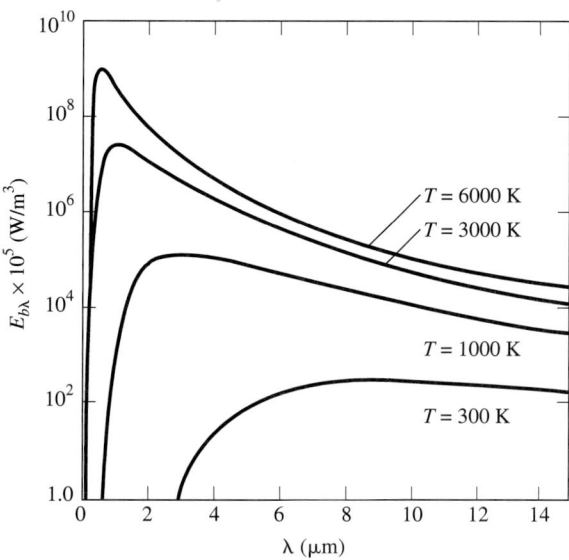

Figure 8.7 Black-body monochromatic emissive power at various temperatures.

increased, the color changes to bright red, then to bright yellow, and finally to white at about 1,300 °C. A solid body appears also brighter at higher temperatures, since a larger portion of its total radiation falls within the visible range.

EXAMPLE 8.4

Reconsider Ex. 8.3. For these radiation sources we wish to determine (a) the maximum monochromatic black-body emissive power and (b) the wavelength corresponding to the maximum emissive power.

(a) Equation (8.36) coupled with Wien's law given by Eq. (8.37),

$$\lambda_{\max} T = 2{,}898 \ \mu\text{m} \cdot \text{K} = 2.898 \times 10^{-3} \text{m} \cdot \text{K},$$

yields, for the maximum of $E_{b\lambda}$

$$\frac{(E_{b\lambda})_{\max}}{T^5} = \frac{3.7415 \times 10^{-16} \text{W} \cdot \text{m}^2}{(2.898 \times 10^{-3})^5 (\text{m} \cdot \text{K})^5 \times (e^{1.4388 \times 10^{-2}/2.898 \times 10^{-3}} - 1)}$$

or

$$\frac{(E_{b\lambda})_{\max}}{T^5} = 1.2865 \times 10^{-5} \text{W/m}^3 \cdot \text{K}^5.$$

Then, the maxima of monochromatic emissive power for the three sources are

$$(E_{b\lambda})_{max,S} = 1.2865 \times 10^{-5} \text{W/m}^3 \cdot \text{K} \times (6{,}000)^5 \text{K}^5 = 10^{14} \text{W/m}^3,$$

$$(E_{b\lambda})_{max,I} = 1.2865 \times 10^{-5} \text{W/m}^3 \cdot \text{K} \times (3{,}000)^5 \text{K}^5 \cong 3.13 \times 10^{12} \text{W/m}^3,$$

and

$$(E_{b\lambda})_{max,E} = 1.2865 \times 10^{-5} \text{W/m}^3 \cdot \text{K} \times (300)^5 \text{K}^5 = 3.13 \times 10^{7} \text{W/m}^3.$$

(b) From Wien's law, [Eq. (8.37)], the peak wavelengths are

$$\lambda_{max,S} = \frac{2{,}898 \mu\text{m} \cdot \text{K}}{6{,}000 \text{K}} \cong 0.48 \mu\text{m},$$

$$\lambda_{max,I} = \frac{2{,}898 \mu\text{m} \cdot \text{K}}{3{,}000 \text{K}} \cong 0.97 \mu\text{m},$$

and

$$\lambda_{max,E} = \frac{2{,}898 \mu\text{m} \cdot \text{K}}{300 \text{K}} \cong 9.66 \mu\text{m}.$$

Note the substantial wavelength difference between the peaks of monochromatic radiation emitted by the sun and earth. ◆

For problems involving real surfaces it is useful to know the fraction of total energy radiated over a wavelength interval $(0, \lambda)$ or (λ_1, λ_2), which is to be designated by $F(0 \to \lambda)$ or $\Delta F(\lambda_1 \to \lambda_2)$, respectively. For an interval $(0, \lambda)$, from

$$F(0 \to \lambda) = \frac{E_b(0, \lambda)}{E_b(0, \infty)} = \frac{\int_0^\lambda E_{b\lambda} d\lambda}{\int_0^\infty E_{b\lambda} d\lambda},$$

where $E_b(0, \infty) = E_b = \sigma T^4$, we have, in terms of λT,

$$F(0 \to \lambda T) = \frac{E_b(0, \lambda T)}{\sigma T^4} = \frac{1}{\sigma} \int_0^{\lambda T} \frac{E_{b\lambda}}{T^5} d(\lambda T). \quad (8.38)$$

This fraction depends on both wavelength and temperature. The fractional function $F(0 \to \lambda T)$ is plotted in Fig. 8.8 and, for computational convenience, tabulated in Table 8.1. For an interval (λ_1, λ_2)

$$\Delta F(\lambda_1 T \to \lambda_2 T) = \frac{E_b(\lambda_1 T, \lambda_2 T)}{\sigma T^4} = \frac{E_b(0, \lambda_2 T)}{\sigma T^4} - \frac{E_b(0, \lambda_1 T)}{\sigma T^4} \quad (8.39)$$

$$= F(0 \to \lambda_2 T) - F(0 \to \lambda_1 T),$$

which is the difference of the ratios obtained from Fig. 8.8 or Table 8.1 for the intervals $(0, \lambda_2)$ and $(0, \lambda_1)$.

Table 8.1 Radiation functions (from Dunkle[5]).

λT	$\dfrac{E_{b\lambda}}{\sigma T^5} \times 10^4$	$F_{0 \to \lambda}(T)$	λT	$\dfrac{E_{b\lambda}}{\sigma T^5} \times 10^4$	$F_{0 \to \lambda}(T)$
$\mu m \cdot K$	$(\mu m \cdot K)^{-1}$		$\mu m \cdot K$	$(\mu m \cdot K)^{-1}$	
500	0.00000672	0	5500	10.340	0.69088
600	0.0003269	0	5600	9.939	0.70102
700	0.004650	0	5700	9.553	0.71077
800	0.03114	0.000016	5800	9.182	0.72013
900	0.1275	0.000087	5900	8.827	0.72914
1000	0.3723	0.000321	6000	8.486	0.73779
1100	0.8550	0.000911	6100	8.159	0.74611
1200	1.646	0.00213	6200	7.845	0.75411
1300	2.774	0.00432	6300	7.544	0.76181
1400	4.223	0.00779	6400	7.256	0.76920
1500	5.934	0.01285	6500	6.980	0.77632
1600	7.827	0.01972	6600	6.716	0.78317
1700	9.811	0.02853	6700	6.463	0.78976
1800	11.799	0.03934	6800	6.220	0.79610
1900	13.716	0.05211	6900	5.988	0.80230
2000	15.501	0.06673	7000	5.766	0.80808
2100	17.113	0.08305	7100	5.553	0.81373
2200	18.524	0.10089	7200	5.348	0.81918
2300	19.720	0.12003	7300	5.153	0.82443
2400	20.698	0.14206	7400	4.966	0.82949
2500	21.465	0.16136	7500	4.786	0.83437
2600	22.031	0.18312	8000	3.995	0.85625
2700	22.412	0.20536	8500	3.354	0.87457
2800	22.626	0.22789	9000	2.832	0.88999
2900	22.692	0.25056	9500	2.404	0.90304
3000	22.627	0.27323	10000	2.052	0.91416
3100	22.450	0.29578	10500	1.761	0.92367
3200	22.178	0.31810	11000	1.518	0.93185
3300	21.827	0.34011	11500	1.315	0.93892
3400	21.411	0.36173	12000	1.145	0.94505
3500	20.942	0.38291	12500	1.000	0.95401
3600	20.432	0.40360	13000	0.8779	0.95509
3700	19.891	0.42376	13500	0.7733	0.95921
3800	19.327	0.44337	14000	0.6837	0.96285
3900	18.748	0.46241	14500	0.6066	0.96607
4000	18.160	0.48087	15000	0.5399	0.96893
4100	17.568	0.49873	15500	0.4821	0.97149
4200	16.976	0.51600	16000	0.4317	0.97377
4300	16.389	0.53268	16500	0.3877	0.97581
4400	15.809	0.54878	17000	0.3492	0.97765
4500	15.240	0.56430	18000	0.2853	0.98081
4600	14.681	0.57926	19000	0.2353	0.98341
4700	14.136	0.59367	20000	0.1958	0.98555
4800	13.606	0.60754	25000	0.0869	0.99217
4900	13.091	0.62089	30000	0.0441	0.99529
5000	12.591	0.63373	35000	0.0247	0.99695
5100	12.108	0.64608	40000	0.0149	0.99792
5200	11.642	0.65795	45000	0.0095	0.99852
5300	11.191	0.66936	50000	0.0063	0.99890
5400	10.758	0.68034	55000	0.0044	0.99917
			∞	0	1.00000

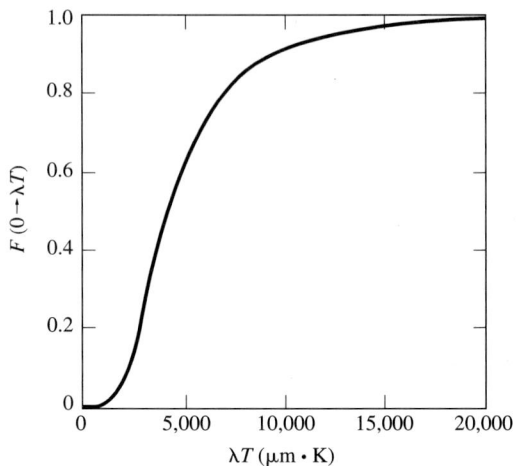

Figure 8.8 Fractional function $F(0 \to \lambda T)$.

Example 8.5

Reconsider Exs. 8.3 and 8.4. For all three sources we wish to determine the percent of the emitted energy that lies in the visible range.

Assume $\lambda = (0.35 - 0.75)\mu\text{m}$ for the visible range. The lower and upper limits of λT are, for the sun,

$$\lambda_1 T_S = 0.35 \mu\text{m} \times 6,000\text{K} = 2,100 \mu\text{m·K},$$

and

$$\lambda_2 T_S = 0.75 \mu\text{m} \times 6,000\text{K} = 4,500 \mu\text{m·K},$$

for the incandescent bulb

$$\lambda_1 T_I = 1,050 \mu\text{m·K}, \quad \lambda_2 T_I = 2,250 \mu\text{m·K},$$

and for the earth

$$\lambda_1 T_E = 105 \mu\text{m·K}, \quad \lambda_2 T_E = 225 \mu\text{m·K}.$$

For a source at a temperature T_0, the fraction of total emitted energy contained in the interval (λ_1, λ_2) is

$$[\Delta F(\lambda_1 T \to \lambda_2 T)]_{T=T_0} = [F_2(0 \to \lambda_2 T) - F_1(0 \to \lambda_1 T)]_{T=T_0},$$

F being available from Table 8.1. Then for the sun, bulb, and earth, respectively,

$$\Delta F_S = 0.5643 - 0.0831 \cong 0.481,$$

$$\Delta F_I = 0.11046 - 0.00062 \cong 0.109,$$

$$\Delta F_E < 10^{-8} \cong 0.$$

This example illustrates the fact that, while 48% and 11% of the emitted energies from sun and an incandescent bulb lie in the visible range, no energy is emitted from earth in the same range. Actually, one could have anticipated these results by inspecting the locations of the maxima in Ex. 8.4. ◆

So far, we have studied the radiation energy associated with black bodies. However, real surfaces do not behave like a black body. The next section is devoted to the surface properties of real surfaces.

8.4 PROPERTIES OF RADIATION ◯

According to electromagnetics, the angle of reflection from an **ideal** (smooth) surface is equal to the angle of incidence [Fig. 8.9(a)]. This reflection is called **specular**. In general, reflection from highly polished surfaces approaches specular behavior. An ordinary mirror reflects specularly in the visible range. Reflection from a **real** (rough) surface is **anisotropic** [Fig. 8.9(b)]. An approximation of the real surface is the **diffuse** surface. Reflection from this surface is **isotropic** [Fig. 8.9(c)]. Industrial (machined, painted, or treated) surfaces may usually be assumed diffuse. Hereafter, unless otherwise specified, we shall deal with diffuse surfaces.

Monochromatic energy G_λ incident a surface is partly absorbed, partly reflected, and partly transmitted (Fig. 8.10). Denoting each part by A_λ, R_λ, and T_λ, respectively, we write

$$G_\lambda = A_\lambda + R_\lambda + T_\lambda$$

or, after dividing each term by G_λ, and introducing the **monochromatic absorptivity** $\alpha_\lambda = A_\lambda/G_\lambda$, **monochromatic reflectivity** $\rho_\lambda = R_\lambda/G_\lambda$ and **monochromatic transmissivity** $\tau_\lambda = T_\lambda/G_\lambda$,

$$1 = \alpha_\lambda + \rho_\lambda + \tau_\lambda. \tag{8.40}$$

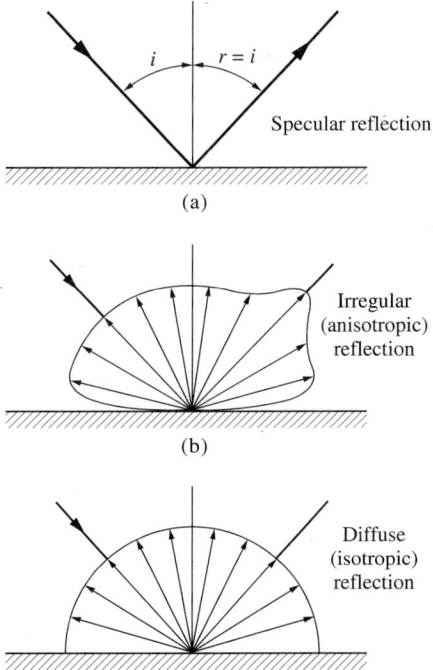

Figure 8.9 Surface types.
(a) Ideal (smooth) surface,
(b) real (rough) surface,
(c) diffuse surface.

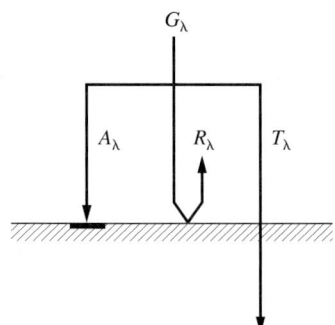

Figure 8.10 Fractions of radiation absorbed, reflected, and transmitted.

For an **opaque** surface, $\tau_\lambda = 0$ and Eq. (8.40) reduces to

$$1 = \alpha_\lambda + \rho_\lambda. \tag{8.41}$$

Glass and many crystals are exceptions and, unless very thick, they are to a degree **transparent** to radiation at certain wavelengths. For a transparent surface, $\rho_\lambda = 0$ and Eq. (8.40) reduces to

$$1 = \alpha_\lambda + \tau_\lambda. \tag{8.42}$$

These properties depend on the wavelength of the incident radiation. Overall properties can be defined by integrating over the whole wave spectrum. For example, the total absorptivity is

$$\alpha = \frac{\int_0^\infty \alpha_\lambda G_\lambda d\lambda}{\int_0^\infty G_\lambda d\lambda} = \frac{\int_0^\infty \alpha_\lambda E_{b\lambda} d\lambda}{E_b} \tag{8.43}$$

if the incident radiation originates from a black body. Note that the overall properties depend on the temperature of the radiation source.

EXAMPLE 8.6[7]

The living room of a modern house has a 4 m × 6 m glass siding. The transmissivity of glass is 0.92 over the interval $\lambda = (3 \times 10^{-7} - 3 \times 10^{-6})$ m and it is opaque at other wavelengths (Fig. 8.11). The interior of the room may be assumed a black body at 300 K. The glass is subject to 400 W/m² from the sun with a black-body temperature of 5,800 K. We wish to determine (a) the total transmissivity of the siding to radiation from the sun, (b) the total transmissivity of the siding to radiation from the room, (c) the energy of sun transmitted through the siding, and (d) the radiant energy from the room transmitted through the siding.

---

[7] The FORTRAN program EX8–6.F is listed in the appendix of this chapter.

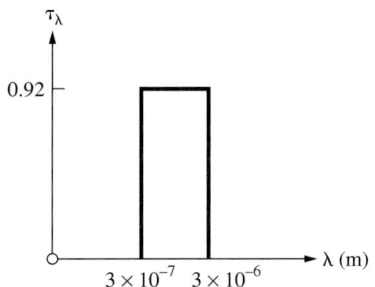

Figure 8.11

(a) The monochromatic transmissivity of the glass is zero over most of the wavelength interval, except between $\lambda_1 = 3 \times 10^{-7}$ m and $\lambda_2 = 3 \times 10^{-6}$ m, where it is 0.92. The total transmissivity to radiation from the sun is

$$\tau_{S \to R} = \frac{\int_0^\infty \tau_\lambda E_{b\lambda} d\lambda}{E_b s} = \frac{0.92 \int_{\lambda_1}^{\lambda_2} E_{b\lambda} d\lambda}{E_b s} \tag{8.44}$$

or

$$\tau_{s \to R} = 0.92 \Delta F(\lambda_1 T_s \to \lambda_2 T_s), \tag{8.45}$$

where $T_S = 5{,}800$ K is the temperature of the sun. The lower and upper bounds of the transmission interval in terms of λT are

$$\lambda_1 T_S = 0.3 \mu\text{m} \times 5{,}800\text{K} = 1{,}740 \mu\text{m·K}$$

and

$$\lambda_2 T_S = 3.0 \mu\text{m} \times 5{,}800\text{K} = 17{,}400 \mu\text{m·K}.$$

Then, the linear interpolation of fractional functions from Table 8.1,

$$[\Delta F(\lambda_1 T \to \lambda_2 T)]_{T=T_S} = [F_2(0 \to \lambda_2 T) - F_1(0 \to \lambda_1 T)]_{T=T_S},$$

gives

$$[\Delta F(\lambda_1 T \to \lambda_2 T)]_{T=T_S} \cong 0.979 - 0.033 = 0.946.$$

The fraction of the sun radiation reaching the room is

$$\text{Total transmissivity}_{S \to R} = \tau_{S \to R} = 0.92 \times 0.946 \cong 0.87 = 87\%.$$

(b) Here we repeat the same procedure as in part (a), but with the room temperature $T_R = 300$ K. For the radiation from the room at a temperature of 300 K,

$$\lambda_1 T_R = 0.3 \mu\text{m} \times 300\text{K} = 90 \mu\text{m·K}$$

and

$$\lambda_2 T_R = 3.0 \mu\text{m} \times 300\text{K} = 900 \mu\text{m·K}.$$

Inserting the fractional functions obtained from Table 8.1 into

$$[\Delta F(\lambda_1 T \to \lambda_2 T)]_{T=T_R} = [F_2(0 \to \lambda_2 T) - F_1(0 \to \lambda_1 T)]_{T=T_R}$$

yields
$$[\Delta F(\lambda_1 T \to \lambda_2 T)]_{T=T_R} = 0.000087 - 0 = 0.0087\%.$$
Then, the fraction of room radiation transmitted through siding is
$$\tau_{R \to \infty} = 0.92 \times 0.000087 \cong 0.00008 = 0.008\%.$$

(c) The sun radiation reaching the room is the amount transmitted through the glass,
$$\dot{Q}^R_{S \to R} = \tau_{S \to R} q^R_{S \to R} A,$$
which, coupled with the result for $\tau_{S \to R}$ from part (a), yields
$$\dot{Q}^R_{S \to R} = 0.87 \times 400\text{W/m}^2 \times 4 \times 6\text{m}^2$$
or
$$\dot{Q}^R_{S \to R} = 8.35\text{kW}.$$

(d) The radiation from the room transmitted through the siding is
$$\dot{Q}^R_{R \to \infty} = \tau_{R \to \infty} q^R_{R \to \infty} A,$$
Using part (b) for $\tau_{R \to \infty}$ and noting that $q^R_{R \to \infty} = \sigma T_R^4$,
$$\dot{Q}^R_{R \to \infty} = 0.00008 \times 5.67\text{W/m}^2 \cdot \text{K}^4 \times \left(\frac{300}{100}\right)^4 \text{K}^4 \times 4 \times 6\text{m}^2$$
or
$$\dot{Q}^R_{R \to \infty} = 0.88\text{W}.$$

Comparison of the foregoing results reveals that glass siding allows the energy of sun to be efficiently transmitted to a room while it practically prevents any radiative losses from the room.[8] The winter advantages and summer disadvantages of this fact should be taken into account in structures involving large glass areas (of buildings and cars).

By the computer program provided, the interested reader may parametrically study the problem for various values of the gases transmissivity. ◆

EXAMPLE 8.7[9]

Heat-reflecting filters, also known as **hot mirrors**, are used for radiant-heat removal. These filters remove the infrared light coming from a source by reflection rather than by absorption but preserve the integrity of the visible light. One such filter has a transmissivity of 70% in the wavelength range $\lambda = (400 - 700)$ nm and is totally reflective outside this range. Such a filter is placed between an observer and a tungsten-halogen lamp having a source temperature of 3,200 K. We wish to determine the fraction of the lamp radiation energy rejected by the filter.

The lower and upper limits of λT for the specified wavelength range and source temperature,
$$\lambda_1 T_L = 400 \times 10^{-3} \mu\text{m} \times 3{,}200\text{K} = 1{,}280 \mu\text{m}\cdot\text{K}$$
and
$$\lambda_2 T_L = 700 \times 10^{-3} \mu\text{m} \times 3{,}200\text{K} = 2{,}240 \mu\text{m}\cdot\text{K},$$

---

[8] This illustrates numerically the greenhouse effect discussed in Section 8.1.

[9] The FORTRAN program EX8–7.F is listed in the appendix of this chapter.

coupled with Table 8.1, yield for the fractional energies, after linear interpolation,

$$F_1(0 \to \lambda_1 T_L) = 0.004 \quad \text{and} \quad F_2(0 \to \lambda_2 T_L) \cong 0.109.$$

The fraction of the total energy contained between two wavelengths is

$$[\Delta F(\lambda_1 T \to \lambda_2 T)]_{T=T_L} = [F_2(0 \to \lambda_2 T) - F_1(0 \to \lambda_1 T)]_{T=T_L}$$

or

$$[\Delta F(\lambda_1 T \to \lambda_2 T)]_{T=T_L} \cong 0.109 - 0.004 = 0.105 = 10.5\%.$$

Combining this result with the spectral transmissivity over the same wavelength range,

$$\tau(0.4\mu m \to 0.7\mu m) = 0.7,$$

gives the fraction of the incident energy which penetrates through the filter,

$$\text{Transmitted fraction} = 0.7 \times 0.105 \cong 0.074 = 7.4\%$$

and

$$\text{Rejected fraction} = (100 - 7.4)\% = 92.6\%.$$

By the computer program provided, the interested reader may parametrically study the problem for various filter transmissivities. ◆

Now, consider an enclosure filled with monochromatic radiation G_λ, and place a monochromatically opaque ($\tau_\lambda = 0$) body into this enclosure (Fig. 8.12). Let the monochromatic reflectivity and emissive power of the body be ρ_λ and E_λ, respectively. Under equilibrium, the first law of thermodynamics for the system enclosing the body gives

$$+G_\lambda - (\rho_\lambda G_\lambda + E_\lambda) = 0,$$

which, in terms of Eq. (8.41), leads to the **Kirchhoff law**,

$$G_\lambda = \frac{E_\lambda}{\alpha_\lambda} = \text{Const.} \tag{8.46}$$

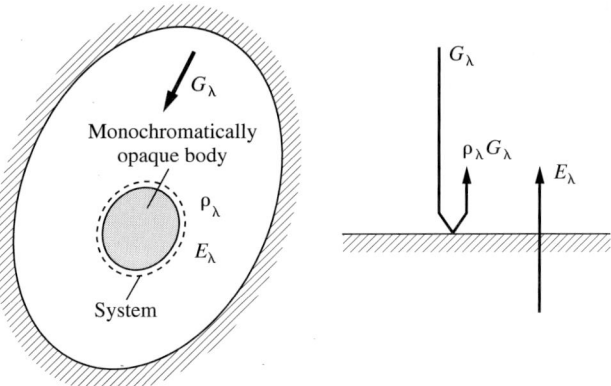

Figure 8.12 The Kirchhoff law.

G_λ is usually called the **irradiation** of an enclosure or the **incident radiation** on a body. For an ideal absorber (black body), $\alpha_\lambda = 1$, and the Kirchhoff law,

$$\frac{E_\lambda}{\alpha_\lambda} = \frac{E_{b\lambda}}{1}, \qquad (8.47)$$

states that, at a given temperature, a black body has the maximum attainable emissive power. Equation (8.47) provides another definition for the monochromatic absorptivity[10] as the monochromatic emissive power of a surface relative to that of the black body,

$$\alpha_\lambda = \frac{E_\lambda}{E_{b\lambda}}, \qquad (8.48)$$

which is valid **only** under **equilibrium** conditions. Furthermore, introducing the **definition** of the monochromatic emissivity of a surface,

$$\epsilon_\lambda = \frac{E_\lambda}{E_{b\lambda}}, \qquad (8.49)$$

we have

$$\alpha_\lambda(\lambda, T) = \epsilon_\lambda(\lambda, T). \qquad (8.50)$$

Since Eq. (8.50) holds only under equilibrium conditions, its use for **nonequilibrium** problems may be justified only under the assumption of **local equilibrium**. As was done with the other properties, a total emissivity (depending on the body temperature) can be defined by integrating over the entire wavelength range.

EXAMPLE 8.8

The SiO-Al coated surface of a satellite is in earth orbit around the sun ($T_S = 5{,}800$ K) and subject in the normal direction to solar flux of 1,400 W/m². The other surface of the satellite is thermally insulated. The spectral absorptivity of the coated surface can be approximated as shown in Fig. 8.13(a). Find the satellite temperature.

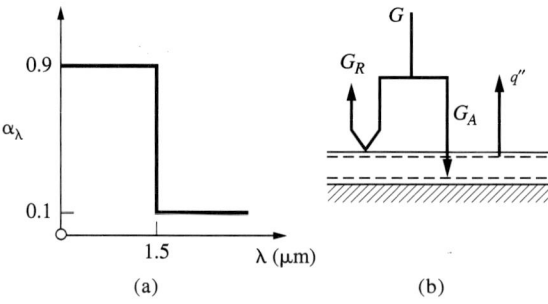

Figure 8.13

---

[10] Recall the definitions leading to Eq. (8.40).

The first law for a system surrounding the satellite [Fig. 8.13(b)] results in a relation between the absorbed incident energy, G_A, and emitted energy, q'',

$$G_A = q''.$$

In view of the spectral behavior of α_λ, two distinct domains ($\alpha = \alpha_1$ for $\lambda < \lambda_1$ and $\alpha = \alpha_2$ for $\lambda > \lambda_1$) need to be considered separately. The absorbed energy, in terms of $\lambda_1 = 1.5\mu m$,

$$G_A = G\left[\alpha_1 F_1(0 < \lambda < 1.5\mu m) + \alpha_2 F_2(1.5\mu m < \lambda < \infty)\right]_{T=T_S};$$

where G is the incident radiation ($G = 1,400 W/m^2$) and $T_S = 5,800$ K. The product

$$\lambda_1 T_S = 1.5\mu m \times 5,800 K = 8,700 \mu m \cdot K,$$

combined with an interpolated fractional function from Table 8.1 gives

$$F_1(0 \to \lambda_1 T_S) = 0.881$$

and, since $F(0 \to \infty) = 1$ by definition,

$$F_2(\lambda_1 T_S \to \infty) = 1 - 0.881 = 0.119.$$

Then, the absorbed incident energy (as well as the emitted energy)

$$G_A = 1,400 W/m^2 \times (0.9 \times 0.881 + 0.1 \times 0.119)$$

or

$$G_A \cong 1,126.7 W/m^2 = q''.$$

In light of Eq. (8.50), which gives for spectral emissivities $\epsilon_1 = \alpha_1$ and $\epsilon_2 = \alpha_2$, the emission may be expressed as

$$q'' = \sigma T_{sat}^4 \left[\epsilon_1 F_1(0 < \lambda < 1.5\mu m) + \epsilon_2 F_2(1.5\mu m < \lambda \infty)\right]_{T=T_{sat}},$$

or

$$q'' = 5.67 \times 10^{-8} \, (W/m^2 \cdot K^4) \, T_{sat}^4 (K^4) \, [0.9 F_1 + 0.1 F_2] = 1,126.7 W/m^2,$$

where T_{sat} is the surface temperature of the satellite. Since T_{sat} is required for F_1 and F_2, the solution is a trial-and-error procedure.

As a first trial, let T_{sat} be 400 K. Then, for

$$\lambda_1 T_{sat} = 1.5\mu m \times 400 K = 600 \mu m \cdot K,$$

Table 8.1 gives

$$F_1 \cong 0.$$

Accordingly,

$$F_2 \cong 1.0,$$

and, the emitted energy based on the assumed T_{sat} becomes

$$q'' = 5.67 \times \left(\frac{400}{100}\right)^4 W/m^2 \cdot K^4 \, [0 + 0.1 \times 1] = 145 W/m^2,$$

which is substantially lower than $1,126.7 \, W/m^2$. Now assuming F_1 and F_2 approximately to retain their initial values at somewhat higher temperatures,

$$q'' = 1,126.7 W/m^2 = 5.67 \times 10^{-8} W/m^2 \cdot K^4 \times T_{sat}^4 K^4 \times 0.1 \times 1,$$

which gives a surface temperature of
$$T_{sat} = 667.7 \text{K}.$$
Then, for
$$\lambda_1 T_{sat} = 1.5 \mu\text{m} \times 667.7\text{K} = 1,001.6 \mu\text{m}\cdot\text{K},$$
Table 8.1 yields
$$F_1 \cong 0.00032$$
and
$$F_2 = 1 - F_1 = 0.99968.$$
In terms of these fractions,
$$q'' = 5.67 \times \left(\frac{667.7}{100}\right)^4 \text{W/m}^2\cdot\text{K}^4 (0.9 \times 0.00032 + 0.1 \times 0.99968)$$
or
$$q'' = 1,129.8 \text{W/m}^2,$$
which is sufficiently close to 1,126.7 W/m². Then the surface temperature of the satellite is, approximately,
$$T_{sat} \cong 668\text{K}.$$

♦

EXAMPLE 8.9

In a combustion experiment of a luminous hydrocarbon flame, it is desired to measure the total irradiation received by a sensing device. The bright yellow color of the flame indicates that the radiation emitted by the small soot particles obeys the Planck distribution. The optical sensing device is a photomultiplier tube (PMT) which converts the photon interaction at the inlet (photocathode) to an electric current at the outlet (anode). The PMT has 3 cm² of active sensor area facing the flame and is placed 2 m away from the flame axis.

Assuming that the flame isotropically radiates a total of 20 kW at a flame temperature of $T_F = 1,800$ K, we wish to determine the energy absorbed by the photocathode surface by considering the typical approximate spectral response curve of the PMT given in Fig. 8.14 and compare the absorbed energy to the typical power of 5 mW for an He-Ne laser beam.

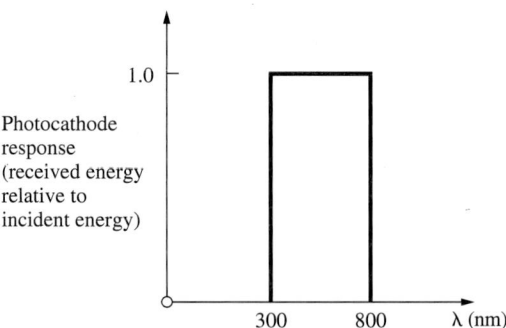

Figure 8.14

For isotropic emission, the radiative heat flux at an arbitrary radius r is

$$q_s^R = \frac{\dot{Q}^R}{4\pi r^2},$$

which gives, for $r = \ell$, ℓ being the distance between the flame and PMT,

$$q_{r=\ell}^R = \frac{\dot{Q}^R}{4\pi \ell^2} = \frac{20{,}000\text{W}}{4\pi \times 2^2 \text{m}^2} \cong 398\text{W/m}^2.$$

In terms of the active sensor area of $A = 3\text{cm}^2$, the total black-body radiation (spectrally integrated from $\ell = 0$ to $\ell \to \infty$) reaching the PMT active surface area is

$$\dot{Q}_{\text{incident}} = q_{r=\ell}^R A = 398\text{W/m}^2 \times 3 \times 10^{-4}\text{m}^2 \cong 0.119\text{W}.$$

In view of the PMT response characteristics, only a fraction of this energy corresponding to the wavelength range 300–800 nm is recognized. The lower and upper limits of λT for the specified source temperature T_F,

$$\lambda_1 T_F = 0.3\mu\text{m} \times 1{,}800\text{K} = 540\mu\text{m}\cdot\text{K}$$

and

$$\lambda_2 T_F = 0.8\mu\text{m} \times 1{,}800\text{K} = 1{,}440\mu\text{m}\cdot\text{K}$$

yield, with interpolated fractional functions from Table 8.1,

$$F_1(0 \to \lambda_1 T_F) \cong 0 \quad \text{and} \quad F_2(0 \to \lambda_2 T_F) \cong 0.0098,$$

which result in

$$[\Delta F(\lambda_1 T \to \lambda_2 T)]_{T=T_F} = [F_2(0 \to \lambda_2 T) - F_1(0 \to \lambda_1 T)]_{T=T_F} \cong 0.0098.$$

Then, the energy absorbed by the photocathode is

$$0.119\text{W} \times 0.0098 \cong 0.00117\text{W} = 1.17\text{mW}$$

or

$$\cong \frac{1}{4} \text{ He-Ne laser power.}$$

Because the level of received black-body emission is comparable to that of the laser, special attention needs to be given to the elimination of this emission in optical diagnostics. For example, in scattering and extinction experiments, narrow-bandpass filters, combined with slits and shields and/or lock-in amplifiers with chopped signals, are common practices. ◆

At a given temperature, if the ratio of the monochromatic emissive power of a surface to the monochromatic emissive power of the black body at the same wavelength remains constant over the entire wavelength spectrum,

$$\epsilon_\lambda(\lambda, T) = \frac{E_\lambda}{E_{b\lambda}} = \epsilon(T) = \text{Const.}, \tag{8.51}$$

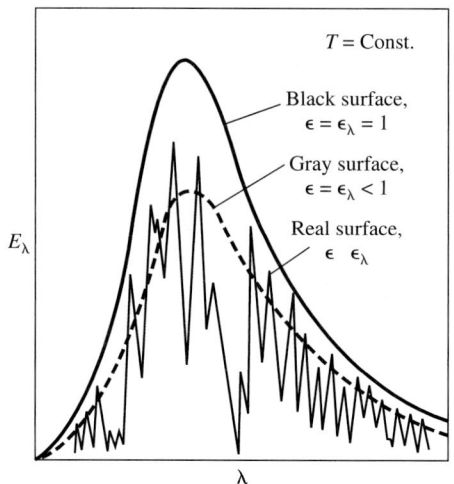

Figure 8.15 Variety of surfaces.

the surface is said to be **gray**. The shape of the monochromatic emissive power of a gray surface is similar to that of the black body at the same temperature, but the value of the emissive power is reduced by the amount of the emissivity (Fig. 8.15).

For a gray surface, Eq. (8.40) may be written in terms of the (total) absorptivity α, reflectivity ρ, and transmissivity τ,

$$1 = \alpha + \rho + \tau, \tag{8.52}$$

which reduces for an opaque gray surface to

$$1 = \alpha + \rho. \tag{8.53}$$

Also, in terms of the (total) emissivity ϵ, the Kirchhoff law becomes

$$\alpha(T) = \epsilon(T). \tag{8.54}$$

For a transparent gray surface, Eq. (8.40) reduces to

$$1 = \alpha + \tau. \tag{8.55}$$

The emissivities of industrially important surfaces for five different temperatures are given in Table 8.2. For more extensive tables see the references cited in Siegel and Howell [3]. Also the effect of wavelength on the monochromatic emissivity of electrical conductors and insulators is demonstrated in Figs. 8.16 and 8.17, respectively. Figure 8.16 shows that polished metal surfaces usually have low monochromatic emissivities, but with oxidation these emissivities appear to assume appreciably increased values (Fig. 8.18). Furthermore, Fig. 8.17 illustrates the fact that electrical insulators exhibit, as a group, a behavior opposite to that of electrical conductors and have high monochromatic emissivities.

Table 8.2 Emissivity of various surfaces (from Brewster [6]).

Material	300 K	500 K	800 K	1,600 K	Material	300 K	500 K	800 K	1,600 K
Metals					**Metals**				
Aluminum					Zinc				
Smooth, polished	0.04	0.05	0.08	0.19	Polished	0.02	0.03		
Smooth, oxidized	0.11	0.12	0.18		Oxidized		0.1		
Rough, oxidized	0.2	0.3			Galvanized	0.02–0.03			
Anodized	0.9	0.7	0.6	0.3					
Brass									
Highly polished		0.03			**Nonmetals**				
Polished	0.1	0.1			Aluminum oxide		0.7	0.6	0.4
Oxidized	0.6				Asbestos	0.95			
Chromium					Asphalt	0.93			
Polished	0.08	0.17	0.26	0.40	Brick				
Copper					Alumina refractory			0.40	0.33
Polished	0.04	0.05	0.18	0.17	Fireclay	0.9		0.8	0.8
Oxidized	0.87	0.83	0.77		Kaolin insulating			0.70	0.53
Gold					Magnesite refractory	0.9			0.4
Highly polished	0.02		0.035		Red, rough	0.9			
Iron and steel					Silica	0.9		0.8	0.8
Iron, polished	0.06	0.08	0.1	0.2	Concrete (rough)	0.94			
Iron, oxidized	0.6	0.7	0.8		Glass	0.95	0.9	0.7	
Stainless, polished	0.1	0.2			Graphite	0.7			0.8
Stainless, oxidized	0.5–0.8				Ice (273 K)				
Mild steel, polished	0.1				Smooth	0.97			
Mild steel, oxidized	0.8	0.8	0.3		Rough	0.99			
Lead					Limestone	0.9	0.8		
Polished	0.05	0.08			Marble (white)	0.95			
Oxidized	0.6	0.6			Mica	0.75			
Mercury					Paints				
Clean	0.1				Aluminized	0.3–0.6			
Magnesium					Most others (incl. white)	0.9			
Polished	0.07	0.13	0.18	0.24	Paper	0.90–0.98			
Nichrome (wire)					Porcelain (glazed)	0.92			
Clean	0.65		0.71		Pyrex	0.82	0.80	0.72	0.6
Oxidized	0.95	0.98			Quartz	0.9		0.6	
Nickel					Rubber				
Polished	0.05	0.07			Hard	0.95			
Oxidized	0.4	0.5			Soft, grey, rough	0.86			
Platinum					Sand (silica)	0.9			
Polished	0.05		0.1		Silicon carbide		0.9		0.8
Oxidized	0.07		0.1		Skin	0.95			
Silver					Snow	0.8–0.9			
Polished	0.01	0.02	0.03		Soil	0.93–0.96			
Oxidized	0.02		0.04		Rocks	0.88–0.95			
Tin					Teflon	0.85	0.92		
Polished	0.05				Vegetation	0.92–0.96			
Tungsten					Water (> 0.1 mm thick)	0.96			
Filament	0.032	0.053	0.088	0.35 (3500 K)	Wood	0.8–0.9			

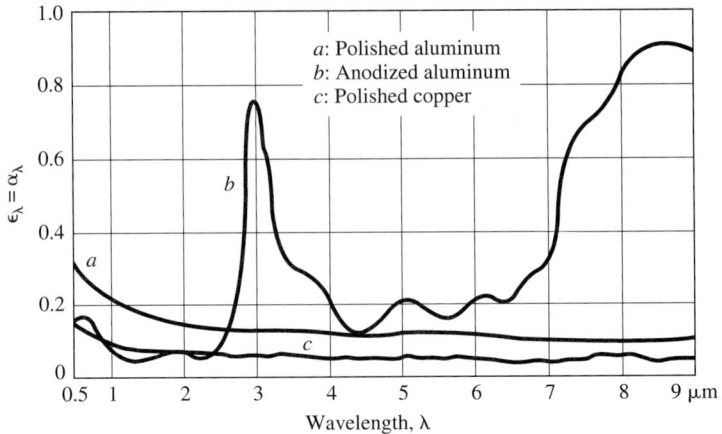

Figure 8.16 Monochromatic emissivity of some electrical conductors (from Kreith and Bohn [7]).

In addition to its variation with wavelength, the monochromatic emissivity of many surfaces is not isotropic and has directional properties. Experimental data on these properties, however, are scarce. Frequently used mean values are $\epsilon/\epsilon_n = 1.2$ for polished metallic surfaces and $\epsilon/\epsilon_n = 0.96$ for insulators. Here ϵ denotes the average hemispherical emissivity and ϵ_n the emissivity normal to the surface.

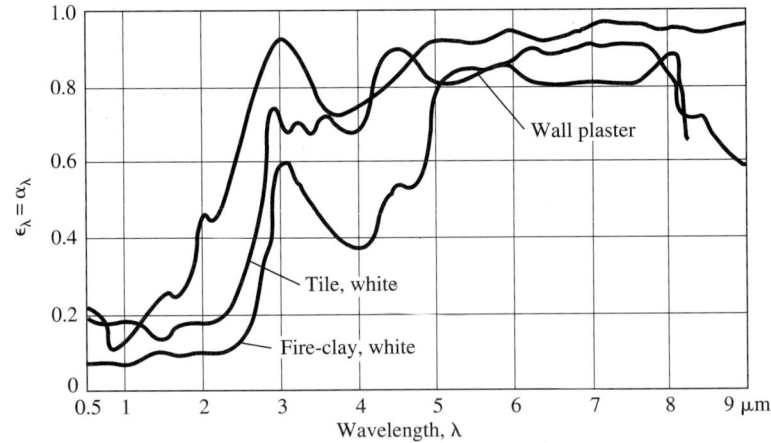

Figure 8.17 Monochromatic emissivity of some electrical insulators (from Kreith and Bohn [7]).

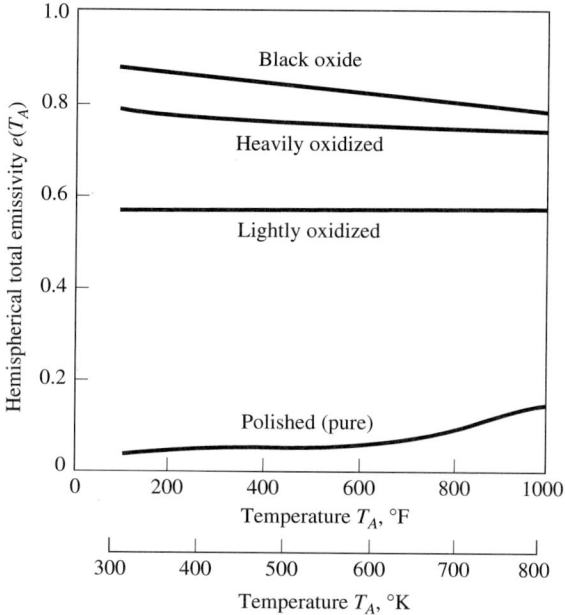

Figure 8.18 Effect of oxidation on emissivity (from Siegel and Howell [3]).

REFERENCES

8.1 M. Planck, *The Theory of Heat Radiation*. Dover, New York, 1959.

8.2 F.K. Richtmyer, E.H. Kennard, and J.N. Cooper, *Introduction to Modern Physics*, 6th ed. McGraw-Hill, New York, 1969.

8.3 R. Siegel and J.R. Howell, *Thermal Radiation Heat Transfer*, 2d ed. Hemisphere/McGraw-Hill, New York, 1981.

8.4 H.C. Hottel and A.F. Sarofim, *Radiative Transfer*. McGraw-Hill, New York, 1967.

8.5 R.V. Dunkle, "Thermal radiation tables and applications," Trans. ASME, **76**, 549, 1954.

8.6 M. Q. Brewster, *Thermal Radiative Transfer and Properties*. Wiley, New York, 1992.

8.7 F. Kreith and M.S. Bohn, *Principles of Heat Transfer*. 4th ed., Harper and Row, New York, 1986.

8.5 APPENDIX

```
C-----------------------------------
C     EX8-6.F (START)
C-----------------------------------
      PROGRAM MAIN
      IMPLICIT REAL*8 (A-H,K-Z)
      PI=4*ATAN(1.)
      WRITE(*,*) 'EXAMPLE 8.6....'
C-----------------------------------
C     INPUT DATA
```

```
C------------------------------------------------
      WRITE(*,*) 'LENGTH: M'
      READ(*,*) L
      WRITE(*,*) 'WIDTH: M'
      READ(*,*) W
      WRITE(*,*) 'TAU: '
      READ(*,*) TAU
      WRITE(*,*) 'LAMBDA1: M'
      READ(*,*) L1
      WRITE(*,*) 'LAMBDA2: M'
      READ(*,*) L2
      WRITE(*,*) 'T_R: K'
      READ(*,*) TR
      WRITE(*,*) 'QS_GLASS: W/m^2'
      READ(*,*) QSG
      WRITE(*,*) 'T_S: K'
      READ(*,*) TS
C------------------------------------------------
C     UNIT CONVERSION
C------------------------------------------------
      L1=1E6*L1
      L2=1E6*L2
C------------------------------------------------
C     CALCULATION
C------------------------------------------------
      L1TS=L1*TS
      L2TS=L2*TS
C------------------------------------------------
C     CHART READINGS
C------------------------------------------------
      WRITE(*,*) 'READ F(0-->LAMBDA*T) FROM TABLE 8.1 WITH'
      WRITE(*,*) 'LAMBDA*T=',L2TS
      WRITE(*,*) 'INPUT F'
      READ(*,*) F2
      WRITE(*,*) 'READ F(0-->LAMBDA*T) FROM TABLE 8.1 WITH'
      WRITE(*,*) 'LAMBDA*T=',L1TS
      WRITE(*,*) 'INPUT F'
      READ(*,*) F1
C------------------------------------------------
C     CALCULATION
C------------------------------------------------
      DTF=F2-F1
      TTAU=TAU*DTF
      DOTQSR=TTAU*QSG*L*W
C------------------------------------------------
C     ANSWER
C------------------------------------------------
      WRITE(*,*) 'TOTAL TRANSMISSIVITY OF THE SIDING'
      WRITE(*,*) 'TO RADIATION FROM THE SUN IS'
      WRITE(*,*) TTAU
      WRITE(*,*) 'RADIATION FROM THE SUN TO THE ROOM IS'
      WRITE(*,*) DOTQSR/1000,' kW'
C------------------------------------------------
C     CALCULATION
C------------------------------------------------
      L1TR=L1*TR
      L2TR=L2*TR
C------------------------------------------------
C     CHART READINGS
```

```
C-------------------------------------------------
      WRITE(*,*) 'READ F(0-->LAMBDA*T) FROM TABLE 8.1 WITH'
      WRITE(*,*) 'LAMBDA*T=',L2TR
      WRITE(*,*) 'INPUT F'
      READ(*,*) F2
      WRITE(*,*) 'READ F(0-->LAMBDA*T) FROM TABLE 8.1 WITH'
      WRITE(*,*) 'LAMBDA*T=',L1TR
      WRITE(*,*) 'INPUT F'
      READ(*,*) F1
C-------------------------------------------------
C     CALCULATION
C-------------------------------------------------
      DTF=F2-F1
      TTAU=TAU*DTF
      DOTQR=TTAU*5.67E-8*TR**4*L*W
C-------------------------------------------------
C     ANSWER
C-------------------------------------------------
      WRITE(*,*) 'TOTAL TRANSMISSIVITY OF THE SIDING'
      WRITE(*,*) 'TO RADIATION FROM THE ROOM IS'
      WRITE(*,*) TTAU
      WRITE(*,*) 'RADIATION OUT FROM THE ROOM IS'
      WRITE(*,*) DOTQR,' W'
      STOP
      END
C----------------------------------------
C     EX8-6.F (END)
C----------------------------------------
C----------------------------------------
C     EX8-7.F (START)
C----------------------------------------
      PROGRAM MAIN
      IMPLICIT REAL*8 (A-H,K-Z)
      PI=4*ATAN(1.)
      WRITE(*,*) 'EXAMPLE 8.7....'
C-------------------------------------------------
C     INPUT DATA
C-------------------------------------------------
      WRITE(*,*) 'TAU: '
      READ(*,*) TAU
      WRITE(*,*) 'LAMBDA1: nM'
      READ(*,*) L1
      WRITE(*,*) 'LAMBDA2: nM'
      READ(*,*) L2
      WRITE(*,*) 'T_L: K'
      READ(*,*) TL
C-------------------------------------------------
C     UNIT CONVERSION
C-------------------------------------------------
      L1=0.001*L1
      L2=0.001*L2
C-------------------------------------------------
C     CALCULATION
C-------------------------------------------------
      L1TL=L1*TL
      L2TL=L2*TL
C-------------------------------------------------
C     CHART READINGS
```

```
C------------------------------------------------
      WRITE(*,*) 'READ F(0-->LAMBDA*T) FROM TABLE 8.1 WITH'
      WRITE(*,*) 'LAMBDA*T=',L2TL
      WRITE(*,*) 'INPUT F'
      READ(*,*) F2
      WRITE(*,*) 'READ F(0-->LAMBDA*T) FROM TABLE 8.1 WITH'
      WRITE(*,*) 'LAMBDA*T=',L1TL
      WRITE(*,*) 'INPUT F'
      READ(*,*) F1
C------------------------------------------------
C     CALCULATION
C------------------------------------------------
      DTF=F2-F1
      TTAU=TAU*DTF
      REFL=1-TTAU
C------------------------------------------------
C     ANSWER
C------------------------------------------------
      WRITE(*,*) 'FRACTION OF ENERGY EJECTED IS'
      WRITE(*,*) REFL*100,'%'
      STOP
      END
C------------------------------------------------
C     EX8-7.F (END)
C------------------------------------------------
```

■ EXERCISES

8.1 Describe the fundamental differences between the conduction and radiation modes of heat transfer.

8.2 What fact originally prompted the electromagnetic nature of radiation?

8.3 Discuss the origin of radiation pressure and the important result this pressure provides.

8.4 What is the relation between optics and electromagnetics? Why are optical concepts such as solid angle and intensity utilized in thermal radiation studies?

8.5 What are the limitations of classical theories, such as electromagnetics, optics, and thermodynamics, for thermal radiation and what fact originally prompted the modern theory of radiation? State briefly the foundations of the modern theory.

8.6 A tungsten filament is heated to 3,000 K. Find the wavelength corresponding to the maximum of the monochromatic radiation energy. Determine the fraction of the total radiation energy in the visible range (0.4 to 0.7 μm), in the infrared range (1 to 20 μm), and in the thermal range (0.1 to 100 μm).

8.7 The spectral emissivity of a paint is approximately 0.2 below 3 μm and 0.8 at longer wavelengths. Determine the total emissivity of a surface covered with this paint at 300 K and 800 K. What is your conclusion?

8.8 The paint of Prob. 8.7 is used to cover a surface which is maintained at 300 K. Let the surface be subjected to (a) solar radiation, and (b) a black source at 800 K. Determine the effective total absorptivity of the surface for these cases.

8.9 Show that Eq. (8.34) may also be expressed in terms of frequency as

$$E_{b\nu} = \frac{2\pi h c \nu^3}{e^{h\nu/kT} - 1}.$$

The monochromatic emissive power in terms of frequency proves convenient for spectral integrations as well as for gas radiation (to be introduced in Chapter 10).

8.10 The design department of a major auto company plans to increase the glass window area of new model cars by 1 m², which creates an additional cooling load due to the greenhouse effect, particularly in hot climates. According to the company's engineers, the present cooling system can provide an additional compressor power of 0.5 kW without major changes in the cooling system. The spectral transmissivity of the window glass is given in Fig. 8P–1. Assuming that the sun and the interior of the car radiate as black-bodies at 5,800 K and 300 K, respectively, and considering that the coefficient of performance β of the cooling system, defined as

$$\beta = \frac{\text{Heat extracted from interior of the car}}{\text{Power input into the compressor}},$$

is 1.5, determine whether or not the additional load remains within the system capacity. Assume that the intensity of solar radiation on earth is 1,000 W/m².

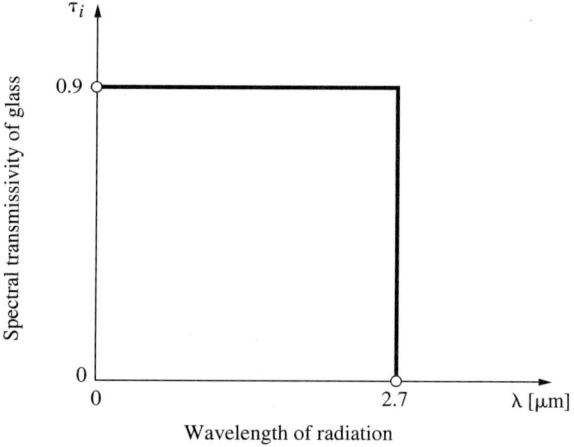

Figure 8P–1

Chapter 9

ENCLOSURE RADIATION

In the preceding chapter we learned the equilibrium aspects of radiation, including emissive power and surface properties. In the present chapter we proceed to the nonequilibrium or transport aspects of radiation.

At low temperatures most monatomic and diatomic gases, as well as air, are electrically neutral. Consequently, in radiation problems involving enclosures filled with neutral gases, the energy exchange among the enclosure surfaces is practically unaffected by the presence of a gas. In other words, most gases at low enough temperatures are transparent and do not influence the radiation-energy transfer. At elevated temperatures gases no longer remain electrically neutral, and they become dissociated or even ionized. They then participate in radiative processes. In this chapter we study radiation problems associated with enclosures in a vacuum or filled with a transparent gas. In the next chapter we will deal with enclosures with a gas participating in the radiation heat transfer.

Before proceeding to a general discussion, we illustrate enclosure radiation in terms of the energy exchange between two closely located, large, parallel, and opaque gray plates (Fig. 9.1). We begin by tracing the radiation energy $\epsilon_1 E_{b1}$ emitted by surface 1, which travels back and forth between the two surfaces until it is finally absorbed. Part $\alpha_2(\epsilon_1 E_{b1})$ of this energy is absorbed by surface 2 and part $\rho_2(\epsilon_1 E_{b1})$ is reflected back to surface 1. There part $\alpha_1 \rho_2(\epsilon_1 E_{b1})$ is absorbed, and part $\rho_1 \rho_2(\epsilon_1 E_{b1})$ is reflected back to surface 2. Again, part $\alpha_2 \rho_1 \rho_2(\epsilon_1 E_{b1})$ is absorbed, part $\rho_1 \rho_2^2(\epsilon_1 E_{b1})$ is reflected back to surface 1, and so on. The radiation energy that has left surface 1 and been

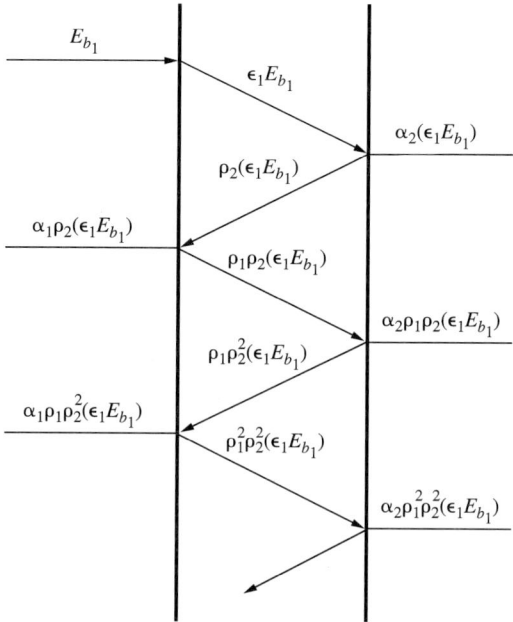

Figure 9.1 Emission, multiple absorption, and reflection.

absorbed by surface 2 is therefore

$$q_{1\to 2} = \lim_{N\to\infty} \alpha_2(\epsilon_1 E_{b1})\left(1 + \rho_1\rho_2 + \rho_1^2\rho_2^2 + \cdots + \rho_1^N\rho_2^N\right), \quad (9.1)$$

where

$$S = \lim_{N\to\infty}\left(1 + \rho_1\rho_2 + \rho_1^2\rho_2^2 + \cdots + \rho_1^N\rho_2^N\right)$$

is the sum of a geometric series. Multiplying this series by $\rho_1\rho_2$ and subtracting the result from the series itself, we get for $\rho_1\rho_2 < 1$,

$$S = \lim_{N\to\infty} \frac{1 - \rho_1^{N+1}\rho_2^{N+1}}{1 - \rho_1\rho_2} \to \frac{1}{1 - \rho_1\rho_2}. \quad (9.2)$$

Inserting Eq. (9.2) into Eq. (9.1), assuming that surface 2 is in local equilibrium and letting $\alpha_2 = \epsilon_2$ in accordance with the Kirchoff law,[1] we obtain

$$q_{1\to 2} = \left(\frac{\epsilon_1\epsilon_2}{1 - \rho_1\rho_2}\right) E_{b1}. \quad (9.3)$$

Similarly, we could trace the radiation energy $\epsilon_2 E_{b2}$ emitted by surface 2, which travels also to and fro between the two surfaces until it is finally absorbed. Accordingly, the

---

[1] Recall Eq. (8.50).

Chap. 9 Enclosure Radiation

interchange of the subscripts in Eq. (9.3) readily gives

$$q_{2\to 1} = \left(\frac{\epsilon_2 \epsilon_1}{1 - \rho_2 \rho_1}\right) E_{b2}. \tag{9.4}$$

The net radiation between surfaces 1 and 2 is then

$$q_{12} = q_{1\to 2} - q_{2\to 1}$$

or

$$q_{12} = \left(\frac{\epsilon_1 \epsilon_2}{1 - \rho_1 \rho_2}\right)(E_{b1} - E_{b2}).$$

Noting for opaque surfaces that $\alpha(=\epsilon) + \rho = 1$, we can express ρ_1 and ρ_2 in terms of ϵ_1 and ϵ_2 and rearrange this equation, after dividing numerator and denominator by $\epsilon_1 \epsilon_2$, as

$$q_{12} = \frac{E_{b1} - E_{b2}}{\dfrac{1}{\epsilon_1} + \dfrac{1}{\epsilon_2} - 1}. \tag{9.5}$$

Here, we utilize this result to demonstrate an important practical fact associated with thermos bottles.

EXAMPLE 9.1[2]

The sides of the two walls facing each other are silvered (Fig. 9.2). The emissivity of silver[3] is $\epsilon \cong 0.02$. The thermos bottle contains hot coffee at temperature $T_1 = 90\,°C$, and the ambient is at a temperature $T_2 = 5\,°C$. We wish to determine the heat loss from a thermos bottle.

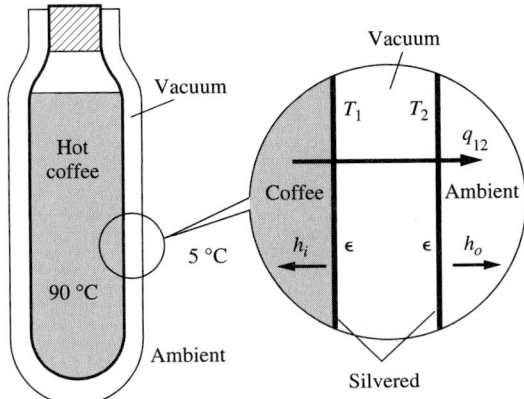

Figure 9.2 Thermos bottle.

---

[2] The FORTRAN program EX9–1.F is listed in the appendix of this chapter.

[3] See Table 8.2.

First, neglecting the inner and outer convective resistances, we assume the inner wall to be at the coffee temperature and the outer wall at the ambient temperature. Next recalling that $E_b = \sigma T^4$ and noting that $\epsilon_1 = \epsilon_2 = \epsilon \ll 1$, we rearrange Eq. (9.5) as

$$q_{12} \cong \frac{1}{2}\epsilon\sigma(T_1^4 - T_2^4),$$

which gives

$$q_{12} = \frac{1}{2} \times 0.02 \times 5.67 \times 10^{-8} \text{ W/m}^2\cdot\text{K}^4 \left[(90+273)^4 - (5+273)^4\right] \text{K}^4$$

or

$$q_{12} \cong 6.5 \text{ W/m}^2.$$

We may now comment on the neglected convective resistances. Letting[4] $h_i \cong 100$ W/m²·K (for stagnant liquids) and $h_0 \cong 10\text{-}100$ W/m²·K (from stagnant to windy air), we find from $q = h\Delta T$ that the temperature drop due to convective resistance does not exceed 1 °C, and we can safely neglect it, as we already have.

By the computer program provided, the interested reader may parametrically study the heat loss depending on the coffee temperature.

To demonstrate the effectiveness of the foregoing radiative insulation, we may evaluate, for example, the conduction across a layer of cork which would have the same insulating effect. Assuming[5] $k \cong 0.04$ W/m·K, we obtain from $q = k\Delta T/\ell$, the required thickness of the cork,

$$\ell = k\Delta T/q = 0.04 \text{ W/m·K} \times \frac{(90-5) \text{ K}}{6.5 \text{ W/m}^2} \cong 0.52 \text{ m!}$$

This result clearly shows how important radiative insulation is. ◆

When we deal with enclosures made of more than two surfaces, the foregoing method of tracing the radiation energy travel to and fro between the surfaces becomes rather involved if not impossible. Two methods that are generally convenient for enclosure radiation are those of **electrical analogy** and **net radiation**. However, before proceeding to these methods, we need to explore the concepts of solid angle, intensity, emissive power, **radiosity**, and **view factor**. We have already discussed the first three of these concepts in Chapter 8. The last two, beginning with radiosity because of its relative simplicity, are developed next.

Consider an opaque gray surface subject to incident radiation G. Employing the definition of (total) emissivity, $\epsilon = E/E_b$, rearrange the **rate of radiation leaving this surface per unit area** as

$$B = \rho G + \epsilon E_b, \tag{9.6}$$

---

[4] See Table 1.2.

[5] See Table 1.1.

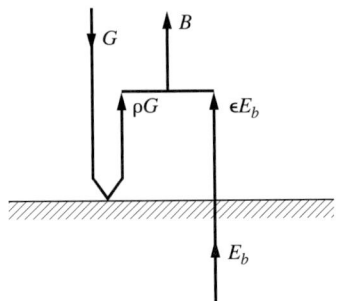

Figure 9.3 Radiosity of an opaque gray surface.

which is called the **radiosity**[6] of an opaque gray surface (Fig. 9.3). For a surface having some transparency, the radiosity also includes the transmitted, as well as reflected and emitted, energy. Having defined the radiosity, we now proceed to the final concept, the view factor.

9.1 VIEW FACTOR

Consider two surfaces A_1 and A_2, as shown in Fig. 9.4. Let the length of the line connecting surface elements dA_1 and dA_2 be r, the angles between r and the normals to dA_1 and dA_2 be ϕ_1 and ϕ_2, and the radiosities and the corresponding intensities of the surfaces be B_1, B_2 and I_1, I_2. We wish to determine the fraction of the radiation energy propagating from A_1 that is intercepted by A_2.

Let us first concentrate on the part of the radiation energy from dA_1 that is intercepted by dA_2. Rearranging Eq. (8.14) as $dq_{1\to 2} = I_1 d\Omega_{12}$, multiplying it by $(dA_n)_1$, and noting $(dA_n)_1 = dA_1 \cos\phi_1$, we have

$$d^2\dot{Q}_{1\to 2} = dq_{1\to 2}(dA_n)_1 = I_1 dA_1 \cos\phi_1 d\Omega_{12},$$

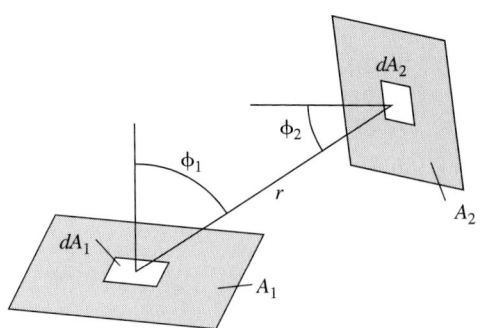

Figure 9.4 View factor.

---

[6] In the literature, J or W also is frequently used to denote the radiosity.

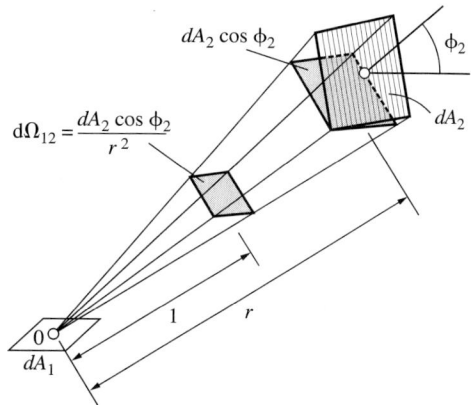

Figure 9.5 Solid angle $d\Omega$-dA relation.

where $d^2 \dot{Q}_{1\to 2}$ is the amount of radiation emitted by dA_1 that is intercepted by dA_2. In terms of $d\Omega_{12} = dA_2 \cos\phi_2 / r^2$ (see Fig. 9.5) and Eq. (8.19) for radiosity[7] $B_1 = \pi I_1$,

$$d^2 \dot{Q}_{1\to 2} = B_1 \frac{\cos\phi_1 \cos\phi_2 \, dA_1 \, dA_2}{\pi r^2}. \tag{9.7}$$

Integrating Eq. (9.7) over A_1 and A_2, and noting that I_1 and B_1 of a diffuse surface are isotropic[8] and independent of the integration, we obtain

$$\dot{Q}_{1\to 2} = B_1 \int_{A_1} \int_{A_2} \frac{\cos\phi_1 \cos\phi_2 \, dA_1 \, dA_2}{\pi r^2}. \tag{9.8}$$

Now we are ready to introduce the concept of **view factor** between surfaces A_1 and A_2, and based on surface A_1, as the fraction F_{12} of the radiation energy $B_1 A_1$ propagating from A_1 that is intercepted by A_2,

$$F_{12} = \frac{\dot{Q}_{1\to 2}}{B_1 A_1}. \tag{9.9}$$

The view factor based on surface A_2, F_{21}, is obtained by changing the subscript in Eq. (9.9). Combining Eqs. (9.8) and (9.9), we get

$$A_1 F_{12} = \int_{A_1} \int_{A_2} \frac{\cos\phi_1 \cos\phi_2 \, dA_1 \, dA_2}{\pi r^2}. \tag{9.10}$$

Interchanging the subscripts in Eq. (9.10), we obtain the important relation

$$\boxed{A_1 F_{12} = A_2 F_{21}}, \tag{9.11}$$

which is known as the **reciprocity rule**.

---

[7] Note that the intensities associated with B and E are different.

[8] Recall Fig. 8.9(c).

Finally, noting that the sum of all fractions based on the same surface should add up to unity, for any surface i of an enclosure made of N surfaces we have

$$\sum_{j=1}^{N} F_{ij} = 1, \quad i = 1, 2, \ldots, N, \qquad (9.12)$$

which is known to be the **summation rule**. Here F_{ii} denotes the fraction of the radiation leaving surface i intercepted by itself. Clearly, $F_{ii} = 0$ if the surface i is flat or convex, and $F_{ii} \neq 0$ if surface i is concave (or can see itself).

In general, a view factor defined by Eq. (9.10) requires the computation of the surface integrals associated with A_1 and A_2. However, considering the radiative symmetry of a particular enclosure, together with Eqs. (9.11) and (9.12), turns out to be sufficient for view factors of a number of enclosures. Here we consider some of these cases:

- For two closely located large parallel plates [Fig. 9.6(a)] $F_{11} = 0$, and from Eq. (9.12) $F_{12} = 1$. This indicates that all of the radiation leaving surface 1 is intercepted by surface 2. The reciprocity rule then states that $F_{12} = F_{21}$, which is an expected result, given the symmetry of the problem.

- For a long cylinder surrounded by another cylinder [Fig. 9.6(b)] we have $F_{11} = 0$ (cylinder 1 does not see itself) and $F_{12} = 1$. Then, from Eq. (9.11) we get $F_{21} = A_1/A_2$, and from Eq. (9.12), or $F_{22} + F_{21} = 1$, we get $F_{22} = 1 - A_1/A_2$ (surface 2 is concave and thus part F_{22} of the radiation leaving surface 2 is intercepted by itself; the rest, F_{21}, is intercepted by surface 1).

- For a long enclosure with an equilateral triangular cross section [Fig. 9.6(c)], noting the symmetry $F_{12} = F_{13}$, and coupling this fact with $F_{12} + F_{13} = 1$, we get $F_{12} = F_{13} = 1/2$.

- For the same enclosure with a right isosceles triangular cross section [Fig. 9.6(d)], noting the symmetry $F_{31} = F_{32} = 1/2$ and coupling this fact with $A_1 F_{13} = A_3 F_{31}$, we get $F_{13} = 1/2(A_3/A_1)$. Inserting this result into $F_{12} + F_{13} = 1$, we obtain $F_{12} = 1 - 1/2(A_3/A_1)$.

- For the same enclosure with a square cross section [Fig. 9.6(e)] we can reduce the problem to the preceding one by introducing an imaginary partition. Then $F_{12} = 1 - 1/2(A_p/A_1)$. Furthermore, noting the symmetry, $F_{12} = F_{13}$, we have $2F_{12} + F_{14} = 1$ and $F_{14} = A_p/A_1 - 1$.

- For the enclosure with the cross section shown in Fig. 9.6(f), again introducing the imaginary partition, we have $F_{13} = F_{1p} = 1/2(A_p/A_1)$ and $F_{12} = 1 - 1/2(A_p/A_1)$.

View factors of many other simple cases may be found in a similar manner. More complicated cases, however, including all of the enclosures of Fig. 9.6, when they have a finite depth, require the use of Eq. (9.10). Computation of the integrals involved with Eq. (9.10) will be illustrated here in terms of an example.

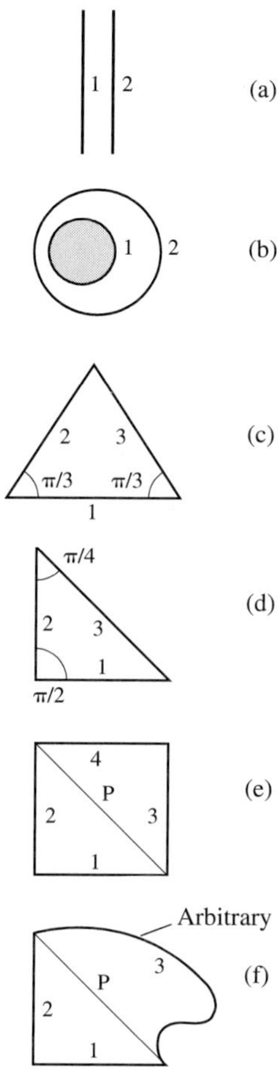

(a) $F_{11} = 0$, $F_{12} = 1$

(b) $F_{11} = 0$, $F_{12} = 1$
$F_{21} = A_1/A_2$, $F_{22} = 1 - A_1/A_2$

(c) $F_{11} = 0$, $F_{12} = F_{13} = 1/2$

(d) $F_{33} = 0$, $F_{31} = F_{32} = 1/2$
$F_{12} = 1 - \frac{1}{2}(A_3/A_1) = 1 - \sqrt{2}/2$
$F_{13} = \frac{1}{2}(A_3/A_1) = \sqrt{2}/2$

(e) $F_{12} = F_{13} = 1 - \frac{1}{2}(A_p/A_1) = 1 - \sqrt{2}/2$
$F_{14} = A_p/A_1 - 1 = \sqrt{2} - 1$

(f) $F_{13} = F_{1p} = \frac{1}{2}(A_p/A_1) = \sqrt{2}/2$
$F_{12} = 1 - \frac{1}{2}(A_p/A_1) = 1 - \sqrt{2}/2$

Figure 9.6 View factors for a long cylinder with various cross sections.

EXAMPLE 9.2

We wish to determine the view factor from a differential area dA_1 to a rectangular area $A_2 = \ell \times L$, dA_1 being parallel to and located at a distance H below one corner of A_2, as shown in Fig. 9.7.

Since one of the areas is differential, the integration will be over A_2 only. Differentiating Eq. (9.10) with respect to A_1 gives

$$F_{d12} = \frac{1}{\pi} \int_{A_2} \frac{\cos\phi_1 \cos\phi_2 dA_2}{r^2},$$

which may be rearranged in terms of $\phi_1 = \phi_2 = \phi$, $\cos\phi = H/r$, $(x^2 + y^2) + H^2 = r^2$, and $dA_2 = dxdy$ as

$$F_{d12} = \frac{H^2}{\pi} \int_0^L \int_0^\ell \frac{dxdy}{(H^2 + y^2 + x^2)^2}$$

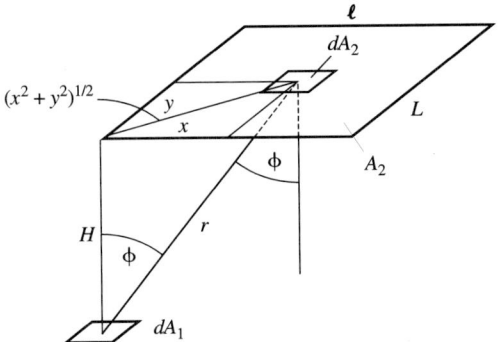

Figure 9.7 View factor F_{d12}.

or, in terms of $H^2 + y^2 = K^2$, as

$$F_{d12} = \frac{H^2}{\pi} \int_0^L \left[\int_0^\ell \frac{dx}{(K^2 + x^2)^2} \right] dy. \tag{9.13}$$

Assuming K constant for the x integration, evaluating the integral in brackets, and integrating the result with respect to y while taking into account the fact that K depends on y, results in

$$F_{d12} = \frac{1}{2\pi} \left[\frac{X}{\sqrt{1 + X^2}} \tan^{-1}\left(\frac{Y}{\sqrt{1 + X^2}}\right) + \frac{Y}{\sqrt{1 + Y^2}} \tan^{-1}\left(\frac{X}{\sqrt{1 + Y^2}}\right) \right], \tag{9.14}$$

where $X = \ell/H$ and $Y = L/H$. Figure 9.8 shows F_{d12} versus X for various values of Y. ◆

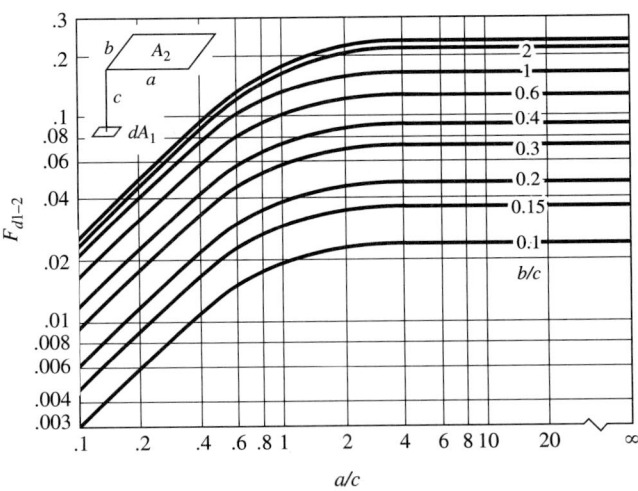

Figure 9.8

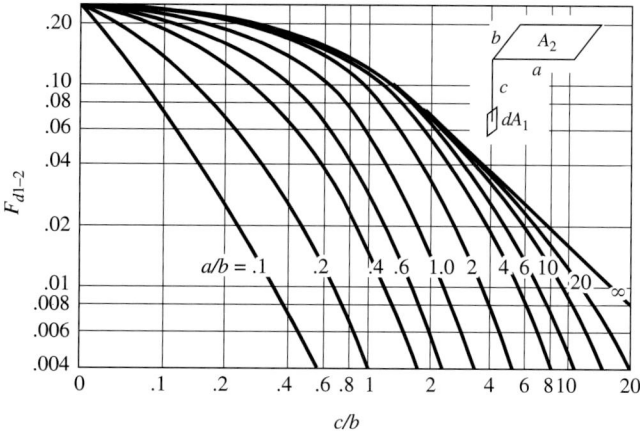

Figure 9.9

We learn with this simple example that the evaluation of a view factor, although it presents no conceptual problem, is usually lengthy and tedious. For later convenience, a number of frequently encountered cases are shown in Figs. 9.8 through 9.13.[9] For more cases the reader is referred to Jakob [1] or Sparrow and Cess [6] and the references cited therein. However, many other cases remain for which no view-factor charts are

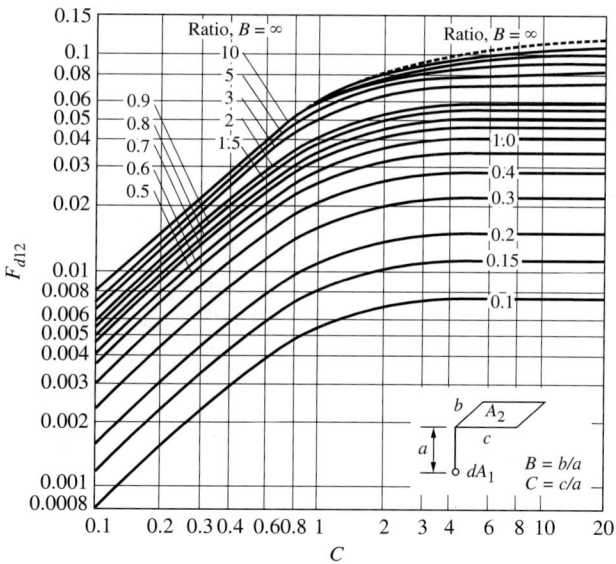

Figure 9.10

---

[9] *Figures 9–8, 9–9, 9–12, 9–13 adapted from E.M. Sparrow and R.D. Cess, *Radiation Heat Transfer*, McGraw-Hill, New York, 1978. Used by permission. Figures 9–10, 9–11 adapted from M. Jakob, *Heat Transfer*, Vol.2, Wiley, New York, 1957. Used by permission.

440 Chap. 9 Enclosure Radiation

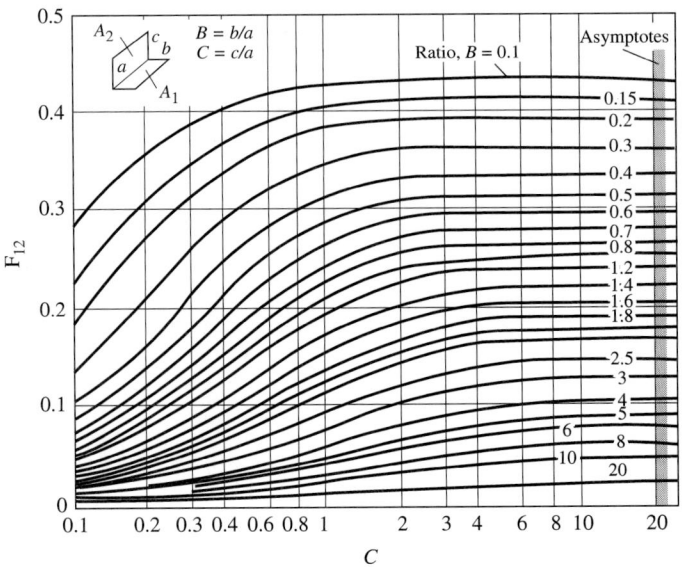

Figure 9.11

available. These may be evaluated by an appropriate superposition of the available charts. The principle behind the evaluation of **view factors by a superposition of charts** is the **conservation of energy**. Here we illustrate the evaluation of some view factors by this approach.

In Fig. 9.14(a), the radiation fraction $dA_1 F_{d1(23)}$ leaving dA_1 and intercepted by $A_2 + A_3$ may be expressed as

$$dA_1 F_{d1(23)} = dA_1 F_{d1(2)} + dA_1 F_{d1(3)}, \qquad (9.15)$$

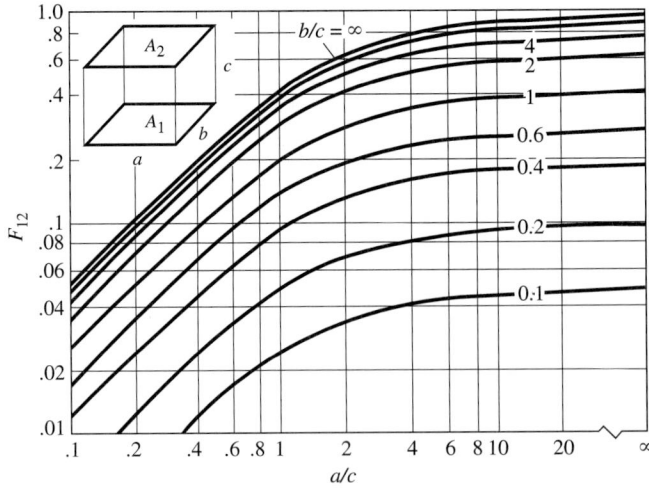

Figure 9.12

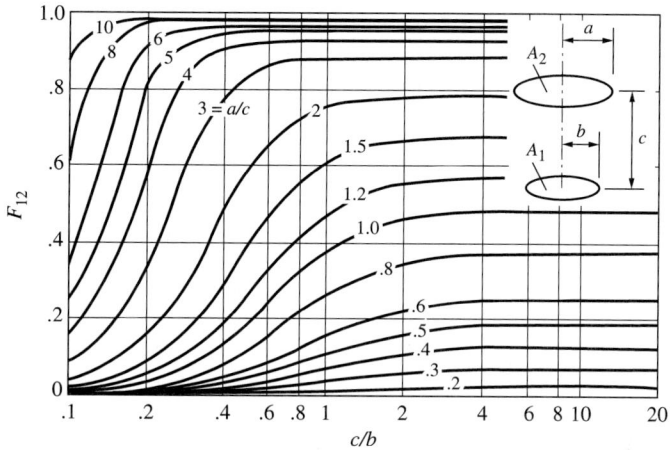

Figure 9.13

where $F_{d1(2)}$ and $F_{d1(3)}$ are available from Fig. 9.8. In Fig. 9.14(b), the radiation fraction $dA_1 F_{d1(2)}$ leaving dA_1 and intercepted by A_2 may be expressed as

$$dA_1 F_{d1(2)} = dA_1 F_{d1(23)} - dA_1 F_{d1(3)},$$

which is Eq. (9.15) solved for $dA_1 F_{d1(2)}$ and where $F_{d1(3)}$ is obtained from Fig. 9.8.

In Fig. 9.15(a), for the radiation fraction $A_1 F_{12}$ leaving A_1 and intercepted by A_2, consider first the radiation $A_2 F_{2(13)}$ leaving A_2 and intercepted by $A_1 + A_3$,

$$A_2 F_{2(13)} = A_2 F_{21} + A_2 F_{23}$$

which may be rearranged by the reciprocity rule as

$$A_{(13)} F_{(13)2} = A_1 F_{12} + A_3 F_{32},$$

or

$$A_1 F_{12} = A_{(13)} F_{(13)2} - A_3 F_{32},$$

where $F_{(13)2}$ and F_{32} are available from Fig. 9.11. In Fig. 9.15(b), for the radiation fraction $A_1 F_{12}$ leaving A_1 and intercepted by A_2, first consider the radiation fraction

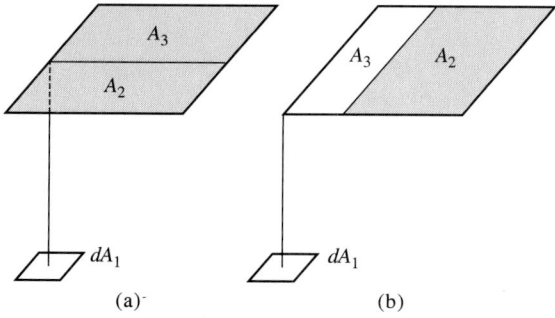

Figure 9.14 Two cases related to Fig. 9.8.

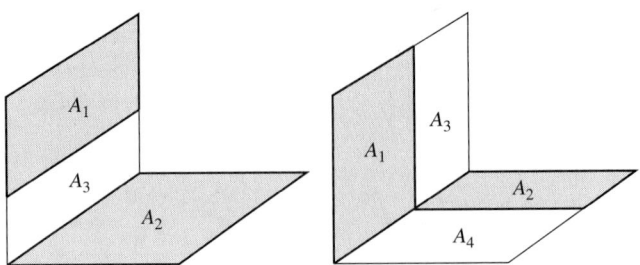

Figure 9.15 Two cases related to Fig. 9.11.

$A_{(13)} F_{(13)(24)}$ leaving $A_1 + A_3$ and intercepted by $A_2 + A_4$,

$$A_{(13)} F_{(13)(24)} = A_1 F_{12} + A_1 F_{14} + A_3 F_{32} + A_3 F_{34}. \tag{9.16}$$

Next consider Eq. (9.10) for F_{12} and F_{34}, and from an inspection of the integral limits involved with F_{12} and F_{34} obtain

$$A_1 F_{12} = A_3 F_{34}. \tag{9.17}$$

Rearranging Eq. (9.16) in terms of Eq. (9.17),

$$A_1 F_{12} = \frac{1}{2} [A_{(13)} F_{(13)(24)} - A_1 F_{14} - A_3 F_{32}], \tag{9.18}$$

where $F_{(13)(24)}$, F_{14}, and F_{32} are available from Fig. 9.11. Equation (9.18) equally applies to F_{12} of Fig. 9.16, provided $F_{(13)(24)}$, F_{14}, and F_{32} are evaluated from Fig. 9.12.

With the evaluation of the view factor, in addition to the concepts of radiosity, solid angle, intensity, and emissive power (the last three from Chapter 8), we complete the concepts needed for enclosure radiation problems. Now we proceed to the solution methods for these problems: electrical analogy and net radiation.

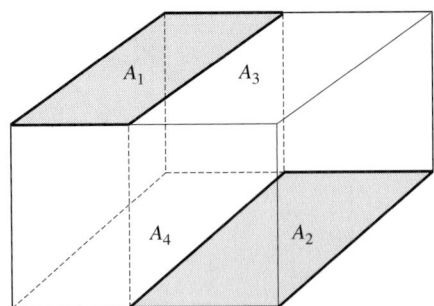

Figure 9.16 A case related to Fig. 9.12.

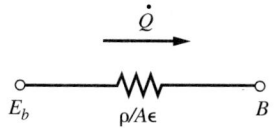

Figure 9.17 Network element associated with surface resistance.

9.2 ELECTRICAL ANALOGY

We have already encountered the application of electrical analogy to conduction. Now we proceed to the use of electrical analogy for radiation. This method is based on two circuit elements. For the first element, reconsider the opaque gray surface of Fig. 9.3. The radiant heat flux from this surface is

$$\frac{\dot{Q}}{A} \equiv q = B - G, \qquad (9.19)$$

where G is the incident radiation. Now recall the definition of radiosity,

$$B = \rho G + \epsilon E_b, \qquad (9.6)$$

and use it to solve for incident radiation G. Insert G into Eq. (9.19) and rearrange the result in terms of $\epsilon = 1 - \rho$ to get

$$\dot{Q} = \frac{E_b - B}{\rho/A\epsilon}, \qquad (9.20)$$

which may be interpreted electrically as the current flow due to the potential drop $E_b - B$ through a resistance $\rho/A\epsilon$. For a black surface, $\rho = 0$ and $E_b = B$. Consequently, $\rho/A\epsilon$ may be assumed as the **surface resistance** (color), which is a measure for the degree of departure from the black surface. The circuit element associated with this resistance is shown in Fig. 9.17.

For the second circuit element, consider the exchange of radiant energy between two surfaces, say 1 and 2. Of the total radiation $B_1 A_1$ leaving surface 1, the amount intercepted by surface 2 is $B_1 A_1 F_{12}$. Similarly, of the total radiation $B_2 A_2$ leaving surface 2, the amount intercepted by surface 1 is $B_2 A_2 F_{21}$. The net exchange of radiant energy between surface 1 and surface 2 is then

$$\dot{Q}^B_{12} = B_1 A_1 F_{12} - B_2 A_2 F_{21},$$

which, in terms of the reciprocity rule, $A_1 F_{12} = A_2 F_{21}$, yields

$$\dot{Q}^B_{12} = \frac{B_1 - B_2}{1/A_1 F_{12}}, \qquad (9.21)$$

where superscript B indicates the heat transfer between radiosities (B). This result may be interpreted electrically as the current flow due to the potential drop $B_1 - B_2$ through a resistance $1/A_1 F_{12}$. Furthermore, $1/A_1 F_{12}$ may be interpreted as the **view resistance**, which indicates the extent to which the surfaces see each other. The circuit element associated with this resistance is shown in Fig. 9.18.

Figure 9.18 Network element associated with view resistance.

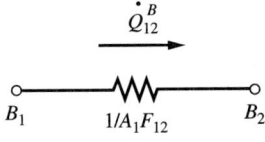

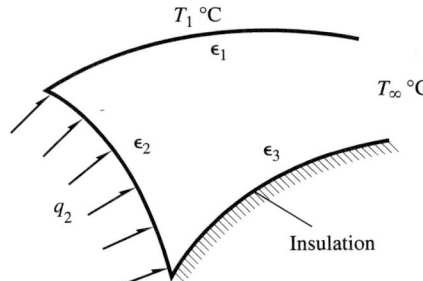

Figure 9.19 An open enclosure.

In terms of the foregoing two circuit elements, we are now ready to solve enclosure problems. The five steps, illustrated in terms of an open enclosure with an arbitrary cross section (Fig. 9.19), are helpful for the solution of these problems (Table 9.1).

Example 9.3

Consider an enclosure made of two concave surfaces with specified temperatures and emissivities (Fig. 9.20). We wish to determine the radiant heat transfer between these surfaces.

Combining in series two surface resistances and one view resistance, as shown in Fig. 9.20, we have

$$\dot{Q}_1 = \dot{Q}_{12}^B = \dot{Q}_2 = \frac{E_{b1} - E_{b2}}{R_{12}}, \tag{9.22}$$

where

$$R_{12} = \frac{\rho_1}{A_1 \epsilon_1} + \frac{1}{A_1 F_{12}} + \frac{\rho_2}{A_2 \epsilon_2}$$

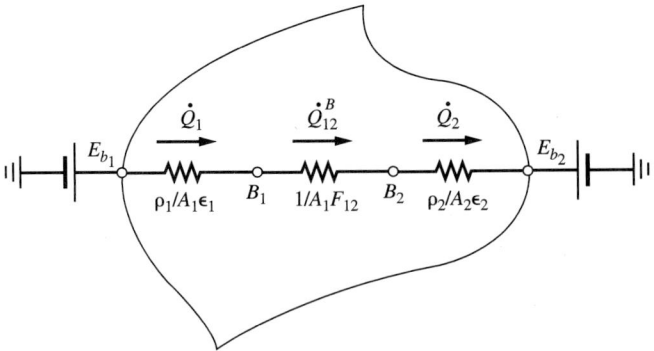

Figure 9.20 Enclosure of two concave surfaces.

Table 9.1 Five steps for electrical analogy

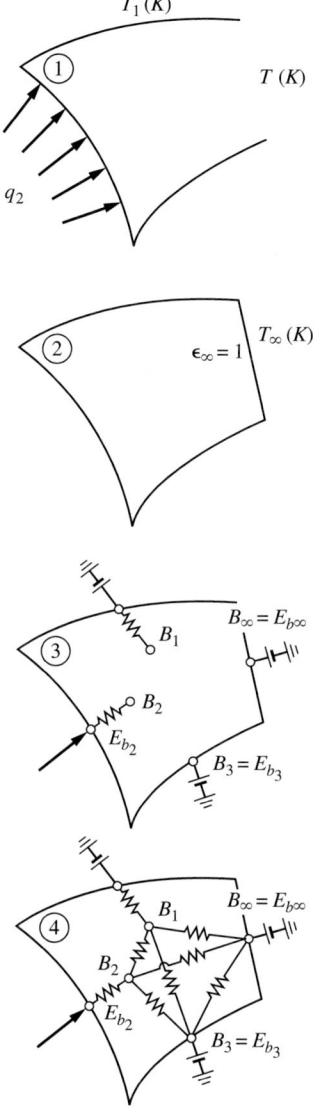

1 Express the temperature of the surfaces with a specified temperature in the absolute scale (Kelvin).

2 Replace the missing surface of an open enclosure with a black surface at the ambient temperature.

3 Attach the appropriate surface (color) resistance to each surface. Note that $B = E_b$ for a black surface and for an insulated surface, so no surface resistance needs to be attached to such surfaces. However, E_b of a black surface is given, while E_b of an insulated surface needs to be determined. Also, note that the color of an insulated surface does not play any role in the problem. Draw circuitry directly on the figure.

4 Connect the radiosity of each surface to the radiosity of all the surfaces that are in direct view of the surface. Use a battery for the E_b of each surface with a specified temperature. Reduce multiple surfaces with identical E_b to a single node.

5 Determine the unknown potentials (emissive powers) and currents (heat fluxes) by the usual solution procedures of electrical circuitry. Use summation and reciprocity rules and choose the most convenient view factors.

---

is the total resistance between emissive powers, E_{b1} and E_{b2}. Note that R_{12} has the unit of reciprocal area. In particular, if the surfaces were parallel and closely located, $F_{12} = 1$ and $A_1 = A_2 = A$, and Eq. (9.22) reduces to Eq. (9.5), as expected. ◆

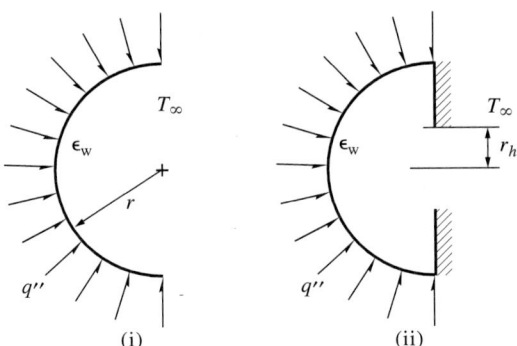

Figure 9.21 Hemispherical container.

EXAMPLE 9.4

A hemispherical container of radius r is subjected to a heat flux q'' in an ambient at temperature T_∞ [Fig. 9.21(a)]. A simple collimator can be constructed by closing the container with a mirror disk having a concentric hole of radius r_h [Fig. 9.21(b)]. **(a)** We wish to determine the increase in the temperature of the container and in the heat flux from the collimator. **(b)** Suppose that the mirror is replaced by a disk of emissivity ϵ_d. What happens to the heat flux through the hole?

(a) First consider the case without disk and replace the opening by a black surface at the ambient temperature. In terms of the circuit shown in Fig. 9.22(a), the radiant heat transfer from the container to the ambient is

$$\dot{Q}_{w\infty} = \frac{E_{bw} - E_{b\infty}}{R_{w\infty}}$$

or, dividing both sides by A_w (the hemispherical area) and in view of $E_b = \sigma T^4$,

$$q'' = \frac{\dot{Q}_{w\infty}}{A_w} = \frac{\sigma(T_w^4 - T_\infty^4)}{A_w R_{w\infty}}, \qquad (9.23)$$

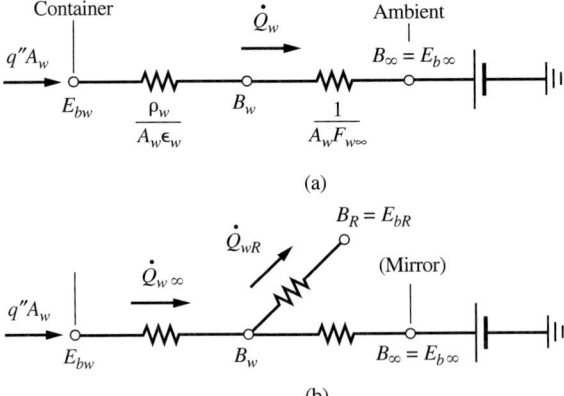

Figure 9.22 Example 9.4.

where $R_{w\infty}$ is the total resistance between the hemispherical wall and the ambient determined from the figure as

$$R_{w\infty} = \frac{\rho_w}{A_w \epsilon_w} + \frac{1}{A_w F_{w\infty}}$$

or, multiplying both sides by A_w and noting that $\rho_w = 1 - \epsilon_w$,

$$A_w R_{w\infty} = \left(\frac{1}{\epsilon_w} - 1\right) + \frac{1}{F_{w\infty}}. \tag{9.24}$$

The reciprocity relation between the wall and the black opening,

$$A_w F_{w\infty} = A_\infty F_{\infty w}, \tag{9.25}$$

gives, in view of $F_{\infty w} = 1$ (all of the radiation leaving the black "disk" is intercepted by the hemisphere),

$$F_{w\infty} = \frac{A_\infty}{A_w}. \tag{9.26}$$

Since $A_\infty = \pi r^2$ and $A_w = 4\pi r^2/2$,

$$\frac{A_\infty}{A_w} = \frac{1}{2}, \tag{9.27}$$

we obtain, in terms of Eq. (9.26), $F_{w\infty} = 1/2$. Then, from Eq. (9.24)

$$A_w R_{w\infty} = \frac{1 + \epsilon_w}{\epsilon_w}, \tag{9.28}$$

and combining with Eq. (9.23),

$$q'' = \left(\frac{\epsilon_w}{1 + \epsilon_w}\right) \sigma \left(T_w^4 - T_\infty^4\right)$$

or, solving for the wall temperature,

$$T_w = \left[\left(\frac{1 + \epsilon_w}{\epsilon_w}\right) \frac{q''}{\sigma} + T_\infty^4\right]^{1/4}. \tag{9.29}$$

In terms of a system surrounding the container, the first law gives for the heat flux through the opening

$$q A_\infty = q'' A_w \tag{9.30}$$

or, in terms of Eq. (9.27),

$$q = 2q''. \tag{9.31}$$

Next we determine the wall temperature and the heat flux when the container is closed with an ideal reflector (mirror surface with $\rho = 1$) having a concentric hole of radius r_h. Following steps 1 through 4 in Table 9.1, the equivalent circuit is constructed in Fig. 9.22(b). For the mirror, $\rho = 1$ and the surface resistance becomes infinity, indicating that, in view of Eq. (9.20), no net heat flows through the reflector. Thus, the branch $B_w - B_R$ does not carry a net energy, Fig. 9.22(b) essentially simplifies to Fig. 9.22(a), and Eqs. (9.23) through (9.26) continue to apply. The difference between the two cases is the modified $F_{w\infty}$ due to the reflector. The area ratio is now

$$\frac{A_\infty}{A_w} = \frac{\pi r_h^2}{4\pi r^2/2} = \frac{1}{2}\left(\frac{r_h}{r}\right)^2, \qquad (9.32)$$

which yields, in terms of Eq. (9.26),

$$F_{w\infty} = \frac{1}{2}\left(\frac{r_h}{r}\right)^2.$$

Substituting this relation into Eq. (9.24),

$$A_w R_{w\infty} = \left(\frac{1}{\epsilon_w} - 1\right) + 2\left(\frac{r}{r_h}\right)^2$$

or

$$A_w R_{w\infty} = \frac{1+\epsilon_w}{\epsilon_w} + 2\left[\left(\frac{r}{r_h}\right)^2 - 1\right], \qquad (9.33)$$

which shows the effect of the reflector on Eq. (9.28). Then, in terms of Eq. (9.23), the wall temperature is

$$T_w = \left\{\left(\frac{1+\epsilon_w}{\epsilon_w} + 2\left[\left(\frac{r}{r_h}\right)^2 - 1\right]\right)\frac{q''}{\sigma} + T_\infty^4\right\}^{1/4}, \qquad (9.34)$$

which is higher than the value of the wall temperature obtained in the absence of a mirror. Combination of Eqs. (9.30) and (9.32) readily gives the heat flux through the hole,

$$q = 2\left(\frac{r}{r_h}\right)^2 q'', \qquad (9.35)$$

exhibiting an enhancement of $(r/r_h)^2$ relative to the case without a reflector (Eq. 9.31).

(b) Radiant energy now reaches also the ambient through the disk. The disk emits radiation on both the inside and outside of the collimator. The equivalent circuit is shown in Fig. 9.23(a), which can be reduced to the one shown in Fig. 9.23(b). The objective of this part is to determine the heat flux through the hole. First we need to calculate $\dot{Q}_h$. At node B_w, the sum of the currents yields

$$\dot{Q}_{w\infty} = \dot{Q}_d + \dot{Q}_h \qquad (9.36)$$

and the resistances connected in parallel between B_w and B_∞ are

$$R_d = \frac{1}{A_w F_{wd}} + 2\frac{\rho_d}{A_d \epsilon_d} + \frac{1}{A_d F_{d\infty}}, \qquad R_h = \frac{1}{A_w F_{w\infty}}.$$

Note that $F_{dw} = 1$, $F_{\infty w} \equiv F_{hw} = 1$, $F_{d\infty} = 1$, and

$$A_w F_{wd} = A_d F_{dw} = A_d, \qquad A_w F_{w\infty} = A_\infty F_{\infty w} = A_\infty \equiv A_h,$$

where $A_d = \pi(r^2 - r_h^2)$ and $A_h = \pi r_h^2$.

Inserting these view factors into R_d and R_h, we get

$$R_d = \frac{1}{A_d}\left(1 + 2\frac{1-\epsilon_d}{\epsilon_d} + 1\right) = \frac{2}{A_d\epsilon_d}, \quad R_h = \frac{1}{A_h}. \tag{9.37}$$

From the equality of potential for the two parallel branches between B_w and $E_{b\infty}$,

$$\dot{Q}_d R_d = \dot{Q}_h R_h$$

or,

$$\dot{Q}_d = \frac{R_h}{R_d}\dot{Q}_h$$

and, in view of Eq. (9.36),

$$\dot{Q}_h = \frac{\dot{Q}_{w\infty}}{\dfrac{R_h}{R_d}+1}. \tag{9.38}$$

From Eq. (9.37)

$$\frac{R_h}{R_d} = \frac{\epsilon_d}{2}\frac{A_d}{A_h} = \frac{\epsilon_d}{2}\left[\left(\frac{r}{r_h}\right)^2 - 1\right],$$

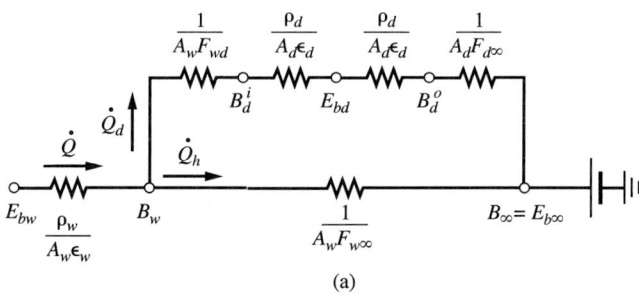

(a)

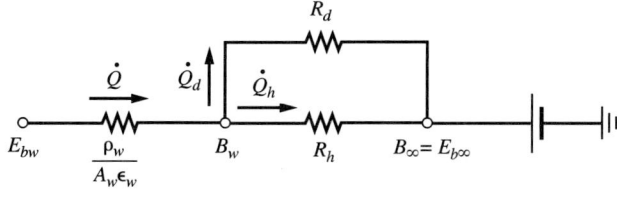

(a)

Figure 9.23 Example 9.4.

which leads, in view of Eq. (9.38), $\dot{Q}_h = qA_h$, and $\dot{Q}_{w\infty} = q''A_w$, to

$$q = \frac{2\left(\dfrac{r}{r_h}\right)^2 q''}{\dfrac{\epsilon_d}{2}\left[\left(\dfrac{r}{r_h}\right)^2 - 1\right] + 1}. \tag{9.39}$$

Equation (9.39) shows, relative to Eq. (9.35), a decrease in the heat flux through the hole. What would happen if the disk were a black surface? How does this compare to the container without a mirror?

◆

EXAMPLE 9.5

Consider two closely located parallel flat plates having temperatures T_1, T_2 and emissivities ϵ_1, ϵ_2. Place between these plates **(a)** a third flat plate with emissivity ϵ_0, **(b)** a flat window screen, η being the total hole area relative to the apparent surface area. Determine the reduction in the radiation heat transfer between the plates.

(a) Flat-plate partition.

The analogous electrical circuit is shown in Fig. 9.24. The resistance equivalent to the six resistances connected in series between potentials E_{b1} and E_{b2} is

$$R_{12} = \frac{1}{A}\left(\frac{\rho_1}{\epsilon_1} + \frac{1}{F_{10}} + 2\frac{\rho_0}{\epsilon_0} + \frac{1}{F_{20}} + \frac{\rho_2}{\epsilon_2}\right),$$

which may be rearranged, in view of $F_{10} = F_{20} = 1$ and $\epsilon + \rho = 1$, as

$$R_{12} = \frac{1}{A}\left[\frac{1}{\epsilon_1} + \frac{1}{\epsilon_2} + 2\left(\frac{1}{\epsilon_0} - 1\right)\right]$$

or, in terms of

$$\frac{1}{\epsilon_{10}} = \frac{1}{\epsilon_1} + \frac{1}{\epsilon_0} - 1 \quad \text{and} \quad \frac{1}{\epsilon_{20}} = \frac{1}{\epsilon_2} + \frac{1}{\epsilon_0} - 1,$$

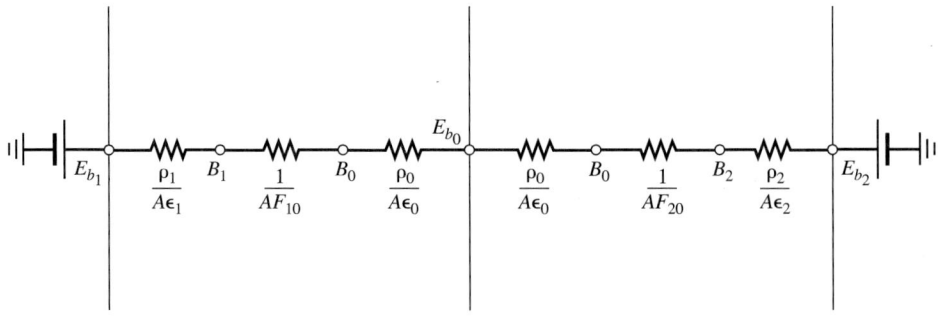

Figure 9.24 Flat-plate partition.

as

$$R_{12} = \frac{1}{A}\left(\frac{1}{\epsilon_{10}} + \frac{1}{\epsilon_{20}}\right) = \frac{1}{A}\left(\frac{\epsilon_{10} + \epsilon_{20}}{\epsilon_{10}\epsilon_{20}}\right).$$

Then

$$\dot{Q}_{1(0)2} = \frac{E_{b1} - E_{b2}}{R_{12}} \tag{9.22}$$

or

$$q_{1(0)2} = \frac{\dot{Q}_{1(0)2}}{A} = \left(\frac{\epsilon_{10}\epsilon_{20}}{\epsilon_{10} + \epsilon_{20}}\right)(E_{b1} - E_{b2}). \tag{9.40}$$

The reduction in the radiation heat transfer resulting from the use of a partition, after introducing $1/\epsilon_{12} = 1/\epsilon_1 + 1/\epsilon_2 - 1$ into Eq. (9.5), may be written as

$$\frac{q_{12} - q_{1(0)2}}{q_{12}} = 1 - \frac{1}{\epsilon_{12}}\left(\frac{\epsilon_{10}\epsilon_{20}}{\epsilon_{10} + \epsilon_{20}}\right). \tag{9.41}$$

For a partition and plates with identical emissivities, $\epsilon_{10} = \epsilon_{20} = \epsilon_{12}$ and Eq. (9.41) reduces to $1 - 1/2 = 1/2$ of the original heat transfer involving no partition.

The temperature of the partition may be obtained by expressing the heat transfer (electric current) between the two plates in terms of the potential drops $E_{b1} - E_{b0}$ and $E_{b0} - E_{b2}$. Thus,

$$\epsilon_{10}(E_{b1} - E_{b0}) = \epsilon_{20}(E_{b0} - E_{b2}),$$

which may be rearranged for E_{b0} to give

$$E_{b0} = \frac{\epsilon_{10}E_{b1} + \epsilon_{20}E_{b2}}{\epsilon_{10} + \epsilon_{20}}$$

or, in terms of $E_b = \sigma T^4$,

$$T_0^4 = \frac{\epsilon_{10}T_1^4 + \epsilon_{20}T_2^4}{\epsilon_{10} + \epsilon_{20}}. \tag{9.42}$$

For identical emissivities, $\epsilon_{10} = \epsilon_{20}$ and Eq. (9.42) reduces to

$$T_0^4 = \frac{1}{2}\left(T_1^4 + T_2^4\right), \tag{9.43}$$

the arithmetic mean of the fourth power of the absolute temperatures. Consequently, the temperature of the partition is closer to that of the plate with higher temperature.

(b) Window-screen partition.

Radiant energy incident on a partition with holes partly escapes the partition and is partly absorbed by and reflected from the partition. Accordingly, the analogous electrical circuit, including the energy that escapes the partition, may be constructed as shown in Fig. 9.25.

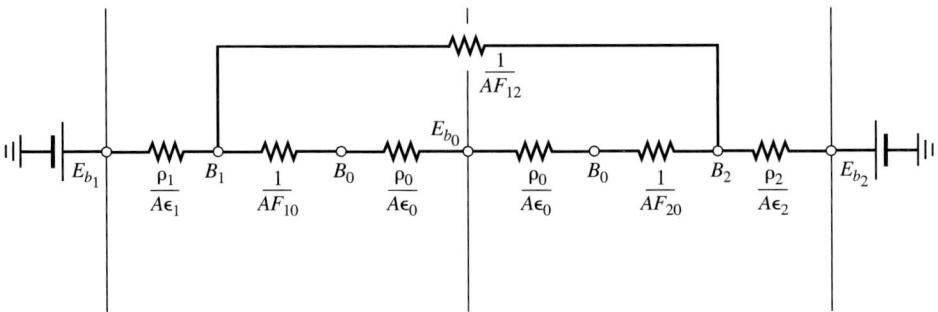

Figure 9.25 Window-screen partition.

The resistance equivalent to that of the lower circuit (involving four resistances connected in series) between potentials B_1 and B_2 is

$$\frac{1}{AF_{10}} + 2\frac{\rho_0}{A_0\epsilon_0} + \frac{1}{AF_{20}},$$

where A_0 is the solid area of the screen. If $A_{hole}/A = \eta$, then $A_0/A = 1 - \eta$. This resistance connected in parallel to the upper circuit (involving resistance $1/AF_{12}$) gives the total resistance between the same potentials,

$$AF_{12} + \frac{1}{(1/AF_{10}) + 2(\rho_0/A_0\epsilon_0) + (1/AF_{20})}.$$

Finally, this resistance connected in series to resistances $\rho_1/A\epsilon_1$ and $\rho_2/A\epsilon_2$ yields the total resistance between potentials E_{b1} and E_{b2},

$$R_{12} = \frac{\rho_1}{A\epsilon_1} + \frac{1}{AF_{12} + \dfrac{1}{(1/AF_{10}) + 2(\rho_0/A_0\epsilon_0) + (1/AF_{20})}} + \frac{\rho_2}{A\epsilon_2}. \tag{9.44}$$

Noting that $F_{12} = \eta$, $F_{10} = F_{20} = 1 - \eta$, and $A_0/A = 1 - \eta$, Eq. (9.44) may be rearranged as

$$R_{12} = \frac{1}{A}\left[\frac{\rho_1}{\epsilon_1} + \frac{1}{\eta + \dfrac{1-\eta}{2(1+\rho_0/\epsilon_0)}} + \frac{\rho_2}{\epsilon_2}\right] \tag{9.45}$$

or, in terms of $1 = \epsilon_0 + \rho_0$, as

$$R_{12} = \frac{1}{A}\left[\frac{\rho_1}{\epsilon_1} + \frac{2}{2\eta + (1-\eta)\epsilon_0} + \frac{\rho_2}{\epsilon_2}\right]. \tag{9.46}$$

Equation (9.45) for $\eta = 1$ reduces to

$$R_{12} = \frac{1}{A}\left(\frac{\rho_1}{\epsilon_1} + 1 + \frac{\rho_2}{\epsilon_2}\right), \tag{9.47}$$

which corresponds to the radiative resistance between two parallel plates in the absence of any partition, and for $\eta = 0$ it reduces to

$$R_{12} = \frac{1}{A}\left(\frac{\rho_1}{\epsilon_1} + \frac{2}{\epsilon_0} + \frac{\rho_2}{\epsilon_2}\right), \tag{9.48}$$

which corresponds to the radiative resistance between two parallel plates with an opaque partition in between. Note that a **glass partition** with negligible reflectivity is a special case of the opaque partition with holes. Letting the transmissivity of the glass be τ_0, and inserting $\rho_0 = 0$ and replacing η with τ_0 in Eq. (9.45), we have

$$R_{12} = \frac{1}{A}\left(\frac{\rho_1}{\epsilon_1} + \frac{2}{1+\tau_0} + \frac{\rho_2}{\epsilon_2}\right). \tag{9.49}$$

Then, the radiation heat transfer between two parallel plates separated by a partition is

$$\dot{Q}_{1(0)2} = \frac{E_{b1} - E_{b2}}{R_{12}}, \tag{9.50}$$

where R_{12} is given by Eq. (9.45) or by its simplified forms [Eqs. (9.47) or (9.48) or (9.49)], depending on the nature of the partition.

We now proceed to the temperature of the partition. First, $\dot{Q}_{1(0)2}$ expressed in terms of the surface resistances of the two plates gives

$$\dot{Q}_{1(0)2} = \frac{E_{b1} - B_1}{\rho_1/A\epsilon_1} = \frac{B_2 - E_{b2}}{\rho_2/A\epsilon_2}. \tag{9.51}$$

Second, the Kirchoff current law applied to the node of potential B_1 yields

$$\dot{Q}_{1(0)2} = \dot{Q}_{12} + \dot{Q}_{10}$$

or, in terms of appropriate potential drops,

$$\dot{Q}_{1(0)2} = \frac{B_1 - B_2}{1/AF_{12}} + \frac{B_1 - E_{b0}}{1/AF_{10} + \rho_0/A_0\epsilon_0}. \tag{9.52}$$

Solving Eq. (9.51) for B_1 and B_2 in terms of $\dot{Q}_{1(0)2}$, inserting B_1 and B_2 into Eq. (9.52), and rearranging the result for E_{b0} leads to the temperature of the partition. Details leading to the explicit form of this temperature are left to the reader.

◆

In terms of the foregoing examples, we have so far tested our knowledge of the fundamental aspects of enclosure radiation. In terms of the following examples we now wish to develop some appreciation on the numerical aspects of enclosure radiation.

EXAMPLE 9.6[10]

A heat flux q'' applied to a hot plate keeps the plate at $T_w = 600$K in an ambient at $T_\infty = 300$K [Fig. 9.26(a)]. The emissivity of the plate surface is $\epsilon_w = 0.9$. **(a)** We wish to evaluate q'' (W/m^2). **(b)** The plate is covered by a hemispherical lid [Fig. 9.26(b)]. The inside and outside surface emissivities of the lid are $\epsilon_0^i = 0.1$ and $\epsilon_0^0 = 0.8$, respectively. The plate temperature is to be kept the same at $T_w = 600$ K. We wish to determine the reduction in q'' and the temperature T_0 of the lid.

(a) Recalling Ex. 9.4(a) and in terms of the circuit shown in Fig. 9.27(a), we have

$$\dot{Q}_{w\infty} = \frac{E_{bw} - E_{b\infty}}{R_{w\infty}}$$

or, in terms of $E_b = \sigma T^4$,

$$q'' = \frac{\dot{Q}_{w\infty}}{A_w} = \frac{\sigma(T_w^4 - T_\infty^4)}{A_w R_{w\infty}}, \qquad (9.53)$$

where, in terms of Eq. (9.24),

$$A_w R_{w\infty} = \left(\frac{1}{\epsilon_w} - 1\right) + \frac{1}{F_{w\infty}},$$

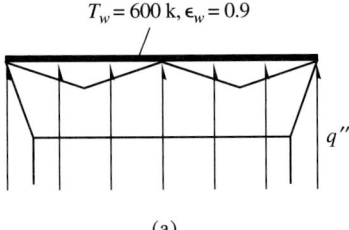

(a)

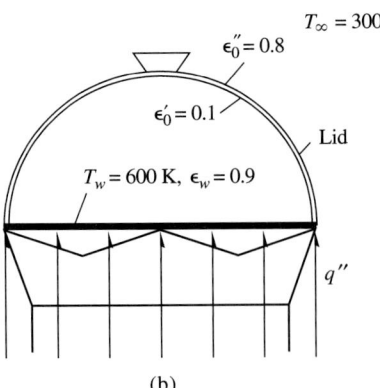

(b)

Figure 9.26 Hot plate.

---

[10] The FORTRAN program EX9–6.F is listed in the appendix of this chapter.

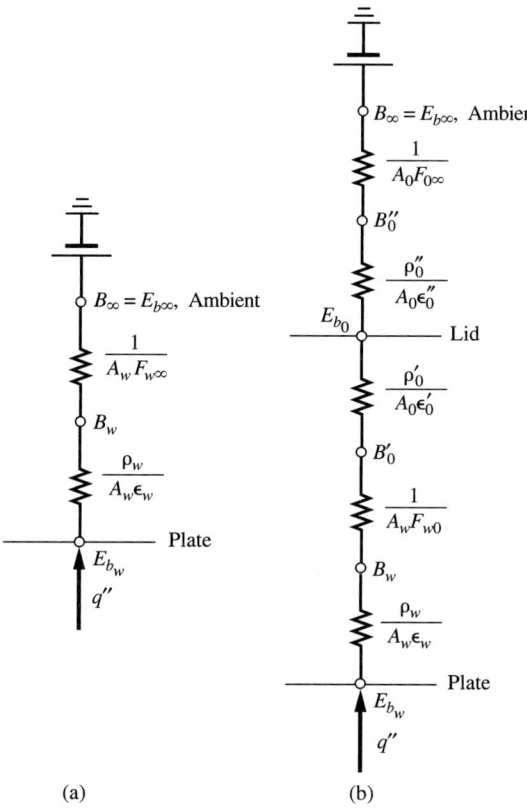

Figure 9.27 Hot plate.

which, in view of $F_{w\infty} = 1$, reduces to

$$A_w R_{w\infty} = 1/\epsilon_w = 1/0.9. \qquad (9.54)$$

Inserting Eq. (9.54) into Eq. (9.53), and noting that

$$\sigma T^4 = 5.67 \times 10^{-8}[\text{W/m}^2\cdot\text{K}^4]T^4[\text{K}^4] = 5.67 \times \left(\frac{T}{100}\right)^4 \text{ W/m}^2,$$

we get

$$q'' = 0.9 \times 5.67 \times (6^4 - 3^4) \cong 6{,}200 \text{ W/m}^2.$$

(b) In terms of the circuit shown in Fig. 9.27(b), we have

$$\dot{Q}_{w\infty} = \frac{E_{bw} - E_{b\infty}}{R_{w\infty}} = \frac{E_{b0} - E_{b\infty}}{R_{0\infty}}, \qquad (9.55)$$

where

$$R_{w\infty} = \frac{\rho_w}{A_w \epsilon_w} + \frac{1}{A_w F_{w0}} + \frac{\rho_0^i}{A_0 \epsilon_0^i} + \frac{\rho_0^0}{A_0 \epsilon_0^0} + \frac{1}{A_0 F_{0\infty}}, \quad R_{0\infty} = \frac{\rho_0^0}{A_0 \epsilon_0^0} + \frac{1}{A_0 F_{0\infty}}.$$

Dividing the first equality of Eq. (9.55) by A_w,

$$q'' = \frac{\dot{Q}_{w\infty}}{A_w} = \frac{\sigma(T_w^4 - T_\infty^4)}{A_w R_{w\infty}},$$

where

$$A_w R_{w\infty} = \left(\frac{1}{\epsilon_w} - 1\right) + \frac{1}{F_{w0}} + \frac{A_w}{A_0}\left(\frac{1}{\epsilon_0^i} - 1 + \frac{1}{\epsilon_0^0} - 1 + \frac{1}{F_{0\infty}}\right),$$

which, in terms of $F_{w0} = 1$, $F_{0\infty} = 1$, and $A_w/A_0 = 1/2$, may be rearranged as

$$A_w R_{w\infty} = \frac{1}{\epsilon_w} + \frac{1}{2}\left(\frac{1}{\epsilon_0^i} + \frac{1}{\epsilon_0^0} - 1\right)$$

or, in view of $\epsilon_w = 0.9$, $\epsilon_0^i = 0.1$, and $\epsilon_0^0 = 0.8$, as

$$A_w R_{w\infty} = \frac{1}{0.9} + \frac{1}{2}\left(\frac{1}{0.1} + \frac{1}{0.8} - 1\right) = 6.236 = 1/0.16.$$

Then, the heat flux is

$$q'' = 0.16 \times 5.67 \text{ W/m}^2\cdot\text{K}^4 \times (6^4 - 3^4) \text{ K}^4 \cong 1{,}102 \text{ W/m}^2$$

and the reduction in q'' is

$$\frac{6{,}200 - 1{,}102}{6{,}200} \times 100 \sim 82\%!$$

Cooking with a covered pot requires less energy input than with an uncovered pot.

For the temperature of the lid, dividing the second equality of Eq. (9.55) by A_w,

$$q'' = \frac{\sigma(T_0^4 - T_\infty^4)}{A_w R_{0\infty}},$$

where

$$A_w R_{0\infty} = \frac{A_w}{A_0}\left(\frac{1}{\epsilon_0^0} - 1 + \frac{1}{F_{0\infty}}\right),$$

or, in view of $F_{0\infty} = 1$,

$$A_w R_{0\infty} = \frac{A_w}{A_0}\frac{1}{\epsilon_0^0}$$

results in, for the heat flux,

$$q'' = \frac{A_0}{A_w}\epsilon_0^0 \sigma(T_0^4 - T_\infty^4),$$

which may be solved for T_0^4 to give

$$T_0^4 = T_\infty^4 + (q''/\epsilon_0^0 \sigma)(A_w/A_0),$$

or

$$\left(\frac{T_0}{100}\right)^4 = 3^4 \text{ K}^4 + \frac{1102 \text{ W/m}^2}{0.8 \times 5.67 \text{W/m}^2 \cdot \text{K}^4} \times \frac{1}{2} \cong 202.5 \text{ K}^4,$$

$$T_0 \cong 377 \text{ K}.$$

How would the results change if inside $\epsilon_0^i = 0.8$ and outside $\epsilon_0^o = 0.1$? What does this say about the best type of pot lid surface?

By the computer program provided, the interested reader may parametrically study the q'' reduction depending on various values of ϵ_0^i and ϵ_0^o. ◆

EXAMPLE 9.7[11]

An electrically heated long solid rod is to be used for the heating of an industrial building. A screen with a ratio of total hole area to total apparent (holes + screen) surface area of $\eta = 0.9$, and inner and outer emissivities of $\epsilon_s^i = 0.08$ and $\epsilon_s^o = 0.9$, is to be placed around the heater to shield the surroundings from the hot surface of the heater (Fig. 9.28). The ambient temperature is $T_\infty = 27\ °\text{C}$. Assume the heater radius to be $r_w = 1$ cm, emissivity $\epsilon_w = 0.8$, the energy generation $u''' = 3$ MW/m^3, and the radius of screen to be $r_s = 3$ cm. Neglecting the effect of convection, we wish to determine **(a)** the increase in the surface temperature of heater and **(b)** the screen temperature.

(a) In the absence of the screen, the heat flow to the ambient from the heater of length ℓ [Fig. 9.29(a)] is

$$\dot{Q}_{w\infty} = u''' A\ell = \frac{E_{bw} - E_{b\infty}}{R_{w\infty}} = q_w A_w,$$

$R_{w\infty}$ being the total resistance. Dividing by A_w and noting $E_b = \sigma T^4$, $A = \pi r_w^2$, and $A_w = 2\pi r_w \ell$, thus $A\ell/A_w = r_w/2$, and the heat flux to ambient is

$$q_w = \frac{u''' r_w}{2} = \frac{\sigma \left(T_w^4 - T_\infty^4\right)}{A_w R_{w\infty}}$$

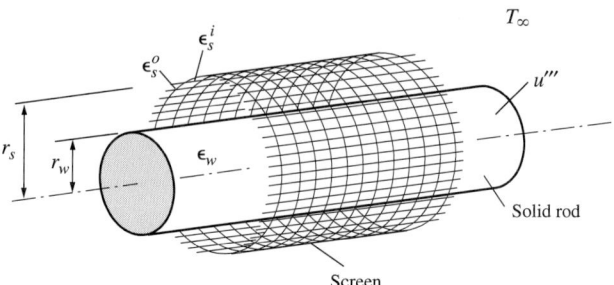

Figure 9.28 Example 9.7.

---

[11] The FORTRAN program EX9–7.F is listed in the appendix of this chapter.

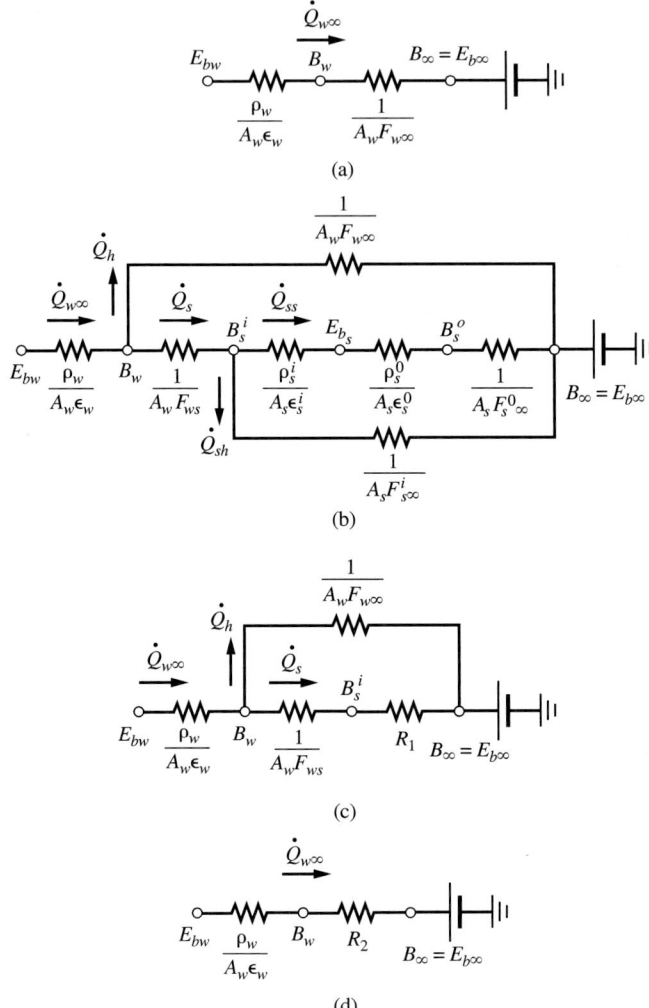

Figure 9.29 Example 9.7.

or, solving for the heater surface temperature,

$$T_w = \left[A_w R_{w\infty} \left(\frac{u''' r_w}{2\sigma} \right) + T_\infty^4 \right]^{1/4}. \tag{9.56}$$

Substituting $F_{w\infty} = 1$ into Eq. (9.24) for the resistance, we get

$$A_w R_{w\infty} = \frac{1}{\epsilon_w}. \tag{9.57}$$

Inserting $A_w R_{w\infty}$ into Eq. (9.56) yields

$$T_w = \left(\frac{1}{\epsilon_w} \frac{u''' r_w}{2\sigma} + T_\infty^4 \right)^{1/4}, \qquad (9.58)$$

which gives, in terms of the given data,

$$T_w = \left(\frac{3 \times 10^6 \text{ W/m}^3 \times 0.01 \text{ m}}{0.8 \times 2 \times 5.67 \times 10^{-8} \text{ W/m}^2 \cdot \text{K}^4} + 300^4 \text{ K}^4 \right)^{1/4}$$

or

$$T_w = 762.9 \text{ K}.$$

Next we proceed to the heater surrounded by the screen. Following the first four steps of Table 9.1, an equivalent electrical circuit is constructed as shown in Fig. 9.29(b). Note that the heater sees the ambient through the holes of screen and that the inner surface of the screen sees the ambient through these holes as well. The outer surface of the screen also sees the ambient. These facts respectively translate to the upper and lower branches of the circuit shown in Fig. 9.29(b).

Combining the resistances for the lower branch (Fig. 9.29c),

$$\frac{1}{R_1} = \frac{1}{\frac{1}{A_s F_{s\infty}^i}} + \frac{1}{\frac{\rho_s^i}{A_s \epsilon_s^i} + \frac{\rho_s^0}{A_s \epsilon_s^0} + \frac{1}{A_s F_{s\infty}^0}}$$

or

$$\frac{1}{R_1} = A_s \left(F_{s\infty}^i + \frac{1}{\frac{1}{\epsilon_s^i} + \frac{1}{\epsilon_s^0} - 2 + \frac{1}{F_{s\infty}^0}} \right), \qquad (9.59)$$

where superscripts i and 0 are used to distinguish the inner and outer surfaces of the screen. Next, combining the resistances between B_w and B_∞ for an equivalent resistance [Fig. 9.29(d)],

$$\frac{1}{R_2} = A_w F_{w\infty} + \frac{1}{R_1 + \frac{1}{A_w F_{ws}}}. \qquad (9.60)$$

Then the total resistance is

$$R_{w\infty} = \frac{\rho_w}{A_w \epsilon_w} + R_2. \qquad (9.61)$$

The three view factors involved with the foregoing resistances are

$$F_{s\infty}^0 = 1, \quad F_{w\infty} = \eta, \quad F_{ws} = 1 - \eta,$$

while $F_{s\infty}^i$ requires some manipulations as follows. The summation rule for the inner surface of the screen gives

$$F_{ss} + F_{sw} + F_{s\infty}^i = 1. \qquad (9.62)$$

The reciprocity rule between the wall and the screen,

$$A_s F_{sw} = A_w F_{ws},$$

yields, in view of $A_w = 2\pi r_w \ell$ and $A_s = 2\pi r_s (1-\eta)\ell$,

$$F_{sw} = \frac{r_w}{r_s}. \tag{9.63}$$

From Eqs. (9.62) and (9.63), assuming that the holes are uniformly distributed over the shell,

$$F^i_{s\infty} = \eta\left(1 - \frac{r_w}{r_s}\right), \quad F_{ss} = (1-\eta)\left(1 - \frac{r_w}{r_s}\right).$$

Then Eqs. (9.59), (9.60), and (9.61) result in

$$R_{w\infty} = \frac{1}{A_w}\frac{1-\epsilon_w}{\epsilon_w} + \cfrac{1}{A_w\eta + \cfrac{1}{A_s\left[\eta\left(1 - \cfrac{r_w}{r_s}\right) + \cfrac{1}{\cfrac{1}{\epsilon_s^i} + \cfrac{1}{\epsilon_s^0} - 1}\right]} + \cfrac{1}{A_w(1-\eta)}}$$

or, after some manipulations,

$$A_w R_{w\infty} = \frac{1}{\epsilon_w} + \cfrac{(1-\eta)\cfrac{r_w}{r_s}}{\eta + \cfrac{1}{\cfrac{1}{\epsilon_s^i} + \cfrac{1}{\epsilon_s^0} - 1}}, \tag{9.64}$$

which shows an increase in the total resistance relative to the case without screen, Eq. (9.57). Inserting $A_w R_{w\infty}$ into Eq. (9.56), we get, for the given data,

$$T_w = \left[\frac{3 \times 10^6 \text{ W/m}^3 \times 0.01 \text{ m}}{2 \times 5.67 \times 10^{-8} \text{ W/m}^2 \cdot \text{K}^4}\left(\frac{1}{0.8} + \cfrac{(1-0.9)/3}{0.9 + \cfrac{1}{\cfrac{1}{0.08} + \cfrac{1}{0.9} - 1}}\right) + 300^4 \text{ K}^4\right]^{1/4},$$

or,

$$T_w \cong 768 \text{ K}.$$

The screen increases the surface temperature of the heater, as expected. Note that for a solid screen ($\eta = 0$), Eq. (9.64) reduces to

$$A_w R_{w\infty} = \frac{1}{\epsilon_w} + \frac{r_w}{r_s}\left(\frac{1}{\epsilon_s^i} + \frac{1}{\epsilon_s^0} - 1\right)$$

and, in view of Eq. (9.56), the surface temperature becomes

$$T_w \cong 1{,}097.5 \text{ K.}$$

In the absence of any screen, $\eta = 1$, and Eq. (9.64) reduces to Eq. (9.57).

(b) To determine the screen temperature T_s we need to know the heat flow $\dot{Q}_{ss}$ through E_{bs} [Fig. 9.29(b)]. However, there are several intermediate steps. First consider the radiation through the holes,[12]

$$\dot{Q}_h = \frac{B_w - E_{b\infty}}{1/A_w F_{w\infty}} = A_w \eta (B_w - E_{b\infty}), \tag{9.65}$$

where B_w is determined from

$$\dot{Q}_{w\infty} = \frac{E_{bw} - B_w}{\rho_w/A_w \epsilon_w} = A_w \left(\frac{\epsilon_w}{1 - \epsilon_w}\right)(E_{bw} - B_w)$$

or, rearranging and noting $\dot{Q}_{w\infty} = u''' A\ell$,

$$B_w = E_{bw} - \left(\frac{1 - \epsilon_w}{\epsilon_w}\right)\frac{u''' r_w}{2},$$

which gives, in terms of the data and T_w from part (a),

$$B_w = 5.67 \times \left(\frac{768}{100}\right)^4 \text{W/m}^2 - \left(\frac{1 - 0.8}{0.8}\right)\frac{3 \times 10^6 \text{ W/m}^3 \times 0.01 \text{ m}}{2}$$

or

$$B_w \cong 15{,}970 \text{ W/m}^2. \tag{9.66}$$

Inserting Eq. (9.66) into (9.65),

$$\dot{Q}_h = 2\pi 0.01\ell(\text{m}^2)0.9\left[15{,}970 - 5.67\left(\frac{300}{100}\right)^4\right](\text{W/m}^2)$$

or

$$\dot{Q}_h \cong 877.1\ell(\text{W}). \tag{9.67}$$

Solving for $\dot{Q}_s$ from [Fig. 9.29(b),(c)],

$$\dot{Q}_{w\infty} = u''' A\ell = \dot{Q}_h + \dot{Q}_s$$

---

[12] Note that, rather than obtaining $\dot{Q}_h$ in terms of B_w from Eq. (9.65), we could have used the equality of potentials for the two branches between B_w and $E_{b\infty}$ in Fig. 9.29(b),(c). However, we avoided this option because of the need for manipulations with a number of resistances.

gives
$$\dot{Q}_s = 3 \times 10^6 \, (\text{W/m}^3) \, \pi (0.01)^2 \ell(\text{m}^3) - 877.1\ell(\text{W})$$

or
$$\dot{Q}_s = 65.38\ell(\text{W}). \tag{9.68}$$

This heat flow reaches the ambient partly through the screen material, $\dot{Q}_{ss}$, and partly through the holes, $\dot{Q}_{sh}$,
$$\dot{Q}_s = \dot{Q}_{ss} + \dot{Q}_{sh}. \tag{9.69}$$

The radiation from the inner surface of the screen to the ambient through the holes is
$$\dot{Q}_{sh} = \frac{B_s^i - E_{b\infty}}{1/A_s F_{s\infty}^i}$$

or, in terms of $A_s = 2\pi r_s \ell (1 - \eta)$ and $F_{s\infty}^i = \eta(1 - r_w/r_s)$,
$$\dot{Q}_{sh} = 2\pi r_s \ell \eta (1 - \eta)(1 - r_w/r_s)\left(B_s^i - E_{b\infty}\right), \tag{9.70}$$

where B_s^i is obtained from
$$\dot{Q}_s = \frac{B_w - B_s^i}{1/A_w F_{ws}},$$

which gives
$$B_s^i = B_w - \frac{1}{2\pi r_w (1-\eta)\ell} \dot{Q}_s$$

and, in terms of B_w and $\dot{Q}_s$ from Eqs. (9.66) and (9.68),
$$B_s^i = 15{,}970 \, \text{W/m}^2 - \frac{1}{2\pi 0.01 \text{m}(1-0.9)} 65.38 \, \text{W/m}$$

or
$$B_s^i = 5{,}564.6 \, \text{W/m}^2. \tag{9.71}$$

Inserting Eq. (9.71) into Eq. (9.70),
$$\dot{Q}_{sh} = 2\pi 0.03(\text{m}) 0.9 (1 - 0.9)(1 - 1/3)\ell(\text{m}) \left(5{,}564.6 - 5.67 \times \left(\frac{300}{100}\right)^4\right) (\text{W/m}^2)$$

or
$$\dot{Q}_{sh} = 57.74\ell(\text{W}). \tag{9.72}$$

Equation (9.69) yields, in terms of Eqs. (9.68) and (9.72),
$$\dot{Q}_{ss} = (65.38 - 57.74)\ell(\text{W}) = 7.64\ell(\text{W}). \tag{9.73}$$

Finally, for T_s, expressing $\dot{Q}_{ss}$ over the resistance between B_s^i and E_{bs},
$$\dot{Q}_{ss} = \frac{B_s^i - E_{bs}}{\rho_s^i/A_s \epsilon_s^i},$$

and, solving for E_{bs},

$$E_{bs} = \sigma T_s^4 = B_s^i - \frac{1}{A_s}\left(\frac{1}{\epsilon_s^i} - 1\right)\dot{Q}_{ss}$$

gives, in view of Eqs. (9.71) and (9.73),

$$E_{bs} = 5{,}564.6 \text{ W/m}^2 - \frac{1}{2\pi \times 0.03\text{m}(1-0.6)}\left(\frac{1}{0.08}-1\right)7.64 \text{ W/m}$$

or

$$E_{bs} = 903.4 \text{ W/m}^2.$$

Then the screen temperature is

$$T_s = \left(\frac{903.4 \text{ W/m}^2}{5.67 \times 10^{-8} \text{ W/m}^2\cdot\text{K}^4}\right)^{1/4}$$

or

$$T_s \cong 355.3\text{K}.$$

If the emissivity of the inner surface of the screen were high—for example, $\epsilon_s^i = \epsilon_s^0 = 0.9$, the screen temperature would become $T_s \cong 483.7$ K, a substantial increase from 355.3 K. The present example illustrates the need of a low inner-surface emissivity for a reasonably low screen temperature. (What would happen to the screen temperature if the outer surface of screen were highly reflective? The variation of the heater emissivity does not affect the screen temperature. Why?)

By the computer program provided, the interested reader may parametrically study the effect of screen size and emissivity. ◆

EXAMPLE 9.8[13]

The surface temperature of a disk of diameter $D = 0.5$ m is kept at $T_w = 230\,°\text{C}$. The emissivity of the disk is $\epsilon_w = 0.8$. **(a)** We wish to determine the radiant heat transfer from one side of the disk to an ambient at a temperature $T_\infty = 25\,°\text{C}$. **(b)** Let the disk be attached to an insulated cylindrical shell ($D/L = 1$), as shown in Fig. 9.30. We wish to evaluate the change in radiant heat transfer from the disk, and the temperature of the shell.

(a) The electrical circuit appropriate for the first part of the problem is identical to that of Ex. 9.6. Consequently, we have from Eqs. (9.53) and (9.54)

$$\dot{Q}_{w\infty} = A_w\epsilon_w\sigma(T_w^4 - T_\infty^4) \tag{9.74}$$

or

$$\dot{Q}_{w\infty} = \frac{\pi}{4} \times 0.5^2 \text{ m}^2 \times 0.8 \times 5.67 \times (5.03^4 - 2.98^4) \text{ W/m}^2,$$

which gives

$$\dot{Q}_{w\infty} \cong 500 \text{ W}.$$

---

[13] The FORTRAN program EX9–8.F is listed in the appendix of this chapter.

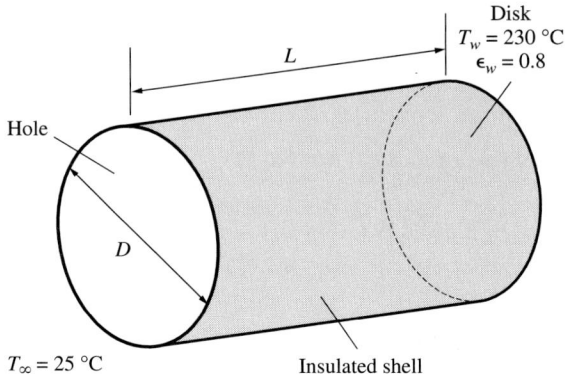

Figure 9.30 Disk and shell.

(b) For the second part of the problem, first we complete an enclosure by replacing the hole in Fig. 9.30 with a black surface at the ambient temperature. Then, from the electric circuit (Fig. 9.31) analogous to the radiation within the enclosure, we have

$$\dot{Q}_{w\infty} = \frac{E_{bw} - E_{b\infty}}{R_{w\infty}}, \qquad (9.75)$$

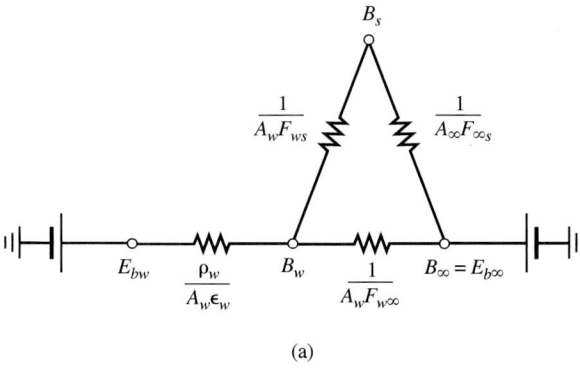

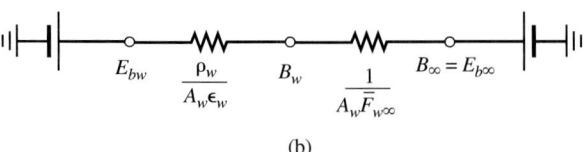

Figure 9.31 Disk and shell. (a) Actual circuit, (b) equivalent circuit.

where
$$R_{w\infty} = \frac{\rho_w}{A_w \epsilon_w} + R_1 \qquad (9.76)$$

and
$$\frac{1}{R_1} = A_w F_{w\infty} + \frac{1}{\frac{1}{A_w F_{ws}} + \frac{1}{A_\infty F_{\infty s}}}, \qquad (9.77)$$

R_1 denoting the effective view resistance. Note that no "color" was attached to the shell, given the fact that it is insulated. Noting that surfaces A_w and A_∞ do not see themselves, employing $F_{w\infty} + F_{ws} = 1$, $F_{\infty w} + F_{\infty s} = 1$, and $A_w F_{w\infty} = A_\infty F_{\infty w}$, Eq. (9.77) may be rearranged as

$$\frac{1}{A_w R_1} = \frac{A_\infty - A_w F_{w\infty}^2}{A_w + A_\infty - 2 A_w F_{w\infty}} \qquad (9.78)$$

or, in view of $A_w = A_\infty$, as

$$\frac{1}{R_1} = \frac{A_w (1 + F_{w\infty})}{2}, \qquad (9.79)$$

which shows an effective increase in the overall view factor. Figure 9.32 illustrates this increase in terms of four configurations. Finally, insertion of Eq. (9.79) into Eq. (9.76), considering $\epsilon + \rho = 1$, gives

$$R_{w\infty} = \frac{1}{A_w} \left(\frac{1}{\epsilon_w} - 1 + \frac{2}{1 + F_{w\infty}} \right), \qquad (9.80)$$

and rearranging Eq. (9.75) in terms of Eq. (9.80) results in

$$\dot{Q}_{w\infty} = A_w \epsilon_w \left(\frac{1 + F_{w\infty}}{1 + \epsilon_w + (1 - \epsilon_w) F_{w\infty}} \right) \sigma (T_w^4 - T_\infty^4). \qquad (9.81)$$

For $F_{w\infty} = 0.17$ (obtained from Fig. 9.13 for $D/L = 1$) and $\epsilon_w = 0.8$, together with A_w, and (T_w, T_∞), Eq. (9.81) yields

$$\dot{Q}_{w\infty} = \frac{\pi}{4} 0.5^2 (\text{m}^2) 0.8 \left(\frac{1 + 0.17}{1 + 0.8 + 0.2 \times 0.17} \right) 5.67 (5.03^4 - 2.98^4) \text{ W/m}^2$$

or
$$\dot{Q}_{w\infty} \cong 319 \text{ W}.$$

(Is this an expected result? Why is there a reduction in the radiation?)

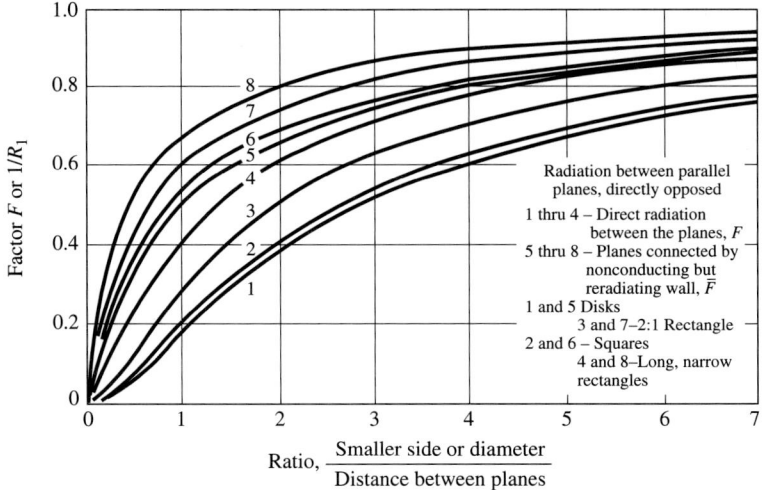

Figure 9.32 View factor F and interchange factor $\overline{F}$ for opposed parallel disks, squares, and Rectangles. Adapted from McAdams, *Heat Transmission*, 3d ed., McGraw-Hill, New York, 1954.

For the reduction in radiation from the disk, employing Eqs. (9.74) and (9.81), we have

$$\left(1 - \frac{1 + F_{w\infty}}{1 + \epsilon_w + (1 - \epsilon_w)F_{w\infty}}\right) \times 100\%$$

or

$$\frac{\epsilon_w(1 - F_{w\infty})}{1 + \epsilon_w + (1 - \epsilon_w)F_{w\infty}} \times 100\%$$

or, in terms of $\epsilon_w = 0.8$ and $F_{w\infty} = 0.17$,

$$\frac{0.8(1 - 0.17)}{1 + 0.8 + 0.2 \times 0.17} \times 100 \cong 36\%.$$

We now proceed to the evaluation of the shell temperature. From the actual circuit of Fig. 9.31 we have

$$\dot{Q}_{w\infty} = \frac{E_{bw} - B_w}{\rho_w/A_w\epsilon_w} = \dot{Q}' + \dot{Q}'', \tag{9.82}$$

where $\dot{Q}'$ is the current from B_w directly to B_∞, and $\dot{Q}''$ is the current from B_w via B_s to B_∞. Expressing these currents in terms of potential drop $B_w - B_\infty$ and $B_w - B_s$, respectively, we get

$$\dot{Q}_{w\infty} = \frac{E_{bw} - B_w}{\rho_w/A_w\epsilon_w} = \frac{B_w - B_\infty}{1/A_w F_{w\infty}} + \frac{B_w - B_s}{1/A_w F_{ws}}. \tag{9.83}$$

From the first equality of Eq. (9.83)

$$B_w = E_{bw} - \left(\frac{\dot{Q}_{w\infty}}{A_w}\right)\frac{\rho_w}{\epsilon_w},$$

which gives

$$B_w = 5.67 \times 5.03^4 \text{ W/m}^2 - \frac{319 \text{W}}{(\pi/4)0.5^2 \text{ m}^2} \times \frac{0.2}{0.8} \cong 3{,}223 \text{ W/m}^2.$$

From the second equality of Eq. (9.83)

$$\dot{Q}_{w\infty} = A_w F_{w\infty}(B_w - B_\infty) + A_w F_{ws}(B_w - B_s),$$

which may be rearranged as

$$B_s = \frac{1}{F_{ws}}\left(B_w - F_{w\infty}B_\infty - \frac{\dot{Q}_{w\infty}}{A_w}\right). \quad (9.84)$$

Noting that $B_\infty = E_{b\infty} = \sigma T_\infty^4$ and $F_{ws} = 1 - F_{w\infty} = 0.83$, we obtain from Eq. (9.84)

$$B_s = \frac{1}{0.83}\left(3{,}223 \text{ W/m}^2 - 0.17 \times 5.67 \times 2.98^4 \text{ W/m}^2 - \frac{319 \text{ W}}{(\pi/4)0.5^2 \text{ m}^2}\right) \cong 1{,}834 \text{ W/m}^2,$$

which, in view of $B_s = E_{bs} = \sigma T_s^4$ for an insulated surface, gives

$$1{,}834 = 5.67\left(\frac{T_s}{100}\right)^4$$

or, the shell temperature,

$$T_s \cong 424 \text{ K} = 151\,°\text{C}.$$

◆

EXAMPLE 9.9[14]

Reconsider the second part of the preceding example. Eliminate the insulation of the shell and assume the inner and outer surfaces of the shell to have the same emissivity, $\epsilon_s = 0.8$. Evaluate the radiant flux from the disk and the temperature of the shell. Compare this flux with that of the preceding example.

---

[14] The FORTRAN program EX9–9.F is listed in the appendix of this chapter.

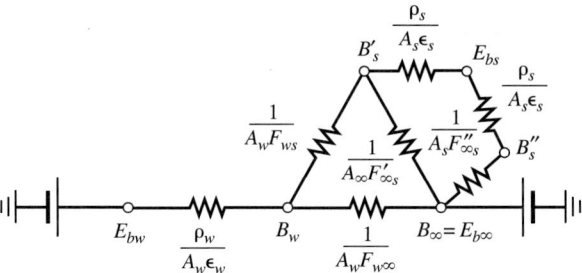

(a) Actual circuit

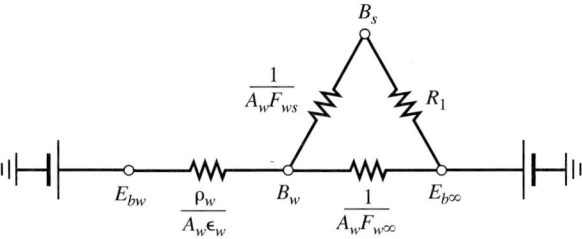

(b) Equivalent circuit

(c) Equivalent second circuit

Figure 9.33 (a) Actual circuit, (b) equivalent first circuit, (c) equivalent second circuit.

The electric circuit analogous to the problem may be obtained by modifying the circuit of Fig. 9.31 as shown in Fig. 9.33. The new circuit involves the inner and outer emissivities of the shell, as well as the view factor between the outer surface of the shell and the ambient. Thus, we have

$$\dot{Q}_{w\infty} = \frac{E_{bw} - E_{b\infty}}{R_{w\infty}}, \quad (9.85)$$

where, in terms of the equivalent second circuit,

$$R_{w\infty} = \frac{\rho_w}{A_w \epsilon_w} + R_2, \quad (9.86)$$

in terms of the equivalent first circuit,

$$\frac{1}{R_2} = A_w F_{w\infty} + \frac{1}{\dfrac{1}{A_w F_{ws}} + R_1}, \quad (9.87)$$

and, in terms of the actual circuit,

$$\frac{1}{R_1} = A_\infty F^i_{\infty s} + \frac{1}{2\dfrac{\rho_s}{A_s \epsilon_s} + \dfrac{1}{A_s F^0_{s\infty}}}, \qquad (9.88)$$

superscripts i and 0 denoting the inner and outer surfaces. Noting that $F^i_{\infty s} = 1 - F_{\infty w} = 1 - F_{w\infty} = 0.83$ and $F^0_{s\infty} = 1$, Eq. (9.88) may be reduced to

$$\frac{1}{R_1} = A_\infty \left[0.83 + \frac{A_s/A_\infty}{(2 - \epsilon_s)/\epsilon_s} \right] \qquad (9.89)$$

and, with $A_s/A_\infty = 4L/D = 4$ and $\epsilon_s = 0.8$, to

$$R_1 = \frac{1}{3.50 A_\infty}.$$

Thus, Eq. (9.87) becomes

$$\frac{1}{R_2} = A_w F_{w\infty} + \frac{1}{\dfrac{1}{A_w F_{ws}} + \dfrac{1}{3.50 A_\infty}},$$

which, in terms of $A_w = A_\infty$ and $F_{ws} = 1 - F_{w\infty}$, may be rearranged as

$$\frac{1}{R_2} = A_w \frac{3.50 + F_{w\infty}(1 - F_{w\infty})}{3.50 + (1 - F_{w\infty})}$$

and with $F_{w\infty} = 0.17$ as

$$R_2 = \frac{1}{0.841 A_w}.$$

Inserting this result into Eq. (9.86), we have

$$R_{w\infty} = \frac{1}{A_w} \left(\frac{\rho_w}{\epsilon_w} + \frac{1}{0.841} \right)$$

or, in terms of $\rho_w = 0.2$ and $\epsilon_w = 0.8$,

$$R_{w\infty} \cong \frac{1}{0.695 A_w}.$$

Finally, from Eq. (9.85),

$$\dot{Q}_{w\infty} = 0.695 A_w \sigma (T_w^4 - T_\infty^4)$$

or

$$\dot{Q}_{w\infty} = 0.695 \frac{\pi}{4} 0.5^2 \text{ m}^2 5.67 (5.03^4 - 2.98^4) \text{ W/m}^2$$

or

$$\dot{Q}_{w\infty} \cong 434 \text{ W}.$$

(Is this an expected result? Explain clearly!). The radiant heat flux relative to that of the preceding example is increased by

$$\frac{434 - 319}{319} \times 100 \cong 36\%.$$

◆

EXAMPLE 9.10[15]

Consider two parallel square plates separated apart by a distance equal to the side of each plate (Fig. 9.34). The temperature and the emissivity of the plates are $T_1 = 800$ K, $T_2 = 600$ K, and $\epsilon_1 = 0.8$, $\epsilon_2 = 0.5$, respectively. The ambient temperature is $T_\infty = 300$ K. We wish to determine the radiant heat exchange between the plates.

The analogous electric circuit is shown in Fig. 9.35. The four black surfaces at the ambient temperature that complete the enclosure are represented by a single node ($B_3 = E_{b3}$). In terms of the Maxwell loop currents[16] $\dot{Q}_1$, $\dot{Q}_2$, and $\dot{Q}_3$ indicated on this figure, the Kirchoff voltage law between potentials 1 and 2 gives

$$E_{b1} - E_{b2} = (\dot{Q}_1 + \dot{Q}_3)\frac{\rho_1}{A_1\epsilon_1} + \dot{Q}_1\left(\frac{1}{A_1 F_{12}}\right) + (\dot{Q}_1 - \dot{Q}_2)\frac{\rho_2}{A_2\epsilon_2}, \quad (9.90)$$

between potentials 2 and 3,

$$E_{b2} - E_{b3} = (\dot{Q}_2 - \dot{Q}_1)\frac{\rho_2}{A_2\epsilon_2} + \dot{Q}_2\left(\frac{1}{A_2 F_{23}}\right), \quad (9.91)$$

and between potentials 3 and 1,

$$E_{b3} - E_{b1} = -\dot{Q}_3\left(\frac{1}{A_3 F_{31}}\right) - (\dot{Q}_1 + \dot{Q}_3)\frac{\rho_1}{A_1\epsilon_1}. \quad (9.92)$$

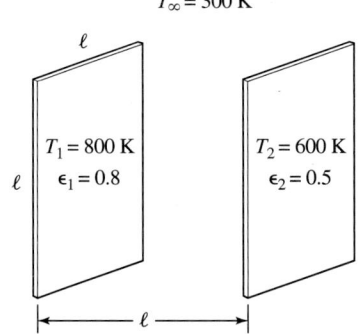

Figure 9.34

---

[15] The FORTRAN program EX9–10.F is listed in the appendix of this chapter.

[16] Details of the solution methods for electrical circuits by the Maxwell loop currents, Δ-Y conversions, and the Kirchoff current and voltage laws are assumed to be known to the student. Otherwise, refer to a standard textbook on electrical circuit theory.

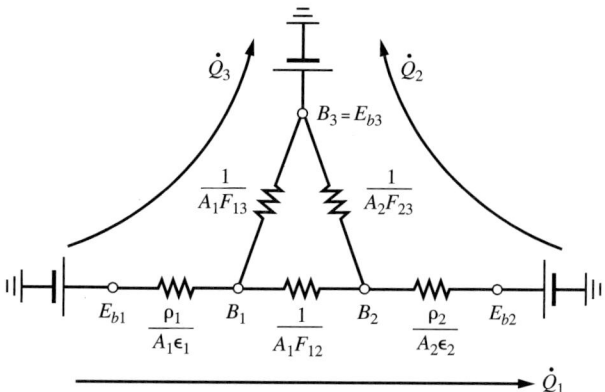

Figure 9.35 Loop currents.

For convenience in algebra, the relation

$$0 = \dot{Q}_1 \left(\frac{1}{A_1 F_{12}} \right) + \dot{Q}_2 \left(\frac{1}{A_2 F_{23}} \right) - \dot{Q}_3 \left(\frac{1}{A_3 F_{31}} \right) \quad (9.93)$$

obtained from the sum of the foregoing three equations is to be considered in place of Eq. (9.90).

Introducing $q_1 = \dot{Q}_1/A_1$, $q_2 = \dot{Q}_2/A_2$, and $q_3 = \dot{Q}_3/A_3$, noting that $A_1 = A_2 = A_3/4$, reading $F_{12} = 0.2$ from Fig. 9.12, and accordingly, $F_{13} = F_{23} = 1 - F_{12} = 0.8$, $F_{31} = (A_1/A_3)F_{13} = 0.8/4 = 0.2$, Eqs. (9.93), (9.91), and (9.92) may be rearranged to give

$$q_1 + 0.25q_2 - q_3 = 0$$
$$-q_1 + 2.25q_2 = 6{,}889 \text{ W/m}^2 \quad (9.94)$$
$$0.25q_1 + 6q_3 = 22{,}765 \text{ W/m}^2$$

This set of algebraic equations are now solved by an iteration method as shown in the FORTRAN program EX9–10.F listed in the Appendix (the general matrix-inversion method to solve this problem is beyond the scope of the current text). Because of the simplicity of the problem, however, an exact solution by the usual elimination procedure is feasible. Thus, we have

$$q_1 = 2{,}627 \text{ W/m}^2, \quad q_2 = 4{,}229 \text{ W/m}^2, \quad q_3 = 3{,}684 \text{ W/m}^2.$$

In terms of these fluxes, the loss from plate 1 is

$$q_1'' = q_1 + 4q_3 = 2{,}627 + 4 \times 3{,}684 = 17{,}363 \text{ W/m}^2$$

and the loss from plate 2 is

$$q_2'' = q_2 - q_1 = 4{,}229 - 2{,}627 = 1{,}602 \text{ W/m}^2.$$

Clearly, plate 2 loses more energy to the ambient than it receives from plate 1.

By the computer program provided, the interested reader may parametrically study the effect of plate separation on the problem. ◆

Having learned the method of electrical analogy and its application to a number of examples, we proceed now to the second method, the method of net radiation, for enclosure radiation problems.

9.3 NET RADIATION

The method is demonstrated here in terms of an enclosure made of N opaque surfaces, some being flat, some convex, and some concave (Fig. 9.36). First, complete an enclosure by replacing any hole with a black surface at the ambient temperature as we have done before for solutions by electrical analogy.

The net radiation from a typical surface, say i, written with the help of Fig. 9.3 and Eq. (9.19) is

$$\frac{\dot{Q}_i}{A_i} \equiv q_i = B_i - G_i, \tag{9.95}$$

which may be rearranged with Eq. (9.6), now expressed for surface i,

$$B_i = \rho_i G_i + \epsilon_i E_{bi}, \tag{9.96}$$

to yield for a surface with specified temperature

$$q_i = \frac{\epsilon_i}{\rho_i}(E_{bi} - B_i) \tag{9.97}$$

and for a surface with specified heat flux

$$E_{bi} = q_i \left(\frac{\rho_i}{\epsilon_i}\right) + B_i. \tag{9.98}$$

For known surface radiosity, Eq. (9.97) gives the radiation flux from a surface with a specified temperature, and Eq. (9.98) gives the temperature of a surface with specified heat flux. Thus, for both cases, the problem is reduced to the evaluation of surface radiosities as follows.

Consider a surface different from surface i, say surface j. The total radiation from this surface is $B_j A_j$. The fraction of this radiation intercepted by surface i is $B_j A_j F_{ji}$, which may be rearranged in terms of the reciprocity relation [Eq. (9.11) for surfaces i and j] $A_j F_{ji} = A_i F_{ij}$ to give $B_j A_i F_{ij}$. Thus, the total radiation leaving all surfaces (including i) and intercepted by surface i is

$$A_i G_i = A_i \sum_{j=1}^{N} B_j F_{ij}, \quad i = 1, 2, \ldots, N, \tag{9.99}$$

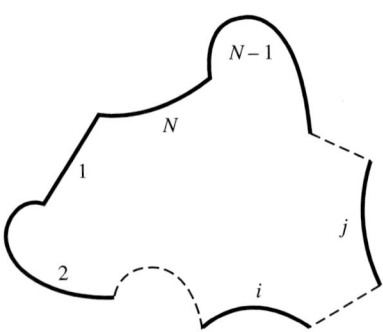

Figure 9.36 N-surface enclosure.

where G_i is the radiation flux incident on surface i. Then, the radiosity of surfaces with specified temperature is found by inserting Eq. (9.99) into Eq. (9.96),

$$B_i = \epsilon_i E_{bi} + \rho_i \sum_{j=1}^{N} B_j F_{ij}, \qquad (9.100)$$

and the radiosity of surfaces with specified heat flux is found by inserting Eq. (9.99) into Eq. (9.95),

$$B_i = q_i + \sum_{j=1}^{N} B_j F_{ij}. \qquad (9.101)$$

Thus, for each case, we end up with a set of N algebraic equations in terms of the unknown radiosities, $B_1, B_2, \ldots, B_N$. These equations can be solved by using a numerical iteration method as shown in the following example. For convenience, the solution procedure for enclosure radiation problems by the method of net radiation is summarized in Table 9.2 in terms of five steps.

Table 9.2 Five steps for net radiation

1. Express the temperature of surfaces with specified temperature in the absolute scale (Kelvin).
2. Replace the missing surface of an open enclosure with a black surface at the ambient temperature.
3. For each surface with specified temperature, write

$$B_i = \epsilon_i E_{bi} + \rho_i \sum_{j=1}^{N} B_j F_{ij}, \qquad (9.100)$$

and for each surface with specified heat flux write

$$B_i = q_i + \sum_{j=1}^{N} B_j F_{ij}, \qquad (9.101)$$

where $q_i = \dot{Q}_i / A_i$.

4. Solve the system of algebraic equations for radiosities given in step 3 by hand (if feasible) or by computational iteration or matrix inversion.
5. Evaluate the radiant flux from a surface with specified temperature from

$$q_i = \frac{\epsilon_i}{\rho_i}(E_{bi} - B_i) \qquad (9.97)$$

and the temperature of a surface with specified flux from

$$E_{bi} = q_i \left(\frac{\rho_i}{\epsilon_i}\right) + B_i. \qquad (9.98)$$

Next, we proceed to an example which illustrates the use of the method of net radiation for enclosure problems.

EXAMPLE 9.11[17]

Reconsider Ex. 9.10. Let the heat loss from plate 2 be given, e.g., $q_2'' = 1{,}602 \text{ W/m}^2$, rather than its temperature. We wish to determine the heat loss from plate 1 and the temperature of plate 2.

Employing Eq. (9.100) for the radiosity of surfaces 1 and 3, and Eq. (9.101) for the radiosity of surface 2, we have

$$B_1 = \epsilon_1 E_{b1} + \rho_1(B_2 F_{12} + B_3 F_{13})$$

$$B_2 = q_2'' + B_1 F_{21} + B_3 F_{23}$$

$$B_3 = E_{b3},$$

which may be rearranged as

$$B_1 - (\rho_1 F_{12}) B_2 = \epsilon_1 E_{b1} + (\rho_1 F_{13}) E_{b3}$$

$$-F_{21} B_1 + B_2 = q_2'' + F_{23} E_{b3},$$

or, in view of $\epsilon_1 = 0.8$, $\rho_1 = 0.2$, $F_{12} = F_{21} = 0.2$, $F_{13} = F_{23} = 0.8$, as

$$B_1 - 0.04 B_2 = 0.8 \times 5.67(8^4 + 0.2 \times 3^4)$$

$$-0.2 B_1 + B_2 = 1{,}602 + 0.8 \times 5.67 \times 3^4,$$

or,

$$B_1 - 0.04 B_2 = 18{,}653$$

$$-0.2 B_1 + B_2 = 1{,}969,$$

which gives

$$B_1 = 18{,}882 \text{ W/m}^2, \quad B_2 = 5{,}745 \text{ W/m}^2.$$

Now, from Eq. (9.97), the radiation from plate 1 is found to be

$$q_1'' = \frac{E_{b1} - B_1}{\rho_1/\epsilon_1} = \frac{5.67 \times 8^4 - 18{,}882}{1/4} = 17{,}369 \text{ W/m}^2,$$

and the temperature of plate 2 from

$$q_2'' = \frac{E_{b2} - B_2}{\rho_2/\epsilon_2}$$

gives

$$E_{b2} = B_2 + q_2''(\rho_2/\epsilon_2)$$

$$E_{b2} = 5{,}745 + 1{,}602 \times 1 = 7{,}347 \text{ W/m}^2$$

or

$$5.67(T_2/100)^4 = 7{,}347$$

or

$$T_2 \cong 600 \text{ K},$$

which recovers the results of Ex. 9.10, as expected. ◆

---

[17] The FORTRAN program EX9–11.F is listed in the appendix of this chapter.

9.4 COMBINED HEAT TRANSFER

The present section deals with a number of examples combining radiation with conduction and/or convection. Most problems involving more than one mode of heat transfer are relatively involved, as they yield nonlinear differential equations and/or boundary conditions whenever radiation is included. They are usually solved after a linearization of the Stefan-Boltzmann law. During this process, however, the quantitative nature of a problem gets lost.

EXAMPLE 9.12

Consider the production of vapor in a boiler tube surrounded by hot furnace gases. Neglecting the effect of curvature, we wish to determine the inner and outer surface temperatures T_0 and T_w (Fig. 9.37).

The large heat transfer coefficient on the evaporating water side provides an inner surface temperature close to the vapor temperature. The small heat transfer coefficient on the gas side suggests dominant radiation. Then, the first law for the system shown in Fig. 9.37 readily yields

$$-q_K + q_R = 0 \tag{9.102}$$

or

$$k \frac{T_w - T_0}{\ell} = \epsilon\sigma \left(T_g^4 - T_w^4\right). \tag{9.103}$$

Equation (9.103) may be reduced, in terms of the Planck number,

$$P_g = \frac{\sigma T_g^4}{kT_g/\ell} \sim \frac{\text{Emission}}{\text{Conduction}}, \tag{9.104}$$

to

$$\frac{T_w - T_0}{T_g - T_0} = \epsilon P_g \frac{1 - (T_w/T_g)^4}{1 - T_0/T_g}. \tag{9.105}$$

Next, linearizing the right side of Eq. (9.103) using a Taylor expansion about T_g,

$$k \frac{T_w - T_0}{\ell} \cong 4\epsilon\sigma T_g^3 \left(T_g - T_w\right),$$

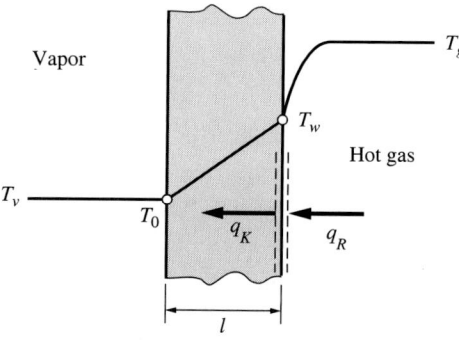

Figure 9.37 Boiler tube wall.

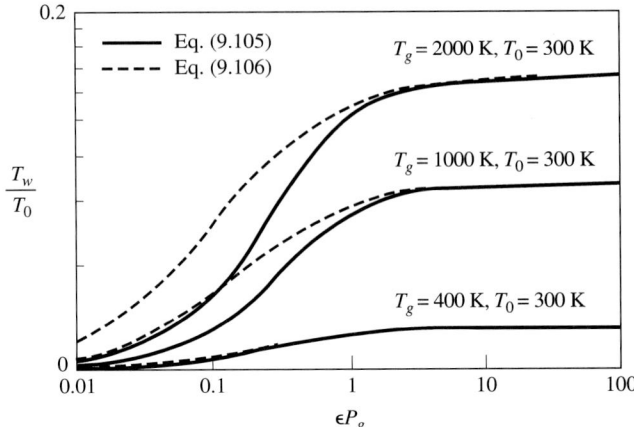

Figure 9.38 Comparison between linearized and nonlinear solutions.

or, rearranging in view of Eq. (9.104), we obtain

$$\frac{T_w - T_0}{T_g - T_0} \cong \frac{4\epsilon P_g}{1 + 4\epsilon P_g}. \tag{9.106}$$

Figure 9.38 compares the nonlinear [Eq. (9.105)] and linear [Eq. (9.106)] solutions for $T_0 = 300$ K and $T_g = 400, 1000,$ and 2000 K. Higher conductivity lowers ϵP_g and allows T_w to approach T_0. Recall Fig. 2.15 and the definition of Bi introduced in Chapter 2. ◆

EXAMPLE 9.13

The thermal boundary layer about a flying object may to a first order be approximated by a Couette flow. Let the velocity of the object be U and its surface emissivity be ϵ_w. The viscosity, thermal conductivity, and temperature of the ambient are μ, k, and T_∞, respectively. Assume the ambient to be transparent and the curvature effects to be negligible. We wish to determine the steady surface temperature of the object.

Newton's law for the control volume shown in Fig. 9.39 gives

$$\frac{d\tau}{dy} = 0, \tag{9.107}$$

where τ is the shear stress in the fluid. The mechanical energy associated with Eq. (9.107) is

$$u\frac{d\tau}{dy} = 0. \tag{9.108}$$

The first law of thermodynamics for the control volume yields

$$-\frac{dq_y^K}{dy} + \frac{d}{dy}(\tau u) = 0. \tag{9.109}$$

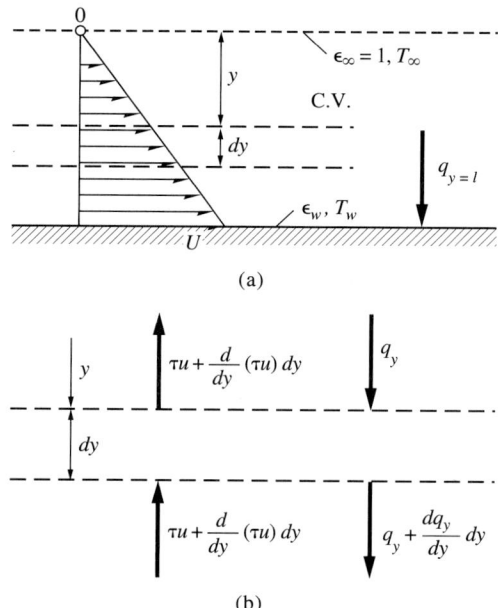

Figure 9.39 (a) Model, (b) balance of total energy.

For a transparent ambient, q_y^R does not contribute to Eq. (9.109). In Chapter 10 we shall study the effect of an absorbing ambient on the problem. Note that

$$\frac{d}{dy}(\tau u) = u\frac{d\tau}{dy} + \tau\frac{du}{dy}, \tag{9.110}$$

where the first term on the righthand side is identical to Eq. (9.108) and is the **displaced** mechanical energy, while the second term is the mechanical energy **dissipated** into heat as a result of deformation, du/dy. This dissipation must be balanced by an **entropy production** in the second law of thermodynamics.

Now, consider the fundamental difference,

$$\text{First law} - \text{Velocity} \times \text{Newton's law}$$

or

$$\text{Total (Thermal + Mechanical) Energy} - \text{Mechanical energy,}$$

by subtracting Eq. (9.108) from Eq. (9.109), which yields the balance of thermal energy,

$$-\frac{dq_y^K}{dy} + \tau\frac{du}{dy} = 0. \tag{9.111}$$

In terms of the Fourier and Newton laws,

$$q_y^K = -k\frac{dT}{dy}, \quad \tau = \mu\frac{du}{dy}, \tag{9.112}$$

Eq. (9.111) leads to

$$\frac{d}{dy}\left(k\frac{dT}{dy}\right) + \mu\left(\frac{du}{dy}\right)^2 = 0, \tag{9.113}$$

which, for constant thermal conductivity, becomes

$$\frac{d^2T}{dy^2} + \frac{\mu}{k}\left(\frac{du}{dy}\right)^2 = 0. \tag{9.114}$$

Also, Eq. (9.107) coupled with Eq. (9.112) gives

$$\frac{d^2u}{dy^2} = 0 \tag{9.115}$$

subject to boundary conditions

$$u(0) = 0, \; u(\ell) = U. \tag{9.116}$$

The solution,

$$u = U\left(\frac{y}{\ell}\right), \tag{9.117}$$

is the well-known velocity profile of the Couette flow. In terms of this velocity, Eq. (9.114) yields

$$\frac{d^2T}{dy^2} + \frac{\mu U^2}{k\ell^2} = 0 \tag{9.118}$$

subject to thermal boundary conditions

$$T(0) = T_\infty, \quad T(\ell) = T_w, \tag{9.119}$$

ℓ being the boundary-layer thickness. The solution of Eq. (9.118) that satisfies Eq. (9.119) is

$$T - T_\infty = (T_w - T_\infty)\frac{y}{\ell} + \frac{\mu U^2}{2k}\frac{y}{\ell}\left(1 - \frac{y}{\ell}\right). \tag{9.120}$$

Clearly, depending on the viscous dissipation, the **heat can be transferred to or from the wall with a temperature higher than the ambient temperature** (Fig. 9.40). Between these cases, there is the case of no heat transfer (insulated wall),

$$\frac{dT(\ell)}{dy} = 0 \tag{9.121}$$

for the case of neglected radiation, which yields the **adiabatic wall** temperature

$$T_a - T_\infty = \frac{\mu U^2}{2k}. \tag{9.122}$$

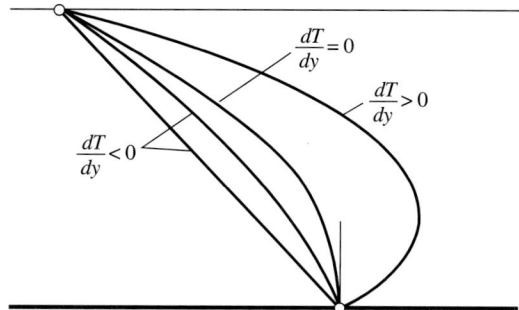

Figure 9.40 Adiabatic wall temperature.

This temperature may be considerably higher than the ambient temperature, depending on high velocity (of space vehicles reaching the melting temperature of the material) or high viscosity (of thrust bearings).

Now, we wish to determine the effect of radiation on T_a. Under the influence of radiation, the condition for insulation given by Eq. (9.121) needs to be replaced by (Fig. 9.41)

$$+q^K_{y=\ell} - q^R = 0 \tag{9.123}$$

or

$$-k\frac{dT(\ell)}{dy} - \epsilon_w \sigma \left(T_a^4 - T_\infty^4\right) = 0. \tag{9.124}$$

Inserting Eq. (9.120) into Eq. (9.124), we get

$$\frac{k}{\ell}\left(T_a - T_\infty - \frac{\mu U^2}{2k}\right) + \epsilon_w \sigma \left(T_a^4 - T_\infty^4\right) = 0. \tag{9.125}$$

For a qualitative answer, linearizing the radiation about a mean temperature T_M,

$$\frac{k}{\ell}\left(T_a - T_\infty - \frac{\mu U^2}{2k}\right) + 4\epsilon_w \sigma T_M^3 (T_a - T_\infty) = 0, \tag{9.126}$$

and introducing the Planck number,

$$P_M = \frac{4\sigma T_M^4}{kT_M/\ell} \sim \frac{\text{Emission}}{\text{Conduction}}, \tag{9.127}$$

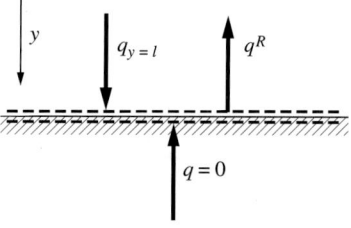

Figure 9.41 Boundary condition.

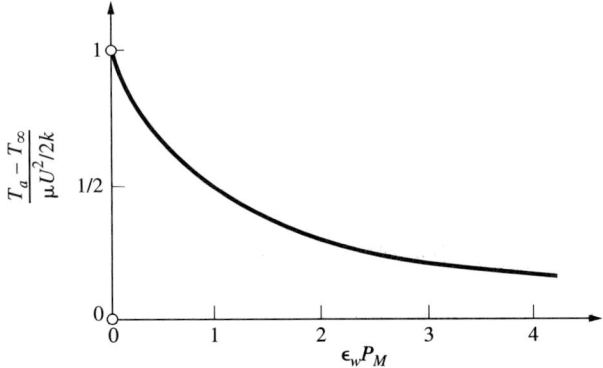

Figure 9.42 Radiation-affected adiabatic wall temperature.

we obtain

$$\frac{T_a - T_\infty}{\mu U^2/2k} = \frac{1}{1 + \epsilon_w P_M}, \qquad (9.128)$$

which shows the **cooling** effect of radiation by lowering T_a (Fig. 9.42). ◆

EXAMPLE 9.14

A semi-infinite vapor layer is being condensed by convective heat loss through a partition separating the vapor from a cold ambient (Fig. 9.43). Assuming the condensate to be transparent and the vapor to be black, we wish to determine the rate of condensation.

In terms of Fig. 9.44, the formulation of the problem is

$$\frac{1}{\alpha}\frac{\partial T}{\partial t} = \frac{\partial^2 T}{\partial x^2} \qquad (9.129)$$

$$T(x, 0) = T(X, t) = T_v \qquad (9.130)$$

$$+k\frac{\partial T(0, t)}{\partial x} = h\left[T(0, t) - T_\infty\right] \qquad (9.131)$$

$$+k\frac{\partial T(X, t)}{\partial x} = \rho h_{fg}\frac{dX}{dt}, \qquad (9.132)$$

where X is the instantaneous thickness of the condensate layer.

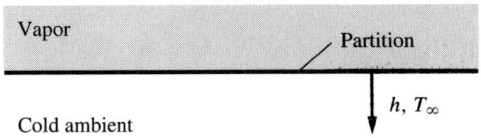

Figure 9.43 Condensation.

Sec. 9.4 Combined Heat Transfer 481

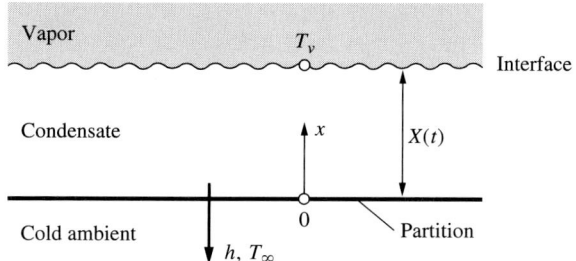

Figure 9.44 Unsteady condensate.

The solution of unsteady problems with moving boundaries is mathematically involved. However, by recognizing the fact that condensation is generally a slow process, neglecting the temporal variation of internal energy (and temperature) in Eq. (9.129), we may obtain a reasonably accurate quasi-steady solution. Thus, Eq. (9.129) is replaced by

$$\frac{\partial^2 T}{\partial x^2} = 0 \tag{9.133}$$

and the unsteadiness is now taken care of only via Equation (9.132).

Equation (9.133) readily integrates to

$$\frac{\partial T}{\partial x} = C_1 \tag{9.134}$$

$$T = C_1 x + C_2. \tag{9.135}$$

From the boundary condition of Eq. (9.130),

$$T_v = C_1 X + C_2, \tag{9.136}$$

and between Eqs. (9.135) and (9.136),

$$T - T_v = C_1(x - X). \tag{9.137}$$

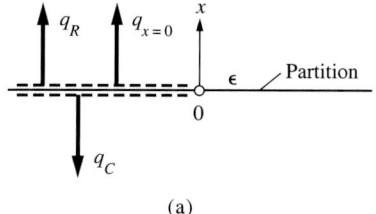

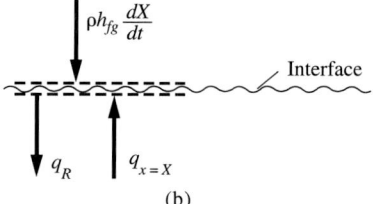

Figure 9.45 Boundary conditions. (a) Partition boundary condition, (b) interface boundary condition.

Also, inserting Eq. (9.137) into Eqs. (9.131) and (9.132) yields

$$C_1 = \frac{T_v - T_\infty}{X + k/h} \tag{9.138}$$

and

$$\frac{k(T_v - T_\infty)}{X + k/h} = \rho h_{fg} \frac{dX}{dt}, \tag{9.139}$$

which may be rearranged as

$$X dX + \left(\frac{k}{h}\right) dX = \frac{k(T_v - T_\infty)}{\rho h_{fg}} dt. \tag{9.140}$$

The integration of Eq. (9.140), in view of $X(0) = 0$, yields

$$\frac{1}{2} X^2 + \left(\frac{k}{h}\right) X - \frac{k(T_v - T_\infty)t}{\rho h_{fg}} = 0 \tag{9.141}$$

whose positive root

$$X = -\frac{k}{h} + \sqrt{\left(\frac{k}{h}\right)^2 + \frac{2k(T_v - T_\infty)t}{\rho h_{fg}}} \tag{9.142}$$

is the instantaneous thickness of the condensate. In terms of the Biot, Jacob, and Fourier numbers defined as

$$B_X = \frac{hX}{k}, \quad J = \frac{c_p(T_v - T_\infty)}{\rho h_{fg}}, \quad F = \frac{\alpha t}{(k/h)^2},$$

Eq. (9.142) becomes

$$B_X = -1 + \sqrt{1 + 2JF} \tag{9.143}$$

or, employing the binomial theorem for a small time t,

$$B_X \cong JF. \tag{9.144}$$

Now we wish to determine the effect of radiation on the problem. For the partition [Fig. 9.45(a)],

$$-q_R - q_{x=0} - q_C = 0, \tag{9.145}$$

or

$$\epsilon \sigma \left[T_v^4 - T^4(0, t)\right] + k \frac{\partial T(0, t)}{\partial x} = h \left[T(0, t) - T_\infty\right], \tag{9.146}$$

and, for the (transparent) interface [Fig. 9.45(b)],

$$q_{x=X} + \rho h_{fg}\frac{dX}{dt} = 0 \qquad (9.147)$$

or

$$-k\frac{\partial T(X,t)}{\partial x} + \rho h_{fg}\frac{dX}{dt} = 0. \qquad (9.148)$$

After linearization about T_M (say T_v for small X, or $T_M^4 = (T_v^4 + \epsilon T_\infty^4)/(1+\epsilon)$ for large h), Eqs. (9.146) and (9.148) become

$$4\epsilon\sigma T_M^3[T_v - T(0,t)] + k\frac{\partial T(0,t)}{\partial x} = h[T(0,t) - T_\infty] \qquad (9.149)$$

and

$$k\frac{\partial T(X,t)}{\partial x} = \rho h_{fg}\frac{dX}{dt}. \qquad (9.150)$$

Now, insertion of Eq. (9.137) into Eqs. (9.149) and (9.150) yields

$$C_1 = \frac{T_v - T_\infty}{(1+R)X + (k/h)} \qquad (9.151)$$

$$kC_1 = \rho h_{fg}\frac{dX}{dt}, \qquad (9.152)$$

where

$$R = 4\epsilon\sigma T_M^3/h \qquad (9.153)$$

is the effect of radiation relative to convection. The elimination of C_1 between Eqs. (9.151) and (9.152) leads to, after some rearrangement,

$$(1+R)XdX + \left(\frac{k}{h}\right)dX = \frac{k(T_v - T_\infty)}{\rho h_{fg}}dt, \qquad (9.154)$$

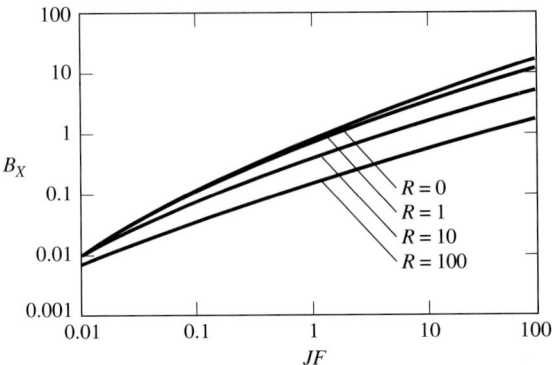

Figure 9.46 Variation of B_X against JF.

which readily integrates to

$$(1 + R)B_X = -1 + \sqrt{1 + 2(1 + R)JF}. \tag{9.155}$$

Figure 9.46 shows the variation of B_X versus JF for various R. For small time, Eq. (9.155) is reduced to Eq. (9.144), as expected (why?). ◆

■ REFERENCES

9.1 M. Jakob, *Heat Transfer*, Vol. 2. Wiley, New York, 1957.

9.2 H.C. Hottel, and A.F. Sarofim, *Radiative Transfer*. McGraw-Hill, New York, 1967.

9.3 W.H. McAdams, *Heat Transmission*. 3d ed. McGraw-Hill, New York, 1954.

9.4 F. Kreith and M.S. Bohn, *Principles of Heat Transfer*. 4th ed. Harper & Row, New York, 1986.

9.5 A.K. Oppenheim, "The network method of radiation analysis," *Trans. ASME*, vol. 78, pp. 725–735, 1956.

9.6 E.M. Sparrow, and R.D. Cess, *Radiation Heat Transfer*. McGraw-Hill, New York, 1978.

9.5 APPENDIX

```
C-----------------------------------
C    EX9-1.F (START)
C-----------------------------------
     PROGRAM MAIN
     IMPLICIT REAL*8 (A-H,K-Z)
     PI=4*ATAN(1.)
     SIGMA=5.67E-8
     WRITE(*,*) 'EXAMPLE 9.1....'
C-------------------------------------------
C    INPUT DATA
```

```
C------------------------------------------------
      WRITE(*,*) 'INPUT THE FOLLOWING DATA...'
      WRITE(*,*) 'E: '
      READ(*,*) E
      WRITE(*,*) 'T_1: C'
      READ(*,*) T1
      WRITE(*,*) 'T_2: C'
      READ(*,*) T2
C------------------------------------------------
C     UNIT CONVERSION
C------------------------------------------------
      T1=T1+273
      T2=T2+273
C------------------------------------------------
C     CALCULATION
C------------------------------------------------
      Q12=E*SIGMA*(T1**4-T2**4)/2
C------------------------------------------------
C     ANSWER
C------------------------------------------------
      WRITE(*,*) 'HEAT LOSS FROM THE BOTTLE IS'
      WRITE(*,*) Q12,' W/m^2'
      STOP
      END
C----------------------------------------
C     EX9-1.F (END)
C----------------------------------------
C----------------------------------------
C     EX9-6.F (START)
C----------------------------------------
      PROGRAM MAIN
      IMPLICIT REAL*8 (A-H,K-Z)
      PI=4*ATAN(1.)
      SIGMA=5.67E-8
      WRITE(*,*) 'EXAMPLE 9.6....'
C--------------------------------------------------
C     INPUT DATA
C--------------------------------------------------
      WRITE(*,*) 'INPUT THE FOLLOWING DATA...'
      WRITE(*,*) 'T_W: K'
      READ(*,*) TW
      WRITE(*,*) 'T_INFTY: K'
      READ(*,*) TINFTY
      WRITE(*,*) 'E_W: '
      READ(*,*) EW
      WRITE(*,*) 'E_O^I: '
      READ(*,*) EOI
      WRITE(*,*) 'E_O^O: '
      READ(*,*) EOO
C--------------------------------------------------
C     CALCULATION
C--------------------------------------------------
      ARW1=1/EW
      Q1=SIGMA*(TW**4-TINFTY**4)/ARW1
      ARW2=1/EW+(1/EOI+1/EOO-1)/2
      Q2=SIGMA*(TW**4-TINFTY**4)/ARW2
      T0=(TINFTY**4+Q2/(EOO*SIGMA*2))**0.25
C--------------------------------------------------
C     ANSWER
```

```
C------------------------------------------------------
      WRITE(*,*) 'HEAT FLUX REQUIRED FOR THE OPEN PLATE IS'
      WRITE(*,*) Q1,' W/m^2'
      WRITE(*,*) 'HEAT FLUX REQUIRED FOR THE COVERED PLATE IS'
      WRITE(*,*) Q2,' W/m^2'
      WRITE(*,*) 'LID TEMPERATURE IS'
      WRITE(*,*) T0,' K'
      STOP
      END
C-----------------------------------
C     EX9-6.F (END)
C-----------------------------------
C-----------------------------------
C     EX9-7.F (START)
C-----------------------------------
      PROGRAM MAIN
      IMPLICIT REAL*8 (A-H,K-Z)
      PI=4*ATAN(1.)
      SIGMA=5.67E-8
      WRITE(*,*) 'EXAMPLE 9.7....'
C--------------------------------------------------
C     INPUT DATA
C--------------------------------------------------
      WRITE(*,*) 'INPUT THE FOLLOWING DATA...'
      WRITE(*,*) 'ETA: '
      READ(*,*) ETA
      WRITE(*,*) 'E_S^I: '
      READ(*,*) ESI
      WRITE(*,*) 'E_S^O: '
      READ(*,*) ESO
      WRITE(*,*) 'T_INFTY: C'
      READ(*,*) TINFTY
      WRITE(*,*) 'R_W: cm'
      READ(*,*) RW
      WRITE(*,*) 'E_W: '
      READ(*,*) EW
      WRITE(*,*) 'U: MW/m^3'
      READ(*,*) U
      WRITE(*,*) 'R_S: cm'
      READ(*,*) RS
C--------------------------------------------------
C     UNIT CONVERSION
C--------------------------------------------------
      TINFTY=TINFTY+273
      RW=0.01*RW
      U=1E6*U
      RS=0.01*RS
C--------------------------------------------------
C     CALCULATION
```

```
C--------------------------------------------------
      ARW1=1/EW
      TW1=(ARW1*U*RW/(2*SIGMA)+TINFTY**4)**0.25
      ARW2=1/EW+(1-ETA)*(RW/RS)/(ETA+1/(1/ESI+1/ESO-1))
      TW2=(ARW2*U*RW/(2*SIGMA)+TINFTY**4)**0.25
      EBW=SIGMA*TW2**4
      BW=EBW-(1-EW)/EW*(U*RW/2)
      AW=2*PI*RW
      EBINFTY=SIGMA*TINFTY**4
      QH=AW*ETA*(BW-EBINFTY)
      QWINFTY=U*PI*RW**2
      QS=QWINFTY-QH
      BSI=BW-QS/(AW*(1-ETA))
      AS=2*PI*RS*(1-ETA)
      FSINFTYI=ETA*(1-RW/RS)
      QSH=(BSI-EBINFTY)*AS*FSINFTYI
      QSS=QS-QSH
      EBS=BSI-(1/ESI-1)*QSS/AS
      TS=(EBS/SIGMA)**0.25
C--------------------------------------------------
C     ANSWER
C--------------------------------------------------
      WRITE(*,*) 'HEATER SURFACE TEMPERATURE WITHOUT SCREEN IS'
      WRITE(*,*) TW1,' K'
      WRITE(*,*) 'HEATER SURFACE TEMPERATURE WITH SCREEN IS'
      WRITE(*,*) TW2,' K'
      WRITE(*,*) 'TEMPERATURE RISE WITH SCREEN (ETA=',ETA,') IS'
      WRITE(*,*) TW2-TW1,' K'
      WRITE(*,*) 'SCREEN TEMPERATURE IS'
      WRITE(*,*) TS,' K'
      STOP
      END
C--------------------------------------------------
C     EX9-7.F (END)
C--------------------------------------------------
C--------------------------------------------------
C     EX9-8.F (START)
C--------------------------------------------------
      PROGRAM MAIN
      IMPLICIT REAL*8 (A-H,K-Z)
      PI=4*ATAN(1.)
      SIGMA=5.67E-8
      WRITE(*,*) 'EXAMPLE 9.8....'
C--------------------------------------------------
C     INPUT DATA
C--------------------------------------------------
      WRITE(*,*) 'INPUT THE FOLLOWING DATA...'
      WRITE(*,*) 'D: m'
      READ(*,*) D
      WRITE(*,*) 'T_W: C'
      READ(*,*) TW
      WRITE(*,*) 'E_W: '
      READ(*,*) EW
      WRITE(*,*) 'T_INFTY: C'
      READ(*,*) TINFTY
C--------------------------------------------------
C     UNIT CONVERSION
C--------------------------------------------------
      TW=TW+273
      TINFTY=TINFTY+273
C--------------------------------------------------
C     CALCULATION
```

```
C-------------------------------------------------
  AW=PI*D**2/4
  QWIN1=AW*EW*SIGMA*(TW**4-TINFTY**4)
  FWINFTY=0.17
  RWIN2=(1/EW-1+2/(1+FWINFTY))/AW
  QWIN2=SIGMA*(TW**4-TINFTY**4)/RWIN2
  EBW=SIGMA*TW**4
  BW=EBW-(QWIN2/AW)*(1-EW)/EW
  BINFTY=SIGMA*TINFTY**4
  FWS=1-FWINFTY
  BS=(BW-FWINFTY*BINFTY-QWIN2/AW)/FWS
  TS=(BS/SIGMA)**0.25
C-------------------------------------------------
C     ANSWER
C-------------------------------------------------
  WRITE(*,*) 'RADIANT HEAT TRANSFER FROM THE DISK TO THE AMBIENT'
  WRITE(*,*) 'WITHOUT THE INSULATED SHELL IS'
  WRITE(*,*) QWIN1,' W'
  WRITE(*,*) 'RADIANT HEAT TRANSFER FROM THE DISK TO THE AMBIENT'
  WRITE(*,*) 'WITH THE INSULATED SHELL IS'
  WRITE(*,*) QWIN2,' W'
  WRITE(*,*) 'SHELL TEMPERATURE IS'
  WRITE(*,*) TS-273,' C'
  STOP
  END
C----------------------------------
C     EX9-8.F (END)
C----------------------------------
C----------------------------------
C     EX9-9.F (START)
C----------------------------------
  PROGRAM MAIN
  IMPLICIT REAL*8 (A-H,K-Z)
  PI=4*ATAN(1.)
  SIGMA=5.67E-8
  WRITE(*,*) 'EXAMPLE 9.9....'
C-------------------------------------------------
C     INPUT DATA
C-------------------------------------------------
  WRITE(*,*) 'INPUT THE FOLLOWING DATA FROM EX. 9.8...'
  WRITE(*,*) 'D: m'
  READ(*,*) D
  WRITE(*,*) 'T_W: C'
  READ(*,*) TW
  WRITE(*,*) 'E_W: '
  READ(*,*) EW
  WRITE(*,*) 'T_INFTY: C'
  READ(*,*) TINFTY
  WRITE(*,*) 'INPUT THE FOLLOWING DATA FROM EX. 9.9...'
  WRITE(*,*) 'E_S: '
  READ(*,*) ES
C-------------------------------------------------
C     UNIT CONVERSION
C-------------------------------------------------
  TW=TW+273
  TINFTY=TINFTY+273
C-------------------------------------------------
C     CALCULATION
```

```
C-----------------------------------------------
      AW=PI*D**2/4
      AINFTY=AW
      AS=4*AINFTY
      FWINFTY=0.17
      FITYSI=1-FWINFTY
      FSITYO=1
      RHOS=1-ES
      R1=1/(AINFTY*FITYSI+1/(2*RHOS/(AS*ES)+1/(AS*FSITYO)))
      FWS=1-FWINFTY
      R2=1/(AW*FWINFTY+1/(1/(AW*FWS)+R1))
      RHOW=1-EW
      RWINFTY=RHOW/(AW*EW)+R2
      EBW=SIGMA*TW**4
      EBINFTY=SIGMA*TINFTY**4
      QWINFTY=(EBW-EBINFTY)/RWINFTY
C-----------------------------------------------
C     ANSWER
C-----------------------------------------------
      WRITE(*,*) 'RADIANT HEAT TRANSFER FROM THE DISK TO THE AMBIENT'
      WRITE(*,*) 'WITH THE NON-INSULATED SHELL IS'
      WRITE(*,*) QWINFTY,' W'
      STOP
      END
C-----------------------------------------------
C     EX9-9.F (END)
C-----------------------------------------------

C-----------------------------------------------
C     EX9-10.F (START)
C-----------------------------------------------
      PROGRAM MAIN
      IMPLICIT REAL*8 (A-H,K-Z)
      REAL A(3,3),Q(3),Q0(3),B(3)
      PI=4*ATAN(1.)
      SIGMA=5.67E-8
      WRITE(*,*) 'EXAMPLE 9.10....'
C-----------------------------------------------
C     INPUT DATA
C-----------------------------------------------
      WRITE(*,*) 'INPUT THE FOLLOWING DATA...'
      WRITE(*,*) 'T_1: K'
      READ(*,*) T1
      WRITE(*,*) 'T_2: K'
      READ(*,*) T2
      WRITE(*,*) 'E_1: '
      READ(*,*) E1
      WRITE(*,*) 'E_2: '
      READ(*,*) E2
      WRITE(*,*) 'T_INFTY: K'
      READ(*,*) TINFTY
C-----------------------------------------------
C     MATRIX: EQS. (9.93), (9.91), AND (9.92) WITH AQ=B
```

```
C-------------------------------------------------------
      A1=1
      A2=1
      A3=4
      F12=0.2
      F13=1-F12
      F23=1-F12
      F31=A1*F13/A3
      A(1,1)=1/F12
      A(1,2)=1/F23
      A(1,3)=-1/F31
      A(2,1)=-(1-E2)/E2*(A1/A2)
      A(2,2)=(1-E2)/E2+1/F23
      A(2,3)=0
      A(3,1)=-(1-E1)/E1
      A(3,2)=0.
      A(3,3)=-1/F31-(1-E1)/E1*(A3/A1)
      EB1=SIGMA*T1**4
      EB2=SIGMA*T2**4
      EB3=SIGMA*TINFTY**4
      B(1)=0
      B(2)=EB2-EB3
      B(3)=EB3-EB1
C-------------------------------------------------
C     ITERATION
C-------------------------------------------------
      DO 10 I=1,3
      Q(I)=1.
      Q0(I)=1.
 10   CONTINUE
      DO 20 I=1,1000
      ERRMAX=1E-10
      Q(1)=(B(1)-A(1,2)*Q(2)-A(1,3)*Q(3))/A(1,1)
      Q(2)=(B(2)-A(2,1)*Q(1)-A(2,3)*Q(3))/A(2,2)
      Q(3)=(B(3)-A(3,1)*Q(1)-A(3,2)*Q(2))/A(3,3)
      DO 30 J=1,3
      IF(Q0(J).NE.0) ERROR=ABS((Q(J)-Q0(J))/Q0(J))
      Q0(J)=Q(J)
      IF(ERROR.GT.ERRMAX) ERRMAX=ERROR
 30   CONTINUE
      IF(ERRMAX.LT.0.0001) GOTO 999
 20   CONTINUE
 999  QP1=(A1*Q(1)+A3*Q(3))/A1
      QP2=(A2*Q(2)-A1*Q(1))/A2
C-------------------------------------------------
C     ANSWER
C-------------------------------------------------
      WRITE(*,*) 'RADIANT HEAT FLUX Q1, Q2, AND Q3 ARE'
      WRITE(*,*) Q(1),Q(2),Q(3),' W/m^2'
      WRITE(*,*) 'RADIANT HEAT LOSS FROM PLATE 1 IS'
      WRITE(*,*) QP1,' W/m^2'
      WRITE(*,*) 'RADIANT HEAT LOSS FROM PLATE 2 IS'
      WRITE(*,*) QP2,' W/m^2'
      STOP
      END
C----------------------------------
C     EX9-10.F (END)
C----------------------------------
C----------------------------------
C     EX9-11.F (START)
```

```fortran
C----------------------------------
      PROGRAM MAIN
      IMPLICIT REAL*8 (A-H,K-Z)
      REAL F(3,3),E(3),B(3),B0(3)
      PI=4*ATAN(1.)
      SIGMA=5.67E-8
      WRITE(*,*) 'EXAMPLE 9.11....'
C--------------------------------------------------
C     INPUT DATA
C--------------------------------------------------
      WRITE(*,*) 'INPUT THE FOLLOWING DATA FROM EX. 9.10...'
      WRITE(*,*) 'T_1: K'
      READ(*,*) T1
      WRITE(*,*) 'E_1: '
      READ(*,*) E1
      WRITE(*,*) 'E_2: '
      READ(*,*) E2
      WRITE(*,*) 'T_INFTY: K'
      READ(*,*) TINFTY
      WRITE(*,*) 'INPUT THE FOLLOWING DATA FROM EX. 9.11...'
      WRITE(*,*) 'Q_2: W/m^2'
      READ(*,*) Q2
C----------------------------------------------------------
C     MATRIX: EQS. (9.100) AND (1.101)
C----------------------------------------------------------
      A1=1
      A2=1
      A3=4
      F(1,1)=0
      F(1,2)=0.2
      F(1,3)=0.8
      F(2,1)=0.2
      F(2,2)=0.
      F(2,3)=0.8
      F(3,1)=F(1,3)*A1/A3
      F(3,2)=F(2,3)*A2/A3
      F(3,3)=0
      E(1)=E1
      E(2)=E2
      E(3)=1
      EB1=SIGMA*T1**4
      EB3=SIGMA*TINFTY**4
C--------------------------------------------------
C     ITERATION
C--------------------------------------------------
      DO 10 I=1,3
      B(I)=1.
      B0(I)=1.
 10   CONTINUE
      DO 20 I=1,1000
      ERRMAX=1E-10
      B(1)=E(1)*EB1+(1-E(1))*(B(1)*F(1,1)+B(2)*F(1,2)+B(3)*F(1,3))
      B(2)=Q2+(B(1)*F(2,1)+B(2)*F(2,2)+B(3)*F(2,3))
      B(3)=E(3)*EB3+(1-E(3))*(B(1)*F(3,1)+B(2)*F(3,2)+B(3)*F(3,3))
      DO 30 J=1,3
      IF(B0(J).NE.0) ERROR=ABS((B(J)-B0(J))/B0(J))
      B0(J)=B(J)
      IF(ERROR.GT.ERRMAX) ERRMAX=ERROR
 30   CONTINUE
      IF(ERRMAX.LT.0.0001) GOTO 999
 20   CONTINUE
 999  QP1=(EB1-B(1))/((1-E(1))/E(1))
      EB2=B(2)+Q2*(1-E(2))/E(2)
      T2=(EB2/SIGMA)**0.25
```

```
C-----------------------------------------------
C     ANSWER
C-----------------------------------------------
      WRITE(*,*) 'RADIOSITY B1, B2, AND B3 ARE'
      WRITE(*,*) B(1),B(2),B(3),' W/m^2'
      WRITE(*,*) 'RADIANT HEAT LOSS FROM PLATE 1 IS'
      WRITE(*,*) QP1,' W/m^2'
      WRITE(*,*) 'SURFACE TEMPERATURE OF PLATE 2 IS'
      WRITE(*,*) T2,' K'
      STOP
      END
C-----------------------------------------------
C     EX9-11.F (END)
C-----------------------------------------------
```

■ EXERCISES

9.1 The cover plate of a solar collector has negligible absorptivity (Fig. 9P–1). Show that the fraction of the energy incident on the upper surface and transmitted through the lower surface is $(1 - \rho)/(1 + \rho)$.

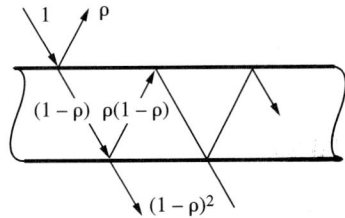

Figure 9P–1

9.2 A fraction of the solar energy incident on the lower surface of a cover plate is transmitted to the absorber plate of a collector (Fig. 9P–2). Show that the fraction of the energy absorbed by the absorber is $\tau\alpha[1 - (1 - \alpha)\rho]$.

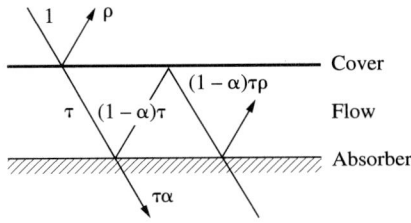

Figure 9P–2

9.3 What provides the effect of radiative insulation in Ex. 9.1? Repeat the example for $\epsilon_1 = \epsilon_2 = \epsilon = 1$. What is your conclusion?

9.4 For an insulated surface, using the definitions of emissivity and radiosity, show that

$$B = G = E_b.$$

What is the effect of ϵ of this surface in an enclosure?

9.5 Reconsider the definition of view factor given by Eq. (9.10). Show that the view factor between a differential surface dA_1 and surface A_2 is reduced to

$$F_{12} = \int_{A_2} \frac{\cos\phi_1 \cos\phi_2 dA_2}{\pi r^2}$$

and the reciprocity relation becomes

$$dA_1 F_{12} = A_2 dF_{21}.$$

9.6 Show that the view factor between two (one differential, one finite) parallel concentric disks (Fig. 9P–3) is given by

$$F_{d12} = 2\ell^2 \int_\ell \frac{\sqrt{\ell^2 + R^2} dr}{r^3} = \frac{R^2}{\ell^2 + R^2}.$$

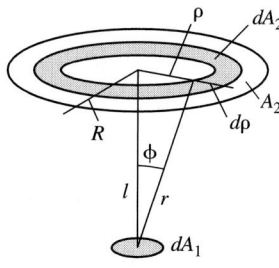

Figure 9P–3

494 Chap. 9 Enclosure Radiation

9.7 Discuss the view factors of the following cases (Fig. 9P–4) in terms of the available charts:

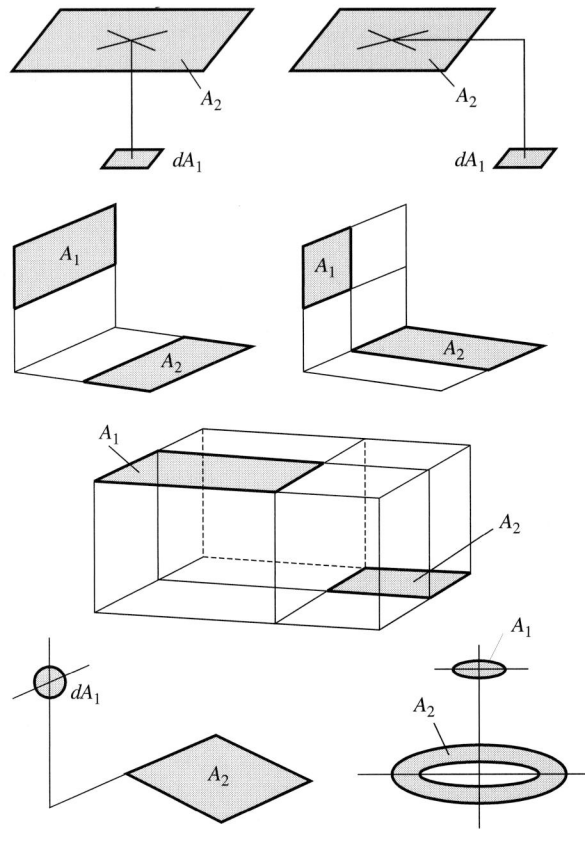

Figure 9P–4

9.8 The view factors between parallel surfaces infinitely long in one direction may readily be obtained by Hottel's "string rule." For two such surfaces (Fig. 9P–5),

$$F_{12} = \frac{(c + d) - (a + b)}{2\ell_1},$$

where c and d are the diagonal (crossed) strings between the extremities of the two surfaces, a and b are the lateral (uncrossed) ones, and ℓ_1 is the length of the first surface. Use this method to determine F_{12} for the following configurations: **(a)** parallel plates, **(b)** perpendicular plates, **(c)** parallel cylinders.

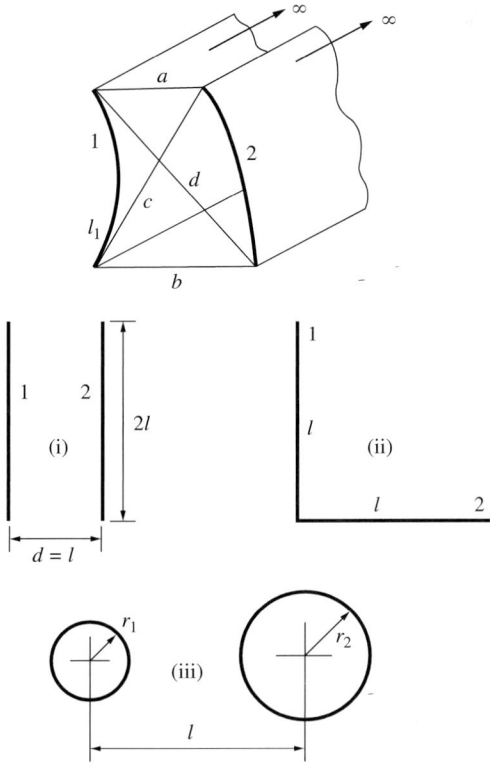

Figure 9P–5

9.9 An optical instrument is modeled as a spherical cavity of diameter D_c inside a metal block at temperature T and surface emissivity ϵ (Fig. 9P–6). The cavity radiates to an ambient at T_∞ through a small opening of diameter D_0 and of negligible length. The effective emissivity of the cavity, ϵ_{eff} is defined by

$$\text{Radiation leaving cavity} \equiv \epsilon_{\text{eff}} A_0 (T^4 - T_\infty^4),$$

where A_0 is the opening cross sectional area. Determine ϵ_{eff} as a function of ϵ, D_c, and D_0. What happens to ϵ_{eff} as D_c/D_0 increases? What is the effect of an odd-shaped cavity on the answer?

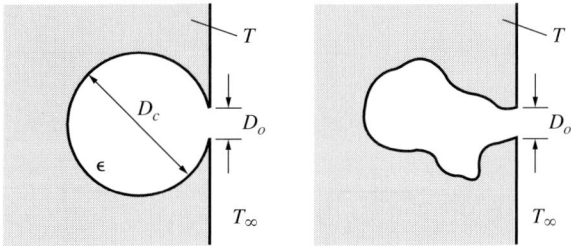

Figure 9P–6

9.10 The radiation from an infinitely long flat plate to an ambient is to be controlled by an infinitely long and electrically heated flat screen ($\eta = A_{\text{holes}}/A_{\text{total area}}$). The plate and ambient temperatures are T_p and T_∞, respectively (Fig. 9P–7). The emissivity of the plate and screen are ϵ_p and ϵ_s, respectively. Show the effect of the screen on the radiant heat transfer from the plate.

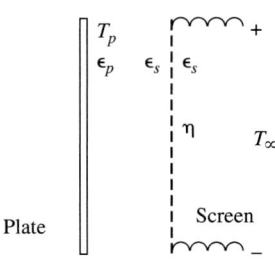

Figure 9P–7

9.11 Consider two large parallel plates with uniform temperature T_1 and T_2 and emissivity ϵ. Place two flat partitions with the same emissivity between the plates. Show that the radiation between the plates is reduced by a factor of 1/3. Generalize this result to the case of n partitions.

9.12 Consider two large parallel plates with uniform temperature T_1 and T_2 (Fig. 9P–8). Place two flat screens ($\eta = A_{\text{holes}}/A_{\text{total area}}$) between the plates. All surfaces have the same emissivity, ϵ. Determine the radiation between the plates relative to that without screens.

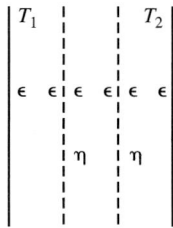

Figure 9P–8

9.13 Calculate the surface temperature of a hemispherical radiant source with $\epsilon = 0.8$ losing energy at the rate $q'' = 1$ kW/m² to an ambient at temperature $T_\infty = 10\,°\text{C}$ (Fig. 9P–9). Find the temperature of the reflector.

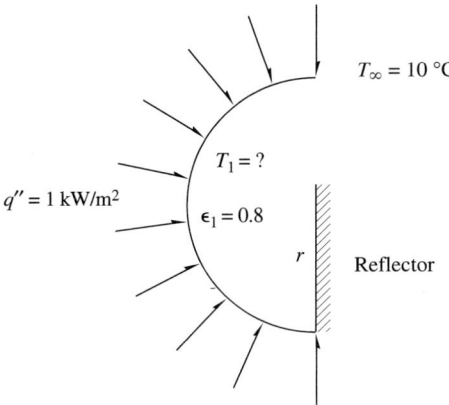

Figure 9P–9

9.14 Consider a cylindrical shell with $D/H = 1$ at a uniform temperature $T_w = 300\,°C$ (Fig. 9P–10). The inside and outside emissivity of the shell is $\epsilon = 0.5$. Calculate the radiant heat transfer from this shell to the ambient at temperature $T_\infty = 20\,°C$.

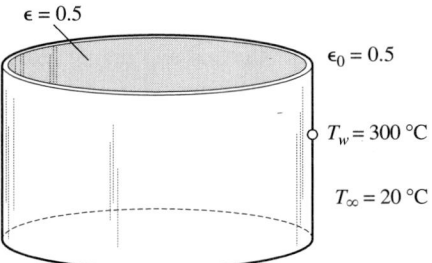

Figure 9P–10

9.15 A thin metal plate of height ℓ is kept at a uniform temperature $T_1 = 800\,°C$ in an ambient at temperature $T_\infty = 10\,°C$ (Fig. 9P–11). The emissivity of the plate is $\epsilon_1 = 0.8$. **(a)** Find the net radiation between one side of the plate and the ambient. **(b)** The plate is surrounded by an insulated semicylinder, as shown in Fig. 9P–11, with $\ell/r = 1$. Find the net radiation between the whole plate and the ambient.

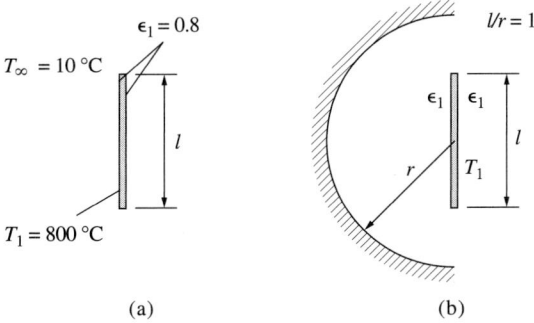

Figure 9P–11

9.16 A heat flux q'' is applied to a disk of diameter D and emissivity ϵ. The ambient temperature is T_∞ (Fig. 9P–12). Neglecting the effect of convection. **(a)** Evaluate the temperature of the plate. **(b)** Repeat part (a) after placing a hemispherical cup of diameter D_0 and emissivity ϵ_0 on top of the disk as shown in Fig. 9P–12.

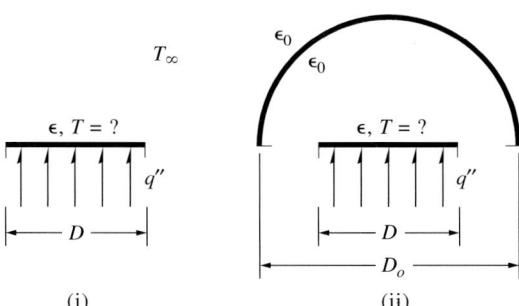

Figure 9P–12

9.17 Electrical power 50 W is supplied to a plate of 0.1 m² area (Fig. 9P–13). The emissivity of the plate is $\epsilon_w = 0.5$. The ambient temperature is $T_\infty = 10\,°\mathrm{C}$. Neglect the effect of convection. **(a)** Find the temperature of the plate, T_w. **(b)** The same amount of power is supplied to a screen of identical apparent area. The wire area of the screen is A_s and $A_s/A = 0.2$. The emissivity of the screen is $\epsilon = 0.5$. Find the temperature of the screen wire.

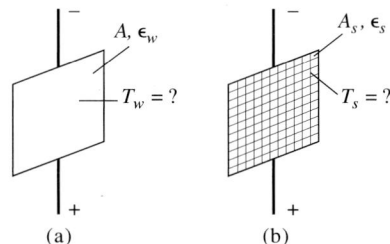

Figure 9P–13

9.18 Hot gases at $T_g = 47\,°\mathrm{C}$ flow in a pipe of 5 cm diameter (Fig. 9P–14). For safety reasons, a radiative insulation is being considered by covering the pipe with a thin concentric metal shell. The emissivities of the pipe and shell are ϵ_p and ϵ_s, respectively. The ambient temperature is $T_\infty = 27\,°\mathrm{C}$. **(a)** Neglecting the effect of convection, find the temperature of the shell and the reduction in heat loss to the ambient for $\epsilon_p = \epsilon_s = 0.1$, and for $\epsilon_p = 1$ and $\epsilon_s = 0.1$. **(b)** For the same reduction of part (a), determine the required thickness of an asbestos insulation.

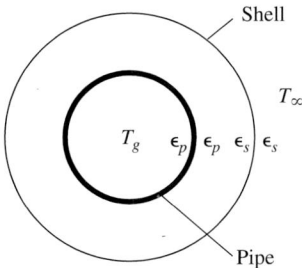

Figure 9P–14

9.19 The inside emissivity of a closed, empty, cylindrical container is desired. A heat source and some insulating material are available. Construct a simple experiment that would help you evaluate this emissivity. Assume that the outside emissivity is known.

9.20 The surface temperature of a disk of diameter $D_1 = 10$ cm is kept at $T_1 = 500$ K. The emissivity of the disk is $\epsilon_1 = 0.8$ (Fig. 9P–15). **(a)** Find the radiant heat transfer from one side of the disk to an ambient at temperature $T_\infty = 300$ K. **(b)** The disk is attached to an insulated cone section ($D_1/D_2 = 1/2$, $L = 10$ cm). Evaluate the change in the heat transfer from the disk.

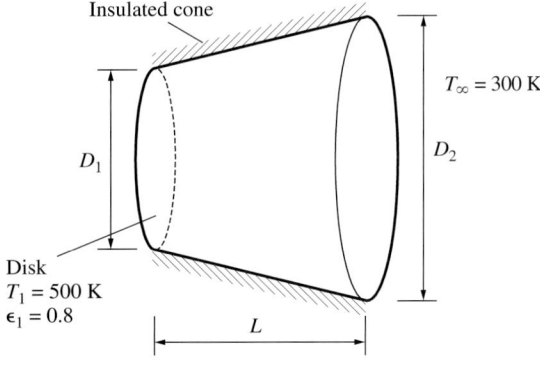

Figure 9P–15

9.21 The surface temperature of a cylinder of diameter $D = 2R$ and emissivity ϵ_w is to be kept at T_w (Fig. 9P–16). **(a)** Find the radiant heat transfer from the cylinder to the ambient. **(b)** Let three different insulated radiation shields separately surround the cylinder. For each case, find the view factors and the radiant heat transfer.

500 Chap. 9 Enclosure Radiation

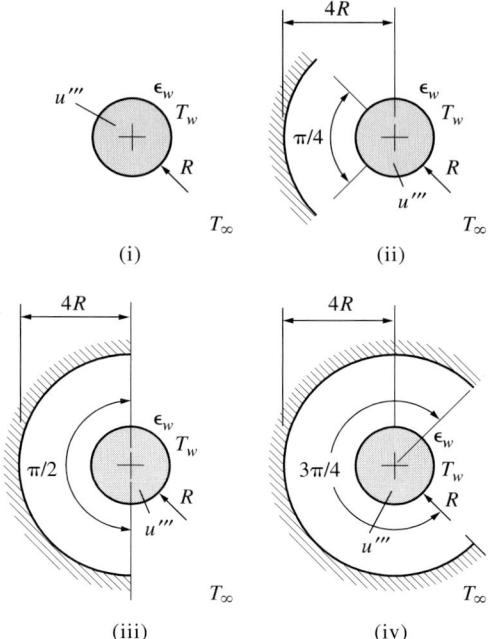

Figure 9P–16

9.22 A long solid cylinder of radius R, emissivity ϵ, and temperature T_w radiates steadily to an ambient at temperature T_∞. **(a)** Find the steady energy loss by radiation from the cylinder. **(b)** Repeat part (a) after placing a radiation shield near the cylinder as shown in Fig. 9P–17.

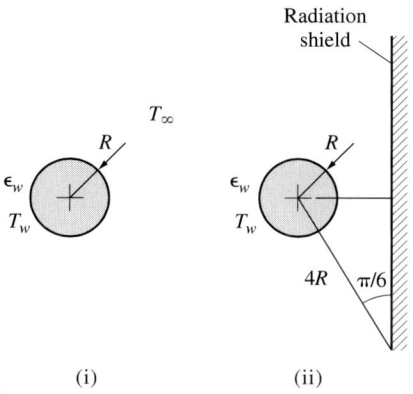

Figure 9P–17

9.23 A spherical satellite of diameter $D_{sat} = 1$ m and an emissivity $\epsilon_{sat} = 0.7$ is placed at a distance of 1.5×10^{11} m from the sun ($D_s = 1.4 \times 10^9$ m). The sun and space are at 5800 K and 0 K, respectively. **(a)** Show that the view factor is

$$F_{sat\text{-}s} = \sin^2 \frac{\phi}{2},$$

where ϕ is the angle between two tangents drawn from the center of the satellite to the sun. **(b)** Replace the sun by a disk ($D_d = D_s$) facing the satellite and having the same diameter as the sun. How does the view factor differ from that of part (a)? **(c)** Find the temperature of the satellite.

9.24 The satellite of Prob. 9.23 is now placed 300 km above the dark side of the earth. Treating the earth as a black body with diameter $D_e = 1.3 \times 10^7$ m and temperature $T_e = 300$ K, **(a)** compare the exact view factor [Prob. 9.23(a)] with that of the disk assumption [Prob. 9.23(b)], **(b)** Find the temperature of the satellite.

9.25 The sketch represents part of a system to be placed in orbit about the earth (Fig. 9P–18). Surfaces 1 and 2 are long concentric cylinders with a vacuum in between. Surface 1 is heated internally by an electrical heater and has an emissivity of $\epsilon_1 = 0.7$. Surface 2 is a thin sheet metal with $\epsilon_2 = 0.9$. The lower half of surface 2 is insulated on the outside, while the upper surface is exposed to space, which may be considered black at 0 K. The uninsulated upper surface of 2 is to be maintained at a temperature of 60 °C. **(a)** Find the power input required per unit length of cylinder 1. **(b)** Find the operating temperature of cylinder 1.

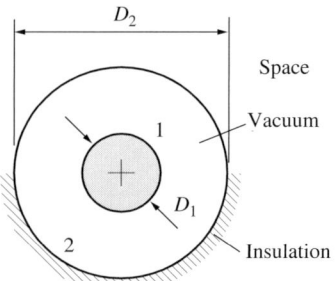

Figure 9P–18

9.26 For a lunar colony to be established on the moon, the intensity of the solar radiation at high noon is a concern (Fig. 9P–19). A "sun shade", consisting of two thin metal plates separated by insulation, is 4 m by 4 m square and is mounted parallel to and 2 m from the lunar surface. The solar radiation is perpendicular to the lunar surface and parallel such that a shadow 4 m × 4 m is cast beneath the shade. The moon's surface acts as a reradiating surface. The irradiation from the sun is $G = 1{,}400$ W/m^2, while it may be assumed that space behaves as a black body at 0 K. Find temperatures T_1, T_2, T_3, and T_4. T_1 is the temperature of the lunar surface far from the shade.

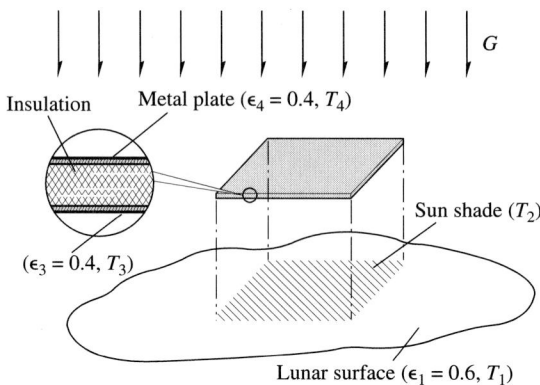

Figure 9P–19

9.27 An oil pipeline (1 m O.D. with 1 cm thick wall) crosses a very large flat desert and is parallel to and 1 m above the ground (Fig. 9P–20). Because of a system failure, the oil stops flowing in the pipe. The effects of any supports may be neglected. **(a)** Find the steady temperature of the sand surface, ignoring the presence of the pipe. **(b)** Assuming the sand behaves as a black-body with a uniform temperature of 97 °C, find the steady temperature of the oil.

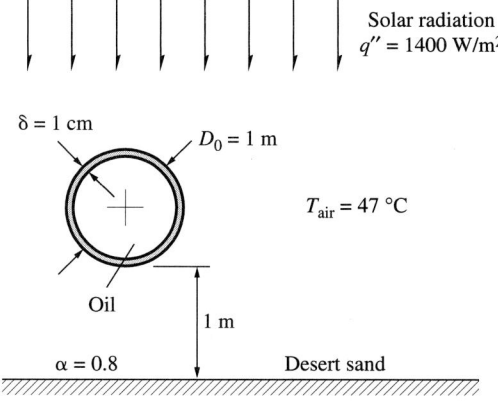

Figure 9P–20

9.28 Large heat-rejection radiators (a heat exchanger or condenser) of advanced design are needed for future systems where large amounts of power will be generated in space (Fig. 9P–21). Because of their weight, the conventional flat plate, tube and fin heat exchangers severely limit maximum power levels and are discouraged from consideration as a space power-generation technique. Accordingly, the development of novel low-mass heat-rejection systems is of major importance for the future of high-power space systems. The moving-belt radiator (MBR) is an example for such systems. Evaluate the heat rejection from an MBR in terms of the nomenclature of the figure.

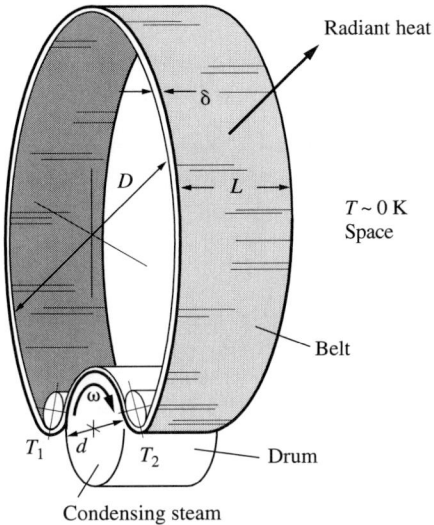

Figure 9P–21

9.29 Energy $u''' = 10^6$ W/m³ is generated in a flat plate of thickness $2\ell = 2$ cm (Fig. 9P–22). This energy is steadily transferred by radiation to an ambient at temperature $T_\infty = 27\,°$C. The thermal conductivity and surface emissivity of the plate are $k = 40$ W/m·K and $\epsilon_w = 0.8$, respectively. **(a)** Evaluate the surface and midplane temperatures of the plate (say T_w and T_0, respectively). **(b)** A solid screen with emissivity $\epsilon_0 = 0.2$ is placed on both sides of the plate. Reevaluate T_w and T_0. **(c)** The screen of part (b) now has holes with $\eta = 0.8 = A_{\text{holes}}/A_{\text{total screen area}}$. Reevaluate T_w and T_0.

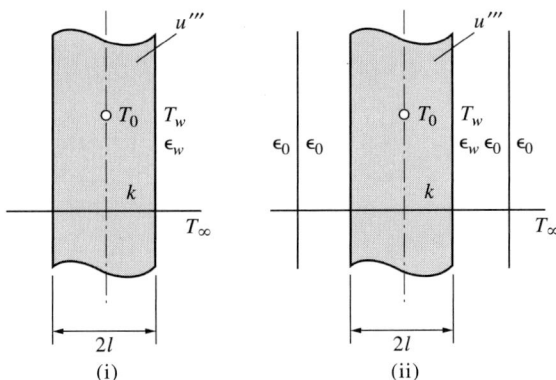

Figure 9P–22

9.30 Air is flowing in a long metal duct 0.4 m in diameter at 5 m/s. The duct has a temperature of 100 °C, and its emissivity is 0.9. A temperature-measuring device in the center of the duct has the shape of a sphere 5 mm in diameter and has an emissivity of 0.7. Its temperature is indicated to be 278 °C. For steady-state conditions, estimate the true temperature of the air in the duct.

9.31 A transparent spherical shell of inner radius r_i and outer radius r_o is filled with water. The shell is suspended such that on one side and away from it is a very large wall of emissivity ϵ_w and temperature T_w. The water is at its freezing point, and the pressure inside the shell is 1 atm. The ambient is at T_a. For $r_i = 0.8$ cm, $r_o = 1.0$ cm, $T_w = -50\,°C$, $\epsilon_w = 0.8$, $T_a = -20\,°C$, $h_{sf} = 333.4$ kJ/kg, determine the rate at which the water freezes **(a)** for stagnant ambient, **(b)** for an air velocity of 10 m/s.

9.32 A square copper plate of thickness δ having the initial temperature T_0 is dropped into an evacuated vertical chamber (Fig. 9P–23) whose walls are maintained at a constant temperature T_w ($\gg T_0$). Using the data given below, compute the temperature of the plate when it reaches the bottom of the chamber. $L_1 = 10$ cm, $L_2 = 10$ m, $\rho_1 = 8{,}000$ kg/m^3, $c_1 = 400$ J/kg·K, $\epsilon_1 = 0.8$, $\epsilon_2 = 0.4$, $\delta = 0.5$ cm, $D_2 = 3$ m, $T_w = 1{,}000\,°C$, $T_0 = 15\,°C$, $g = 9.81$ m/s^2.

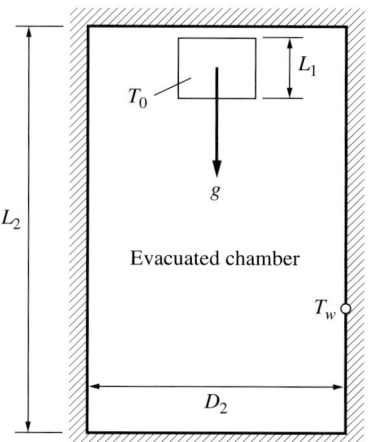

Figure 9P–23

9.33 Consider an ordinary light bulb (Fig. 9P–24). For $P = 200$ W, $\rho = 2{,}500$ kg/m^3, $c_p = 800$ J/kg·K, $\delta = 0.4$ mm, $\alpha = 0.10$ (emissivity), $T_\infty = 20\,°C$, $D = 8$ cm. **(a)** Calculate the steady surface temperature of the lighted bulb. **(b)** Find the time required for the surface temperature of the bulb to reach this steady value. **(c)** How would you modify your analysis for a fluorescent lamp?

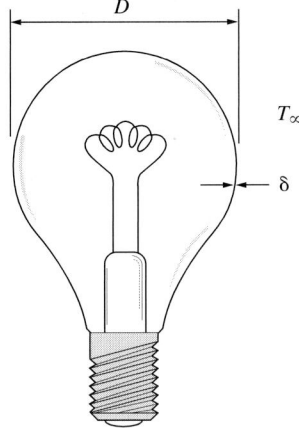

Figure 9P–24

9.34 Two thin-walled glossy (shiny) metal cylindrical containers 10 cm in diameter by 12 cm long are suddenly filled with water at $T = 95\ °C$. Assume that the temperature of the water in each container always remains uniform. One container is covered with a layer of asbestos 1/2 mm thick ($k = 0.151$ W/m·K). The ambient is at $20\ °C$. Find the temperature in each container after 5 minutes.

CHAPTER 10

GAS RADIATION

In the preceding chapter we discussed radiation in enclosures filled with gases at low temperatures. Recognizing that most monatomic and diatomic gases, as well as air, are transparent at these temperatures, we assumed the radiation energy exchange among the enclosure surfaces to be unaffected by the presence of these gases. At elevated temperatures, gases no longer remain transparent and become to some degree opaque. Then they start participating in the energy-exchange process by absorbing, and emitting (and sometimes scattering) this energy. Important technological problems such as the furnaces of steam boilers, diesel engines, nuclear explosions, plasma generators for nuclear fusion, rocket propulsion, hypersonic shock at elevated temperatures, and ablating systems involve the effect of gas radiation. Although some of these problems are relatively new, the absorption of solar radiation in the atmosphere has been receiving attention for about a century and is an example of gas radiation. The technological importance of gas radiation was first recognized during the early part of this century in connection with the heat transfer inside boiler furnaces. The radiation energy emitted by flames in furnaces and in diesel-engine combustion chambers depends not only on the gaseous emission but also on the heated carbon (soot) particles formed within flames.

Both technological and atmospheric radiation are mostly associated with water vapor and carbon dioxide, which are significant emitters and absorbers. Other examples of gases of significant emission and absorption properties are carbon monoxide, sulfur dioxide, ammonia, and hydrogen chloride. In general, radiation occurs over a number of discrete bands of the electromagnetic spectrum (recall Fig. 8.2). However,

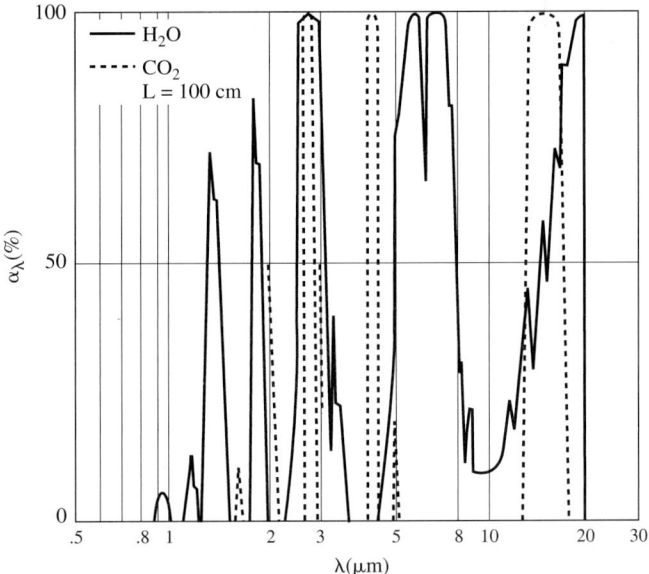

Figure 10.1 Absorption bands of CO_2 and H_2O vapor (from Rohsenow and Choi [9]). Used by permission.

there is a difference between the spectral behavior of opaque solids and gases. As shown in Figs. 8.16 and 8.17, the wavelength dependence of radiation properties for these solids is quite smooth, except for some cases which may be somewhat irregular. Gas properties, on the other hand, exhibit a very irregular wavelength dependence. Figure 10.1 illustrates this fact in terms of the absorption bands of carbon dioxide and water vapor. Actually, the radiation emitted or absorbed by a solid happens within the volume rather than the surface of the solid. The physics of radiation has consequently a common foundation for all media. The spectral differences are caused by the variety of energy transitions, which can best be explained in terms of quantum mechanics. This goes, however, beyond the scope of this text.

For the radiation properties of gases, and for the effect of radiation on the balance of thermal energy, we need to know the equation governing the balance of radiation energy. The next section is devoted to the development of this equation.

10.1 BALANCE OF RADIATION ENERGY

Problems involving gas radiation are an order of magnitude more complicated than those involving conduction, convection, and/or enclosure radiation. Consequently, the following discussion is restricted to one-dimensional problems.

Consider a one-dimensional optical system involving an absorbing and emitting medium (Fig. 10.2). The effect of scattering, which may be important in combustion

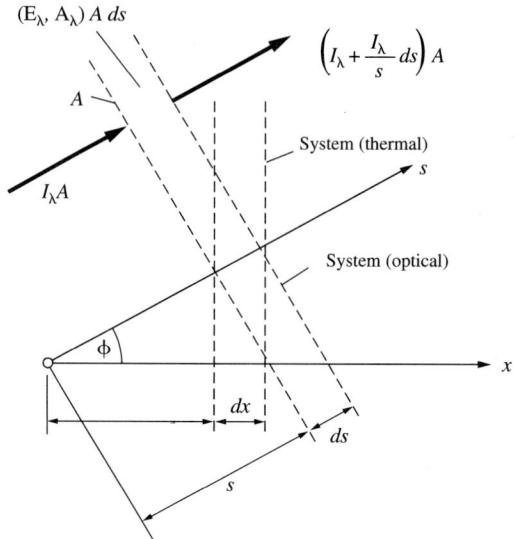

Figure 10.2 Balance of one-dimensional monochromatic radiation energy.

problems, is neglected. The balance of monochromatic radiation energy for the optical system gives

$$\frac{\partial I_\lambda}{\partial s} \cong E_\lambda - A_\lambda, \tag{10.1}$$

where I_λ is the monochromatic intensity and E_λ and A_λ denote respectively the monochromatic emission and absorption of radiation per unit volume. Equation (10.1) applies to unsteady as well as the steady problems. The effect of unsteadiness on Eq. (10.1), being inversely proportional to the velocity of light, is negligible. Details of this fact, however, are beyond the scope of this text.

Introducing the monochromatic absorption coefficient,

$$\kappa_\lambda = A_\lambda / I_\lambda, \tag{10.2}$$

Eq. (10.1) may be rearranged as

$$\frac{\partial I_\lambda}{\partial s} = E_\lambda - \kappa_\lambda I_\lambda. \tag{10.3}$$

Under radiative equilibrium, the radiation is uniform in space and $\partial I_\lambda / \partial s = 0$. Then Eq. (10.3) reduces to

$$E_\lambda = \kappa_\lambda I_\lambda^o, \tag{10.4}$$

where $I_\lambda^o = E_{b\lambda}/\pi$ according to Eq. (8.19), and $E_{b\lambda}$ is given by Eq. (8.34).

For nonequilibrium problems involving radiation, assuming **local equilibrium** for the emission of radiation and inserting Eq. (10.4) into Eq. (10.3) yields

$$\frac{\partial I_\lambda}{\partial s} = \kappa_\lambda \left(I_\lambda^o - I_\lambda\right), \qquad (10.5)$$

which is the balance of monochromatic radiation energy (this balance is often referred to as the transfer equation). Equation (10.5) plays a pivotal role in radiation problems. It is useful for the evaluation of radiative properties and for the evaluation of the radiation heat flux by which the thermal energy balance is modified.

10.2 RADIATION PROPERTIES OF GASES

The radiation properties of weakly absorbing fluids are usually determined by the experimental setup shown in Fig. 10.3. The fluid fills a container with two windows. One of the windows faces a known radiation source (say a black body) and the other faces a spectroscope which resolves the radiation beam. The container at a uniform pressure and temperature provides a uniform density and absorption coefficient.

Under steady conditions, the integration of Eq. (10.5) results in

$$I_\lambda^o - I_\lambda(s) = Ce^{-\kappa_\lambda s}. \qquad (10.6)$$

Letting $I_\lambda = I_\lambda(0)$ for $s = 0$,

$$I_\lambda^o - I_\lambda(0) = C,$$

and using this result in Eq. (10.6) yields, after some rearrangement,

$$I_\lambda(s) = I_\lambda^o \left(1 - e^{-\kappa_\lambda s}\right) + I_\lambda(0)e^{-\kappa_\lambda s}. \qquad (10.7)$$

Now, inserting $s = L$ into Eq. (10.7) gives the monochromatic intensity of the radiation at the container exit,

$$I_\lambda(L) = I_\lambda^o \left(1 - e^{-\kappa_\lambda L}\right) + I_\lambda(0)e^{-\kappa_\lambda L}, \qquad (10.8)$$

where the first and second righthand terms respectively show the emission and absorption of the radiation within the container (why?). If the temperature of the black body is small compared to that of the container fluid, $I_\lambda(0) \ll I_\lambda^o$ and Eq. (10.8) reduces to its emission term,

$$I_\lambda(L) \cong I_\lambda^o \left(1 - e^{-\kappa_\lambda L}\right). \qquad (10.9)$$

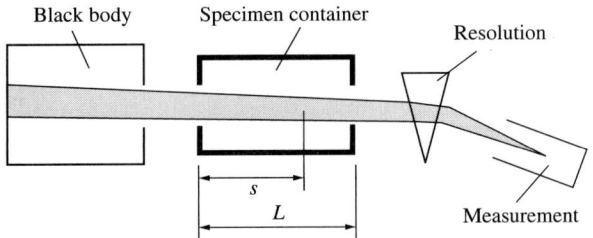

Figure 10.3 Setup for measurement of monochromatic radiation.

If the temperature of the container fluid is small compared with the temperature of the black body, $I_\lambda^o \ll I_\lambda(0)$ and Eq. (10.8) reduces to its absorption term,

$$I_\lambda(L) \cong I_\lambda(0)e^{-\kappa_\lambda L}. \tag{10.10}$$

Defining the monochromatic emissivity ϵ_λ as the ratio of the monochromatic intensity of radiation leaving the fluid relative to the monochromatic intensity of the black body at the same temperature, we have from Eq. (10.9)

$$\epsilon_\lambda = I_\lambda(L)/I_\lambda^o = 1 - e^{-\kappa_\lambda L}. \tag{10.11}$$

Also, defining monochromatic transmissivity τ_λ as the ratio of the monochromatic intensity of radiation leaving the fluid relative to the monochromatic intensity of radiation entering it, we get from Eq. (10.10)

$$\tau_\lambda = I_\lambda(L)/I_\lambda(0) = e^{-\kappa_\lambda L}. \tag{10.12}$$

Clearly, from Eqs. (10.11) and (10.12),

$$\epsilon_\lambda = 1 - \tau_\lambda = \alpha_\lambda, \tag{10.13}$$

where α_λ is the monochromatic absorptivity of the fluid, and the Kirchhoff law continues to hold for the fluid.

Since gases are the most frequently encountered absorbing media in engineering and environmental problems, further discussion on the radiation properties of fluids is confined to gases. We are mainly interested in the total effect of radiation, which is obtained by a summation of each wavelength of significance. Introducing the monochromatic mass-absorption coefficient $\kappa_\lambda' = \kappa_\lambda/\rho$, and noting that $\rho \sim p$ for gases at uniform temperature,

$$\kappa_\lambda = \rho\kappa_\lambda' = p\kappa_\lambda''.$$

Then, we have from Eqs. (10.12) and (10.13) the total emissivity of the gas component at partial pressure p,

$$\epsilon = 1 - \frac{\sum_\lambda I_\lambda(0)e^{-\kappa_\lambda'' pL}}{\sum_\lambda I_\lambda(0)}. \tag{10.14}$$

Hottel[1] has developed a series of charts from experimental data for the total emissivity of various gases. Figures 10.4 and 10.6 show those for CO_2 and H_2O vapor at 1 atm total pressure and negligible partial pressure. Similar charts are available for CO, NH_3 and SO_2. The emissivities obtained from these figures are multiplied by the factors given in Figs. 10.5 and 10.7, which correct for departures from 1 atm total pressure. Note that these figures employ English (°F, ft) rather than SI (metric) units.

---

[1] Figures 10.4 and 10.6 are adapted from H. C. Hottel and R. B. Egbert, *AICHE Trans.* **38**, (1942). Used by permission. Figures 10.5, 10.7, and 10.8 are adapted from H. C. Hottel, Ch. 4 of *Heat Transmission*, 3d ed. By W. H. McAdams, McGraw-Hill, New York, 1954. Used by permission.

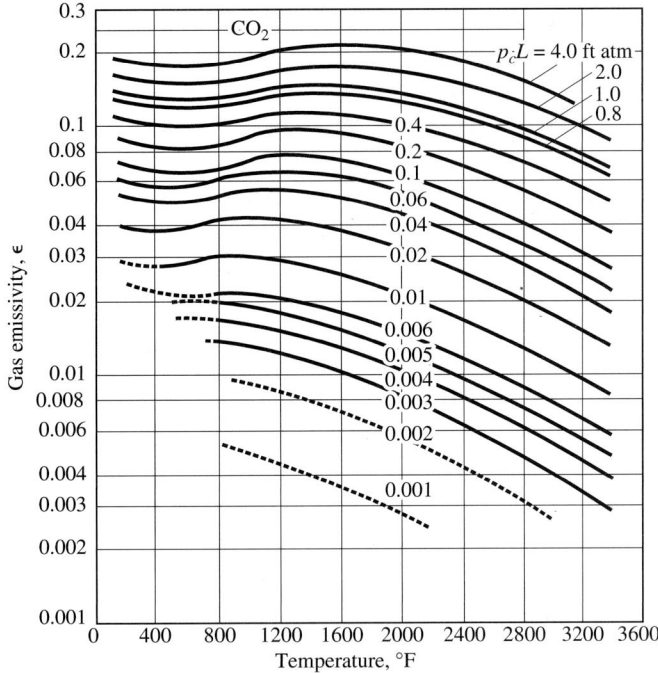

Figure 10.4 Emissivity of CO_2 at 1 atm total pressure and small partial pressure. From McAdams [25].

Now, consider a gas of thickness L and temperature T_g exchanging radiant energy with a black surface at temperature T_s. The gas emissivity ϵ_g is obtained from Figs. 10.4 and 10.5 for CO_2 and Figs. 10.6 and 10.7 for H_2O vapor. For example, the value of ϵ_g ($T_g, p_g L$) from Fig. 10.4 multiplied by the correction factor C_g from Fig. 10.5 includes the effect of CO_2 pressure on the gas emissivity.

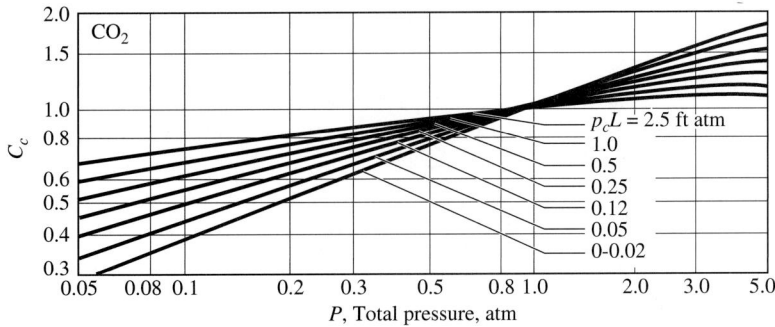

Figure 10.5 Correction factor for converting emissivity of CO_2 at 1 atm total pressure to emissivity at P atm total pressure. From McAdams [25].

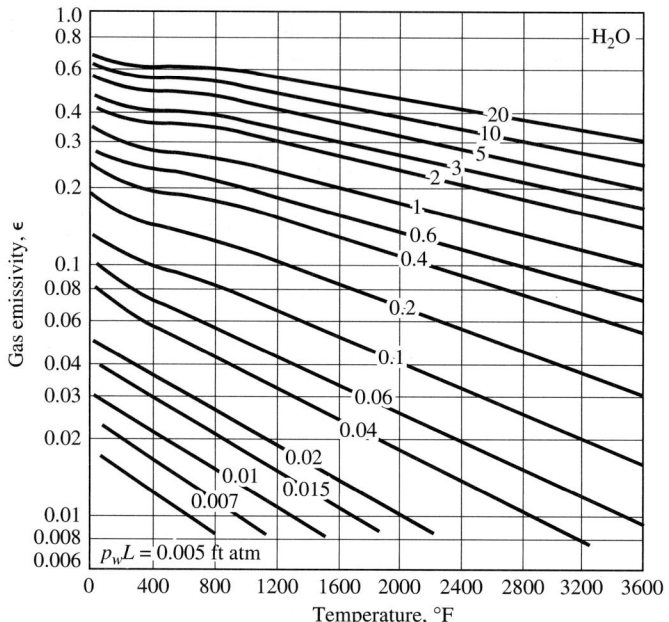

Figure 10.6 Emissivity of H_2O vapor at 1 atm total pressure and small partial pressure. From McAdams [25].

Under equilibrium, $T_g = T_s$ and $\alpha_g = \epsilon_g$. Under nonequilibrium, $T_g \neq T_s$ and the following empirical relation holds:

$$\alpha_{cs} = C_c \left(\frac{T_g}{T_s}\right)^{0.65} \epsilon_c \left(T_s, p_c L \frac{T_s}{T_g}\right) \tag{10.15}$$

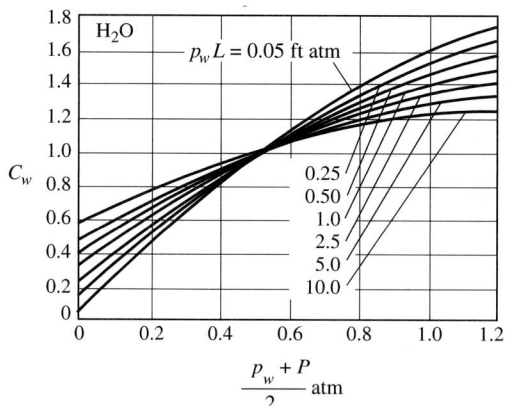

Figure 10.7 Correction factor for converting emissivity of H_2O vapor at 1 atm total pressure to emissivity at P atm total pressure. From McAdams [25].

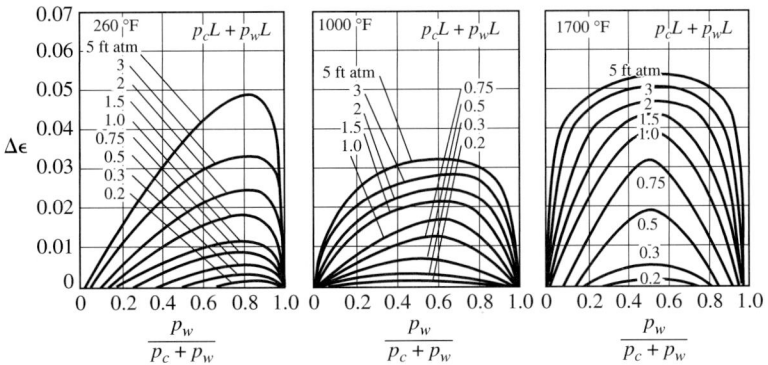

Figure 10.8 Correction to gas emissivity resulting from spectral overlap of CO_2 and H_2O vapor. From McAdams [25].

for CO_2, and

$$\alpha_{ws} = C_w \left(\frac{T_g}{T_s}\right)^{0.45} \epsilon_w \left(T_s, p_w L \frac{T_s}{T_g}\right) \quad (10.16)$$

for H_2O, where ϵ_c and ϵ_w are obtained from Figs. 10.4 and 10.6 at T_s, and C_c, C_w are the correction factors from Figs. 10.5 and 10.7. When both CO_2 and H_2O vapor exist in a mixture, the emissivity is less than the sum of emissivities evaluated for each gas alone because of the overlap in bands. Figure 10.8 gives the amount of correction $\Delta\epsilon$ to be subtracted from the sum of emissivities. Thus, for a mixture of CO_2 and H_2O vapor,

$$\epsilon_g = \epsilon_c + \epsilon_w - \Delta\epsilon. \quad (10.17)$$

EXAMPLE 10.1

The exhaust gases from a combustor burning a hydrocarbon fuel are at $p = 4$ atm and $T_g = 1200$ K. The exhaust-gas composition is 15% CO_2, 7.5% H_2O and 77.5% N_2. We wish to determine the emissivity for a path length of $L = 1$ m and the absorptivity for radiation from a black wall at $T_s = 600$ K. Assume that the nitrogen is transparent and has no effect on the radiation.

For $T_g = 1200$ K $\cong 1700\,°F$, the partial pressure of CO_2 is $p_c = 0.15 \times 4 = 0.6$ atm,

$$p_c L = 0.6 \text{atm} \times 1\text{m} = 0.6 \text{m·atm} \cong 2 \text{ft·atm},$$

for the water vapor, $p_w = 0.075 \times 4 = 0.3$ atm,

$$p_w L = 0.3 \text{atm} \times 1\text{m} = 0.3 \text{m·atm} \cong 1 \text{ft·atm}.$$

In order to use Figs. 10.4–10.8 we need

$$(p_c + p_w)L \cong 3\text{ft·atm}$$

$$\frac{p_c + p_w}{p} = \frac{0.6 + 0.3}{4} = 0.225$$

$$\frac{p_w}{p_c + p_w} = \frac{0.3}{0.6 + 0.3} = 0.333$$

$$\frac{p_w + p}{2} = \frac{0.3 + 4}{2} = 2.15,$$

from Figs. 10.4 through 10.8,

$$\epsilon_c = 0.18, \quad C_c \cong 1.1$$

$$\epsilon_w = 0.19, \quad C_w \cong 1.6$$

and

$$\Delta\epsilon \cong 0.05.$$

Then

$$\epsilon_g = C_c\epsilon_c + C_w\epsilon_w - \Delta\epsilon$$

or

$$\epsilon_g = 1.1 \times 0.18 + 1.6 \times 0.19 - 0.05 \cong 0.452.$$

In order to get the absorptivity, we need to use Eqs. (10.15) and (10.16), then correct the total absorptivity using $\Delta\epsilon$ from Fig. 10.8. For

$$p_c L(T_s/T_g) = 1\text{ft·atm}, \quad p_w L(T_s/T_g) = 0.5\text{ft·atm}$$

and

$$\left(\frac{T_g}{T_s}\right)^{0.65} = 2^{0.65} = 1.57, \quad \left(\frac{T_g}{T_s}\right)^{0.45} = 2^{0.45} = 1.37$$

from Figs. 10.4 through 10.8,

$$\epsilon_c = 0.15, \quad \epsilon_w = 0.14$$

and

$$\alpha_{gs} = 1.1 \times 0.15 \times 1.57 + 1.6 \times 0.14 \times 1.37 - 0.05 = 0.516.$$

♦

The foregoing charts are based on gases in a hemispherical container of radius L radiating to differential area at the center of the base (Fig. 8.3). In terms of an equivalent mean beam length, however, the use of these charts may be extended to other gas shapes. It can be shown for irregular gas shapes that the mean length is

$$L = 4\frac{V}{A}, \tag{10.18}$$

where V is the volume and A is the peripheral area of gas. Details of this result are beyond the scope of this text. Equation (10.18) is for gases at a uniform temperature

Table 10.1 (from Rohsenow and Hartnett [24]).

Sphere	$\frac{2}{3}$ (diameter)
Infinite cylinder	Diameter
Infinite parallel planes	2 (distance between planes)
Semi-infinite cylinder radiating to center of base	Diameter
Right circular cylinder height equal to diameter	
Radiating to center of base	Diameter
Radiating to whole surface	$\frac{2}{3}$ (diameter)
Infinite cylinder of half-circular cross section radiating to spot in middle of flat side	Radius
Rectangular parallelpipeds	
-Cube	$\frac{2}{3}$ (edge)
radiating to 1 × 4 face	0.9 (shortest edge)
radiating to 1 × 1 face	0.86 (shortest edge)
radiating to all faces	0.891 (shortest edge)
Space outside infinite bank of tubes with conters on equilateral triangles	
Tube diameters — clearance	3.4 (clearance)
Tube diameter — 1/2 clearance	4.44 (clearance)

and small pL product. For not so small pL, the actual L is somewhat smaller than the L obtained from Eq. (10.18). For most practical cases the actual L is about 85% of Eq. (10.18). Table 10.1 gives values of L for various gas shapes.

So far, we have studied the evaluation of mean radiation properties. We may employ these properties in enclosure problems involving lumped gas radiation. For an opaque gray gas

$$\epsilon_\lambda = \alpha_\lambda = \alpha = \epsilon$$

and

$$\tau = 1 - \epsilon.$$

Now, assume this gas isothermally fills an enclosure made of gray surfaces. The resulting enclosure problems can be treated by replacing the gas with a transparent solid partition with negligible reflectivity.

Example 10.2

Two parallel black plates at temperature $T_1 = 1,000$ K and $T_2 = 600$ K are a distance $\ell = 15$ cm (∼ 6 inch) apart in CO_2 at 1 atm. We wish to determine the radiant heat transfer between the plates.

The problem is a special case of Ex. 9.5. Replacing the glass transmissivity τ_0 of Ex. 9.5 with a gas transmissivity $\tau_{1(g)2}$, we have from Eq. (9.49), for black walls,

$$R_{12} = \frac{1}{A}\left(\frac{2}{1 + \tau_{1(g)2}}\right) \tag{10.19}$$

and, from Eq. (9.50),

$$\frac{\dot{Q}_{1(g)2}}{A} = \frac{1}{2}(1 + \tau_{1(g)2})(E_{b1} - E_{b2}). \tag{10.20}$$

Thus, the problem is reduced to finding $\tau_{1(g)2}$.

For the gas temperature, as a first trial, assume equilibrium and let $\epsilon_{1g} \cong \epsilon_{g2}$ (gas emissivities on the side of each wall). This gives

$$T_g^4 = \frac{1}{2}(T_1^4 + T_2^4) = \frac{1}{2}(1{,}000^4 + 600^4)$$

or

$$T_g = 867\text{K}(\sim 1{,}100\,°\text{F}).$$

Also, the mean beam length is

$$L = 0.85 \times 4 \times \frac{V}{A} = 0.85 \times 4 \times \frac{A \times \ell/2}{A}$$

or

$$L = 0.85 \times 4 \times \frac{\ell}{2} = \frac{0.85 \times 4 \times 6}{12 \times 2} = 0.85\,\text{ft},$$

which gives

$$p_g L = 0.85\,\text{ft}\cdot\text{atm}.$$

Under equilibrium, for $T_g \cong 1{,}100\,°\text{F}$ and $p_g L = 0.85$ ft·atm, we have from Fig. 10.4,

$$\epsilon_g \cong 0.14 \quad \text{and} \quad \tau_g \cong 0.86.$$

Then, Eq. (10.20) yields

$$\frac{\dot{Q}_{1(g)2}}{A} = \frac{1}{2}(1 + 0.86) \times 5.67 \times 10^{-8}\,\text{W/m}^2\cdot\text{K}^4 \times (1{,}000^4 - 600^4)\text{K}^4$$

or

$$\frac{\dot{Q}_{1(g)2}}{A} = 45.9\,\text{kW/m}^2. \tag{10.21}$$

In the actual case, $\epsilon_{1g} \neq \epsilon_{g2}$, ϵ_{1g} between the hot wall and gas, and ϵ_{g2} between the gas and cold wall need to be evaluated separately. However, the difference between the emissivities is not expected to significantly affect Eq. (10.21). Demonstration of this fact is left to reader. ◆

Having reduced the problems of lumped gas radiation to enclosure problems with transparent partitions, we proceed next to problems of distributed gas radiation.

10.3 DISTRIBUTED GAS RADIATION

In this section we restrict ourselves to gray gases, as we did in the foregoing brief discussion on the lumped gas radiation. Spectral considerations, although important for the quantitative effects of radiation, are beyond the scope of this text. Consequently, our objective is to gain some knowledge only on the qualitative foundations of gas radiation.

Our starting point is the balance of radiation energy (transfer equation), obtained from the average of Eq. (10.5) over the wavelength spectrum,

$$\frac{\partial I}{\partial s} = \kappa (I_0 - I). \tag{10.22}$$

In terms of Fig. 10.9, the integration of this equation with respect to s yields

$$I(s) = I(0)e^{-\tau(s,0)} + \int_0^s I_0(s')e^{-\tau(s,s')}\kappa \, ds', \tag{10.23}$$

where $I(0)$ is the intensity at location $s = 0$, and $\tau(s, s')$ is the **optical thickness** of the medium

$$\tau(s, s') = \int_{s'}^{s} \kappa(s'')\,ds''. \tag{10.24}$$

Clearly, Eq. (10.23) expresses the fact that the intensity at point s and in direction $\mathbf{m}$ (Fig. 10.9) results from the emission of all the interior points such as s' and from the emission of the boundary $s = 0$, respectively reduced by factors $e^{-\tau(s,s')}$ and $e^{-\tau(s,0)}$ to allow for the absorption by the intervening matter.

To simplify our discussion, we neglect for the time being the effect of the boundaries by eliminating $I(0)e^{-\tau(s,0)}$ from Eq. (10.23). Then we have

$$I(s) = \int_0^s I_0(s')e^{-\tau(s,s')}\kappa \, ds'. \tag{10.25}$$

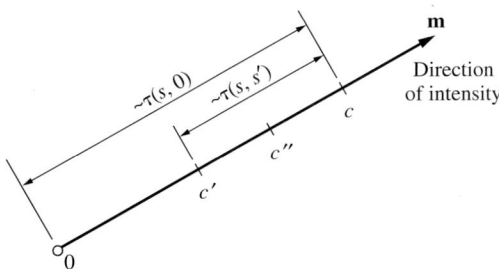

Figure 10.9 Optical coordinate.

Furthermore, we restrict ourselves to two limiting cases corresponding to negligible and rapid decay of $e^{-\tau(s,s')}$, i.e., thin and thick gases.

10.3.1 Thin Gas

In the first case, $e^{-\tau(s,s')} \sim 1$ and the absorption is negligible, so Eq. (10.25) becomes

$$I(s) = \int_0^s I_0(s')\kappa\, ds'. \tag{10.26}$$

Clearly, the assumption of

$$I_0 \gg I \tag{10.27}$$

reduces to

$$\frac{\partial I}{\partial s} = \kappa I_0, \tag{10.28}$$

whose integration leads also to Eq. (10.26). The condition stated by Eq. (10.27), that is,

$$\boxed{\text{Emission} \gg \text{Absorption}}$$

is the definition of a **thin gas**.

Next we consider the heat flux associated with this gas. The derivative $\partial/\partial s$ in Eq. (10.28) is the gradient in the direction of the propagation of radiation. In terms of a cartesian coordinate, say x, noting from (Fig. 10.2) that

$$\frac{\partial}{\partial s} = \frac{\partial}{\partial x}\frac{\partial x}{\partial s} = \frac{\partial}{\partial x}\cos\phi, \tag{10.29}$$

Eq. (10.28) may be rearranged as

$$\frac{\partial I}{\partial x}\cos\phi = \kappa I_0. \tag{10.30}$$

On the other hand, recalling Eq. (8.18),

$$E = q^R = \int_\Omega I \cos\phi\, d\Omega, \tag{8.18}$$

differentiating this equation with respect to x and changing the order of differentiation and integration yields

$$\frac{\partial q^R}{\partial x} = \int_\Omega \frac{\partial I}{\partial x}\cos\phi\, d\Omega. \tag{10.31}$$

This result may be rearranged with the help of Eq. (10.30) to give

$$\frac{\partial q^R}{\partial x} = \int_\Omega \kappa I_0(s')\, d\Omega. \tag{10.32}$$

Finally, noting that I_0 is independent of the solid angle and that $\pi I_0(s') = E_b$, Eq. (10.32) may be reduced to

$$\boxed{\frac{\partial q^R}{\partial x} = 4\kappa E_b}, \qquad (10.33)$$

the heat flux associated with a thin gray gas. In terms of a characteristic length ℓ and optical thickness $\tau = \kappa \ell$, Eq. (10.33) may be rearranged on dimensional grounds to give

$$\boxed{q^R \sim \tau E_b}, \qquad (10.34)$$

which will prove convenient later.

So far, we have considered one extreme case corresponding to the negligible decay of $e^{-\tau(s,s')}$, which led us to the thin-gas limit. We proceed now to the other extreme case corresponding to the rapid decay of $e^{-\tau(s,s')}$.

10.3.2 Thick Gas

In terms of a constant absorption coefficient κ and $\tau = \kappa s$ and $\tau' = \kappa s'$, first rearrange Eq. (10.25) as

$$I(\tau) = \int_0^\tau I_0(\tau') e^{-(\tau-\tau')} d\tau'. \qquad (10.35)$$

Next, assume that the absorption occurs over a short distance because of large κ and that the equilibrium intensity at τ' may be approximated by its tangent about τ (Fig. 10.10),

$$I_0(\tau') = I_0(\tau) + \left(\frac{\partial I_0}{\partial \tau}\right)_{\tau=\tau'} (\tau' - \tau), \qquad (10.36)$$

which is the two-term Taylor expansion about $\tau' = \tau$. Inserting Eq. (10.36) into Eq. (10.35), and noting that $I_0(\tau)$ and $(\partial I_0/\partial \tau')_{\tau'=\tau}$ are constants with respect to the integration in τ',

$$I(\tau) = I_0(\tau) \int_0^\tau e^{-(\tau-\tau')} d\tau' + \left(\frac{\partial I_0}{\partial \tau}\right) \int_0^\tau (\tau' - \tau) e^{-(\tau-\tau')} d\tau'. \qquad (10.37)$$

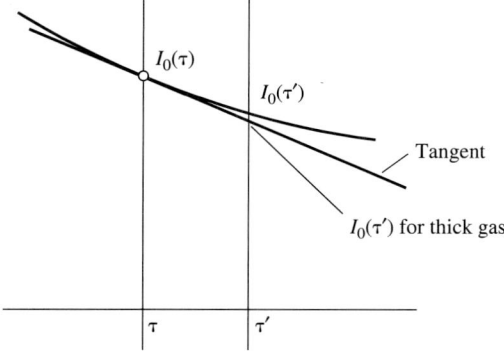

Figure 10.10 Thick-gas approximation.

By the help of an integral table,

$$\int_0^\tau e^{-(\tau-\tau')}d\tau' = e^{-\tau}\int_0^\tau e^{\tau'}d\tau' = 1$$

$$\int_0^\tau (\tau'-\tau)e^{-(\tau-\tau')}d\tau' = e^{-\tau}\left(\int_0^\tau \tau' e^{\tau'}d\tau' - \tau\int_0^\tau e^{\tau'}d\tau'\right) = -1,$$

Eq. (10.37) may be rearranged as

$$I(\tau) = I_0(\tau) - \frac{\partial I_0(\tau)}{\partial \tau} \qquad (10.38)$$

or, on dimensional grounds,

$$I \sim I_0\left(1 - \frac{1}{\tau} + \cdots\right),$$

which implies for $\tau \gg 1$

$$I \sim I_0. \qquad (10.39)$$

The condition stated by this result, that is,

$$\boxed{\text{Emission} \sim \text{Absorption}},$$

is the definition of a **thick gas**.

Now, we consider the heat flux associated with this gas. Inserting Eq. (10.38) into Eq. (8.18) gives

$$q^R = I_0 \int_\Omega \cos\phi\, d\Omega - \frac{1}{\kappa}\left(\frac{\partial I_0}{\partial x}\right)\int_\Omega \cos^2\phi\, d\Omega. \qquad (10.40)$$

Recalling $d\Omega = d\theta \sin\phi\, d\phi$ from Fig. 8.3, the integrals of Eq. (10.40) may be rearranged and evaluated. The result is

$$\int_\Omega \cos\phi\, d\Omega = \int_0^{2\pi} d\theta \int_0^{2\pi} \cos\phi \sin\phi\, d\phi = 0$$

$$\int_\Omega \cos^2\phi\, d\Omega = \int_0^{2\pi} d\theta \int_0^{2\pi} \cos^2\phi \sin\phi\, d\phi = \frac{4}{3}\pi.$$

Employing this result and $\pi I_0 = E_b$, we obtain from Eq. (10.40)

$$\boxed{q^R = -\frac{4}{3\kappa}\frac{\partial E_b}{\partial x}}, \qquad (10.41)$$

the heat flux associated with a thick gray gas. In terms of a characteristic length ℓ and optical thickness $\tau = \kappa\ell$, Eq. (10.41) may be rearranged on dimensional grounds to give

$$\boxed{q^R \sim \frac{E_b}{3\tau}}. \qquad (10.42)$$

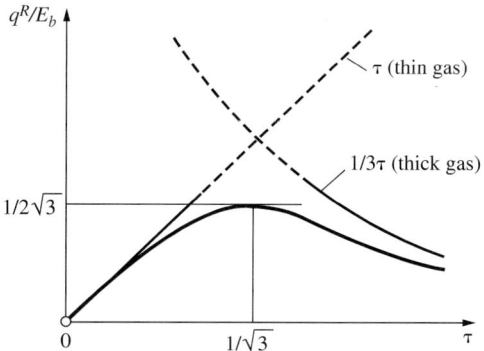

Figure 10.11 Radiant heat flux for thin gas, thick gas, and for any optical thickness.

To find the heat flux resulting from radiating gases with an arbitrary optical thickness, consider the heat flux given by Eqs. (10.34) and (10.42) (Fig. 10.11). It can be shown for an arbitrary optical thickness that the heat flux, based on the assumption of isotropic radiation stress (pressure), is

$$q^R \sim \frac{\tau E_b}{1 + 3\tau^2}. \qquad (10.43)$$

Details leading to this equation are beyond the scope of this text. Note that the limits of Eq. (10.43) for $\tau \to 0$ and $\tau \to \infty$ respectively lead to Eqs. (10.34) and (10.42), as expected.

Finally, we may summarize what we have learned so far on the heat flux associated with radiating gases. Figure 10.12 shows separately the effect of emission (a measure of the hotness) and that of absorption (a measure of the optical thickness).

10.3.3 Effect of Boundaries

In the foregoing discussion we neglected the effect of boundaries. A knowledge of this effect is needed before we can proceed to some illustrative examples. The behavior of a thick gas near a boundary is beyond the scope of this text (the interested reader may refer to Arpacı, [15]; Arpacı and Larsen, [16]; Lord and Arpacı [17]. However, most fluids of technological importance may be adequately described by the assumption of thin gas, and the behavior of thin gases near a boundary is the knowledge we need. Because of its negligible absorption, the thin gas is usually influenced by the geometry of the enclosures, and its behavior near a boundary depends on this geometry as well as on the boundary itself.

Here, after some general remarks, we demonstrate the boundary and geometry effects in terms of a thin gas between two parallel plates. First, reconsider the radiation energy balance given by Eq. (10.22). The one-dimensional cartesian form of this balance, obtained with the help of Eq. (10.29), is

$$\frac{\partial I}{\partial x} \cos \phi = \kappa (I_0 - I). \qquad (10.44)$$

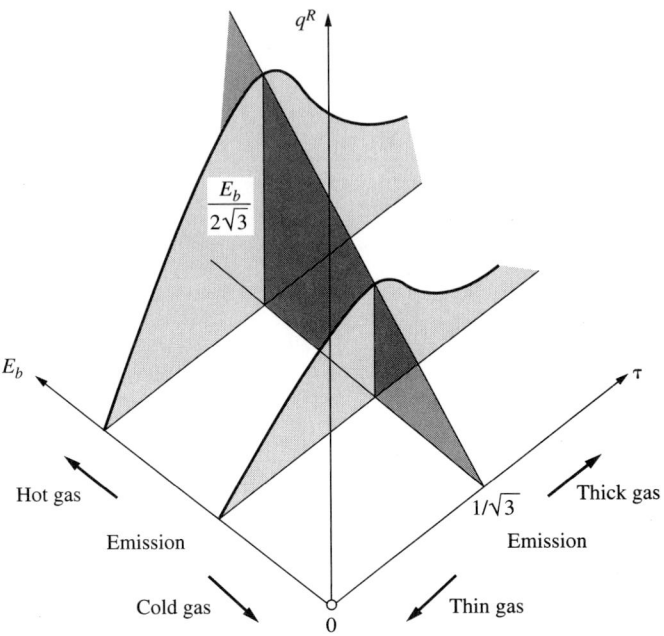

Figure 10.12 Emission and absorption effects on radiant heat flux.

Integrating this equation over the solid angle, employing Eq. (8.18), and noting that I_0 is isotropic and independent of the solid angle and that $\pi I_0 = E_b$, yields

$$\frac{\partial q^R}{\partial x} = 4\kappa E_b - \kappa \int_\Omega I\,d\Omega. \tag{10.45}$$

Clearly, the difference between Eqs. (10.33) and (10.45),

$$\int_\Omega I\,d\Omega, \tag{10.46}$$

represents the effect of boundaries. Now, assume **hemispherical isotropy**, and replace the actual intensity with **two-stream** (outward and inward) intensities as shown in Fig. 10.13. Then, Eq. (10.46) may be rearranged as

$$\int_\Omega I\,d\Omega = 2\pi(I^+ + I^-). \tag{10.47}$$

Note that, for (spherical) isotropy, $I^+ = I^- = I_0$, and Eq. (10.47) gives $4\pi I_0$, as expected.

After the foregoing general considerations, we now consider two parallel plates. In a manner similar to the development leading to Eq. (9.6), we obtain (see Fig. 10.14)

$$I_1^+ = \epsilon_1 I_{01} + \rho_1 I_1^-, \tag{10.48}$$

$$I_2^- = \epsilon_2 I_{02} + \rho_2 I_2^+. \tag{10.49}$$

Sec. 10.3 Distributed Gas Radiation

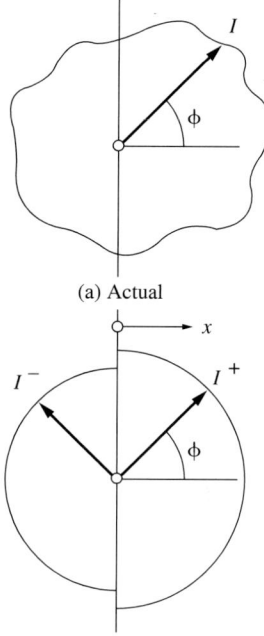

Figure 10.13 Hemispherical approximation of actual intensity. (a) Actual, (b) hemispherical.

For a thin gas the absorption is negligible, and I^+ and I^- remain uniform across the thickness of the gas between the plates (Fig. 10.15). Consequently, I^+ and I^- of surface 1 become identical respectively to I^+ and I^- of surface 2. Now, solving Eqs. (10.48) and (10.49) for I^+ and I^- yields

$$I^+ = \frac{\epsilon_1 I_{01} + \rho_1 \epsilon_2 I_{02}}{1 - \rho_1 \rho_2}, \quad I^- = \frac{\epsilon_2 I_{02} + \rho_2 \epsilon_1 I_{01}}{1 - \rho_1 \rho_2}. \tag{10.50}$$

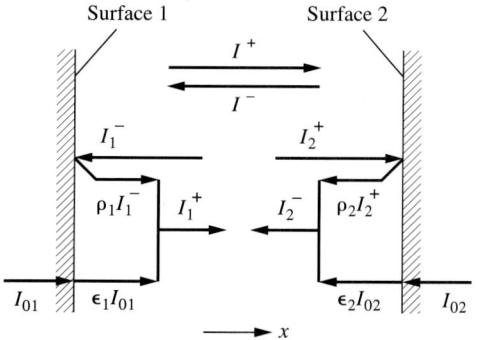

Figure 10.14 Boundary intensities for two parallel plates.

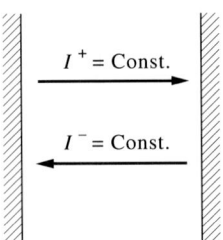

Figure 10.15 Hemispherical intensities for thin gas between parallel plates.

Summing these intensities, noting that $\rho + \epsilon = 1$ and that $\pi I_0 = E_b$, we get

$$2\pi(I^+ + I^-) = 4\frac{\left(\frac{1}{\epsilon_2} - \frac{1}{2}\right)E_{b1} + \left(\frac{1}{\epsilon_1} - \frac{1}{2}\right)E_{b2}}{\left(\frac{1}{\epsilon_2} - \frac{1}{2}\right) + \left(\frac{1}{\epsilon_1} - \frac{1}{2}\right)} = 4E_{bM}, \quad (10.51)$$

where E_{bM} is an average emissive power, introduced for later convenience. For surfaces with identical emissivities,

$$E_{bM} = \frac{1}{2}(E_{b1} + E_{b2}),$$

as expected. Finally, inserting Eq. (10.51) into Eq. (10.47) and the result into Eq. (10.45) gives

$$\boxed{\frac{\partial q^R}{\partial x} = 4\kappa(E_b - E_{bM})}. \quad (10.52)$$

The effect of the boundaries on a thin gas in a slot turns out to be E_{bM} [recall Eq. (10.33)]. Now we proceed to a couple of illustrative examples involving thin gases.

EXAMPLE 10.3

Consider a stagnant thin gas between two vertical plates separated by a distance ℓ. The temperature and emissivity of the plates are T_1, T_2 and ϵ_1, ϵ_2, respectively. The conductivity and absorptivity of the gas are k and κ. We wish to determine the temperature distribution in the gas.

Consider the differential system shown in Fig. 10.16. The first law of thermodynamics (conservation of total energy) applied to this system yields

$$-\frac{d}{dx}(q_x^K + q_x^R) = 0, \quad (10.53)$$

where q_x^K and q_x^R are the heat fluxes associated with conduction and radiation, respectively.

Sec. 10.3 Distributed Gas Radiation

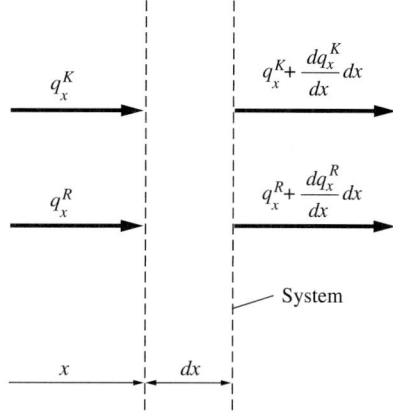

Figure 10.16 Balance of thermal (conduction + radiation) energy.

To proceed further, we need the particular laws relating the heat fluxes to temperature. We know from Chapter 1 and from the development preceding the present example that

$$q_x^K = -k\frac{dT}{dx} \tag{10.54}$$

$$\frac{dq_x^R}{dx} = 4\kappa(E_b - E_{bM}). \tag{10.55}$$

Inserting Eqs. (10.54) and (10.55) into Eq. (10.53) and noting that $E_b = \sigma T^4$ gives, for constant properties, the governing equation,

$$k\frac{d^2T}{dx^2} - 4\kappa\sigma(T^4 - T_M^4) = 0, \tag{10.56}$$

subject to the boundary conditions,

$$T(0) = T_1 \quad \text{and} \quad T(\ell) = T_2. \tag{10.57}$$

There is no difficulty in the numerical integration of Eq. (10.56), and the mathematics of Chapter 4 may easily be used here. However, rather than a numerical approach, we proceed by assuming $\Delta T = T_1 - T_2$ is small, and expand T about T_M into a Taylor series,

$$T^4 = T_M^4 + 4T_M^3(T - T_M) + \cdots. \tag{10.58}$$

In terms of this approximation, Eq. (10.56) reduces to the governing equation of extended surfaces [Eq. (2.109) with $u''' = 0$],

$$k\frac{d^2T}{dx^2} - 16\kappa\sigma T_M^3(T - T_M) = 0. \tag{10.59}$$

Introducing a dimensionless distance $\xi = x/\ell$, optical thickness $\tau = \kappa\ell$, and the **Planck number** based on temperature T_M,

$$P_M = \frac{4\sigma T_M^3}{k/\ell} \sim \frac{\text{Radiation}}{\text{Conduction}}, \tag{10.60}$$

and $4\tau P_M = R^2$, Eqs. (10.59) and (10.57) may be rearranged to yield

$$\frac{d^2 T}{d\xi^2} - R^2(T - T_M) = 0 \tag{10.61}$$

with

$$T(0) = T_1, \quad T(1) = T_2. \tag{10.62}$$

Expressing the general solution of Eq. (10.61) in terms of hyperbolic functions (recall Ex. 2.10), we have

$$T - T_M = C_1 \cosh R\xi + C_2 \sinh R\xi. \tag{10.63}$$

The use of the boundary conditions gives

$$T_1 - T_M = C_1$$

$$T_2 - T_M = C_1 \cosh R + C_2 \sinh R.$$

Obtaining C_1 and C_2 from these expressions and inserting them into Eq. (10.63) yields, after some rearrangement, the temperature distribution in the gas

$$\frac{T - T_M}{T_1 - T_M} = \frac{\sinh R(1 - \xi)}{\sinh R} + \left(\frac{T_2 - T_M}{T_1 - T_M}\right) \frac{\sinh R\xi}{\sinh R}. \tag{10.64}$$

For $\tau \to 0$ or $P_M \to 0$ (which implies $T_M \to 0$), all of the hyperbolic functions approach their arguments, and Eq. (10.64) reduces to the temperature distribution in the gas resulting from conduction alone,

$$\frac{T - T_2}{T_1 - T_2} = 1 - \xi. \tag{10.65}$$

Next, we proceed to another example, the so-called problem of aerodynamic heating. We learned in Chapter 9 the effect of enclosure radiation on this problem. ◆

EXAMPLE 10.4

The thermal boundary layer about a flying object may to a first order be approximated by a Couette flow. Let the velocity of the object be U and its surface emissivity be ϵ_w. The viscosity, thermal conductivity, absorption, and temperature of the ambient are μ, k, κ, and T_∞, respectively. Curvature effects are negligible. We wish to determine the steady surface temperature of the object.

The first law of thermodynamics applied to the differential system shown in Fig. 10.17, including effects of conduction, radiation, and shear work done on the system, gives

$$-\frac{d}{dy}(q_y^K + q_y^R) + \frac{d}{dy}(\tau_1 u) = 0, \tag{10.66}$$

where τ_1 is the longitudinal shear stress. Noting that

$$\frac{d}{dy}(\tau_1 u) = \tau_1 \frac{du}{dy} + u \frac{d\tau_1}{dy},$$

and that the longitudinal momentum is

$$\frac{d\tau_1}{dy} = 0,$$

Eq. (10.66) may be reduced to

$$-\frac{d}{dy}(q_y^K + q_y^R) + \tau_1 \frac{du}{dy} = 0. \tag{10.67}$$

Particular laws associated with momentum and conduction are

$$\tau_1 = \mu \frac{du}{dy} \tag{10.68}$$

and

$$q_y^K = -k\frac{dT}{dy}. \tag{10.69}$$

However, the particular law for radiation needs some attention. Replacing the fluid beyond the boundary layer with a black surface at the edge of boundary layer (Fig. 10.17), rearranging the notation of Eq. (10.51) with $\epsilon_1 = \epsilon_\infty = 1$, $\epsilon_2 = \epsilon_w$, $E_{b1} = E_{b\infty}$ and $E_{b2} = E_{bw}$, and following some manipulations, we have

$$E_{bM} = E_{b\infty} - \frac{\epsilon_w}{2}(E_{bw} - E_{b\infty}). \tag{10.70}$$

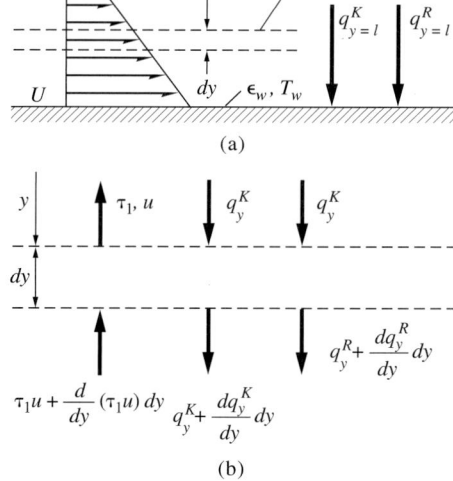

Figure 10.17 Example 10.4. (a) Model, (b) balance of total energy.

Now, inserting Eq. (10.70) into Eq. (10.47) and the result into Eq. (10.45), and employing $E_b = \sigma T^4$, yields the particular radiation law

$$\frac{dq^R}{dy} = 4\kappa\sigma(T^4 - T_\infty^4) - 2\kappa\epsilon_w\sigma(T_w^4 - T_\infty^4). \tag{10.71}$$

Finally, in terms of Eqs. (10.68), (10.69), and (10.71) and the Couette velocity $u = U(y/\ell)$, Eq. (10.67) may be rearranged to give the governing equation

$$k\frac{d^2T}{dy^2} + \mu\left(\frac{U}{\ell}\right)^2 - 4\kappa\sigma(T^4 - T_\infty^4) + 2\kappa\epsilon_w\sigma(T_w^4 - T_\infty^4) = 0 \tag{10.72}$$

subject to

$$T(0) = T_\infty \quad \text{and} \quad T(\ell) = T_w. \tag{10.73}$$

To proceed further, we assume that $T_w - T_\infty$ is small, and we expand both T and T_w about T_∞,

$$T^4 = T_\infty^4 + 4T_\infty^3(T - T_\infty) + \cdots,$$

$$T_w^4 = T_\infty^4 + 4T_\infty^3(T_w - T_\infty) + \cdots.$$

Employing these linearizations, and the dimensionless distance $\eta = y/\ell$, optical thickness $\tau = \kappa\ell$, Planck number $P_\infty = 4\sigma T_\infty^3/(k/\ell)$ and $4\tau P_\infty = R^2$, Eq. (10.72) may be further rearranged as

$$\frac{d^2T}{d\eta^2} - R^2\left(T - \left[T_\infty + \frac{1}{2}\epsilon_w(T_w - T_\infty) + \frac{\mu U^2/k}{4\tau P_\infty}\right]\right) = 0$$

or, introducing

$$T_M = T_\infty + \frac{1}{2}\epsilon_w(T_w - T_\infty) + \frac{\mu U^2/k}{R^2}, \tag{10.74}$$

as

$$\frac{d^2T}{d\eta^2} - R^2(T - T_M) = 0 \tag{10.75}$$

$$T(0) = T_\infty \quad \text{and} \quad T(1) = T_w. \tag{10.76}$$

Clearly, except for the difference in notation, the formulation of the preceding example given by Eq. (10.61) subject to Eq. (10.62) is identical to the foregoing formulation. Consequently, the solution of this formulation may be readily obtained from Eq. (10.64) by a notation change involving $\xi \to \eta$, $T_1 \to T_\infty$, $T_2 \to T_w$, and $P_w \to P_\infty$. Thus, we have

$$\frac{T - T_M}{T_\infty - T_M} = \frac{\sinh R(1 - \eta)}{\sinh R} + \left(\frac{T_w - T_M}{T_\infty - T_M}\right)\frac{\sinh R\eta}{\sinh R}, \tag{10.77}$$

the ambient temperature in terms of the surface temperature of the projectile. ◆

Actually, we wish to know the increase in the surface temperature of the projectile resulting from viscous dissipation. Employing Eq. (10.77) with the **steady** condition of zero heat flux across the projectile surface gives this increase. However, note that due to radiation, the vanishing surface gradient of the fluid temperature no longer is the condition of zero heat flux. We need to develop this condition for the combined case of conduction and radiation. First reconsider Eq. (8.18),

$$q^R = \int_\Omega I \cos\phi \, d\Omega \tag{8.18}$$

or, in terms of $d\Omega = \sin\phi \, d\phi \, d\theta$ (recall Fig. 8.3),

$$q^R = \int_\Omega I \sin\phi \cos\phi \, d\phi \, d\theta.$$

Next, assume hemispherical isotropy, and replace the actual intensity with two-stream intensities (recall Fig. 10.13). The explicit use of the θ and ϕ limits leads to

$$q^R = \int_0^{2\pi} d\theta \int_0^\pi I \sin\phi \cos\phi \, d\phi$$

or, in terms of $\cos\phi = \mu$,

$$q^R = -2\pi \int_1^{-1} I\mu \, d\mu,$$

or

$$q^R = -2\pi \left(I^+ \int_1^0 \mu \, d\mu + I^- \int_0^{-1} \mu \, d\mu \right).$$

Finally, carrying out the simple integrals, we get

$$\boxed{q^R = \pi(I^+ - I^-)} \tag{10.78}$$

or, in terms of the notation of the present example,

$$I^+ - I^- = \pi\epsilon_w(I_{0\infty} - I_{0w})$$

or

$$q^R = \epsilon_w(E_{b\infty} - E_{bw}). \tag{10.79}$$

Now, the condition of zero heat flux (insulated surface) may be written as

$$q^K_{y=\ell} + q^R_{y=\ell} = 0$$

or, in terms of Eqs. (10.69) and (10.79), and $E_b = \sigma T^4$,

$$-k\frac{dT(\ell)}{dy} + \epsilon_w \sigma(T_\infty^4 - T_w^4) = 0. \tag{10.80}$$

Linearizing the radiation flux as before, using $\eta = y/\ell$ and $P_\infty = 4\sigma T_\infty^3/(k/\ell)$ yields

$$-k\frac{dT(1)}{d\eta} = \epsilon P_\infty(T_w - T_\infty), \tag{10.81}$$

which is identical in form to a convection boundary condition in terms of a heat transfer coefficient.

Inserting Eq. (10.77) into Eq. (10.81), recalling Eq. (10.74), and following some algebraic manipulations, we get the temperature increase above the ambient temperature resulting from viscous dissipation,

$$T_w - T_\infty = \frac{\mu U^2}{2k} \left[\frac{\frac{2}{R^2}(\cosh R - 1)}{\frac{1}{2}\epsilon_w + \left(1 - \frac{1}{2}\epsilon_w\right)\cosh R + \epsilon_w P_\infty \frac{\sinh R}{R}} \right], \quad (10.82)$$

where the terms in brackets show the effect of thin gas and enclosure radiation. In Fig. 10.18, $(T_w - T_\infty)/(\mu U^2/2k)$ is plotted against τ for some values of ϵ_w and P_∞.

A gas becomes transparent as $\tau \to 0$, or cold as $P_\infty \to 0$, and

$$\sinh R \to R, \quad \cosh R \to 1 + \frac{R^2}{2},$$

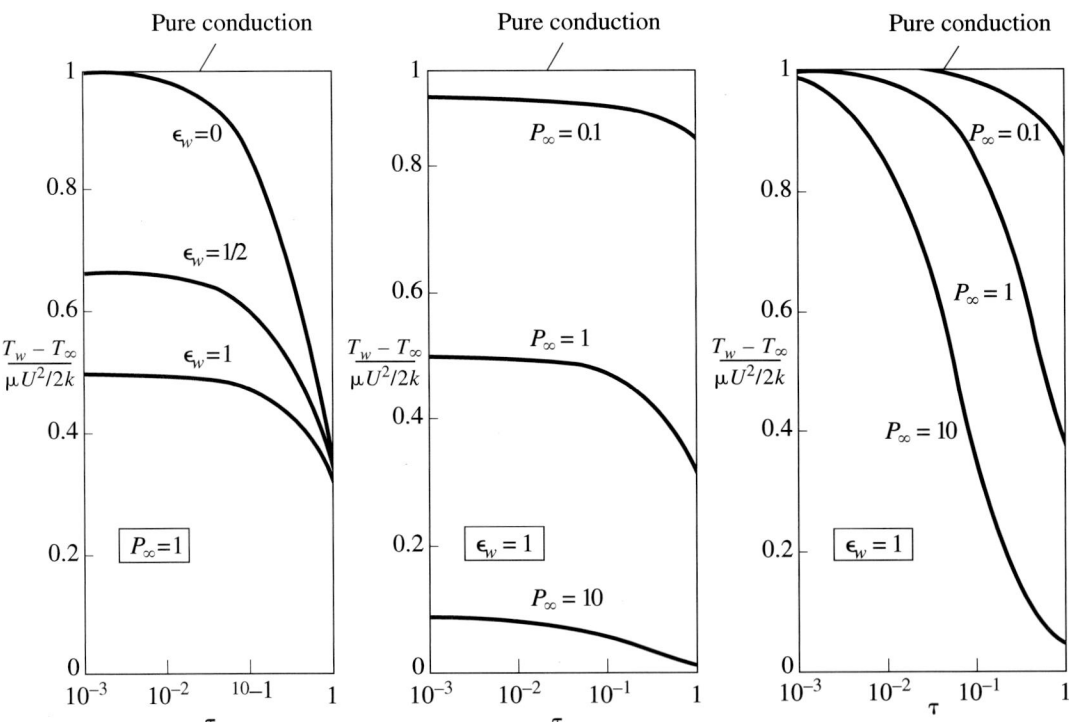

Figure 10.18 Effect of gas emission and gas absorption on the temperature of (a) adiabatic black surface, (b) adiabatic mirror surface.

and Eq. (10.82) reduces to

$$T_w - T_\infty = \frac{\mu U^2}{2k}\left(\frac{1}{1 + \epsilon_w P_\infty}\right), \qquad (10.83)$$

where the terms in parentheses show the effect of enclosure radiation. Furthermore, $\epsilon_w \to 0$ for a mirror surface, and Eq. (10.83) reduces to

$$T_w - T_\infty = \frac{\mu U^2}{2k}, \qquad (10.84)$$

which involves only the effect of conduction. In the literature, Eq. (10.84) is often rearranged as

$$\frac{\mu U^2}{k(T_w - T_\infty)} = \left(\frac{\mu c_p}{k}\right)\frac{U^2}{c_p(T_w - T_\infty)} = \frac{1}{2}, \qquad (10.85)$$

where $\mu c_p/k$ is the Prandtl number and

$$\frac{U^2}{c_p(T_w - T_\infty)} = \mathrm{Ec}$$

is the Eckert number. On dimensional grounds, Eq. (10.85) may be written as

$$\frac{\mu(U/\ell)^2}{k(T_w - T_\infty)/\ell^2} \sim \frac{\text{Dissipation}}{\text{Conduction}}$$

which shows the importance of viscous dissipation relative to conduction.

EXAMPLE 10.5

Some of the features of a flame in a premixed combustible gas stream may be elucidated by considering a fluid at temperature T_∞ flowing with velocity V into a porous plug (approximating the flame front) held at temperature T_w. To eliminate the attenuation in thin gas, let a black porous plug with temperature T_∞ be placed at a distance ℓ from the flame. We wish to determine the temperature distribution in the flame.

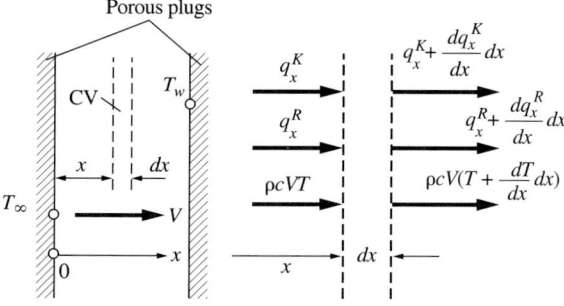

Figure 10.19 The first law applied to the control volume.

532 Chap. 10 Gas Radiation

The governing equation (adding enthalpy flow to, excluding viscous dissipation from, and replacing y with x coordinate in the preceding example) can be written as

$$\frac{d^2T}{d\xi^2} - 2\Pi \frac{dT}{d\xi} - R^2(T - T_M) = 0, \tag{10.86}$$

where $\xi = x/\ell$, $\Pi = V\ell/2\alpha$,

$$T_M = T_\infty + \frac{1}{2}\epsilon_w(T_w - T_\infty), \tag{10.87}$$

and the remaining definitions are from the preceding example. The boundary conditions are

$$T(0) = T_\infty \quad \text{and} \quad T(1) = T_w. \tag{10.88}$$

The solution of the problem, following the usual procedure and after some tedious manipulations, is found to be

$$\frac{T - T_M}{T_\infty - T_M} = e^{\Pi\xi}\frac{\sinh H(1-\xi)}{\sinh H} - e^{-\Pi(1-\xi)}\left(\frac{T_w - T_M}{T_\infty - T_M}\right)\frac{\sinh H\xi}{\sinh H}, \tag{10.89}$$

where $H = (\Pi^2 + R^2)^{1/2}$. Equation (10.88) reduces, for $\Pi \to 0$, to Eq. (10.64), and for $R \to 0$, to

$$\frac{T - T_\infty}{T_w - T_\infty} = \frac{1 - e^{2\Pi\xi}}{1 - e^{2\Pi}}, \tag{10.90}$$

and for $\Pi \to 0$ and $R \to 0$ to Eq. (10.65), as expected. Elimination of T_M between Eqs. (10.87) and (10.89) yields

$$\frac{T - T_\infty}{T_w - T_\infty} = \frac{\epsilon_w}{2}\left[1 - e^{\Pi\xi}\frac{\sinh H(1-\xi)}{\sinh H}\right] - \left(\frac{\epsilon_w}{2} - 1\right)e^{\Pi(1-\xi)}\frac{\sinh H\xi}{\sinh H}. \tag{10.91}$$

Figure 10.20 shows the effect of thin gas and enclosure radiation on the gas flow between the two porous plugs. ◆

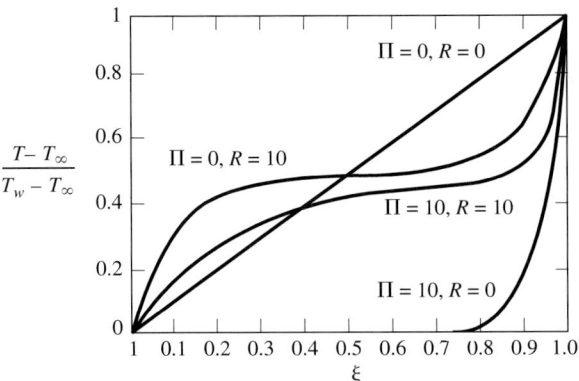

Figure 10.20 Effect of thin gas and enclosure radiation on the gas flow.

An attempt has been made in this chapter to introduce the reader to gas radiation, which is vitally important for current technological and environmental problems. However, because of the complexity of the subject, some of its aspects—such as thick gases near boundaries, intermediate optical thicknesses, spectral effects, weighted nongrayness, multidimensional effects, and scattering—are left untreated. Readers interested in these topics should consult the references of this chapter and the current literature.

■ REFERENCES

10.1 E.M. Sparrow and R.D. Cess, *Radiation Heat Transfer*. Augmented edition. Hemisphere/McGraw-Hill, New York, 1978.

10.2 W.G. Vincenti and C.H. Kruger, *Introduction to Physical Gas Dynamics*. Wiley, New York, 1965.

10.3 S.S. Penner and D.B. Olfe, *Radiation and Reentry*. Academic Press, New York, 1968.

10.4 R. Siegel and J.R. Howell, *Thermal Radiation Heat Transfer*, 2d ed. Hemisphere/McGraw-Hill, New York, 1981.

10.5 M.N. Özışık, *Radiative Transfer*. Wiley-Interscience, New York, 1973.

10.6 S.S. Penner, *Quantitative Molecular Spectroscopy and Gas Emissivities*. Addison-Wesley, Reading, MA, 1959.

10.7 H.C. Hottel and A.F. Sarofim, *Radiative Transfer*. McGraw-Hill, New York, 1967.

10.8 E.R.G. Eckert and R.M. Drake, *Analysis of Heat and Mass Transfer*. McGraw-Hill, New York, 1972.

10.9 W.M. Rohsenow and H. Choi, *Heat, Mass and Momentum Transfer*. Prentice-Hall, Englewood Cliffs, NJ, 1961.

10.10 R.D. Cess, *Advances in Heat Transfer*, 1, 1964.

10.11 R. Viskanta, *Advances in Heat Transfer*, 3, 1966.

10.12 C.L. Tien, *Advances in Heat Transfer*, 5, 1968.

10.13 R.D. Cess and S.N. Tiwari, *Advances in Heat Transfer*, 8, 1972.

10.14 D.K. Edwards, *Advances in Heat Transfer*, 12, 1976.

10.15 V.S. Arpacı, *Int. J. Heat Mass Transfer*, 11, 871 (1968).

10.16 V.S. Arpacı, and P.S. Larsen, *AIAA J.*, 7, 602 (1969).

10.17 H.A. Lord and V.S. Arpacı, *Int. J. Heat Mass Transfer*, 13, 1737 (1970).

10.18 W.G. Vincenti and S.C. Traugott, *Annual Review of Fluid Mechanics*, 3, (1971).

10.19 S.E. Gilles, A.C. Cogley, and W.G. Vincenti, *Int. J. Heat Mass Transfer*, 12, 445 (1969).

10.20 V.S. Arpacı, and R.J. Tabaczynski, 1982, "Radiation-affected flame propagation," *Combust. Flame*, 46, 315.

10.21 V.S. Arpacı, and R.J. Tabaczynski, 1984, "Radiation-affected laminar flame quenching," *Combust. Flame*, 57, 169.

10.22 V.S. Arpacı, and S. Troy, 1990, "On the attenuating thin gas," *J. Thermophysics Heat Transfer*, 4, 407.

10.23 V. S. Arpacı, 1991, "Radiative entropy production-heat loss into entropy," *Advances in Heat Transfer*, vol. 21, p. 239 (ed. J.P. Hartnett, and T.F. Irvine). Academic Press, New York.

10.24 W.M. Rohsenow, and J.P. Hartnett eds., *Handbook of Heat Transfer*, McGraw-Hill, New York, 1973.

10.25 W.H. McAdams, *Heat Transmission*, 3d ed. McGraw-Hill, New York, 1954.

■ EXERCISES

10.1 Reconsider Ex. 10.1 for $T_g = 600$ K and $T_s = 1{,}200$ K.

10.2 Determine the radiant heat transfer through atmospheric CO_2 between two large parallel black plates 10 cm apart at temperatures of 1,200 K and 600 K.

10.3 Repeat Prob. 10.2 for plates with emissivity $\epsilon = 0.5$.

10.4 Repeat Prob. 10.2 for CO_2 at pressure $p = 3$ atm.

10.5 Arpacı and Troy [22] approximates the effect of thin-gas attenuation on Eq. (10.71) by

$$\frac{dq^R}{dy} = 4\kappa\sigma(T^4 - T_\infty^4) - 2\epsilon_w\kappa\sigma(T_w^4 - T_\infty^4)e^{-\sqrt{3}\kappa y}.$$

Neglecting conduction, show for a semi-infinite thin gas next to a wall at temperature T_w that

$$\frac{E_b - E_{b\infty}}{E_{bw} - E_{b\infty}} = \frac{\epsilon_w}{2}e^{-\sqrt{3}\kappa y},$$

T_∞ being the gas temperature far from the wall. Note that $E_b(0) \neq E_{bw}$, indicating to a **temperature jump** on the wall in the absence of conduction.

10.6 Reconsider Ex. 2.1 for attenuating thin gas.

10.7 Reconsider Ex. 2.3 for attenuating thin gas.

CHAPTER 11

PHASE CHANGE

In Chapters 5 and 6 we studied convection heat transfer without phase change. We now proceed to convection with phase change. In particular, we consider convection with melting-solidification, and condensation-evaporation (boiling). Many environmental processes as well as engineering problems involve phase change. Examples are the freezing of lakes and the melting of snow in nature, evaporation and condensation in a closed-loop power cycle of a conventional or nuclear power plant, closed-loop refrigeration cycle of a home or commercial cooling system, etc.

The important difference between single-phase and two-phase flows is the interface latent heat effect involved with the latter. Also, in cases with strong curvature effects, surface tension needs to be taken into account. Because of latent heat, the heat transfer rates in two-phase problems are an order of magnitude larger than those in single phase.

Our approach to two-phase problems begins with an illustrative example dealing with solidification in a stagnant liquid.

AN ILLUSTRATIVE EXAMPLE

Consider a liquid initially at temperature T_∞ suddenly brought into contact with a plane wall at constant temperature T_w. Here we consider the case of a liquid at the melting point solidifying on an isothermal subcooled wall (Fig. 11.1). We wish to find the thickness of the solidified layer, $\delta(t)$.

536 Chap. 11 Phase Change

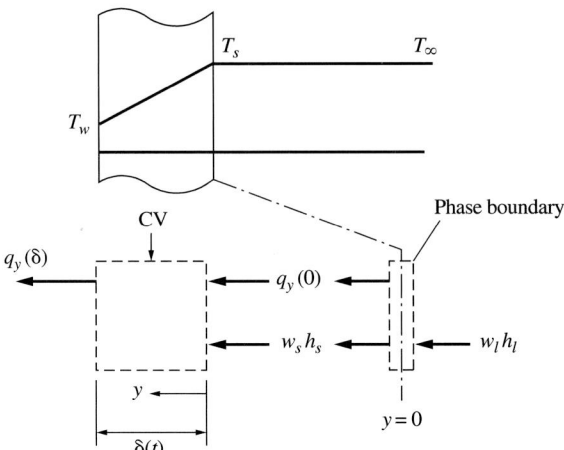

Figure 11.1 Configuration and control volume of a solidification process at plane wall.

The problem is in terms of a clearly defined penetration depth (instantaneous thickness of the solidified liquid), and it is well suited for an integral formulation. For an expanding control volume which encloses the solid, the conservation of mass becomes

$$\frac{d}{dt}\int_0^\delta \rho\,dy - w_s = 0, \tag{11.1}$$

where w_s is the mass flow per unit area of solidified material entering the control volume at $y = 0$. The conservation of mass and balance of thermal energy for the control volume enclosing the phase boundary give

$$w_s - w_\ell = 0; \quad w_s h_{s\ell} = -k\frac{\partial T(0,t)}{\partial y}, \tag{11.2}$$

where $h_{s\ell} = h_\ell - h_s$ denotes the heat of fusion. In terms of w_s, obtained by carrying out the integration in Eq. (11.1),

$$\rho\frac{d\delta}{dt} = w_s, \tag{11.3}$$

Eq. (11.2) becomes

$$\rho h_{s\ell}\frac{d\delta}{dt} = -k\left(\frac{\partial T}{\partial y}\right)_{y=0}. \tag{11.4}$$

A first-order polynomial profile for the solid satisfying the boundary conditions

$$T(0,t) = T_s; \quad T(\delta,t) = T_w, \tag{11.5}$$

yields

$$\frac{T - T_s}{T_w - T_s} = \frac{y}{\delta}. \tag{11.6}$$

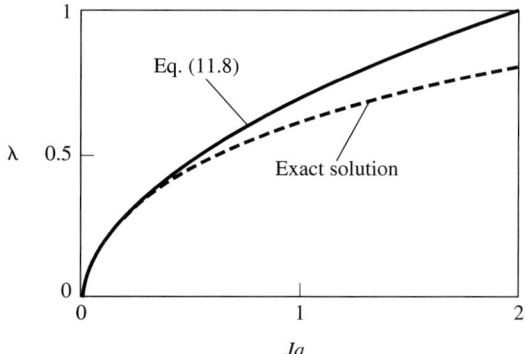

Figure 11.2 Comparison between the quasi-steady solution and the exact solution.

Inserting Eq. (11.6) into Eq. (11.4) leads to a formulation in terms of $\delta(t)$,

$$\delta \frac{d\delta}{dt} = \alpha \text{Ja}, \quad \delta(0) = 0, \tag{11.7}$$

where $\text{Ja} = c(T_s - T_w)/h_{s\ell}$ denotes the *Jakob number* and $\alpha = k/\rho c$ the thermal diffusivity. The solution of Eq. (11.7) is

$$\lambda = \frac{\delta(t)}{2\sqrt{\alpha t}} = \left(\frac{\text{Ja}}{2}\right)^{1/2}. \tag{11.8}$$

A comparison of our first-order approximation with the exact solution is shown in Fig. 11.2.

The rest of the chapter follows our approach to Chapters 5 and 6. First, we study laminar two-phase flow by approximate analytical means. Next, we consider the dimensionless numbers appropriate for two-phase. Finally, we correlate experimental data on turbulent two-phase in terms of these numbers.

11.1 LAMINAR TWO-PHASE

In the foregoing illustrative example we considered a two-phase problem with an unsteady interface. Here, we deal with a problem involving a spatially developing interface.

EXAMPLE 11.1

Filmwise condensation (or evaporation) involves heat transfer to the liquid-vapor interface as well as convective flow in either or both phases. Here, we consider steady vapor condensation at the saturation temperature T_s which forms a liquid film while flowing down next to a vertical isothermal wall at $T_w < T_s$ (Fig. 11.3). Assuming a continuous smooth film starting at $x = 0$, we wish to determine the variation in film thickness $\delta(x)$ and the local Nusselt number.

538 Chap. 11 Phase Change

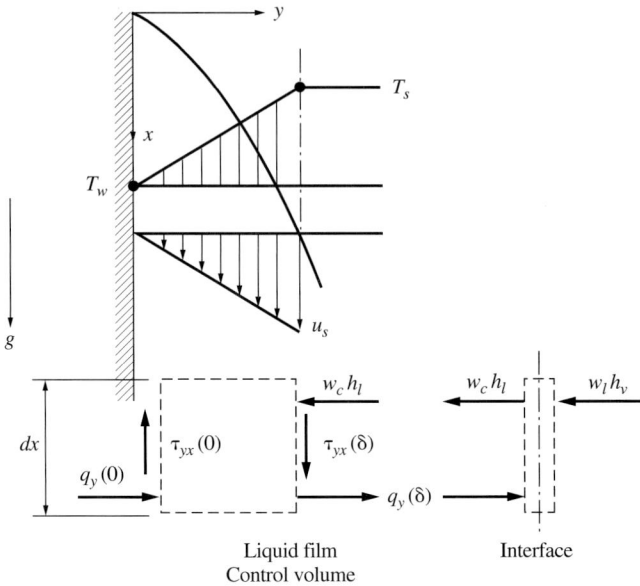

Figure 11.3 Filmwise condensation of saturated vapor along a vertical isothermal wall.

The control volume enclosing the liquid film (but excluding the interface) increases in thickness with x as vapor condenses at mass flow w_c. An integral balance for the conservation of mass and balance of momentum yields

$$\frac{d}{dx} \int_0^\delta \rho u \, dy - w_c = 0 \tag{11.9}$$

$$\left(1 - \frac{\rho_v}{\rho}\right) \rho g \delta = \tau_{yx}\Big|_{y=0} = \mu \frac{\partial u}{\partial y}\Big|_{y=0}, \tag{11.10}$$

and the balance of thermal energy of the interface gives

$$-w_c h_{\ell v} = (q_y)_\delta, \tag{11.11}$$

where $h_{\ell v} = h_v - h_\ell$ denotes the heat of evaporation. The remaining boundary conditions in y are

$$u(x, 0) = 0; \quad u(x, \delta) = u_s(x); \quad T(x, 0) = T_w; \quad T(x, \delta) = T_s, \tag{11.12}$$

where the upstream condition is $\delta(0) = 0$ and $u_s(x)$ is to be determined.

For a first-order solution, assuming linear velocity and temperature profiles satisfying Eq. (11.12), we have

$$\frac{u}{u_s} = \eta; \quad \frac{T - T_s}{T_w - T_s} = 1 - \eta, \tag{11.13}$$

with the notation $u(x, \delta) = u_s(x)$, $T(x, \delta) = T_s$, and $\eta = y/\delta$. Using the interfacial shear stress from Eq. (11.10), we get

$$u_s = \left(1 - \frac{\rho_v}{\rho}\right)\frac{g\delta^2}{\nu}, \tag{11.14}$$

and from Eq. (11.11),

$$w_c = \frac{\rho\alpha}{\delta}\text{Ja}, \tag{11.15}$$

where $\text{Ja} = c(T_s - T_w)/h_{\ell v}$ denotes the Jakob number and $\alpha = k/\rho c$ the thermal diffusivity of the liquid. Employing Eq. (11.13) in Eq. (11.9), carrying out the integration, and rearranging, we obtain

$$\delta(x) = \left[\frac{(8/3)\alpha\nu x\text{Ja}}{(1 - \rho_v/\rho)g}\right]^{1/4} \tag{11.16}$$

and the local heat transfer,

$$\text{Nu}_x = \frac{hx}{k} = \frac{(\partial T/\partial\eta)_{\eta=0}\, x}{(T_s - T_w)\delta} = \frac{x}{\delta} \tag{11.17}$$

or

$$\text{Nu}_x = \left[\frac{\text{Ra}_x}{(8/3)\text{Ja}}\right]^{1/4}, \tag{11.18}$$

where $\text{Ra}_x = (\Delta\rho/\rho)gx^3/\nu\alpha$, with $\Delta\rho = \rho - \rho_v$, denotes the local Rayleigh number. Note that Eq. (11.18) is valid only when $\text{Ja}/\text{Pr} \ll 1$ for the case of viscous oils, i.e., for large Pr, since $\text{Ja} \ll 1$ for most practical cases.

For liquid metals, on the other hand, the rate of condensation is high, and the interfacial shear stress causes a noticeable decrease in the Nusselt number as the parameter Ja/Pr increases. For this case, the integral balance of momentum is

$$\frac{d}{dx}\int_0^\delta \rho uu\, dy = \left(1 - \frac{\rho_v}{\rho}\right)\rho g, \tag{11.19}$$

and the velocity profile may approximately be assumed a constant across the film, i.e., $u = u_s(x)$. Substituting Eq. (11.13) with $u/u_s = 1$ into (11.19) and carrying out the integration yields

$$\frac{d}{dx}(u_s^2\delta) = \left(1 - \frac{\rho_v}{\rho}\right)g\delta. \tag{11.20}$$

Next, on the basis of our experience on Ex. 5.4, we try a solution of the form

$$\delta(x) = C_1 x^m; \quad u_s(x) = C_2 x^n, \tag{11.21}$$

which, in terms of Eqs. (11.19) and (11.20), gives

$$m = \frac{1}{4}; \quad n = \frac{1}{2}, \tag{11.22}$$

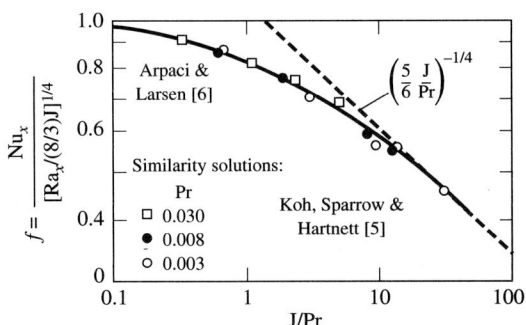

Figure 11.4 Filmwise condensation departure from the Nusselt theory for liquid metals.

while the coefficients C_1 and C_2 satisfy the algebraic equations,

$$\frac{3}{4}C_1 C_2 = \alpha \frac{\text{Ja}}{C_1}, \tag{11.23}$$

$$\frac{5}{4}C_1 C_2^2 = \left(1 - \frac{\rho_v}{\rho}\right) g C_1, \tag{11.24}$$

leading to

$$C_1 = \left[\frac{(8/3)\alpha \nu \text{Ja}}{(1 - \rho_v/\rho)g} \frac{5\text{Ja}}{6\text{Pr}}\right]^{1/4} \tag{11.25}$$

$$C_2 = \left[\frac{4}{5}(1 - \frac{\rho_v}{\rho})g\right]^{1/2}. \tag{11.26}$$

The local Nusselt number for liquid metals is then

$$\text{Nu}_x = \left[\frac{\text{Ra}_x}{(8/3)\text{Ja}}\right]^{1/4} f, \tag{11.27}$$

where

$$f = \left(\frac{5}{6}\frac{\text{Ja}}{\text{Pr}}\right)^{-1/4}.$$

In Fig. 11.4 Eqs. (11.18) and (11.27) are compared with more accurate results by Arpacı and Larsen (1984) and a similarity solution by Koh, Sparrow, and Hartnett [5] for $\text{Pr} = 0.03, 0.008,$ and 0.003. ◆

The present example shows how a strong coupling among the conservation of mass, balance of momentum, and energy equations arises in problems involving phase change. Next we proceed to a dropwise film-boiling problem.

Example 11.2

Consider dropwise film boiling on an isothermal horizontal plate. We wish to evaluate the heat transfer coefficient.

Experiments show that a drop of liquid evaporates rapidly when placed on a horizontal surface heated to a few degrees above the saturation temperature; the evaporation is slow, however, if the surface is well above saturation. In the first case the liquid can wet the surface, creating nucleate boiling with bubble formation, liquid film evaporation, and high rates of heat transfer. In the second case the drop remains suspended on a poorly conducting vapor film, which prevents direct contact between the liquid and the hot surface. The latter is the Leidenfrost problem of film boiling and is our concern here.

This is a difficult problem, involving complex geometry, unsteadiness, and an interface boundary whose location is not known a priori. We make several assumptions but retain the essential physics. First, we consider the evaporation to be quasi-steady. The weight of the drop is then balanced by the surface integral of pressure forces. The excess pressure in the vapor film arises from that required to maintain the outward viscous flow of vapor escaping from underneath the drop. In addition, we use a quasi-developed integral formulation for the flow of vapor which is generated from the drop by conduction through the vapor film. Since heat transfer, vapor flow, and excess pressure increase with decreasing film thickness, the drop tends to settle at a quasi-steady film thickness to be determined as a part of the solution. We approximate the geometry of an actual drop, which is determined by gravity, surface tension, and pressure distribution, by a hemisphere (Fig. 11.5). The liquid is at rest and isothermal at the saturation temperature, T_s. The heating surface is isothermal at T_w, which exceeds the Leidenfrost temperature.

A quasi-steady integral balance for the mass, momentum, and energy of the vapor film is

$$\frac{1}{r}\frac{\partial}{\partial r}\left(r\int_0^\delta \rho u \, dz\right) - w_\delta = 0 \tag{11.28}$$

$$\frac{1}{r}\frac{\partial}{\partial r}\left(r\int_0^\delta \rho u^2 \, dz\right) - u(r,\delta)w_\delta = -\frac{\delta}{r}\frac{\partial}{\partial r}(rp) + \tau_{zr}|_{z=\delta} - \tau_{zr}|_{z=0} \tag{11.29}$$

$$\frac{1}{r}\frac{\partial}{\partial r}\left(r\int_0^\delta \rho c_p u T \, dz\right) - c_p T(r,\delta) w_\delta = -q_z|_{z=\delta} + q_z|_{z=0} \tag{11.30}$$

w_δ denoting the evaporating mass flow per unit area, subject to the boundary conditions

$$u(r,0) = 0; \quad u(r,\delta) = 0; \quad T(r,0) = T_w; \quad T(r,\delta) = T_s \tag{11.31}$$

$$\tau_{zr} = \mu\frac{\partial u}{\partial z}; \quad q_z = -k\frac{\partial T}{\partial z}. \tag{11.32}$$

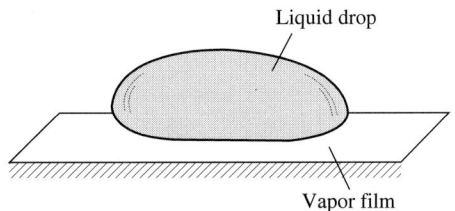

Figure 11.5 Liquid-drop modeling.

Also, the balance of thermal energy at the liquid-vapor interface is

$$h_{fg} w_\delta = -k \frac{\partial T(r,z)}{\partial z}, \tag{11.33}$$

where h_{fg} denotes the heat of evaporation.

For a quasi-developed solution, ignoring the left side of Eq. (11.30), we assume a linear temperature distribution,

$$\frac{T - T_s}{T_w - T_s} = 1 - \frac{z}{\delta}, \tag{11.34}$$

also, neglecting the radial variation of the vapor film, assume a parabolic velocity distribution,

$$\frac{u}{u_m} = 6\frac{z}{\delta}\left(1 - \frac{z}{\delta}\right), \tag{11.35}$$

where $u_m(r)$ denotes the mean vapor velocity. A quasi-developed balance for both the momentum and energy is justified by the Reynolds analogy, provided that the Prandtl number is of order unity and the pressure gradient is zero.

Using Eqs. (11.34) and (11.35), eliminating w_δ by Eq. (11.33), we get from Eqs. (11.28) and (11.29),

$$\frac{1}{r}\frac{\partial}{\partial r}(ru_m) = \frac{\alpha \mathrm{Ja}}{\delta} \tag{11.36}$$

$$\frac{6}{5}\frac{1}{r}\frac{\partial}{\partial r}(ru_m^2) = -\frac{\delta}{\rho}\frac{1}{r}\frac{\partial}{\partial r}(rp) - \frac{12\nu u_m}{\delta}, \tag{11.37}$$

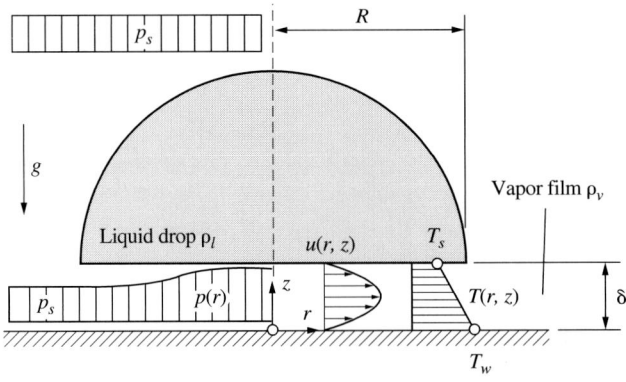

Figure 11.6 Idealized drop evaporation.

where Ja $= c_p(T_w - T_s)/h_{fg}$ denotes the Jacob number. Using symmetry and denoting by p_s the pressure outside the drop, the boundary conditions become

$$u_m(0) = 0; \quad \frac{dp(0)}{dr} = 0; \quad p(R) = p_s; \quad \delta = \text{Const.} \tag{11.38}$$

Introducing $u_m = \alpha r \text{Ja}/2\delta$, obtained from Eq. (11.36), into Eq. (11.37) and integrating once yields

$$p(r) - p_s = 2\mu\alpha\text{Ja}\left(\frac{R^2}{\delta^4}\right)\left(1 + \frac{3}{20}\frac{\text{Ja}}{\text{Pr}}\right)\left[1 - \left(\frac{r}{R}\right)^2\right], \tag{11.39}$$

where $\text{Pr} = \nu/\alpha$ and the factor $(3/20)\text{Ja}/\text{Pr}$ denotes the inertial effect.

Now, the vertical balance of momentum for the drop, ignoring the unsteadiness and momentum flow due to evaporation, gives

$$0 = \int_0^R 2\pi r\,[p(r) - p_s]\,dr - (\rho_\ell - \rho)g\frac{2}{3}\pi R^3, \tag{11.40}$$

where the subscript ℓ refers to the liquid. Inserting Eq. (11.39) into Eq. (11.40), and using $D = 2R$ as the characteristic length, yields the film thickness

$$\delta = \left[\frac{3}{4}\frac{\alpha\mu\text{Ja}D}{g(\rho_\ell - \rho)}\left(1 + \frac{3}{20}\frac{\text{Ja}}{\text{Pr}}\right)\right]^{1/4}, \tag{11.41}$$

and the mean instantaneous heat transfer coefficient,

$$\bar{h}(t) = \frac{1}{T_w - T_s}\frac{1}{\pi R^2}\int_0^R 2\pi r q_y(r, 0)\,dr \tag{11.42}$$

or

$$\text{Nu} = \frac{\bar{h}(t)D}{k} = \left[\frac{4}{3}\frac{g(\rho_\ell - \rho)D^3}{\mu\alpha\text{Ja}\left(1 + \frac{3}{20}\frac{\text{Ja}}{\text{Pr}}\right)}\right]^{1/4} \tag{11.43}$$

which may be rearranged in terms of

$$\Pi_2 = \frac{\text{Ra}/\text{Ja}}{1 + \frac{3}{20}\text{Ja}/\text{Pr}} \tag{11.44}$$

as

$$\text{Nu} = \left(\frac{4}{3}\Pi_2\right)^{1/4}. \tag{11.45}$$

◆

So far we have studied a number of illustrative examples for two-phase laminar heat transfer following the analytical approach we used in Chapter 5. For two-phase turbulent heat transfer we use an approach based on two-length scale dimensional analysis and the correlation of experimental data in terms of dimensionless numbers resulting from this analysis.

11.2 A DIMENSIONLESS NUMBER

Consider the control volume shown in Fig. 11.7. The balance between body and surface forces acting on this control volume yields

$$F_I + F_V \sim F_B, \tag{11.46}$$

F_I being the inertial force, F_V the viscous force, and F_B the buoyant force.

The balance of thermal energy for the same control volume gives

$$\dot{Q}_H + \dot{Q}_2 \sim \dot{Q}_K, \tag{11.47}$$

$\dot{Q}_H$ being the longitudinal net enthalpy flow, $\dot{Q}_2$ the enthalpy flow across the two-phase interface, and $\dot{Q}_K$ the conduction. In situations involving phase change, the Jacob number denoting the ratio of sensible heat to latent heat,

$$\frac{\dot{Q}_H}{\dot{Q}_2} \sim \frac{c_p \Delta T}{h_{fg}} = \text{Ja} \ll 1, \tag{11.48}$$

c_p being the specific heat at constant pressure, ΔT a longitudinal temperature difference, and h_{fg} the latent heat, is customarily utilized. In view of Eq. (11.48), however, $\dot{Q}_H \ll \dot{Q}_2$ and Eq. (11.47) is reduced to

$$\dot{Q}_2 \sim \dot{Q}_K. \tag{11.49}$$

Recalling Chapter 5, a dimensionless number resulting from Eq. (11.46) is

$$\frac{F_B}{F_I + F_V} \tag{11.50}$$

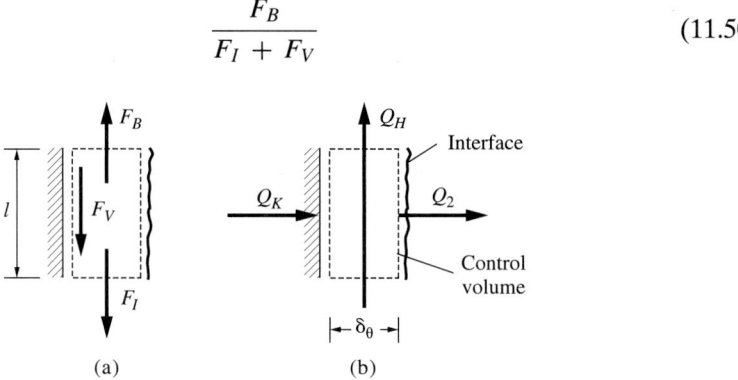

Figure 11.7 Configuration of a two-phase control volume.

or
$$\frac{F_B/F_V}{F_I/F_V + 1}. \tag{11.51}$$

Also, a dimensionless number associated with Eq. (11.49) is
$$\dot{Q}_2/\dot{Q}_K. \tag{11.52}$$

On dimensional grounds
$$\frac{F_B}{F_V} \sim \frac{g(\Delta\rho)\ell^2}{\mu V}, \quad \frac{F_I}{F_V} \sim \frac{\rho V \ell}{\mu}, \quad \frac{\dot{Q}_2}{\dot{Q}_K} \sim \frac{\rho V \ell h_{fg}}{k\Delta T},$$

where ℓ is a characteristic length and the rest of the notation is conventional. However, the foregoing nondimensionalizations in terms of a characteristic velocity V are not appropriate for buoyancy-driven flows. Velocity is now a dependent variable and should not appear in the dimensionless numbers describing these flows. Accordingly, one needs to combine Eq. (11.52) with Eq. (11.51) for a result independent of V. The only combination which eliminates velocity between Eq. (11.51) and (11.52) is

$$\Pi_2 \sim \frac{(F_B/F_V)(\dot{Q}_2/\dot{Q}_K)}{(F_I/F_V)(\dot{Q}_K/\dot{Q}_2) + 1} \sim \frac{Ra_2}{Pr_2^{-1} + 1} \tag{11.53}$$

and its limit
$$\lim_{Pr_2 \to \infty} \Pi_2 \sim Ra_2, \tag{11.54}$$

where
$$Pr_2 \sim \left(\frac{F_V}{F_I}\right)\left(\frac{\dot{Q}_2}{\dot{Q}_K}\right) \sim \frac{\mu h_{fg}}{k\Delta T}, \tag{11.55}$$

$$Ra_2 \sim \left(\frac{F_B}{F_V}\right)\left(\frac{\dot{Q}_2}{\dot{Q}_K}\right) \sim \frac{g(\Delta\rho)\rho h_{fg}\ell^3}{\mu k(\Delta T)} \tag{11.56}$$

which clearly demonstrate that Π_2 does not include the Jacob number. This fact should not be surprising in view of Eq. (11.48). Actually, either Pr_2 or Ra_2 would have been a more appropriate definition for the Jacob number.

The heat transfer across the a two-phase film is then represented by
$$Nu = f(\Pi_2), \tag{11.57}$$

or its limit
$$\lim_{Pr_2 \to \infty} Nu = f(Ra_2). \tag{11.58}$$

Note that a "two-phase specific heat,"
$$(c_p)_2 = h_{fg}/\Delta T, \tag{11.59}$$

may be defined as the natural limit of $(\partial h/\partial T)_p$, ΔT being the temperature jump across the interface. In terms of this definition,

$$\alpha_2 = \frac{k}{\rho(c_p)_2}, \quad \mathrm{Pr}_2 = \frac{\nu}{\alpha_2}, \quad \mathrm{Ra}_2 = \frac{g}{\nu\alpha_2}\left(\frac{\Delta\rho}{\rho}\right)\ell^3, \tag{11.60}$$

and Pr_2 and Ra_2 assume their conventional forms.

Here, for customary reasons only, Eqs. (11.55) and (11.56) are rearranged by $\dot{Q}_H$,

$$\mathrm{Pr}_2 \sim \left(\frac{F_V}{F_I}\right)\left(\frac{\dot{Q}_H}{\dot{Q}_K}\right)\left(\frac{\dot{Q}_2}{\dot{Q}_H}\right) \tag{11.61}$$

and

$$\mathrm{Ra}_2 \sim \left(\frac{F_B}{F_V}\right)\left(\frac{\dot{Q}_H}{\dot{Q}_K}\right)\left(\frac{\dot{Q}_2}{\dot{Q}_H}\right), \tag{11.62}$$

which, in terms of the usual Prandtl and Rayleigh numbers,

$$\mathrm{Pr} \sim \left(\frac{F_V}{F_I}\right)\left(\frac{\dot{Q}_H}{\dot{Q}_K}\right) \sim \frac{\nu}{\alpha} \tag{11.63}$$

and

$$\mathrm{Ra} \sim \left(\frac{F_B}{F_V}\right)\left(\frac{\dot{Q}_H}{\dot{Q}_K}\right) \sim \frac{g}{\nu\alpha}\left(\frac{\Delta\rho}{\rho}\right)\ell^3, \tag{11.64}$$

become

$$\mathrm{Pr}_2 = \mathrm{Pr}/\mathrm{Ja}, \quad \mathrm{Ra}_2 = \mathrm{Ra}/\mathrm{Ja}. \tag{11.65}$$

Then, Eq. (11.53) gives

$$\Pi_2 \sim \frac{\mathrm{Ra}/\mathrm{Ja}}{\mathrm{Ja}/\mathrm{Pr} + 1} \tag{11.66}$$

and

$$\lim_{\mathrm{Pr}\to\infty} \Pi_2 = \mathrm{Ra}/\mathrm{Ja}. \tag{11.67}$$

The foregoing dimensional analysis based on one-length scale arguments leads to implicit relations such as Eqs. (11.57) and (11.58). For explicit relations such as Eqs. (11.18), (11.27), and (11.45) we proceed to a dimensional analysis, to be illustrated first for laminar flows and extended later to turbulent flows.

11.2.1 A Dimensional Approach

In terms of a flow scale ℓ (or x) and a diffusion scale δ, an explicit dimensionless form of Eq. (11.46) is

$$u\frac{u}{\ell} + v\frac{u}{\delta^2} \sim g\left(\frac{\Delta\rho}{\rho}\right) \tag{11.68}$$

and an explicit dimensionless form of Eq. (11.49) is

$$\rho \left(\frac{u\delta_\theta}{\ell} \right) h_{fg} \sim k \frac{\Delta T}{\delta_\theta}, \qquad (11.69)$$

where $u\delta_\theta/\ell$ is the interface velocity of the transversal mass flow expressed in terms of the longitudinal mass flow.

Noting that the thickness of the momentum and thermal boundary layers is about the same,

$$\delta \sim \delta_\theta, \qquad (11.70)$$

and rearranging Eq. (11.68) in terms of Eq. (11.70) yields

$$\frac{u}{\delta_\theta^2} \left(1 + \frac{u\delta_\theta^2}{\nu\ell} \right) \sim \frac{g}{\nu} \left(\frac{\Delta\rho}{\rho} \right). \qquad (11.71)$$

Separately, Eq. (11.69) gives

$$\frac{u\delta_\theta^2}{\ell} \sim \frac{k\Delta T}{\rho h_{fg}}, \qquad (11.72)$$

which may be rearranged in terms of Eqs. (11.59) and (11.60) as

$$\frac{u\delta_\theta^2}{\ell} \sim \frac{k}{\rho(c_p)_2} = \alpha_2 \qquad (11.73)$$

or

$$u \sim \frac{\alpha_2 \ell}{\delta_\theta^2}. \qquad (11.74)$$

Insertion of Eq. (11.74) into Eq. (11.71) leads to

$$\frac{\ell}{\delta_\theta^4} \sim \frac{(g/\nu\alpha_2)(\Delta\rho/\rho)}{1 + \mathrm{Pr}_2^{-1}}, \qquad (11.75)$$

or, in terms of Eq. (11.53), to

$$\frac{\ell}{\delta_\theta} \sim \Pi_2^{1/4} \sim \mathrm{Nu}. \qquad (11.76)$$

As Pr_2 (or Pr) $\to \infty$, the inertial effect becomes negligible and Eq. (11.76) is reduced to

$$\mathrm{Nu} \sim \mathrm{Ra}_2^{1/4}. \qquad (11.77)$$

Clearly, Eq. (11.76) is identical to Eq. (11.45) except for a numerical constant, and Eq. (11.77) to Eq. (11.18) in the same manner. Dimensional arguments lead, for the case of turbulent two-phase films, to

$$\mathrm{Nu} \sim \Pi_2^{1/3} \qquad (11.78)$$

and to its limit for $\mathrm{Pr} \to \infty$,

$$\mathrm{Nu} \sim \mathrm{Ra}_2^{1/3}. \qquad (11.79)$$

The details of the development leading to these results, however, are beyond the intended scope of this text.

For film boiling of cryogenic liquids at atmospheric pressure, correlating experimental data for liquid nitrogen, Frederkin and Clark [9] recommend

$$\bar{h} = 0.15 \left[\frac{(\rho_\ell - \rho_v) g h'_{fg} k_v^2}{\nu_v (T_w - T_{\text{sat}})} \right]^{1/3}, \quad \frac{(\rho_\ell - \rho_v) g h'_{fg} L^3}{k_v \nu_v (T_w - T_{\text{sat}})} > 5 \times 10^7, \quad (11.80)$$

where $h'_{fg} = h_{fg} + 0.5 c_{pv}(T_w - T_{\text{sat}})$ and all properties are evaluated at the film temperature. This correlation can readily be expressed in terms of Eq. (11.79), leading to

$$\text{Nu} = 0.15 \text{Ra}_2^{1/3}, \quad \text{Ra}_2 > 5 \times 10^7. \quad (11.81)$$

So far we have dealt with cases of two-phase involving a continuous film. Next we proceed to cases involving nucleation.

11.3 REGIMES OF BOILING

To identify different regimes of pool boiling, consider an electrically heated horizontal wire submerged in a pool of liquid at saturation temperature T_{sat}. The heat flux through the surface rather than the wall temperature is controlled and can be determined by measuring the electrical current and potential drop through the wire. The temperature of the wire is obtained by measuring its electrical resistance, which depends on temperature. Figure 11.8 represents the typical boiling curve obtained from such experiments. Initially, the wire temperature increases above the saturation temperature, and as long as it is within 5 °C, i.e., $\Delta T_e \leq 5\,°C$, no vapor bubbles are seen (**free convection boiling**). When ΔT_e is increased further, vapor bubbles grow and rise rapidly from nuclei at favored spots on the metal surface and collapse as they project into the bulk subcooled liquid. With further increases in ΔT_e, larger and more numerous bubbles are formed and escape as jets or columns, which subsequently merge into slugs of vapor (**nucleate boiling**). Additional heating causes the vapor stream upward so fast that the liquid downflow to the surface is unable to sustain a higher evaporation rate and gives rise to a **peak heat flux** q_{max}. Finally, an unstable film forms around the wire, and large bubbles originate at the outer upper surface of the film (**transition boiling regime**). The heat flux is reduced because of the poor conductivity of the vapor around the wire, and there is a local minimum value q_{min}, referred to as the Leidenfrost point (recall Ex. 11.2). With more heating, the wire surface is completely covered by a vapor blanket, which is stable in the sense that it does not collapse and reform repeatedly (**film boiling**). Since the temperature level is very high, radiation heat transfer across the vapor film becomes significant. Transition boiling is difficult to obtain with electrical heating. Further heating after the input heat flux reaches q_{max} causes a large surface-temperature rise toward point M and melts the wire. For this reason, point B is sometimes called the "burn-out point."

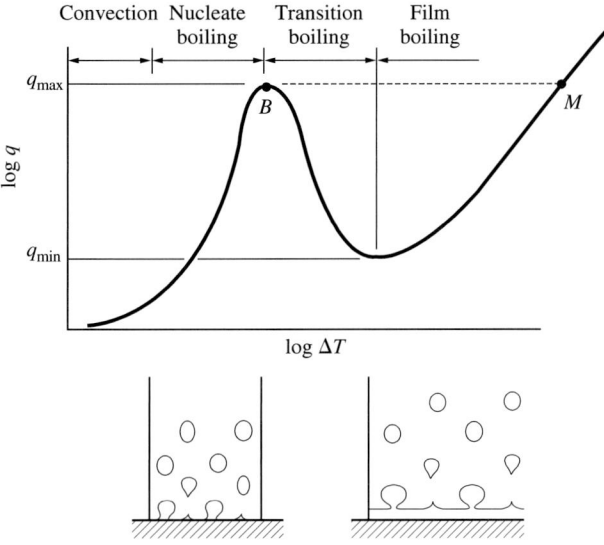

Figure 11.8 Boiling curve.

A Model (the Rohsenow Correlation). Experimental studies show that the relation between q/A and ΔT, depending on the cavity-size distribution, eventually leads, for most commercially available surfaces, to[1]

$$\frac{q}{A} \sim (\Delta T)^3. \tag{11.82}$$

For a dimensionless interpretation of this relation, consider the momentum balance

$$F_I \sim F_V \tag{11.83}$$

and the thermal energy balance

$$\dot{Q}_2 \sim \dot{Q}_K. \tag{11.84}$$

On dimensional grounds, Eqs. (11.83) and (11.84) give

$$\rho V^2 D^2 \sim \mu V D \tag{11.85}$$

and

$$\rho h_{fg} V \sim k \frac{\Delta T}{D}. \tag{11.86}$$

Elimination of V between Eqs. (11.85) and (11.86) yields

$$k \Delta T \sim \mu h_{fg} \tag{11.87}$$

or, a dimensionless number,

$$\frac{k \Delta T}{\mu h_{fg}}, \tag{11.88}$$

---

[1] q/A, the usual notation used in boiling, corresponds to $\dot{Q}/A$ consistently used in this text.

which, after multiplying its numerator and denominator by c_p, leads to

$$\frac{kc_p\Delta T}{\mu c_p h_{fg}} \equiv \frac{\text{Ja}}{\text{Pr}} = \frac{1}{\text{Pr}_2}. \tag{11.89}$$

Then, Eq. (11.82) may be nondimensionalized as

$$\frac{q/A}{\dot{Q}_K} \sim \left(\frac{\text{Ja}}{\text{Pr}}\right)^3 \tag{11.90}$$

or, in terms of Eqs. (11.86) and (11.87), as

$$\frac{q}{A} \sim \frac{\mu h_{fg}}{D}\left(\frac{\text{Ja}}{\text{Pr}}\right)^3 \tag{11.91}$$

where D is the unknown bubble diameter (to be distinguished by D_b) at its departure from the heating surface. Actually, the rise velocity (say V_b) as well as D_b of bubbles are unknowns of nucleation and need to be determined depending on ΔT (or $\Delta \rho$ for an isothermal model).

The dynamics of a bubble is governed by

$$F_B + F_S \sim F_I + F_V, \tag{11.92}$$

F_S being the surface tension force. Before departure (or under equilibrium)

$$F_B \sim F_S \tag{11.93}$$

or, on dimensional grounds,

$$g\Delta\rho D_b^3 \sim \sigma D_b, \tag{11.94}$$

which leads to an experimentally supported diameter,

$$D_b \sim \left(\frac{\sigma}{g\Delta\rho}\right)^{1/2}. \tag{11.95}$$

In terms of this result, the Rohsenow correlation for nucleate pool boiling becomes

$$\frac{q}{A} \sim \mu h_{fg}\left[\frac{g(\rho_\ell - \rho_v)}{\sigma}\right]^{1/2}\left(\frac{\text{Ja}_\ell}{\text{Pr}_\ell}\right)^3 \tag{11.96}$$

or, with a slight Pr correction for some fluid-surface combinations,

$$\frac{q}{A} = \mu h_{fg}\left[\frac{g(\rho_\ell - \rho_v)}{\sigma}\right]^{1/2}\left[\frac{c_{p,\ell}(T_w - T_{\text{sat}})}{C_{s,f}h_{fg}\text{Pr}_\ell^n}\right]^3, \tag{11.97}$$

where T_w is the surface temperature, and $C_{s,f}$ and n depend on the surface-liquid combination. All of the properties in Eq. (11.97) are evaluated at the saturation temperature. Table 11.1 gives the numerical values for some C_{sf} and n combinations. Table 11.2 is for surface tension. More extensive tables are available in the literature (see Rohsenow and Hartnett, [4]).

Table 11.1 Fluid-Surface Combination

Fluid Surface Combination	$C_{s,f}$	n
Water–copper		
Scored	0.0068	1.0
Polished	0.0130	1.0
Water–stainless steel		
Chemically etched	0.0130	1.0
Mechanically polished	0.0130	1.0
Ground and polished	0.0060	1.0
Water–brass	0.0060	1.0
Water–nickel	0.006	1.0
Water–platinum	0.0130	1.0
n-Pentane–copper		
Polished	0.0154	1.7
Lapped	0.0049	1.7
Benzene–chromium	0.101	1.7
Ethyl alcohol–chromium	0.0027	1.7

EXAMPLE 11.3

Water at 1 atm boils under nucleate conditions on a mechanically polished flat stainless steel plate kept at temperature $T_s = 395$ K. We wish to determine the heat flux.

From Appendix B–3 and Table 11.2, the thermomechanical water properties at saturation temperature are

Water at $T_g = 373$ K

$c_{p\ell} = 4.217$ kJ/kg·K
$h_{fg} = 2.257 \times 10^{-3}$ kJ/kg
$\sigma = 58.9 \times 10^{-3}$ kg/m³
$\rho_\ell = 958$ kg/m³
$\mu_\ell = 279 \times 10^{-6}$ N·s/m²
$Pr_\ell = 1.76$

and from Table 11.1, for water boiling on a mechanically polished surface,

$$C_{s,f} = 0.013, \quad n = 1.$$

With these values,

$$\frac{c_{p\ell}(T_w - T_{\text{sat}})}{C_{s,f} h_{fg} Pr_\ell^n} = \frac{4.217 \text{ kJ/kg·K} \times (395 - 373) \text{ K}}{0.013 \times 2.257 \times 10^3 \text{ kJ/kg} \times 1.76} = 1.8$$

Table 11.2 Surface tension

Fluid	T K	σ × 10³ N/m	Fluid	T K	σ × 10³ N/m
Water	275	75.3	Refrigerant–12	180	25.6
	280	74.8		190	24.0
	290	73.7		200	22.4
	300	71.7		210	20.9
	310	70.0		220	19.4
	320	68.3		230	17.9
	330	66.6		240	16.5
	340	64.9		250	15.0
	350	63.2		260	13.6
	360	61.4		270	12.3
	370	59.5		280	10.9
	373.15	58.9		290	9.66
	380	57.6		300	8.39
	390	55.6		310	7.15
	400	53.6		320	5.97
	420	49.4	Mercury	300	470
	440	45.1		400	450
	460	40.7		500	430
	480	36.2		600	400
	500	31.6		700	380
	550	19.7	Potassium	400	110
	600	8.4		500	105
	647.30	0.0		600	97
				700	90
				800	83
Ammonia	220	39		900	76
	240	34			
	260	30	Sodium	500	175
	280	25		700	160
	300	20		900	140
	320	16		1100	120

and

$$\frac{g(\rho_\ell - \rho_v)}{\sigma} \cong \frac{9.81 \text{ m/s}^2 \times 958 \text{ kg/m}^3}{58.9 \times 10^{-3} \text{ N/m}} = 159 \times 10^3 \text{ m}^{-2}.$$

Then, from Eq. (11.97),

$$\frac{q}{A} = 279 \times 10^{-6} \text{ N·s/m}^2 \times 2.257 \times 10^3 \text{ kJ/kg} \times (159 \times 10^3 \text{ m}^{-2})^{1/2} \times (1.8)^3$$

or

$$\frac{q}{A} = 1{,}464 \text{ kW/m}^2.$$

Also,
$$h = \frac{q/A}{T_w - T_{sat}} = \frac{1{,}464 \text{ kW/m}^2}{(395 - 373) \text{ K}} = 66{,}545 \text{ W/m}^2 \cdot \text{K}$$

which, in Table 1.2, corresponds to the upper bound of boiling water (note the ±30% uncertainty involved usually with these coefficients). ◆

On the boiling curve (Fig. 11.8), the upper limit of nucleate boiling, $(q/A)_{max}$, and the lower limit of film boiling, $(q/A)_{min}$, are important limits. Next, we review the existing models for these limits.

Models for $(q/A)_{max}$ and $(q/A)_{min}$ (Kutatelatze, Zuber Correlations).

For the bubble-rise velocity V_b, reconsider Eq. (11.92) under the assumption of $F_B \gg F_S$ and $F_I \gg F_V$,

$$F_B \sim F_I \tag{11.98}$$

or, on dimensional grounds,

$$g \Delta \rho D_b^3 \sim \rho V_b^2 D_b^2, \tag{11.99}$$

which gives

$$V_b \sim \left[g \left(\frac{\Delta \rho}{\rho} \right) D_b \right]^{1/2} \tag{11.100}$$

or, in terms of the D_b obtained from Eq. (11.95),

$$V_b \sim \left(\frac{\sigma g \Delta \rho}{\rho^2} \right)^{1/4}. \tag{11.101}$$

Then, the energy balance between the applied heat flux and the enthalpy flow of evaporation,

$$\frac{q}{A} \sim \rho h_{fg} V_b, \tag{11.102}$$

coupled with Eq. (11.101) yields

$$\frac{q}{A} \sim \rho h_{fg} \left(\frac{\sigma g \Delta \rho}{\rho^2} \right)^{1/4} \tag{11.103}$$

or, with the experimentally determined numerical coefficients,

$$\left(\frac{q}{A} \right)_{max} = 0.149 \rho_v h_{fg} \left[\frac{\sigma g (\rho_\ell - \rho_v)}{\rho_v^2} \right]^{1/4} \tag{11.104}$$

and

$$\left(\frac{q}{A} \right)_{min} = 0.09 \rho_v h_{fg} \left[\frac{\sigma g (\rho_\ell - \rho_v)}{(\rho_\ell + \rho_v)^2} \right]^{1/4}. \tag{11.105}$$

■ REFERENCES

11.1 W.H. McAdams, *Heat Transmission*, 3d ed. McGraw-Hill, New York, 1954.

11.2 M. Jacob, *Heat Transfer*, vol. 1. Wiley, New York, 1949.

11.3 W.M. Rohsenow, and H.Y. Choi, *Heat, Mass and Momentum Transfer*, Prentice Hall, Englewood Cliffs, NJ, 1961.

11.4 W.M. Rohsenow and J.P. Hartnett, Handbook of Heat Transfer, ch. 13. McGraw-Hill, New York, 1973.

11.5 J.C.Y. Koh, E.M. Sparrow, and J.P. Hartnett, "Two Phase Boundary Layer in Laminar Film Condensation," *Int. J. Heat Mass Transfer*, 2, 1961.

11.6 V.S. Arpacı, and P.S. Larsen, *Convection Heat Transfer*. Prentice Hall, Englewood Cliffs, NJ, 1984.

11.7 A.F. Mills, *Heat Transfer*. 2nd ed., Prentice-Hall, Englewood Cliffs, NJ., 1999.

11.8 V.S. Arpacı, *Microscales of Turbulence-Heat and Mass Transfer Correlations*. Gordon and Breach Science Publishers, Amsterdam-The Netherlands, 1997.

11.9 T.H.K. Frederkin, and J.A. Clark, *Advances in Cryogenic Engineering*, Vol. 8, Plenum, New York, 1962.

■ EXERCISES

11.1 A reactor core element is simulated by a coaxial sodium flow over a solid fuel rod (Fig. 6P–1). In terms of the following data, $u''' = 35$ MW/m^3, $D_i = 25$ mm, $D_0 = 37.5$ mm, $\ell = 6$ m, $T_i = 90\ °\text{C}$, $V = 1.2$ m/s, and $k_{\text{fuel rod}} = 9$ W/m·K, evaluate (a) the exit temperature of sodium, (b) the surface temperature of the fuel rod, (c) the center-line temperature of the fuel rod.

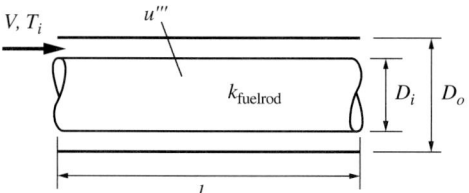

Figure 11P–1

Appendix A

CORRELATIONS

Fluid mechanics and heat transfer correlations involved with internal and external forced convection and natural convection are summarized in this appendix. These correlations are organized in the following manner:

First, whenever possible, a schematic configuration is shown. Features such as geometry, flow condition, flow regime, and working fluid are categorized to help the reader find the equation number—(A.1), (A.2) etc.—corresponding to the friction factor or the Nusselt number for the particular problem of interest.

Second, a numbered table follows each schematic configuration. Each table has five columns:

1. **Eq.:** the equation number found in the schematic.
2. **Correlation:** Correlations for the friction factor or the Nusselt number.
3. **Condition:** the condition to which the correlation applies.
4. **Property:** the temperature to be used to evaluate properties such as density ρ, viscosity μ, etc.
5. **Ref.:** the number of the reference in which further details can be found.

Further details on using the appendix can be found in Chapter 6.

Forced Convection in Circular Pipes

GEOMETRY							
FLOW CONDITION		Developing flow, developing temperature		Developed flow, developing temperature		Developed flow, developed temperature	
FRICTION FACTOR		Laminar	Turbulent	Laminar	Turbulent	Laminar	Turbulent
Friction Factor	Smooth wall					(A.1)	(A.2, 3, 4)
	Rough wall					(A.1)	(A.5)
Nusselt number	Constant T_w	(A.9)		(A.8)		(A.7)	smooth wall (A.10–13) / rough wall (A.14)
	Constant q_w					(A.7)	smooth wall (A.13) / rough wall

Gases, water, viscous oils

FLOW CONDITION		Developing flow, developing temperature		Developed flow, developing temperature		Developed flow, developed temperature	
FRICTION FACTOR		Laminar	Turbulent	Laminar	Turbulent	Laminar	Turbulent
Nusselt number	Constant T_w						(A.15, 18)
	Constant q_w						(A.15, 16, 17, 18)

Liquid metals

Convection used in the tables below is as follows:

$$\text{Re}_D = \frac{UD}{\rho}, \quad U = \frac{\dot{m}}{\rho A}, \quad T_f = \frac{T_w + T_b}{2}, \quad T_b = \frac{\int_0^R u(r)T(r)dr}{\int_0^R u(r)dr}$$

$$f = \frac{\tau_w}{\frac{1}{2}\rho U^2} = \frac{1}{4}\frac{-(dP/dx)D_H}{\frac{1}{2}\rho U^2}$$

$$Gz = \frac{D}{L}\text{Re}_D \text{Pr}$$

$$D_H = 4 \times \frac{\text{Cross-sectional area}}{\text{Wetted perimeter}} = 4 \times \frac{A}{P}$$

ϵ: Surface roughness

Table A.1 Correlation for forced convection

Forced Convection in Circular Pipes

Eq.	Correlation	Condition	Property	Ref.
(A.1)	$f = \dfrac{16}{\text{Re}_D}$	$0 < \text{Re}_D < 2300$	T_f	[1]
(A.2)	$f = \dfrac{0.0791}{\text{Re}_D^{1/4}}$	$2300 < \text{Re}_D \leq 10^5$	T_f	[1]
(A.3)	$f = \dfrac{0.046}{\text{Re}_D^{1/5}}$	$\text{Re}_D \geq 10^5$	T_f	[1]
(A.4)	$\left(\dfrac{2}{f}\right)^{1/2} = 2.5 \ln\left[\text{Re}_D \left(\dfrac{f}{2}\right)^{1/2}\right] + 0.3393$	$\text{Re}_D > 2300$	T_f	[2]
(A.5)	$f = \dfrac{0.1250}{\left[\log\left(\dfrac{5.74}{\text{Re}_D^{0.9}} + \dfrac{\epsilon/D}{3.7}\right)\right]^2}$	$\text{Re}_D > 2300$	T_f	

Eq.	Correlation	Condition	Property	Ref.
(A.6)	$\mathrm{Nu}_D = 4.36$	$\mathrm{Pr} \geq 0.6$	T_b	[3]
(A.7)	$\mathrm{Nu}_D = 3.66$	$\mathrm{Pr} \geq 0.6$	T_b	[3]
(A.8)	$\mathrm{Nu}_D = 3.66 + \dfrac{0.0668\mathrm{Gz}}{1 + 0.04\mathrm{Gz}^{2/3}}$	$\mathrm{Pr} \geq 1$, or an unheated starting length $\mathrm{Gz} = \dfrac{D}{L}\mathrm{Re}_D \mathrm{Pr}$	T_b	[4]
(A.9)	$\mathrm{Nu}_D = 1.86 \mathrm{Gz}^{1/3} \left(\dfrac{\mu}{\mu_w}\right)^{0.14}$	$0.48 < \mathrm{Pr} < 16700$ $0.0044 < \dfrac{\mu}{\mu_w} < 9.75$ $\mathrm{Gz}\left(\dfrac{\mu}{\mu_w}\right)^{0.14} \geq 2$	T_b	[5]
(A.10)	$\mathrm{Nu}_D = 0.023 \mathrm{Re}_D^{4/5} \mathrm{Pr}^{1/3}$	$0.7 \leq \mathrm{Pr} \leq 160$ $\dfrac{L}{D} \geq 60$ $\mathrm{Re} \geq 10^4$	T_b	[6]
(A.11)	$\mathrm{Nu}_D = 0.023 \mathrm{Re}_D^{4/5} \mathrm{Pr}^n$ $T_w > T_b, \quad n = 0.4$ $T_w < T_b, \quad n = 0.3$	$0.7 \leq \mathrm{Pr} \leq 160$ $\dfrac{L}{D} \geq 60$ $\mathrm{Re} \geq 10^4$	T_b	[7]
(A.12)	$\mathrm{Nu}_D \doteq 0.027 \mathrm{Re}_D^{4/5} \mathrm{Pr}^{1/3} \left(\dfrac{\mu}{\mu_w}\right)^{0.14}$	$0.7 \leq \mathrm{Pr} \leq 16{,}700$ $\dfrac{L}{D} \geq 60$ $\mathrm{Re} \geq 10^4$	T_b (μ_w at T_w)	[5]
(A.13)	$\mathrm{St}_D = \dfrac{f/2}{1.07 + 12.7(f/2)^{1/2}(\mathrm{Pr}^{2/3} - 1)} \left(\dfrac{\mu_b}{\mu_w}\right)^n$ $T_w > T_b, \quad n = 0.11$ $T_w < T_b, \quad n = 0.25$ $q_w = \text{constant or gases}, \quad n = 0$	$0.5 < \mathrm{Pr} < 2000$ $10^4 < \mathrm{Re} < 5 \times 10^6$ $0 < \dfrac{\mu_b}{\mu_w} < 40$	T_f (μ_b at T_b) (μ_w at T_w)	[8]
(A.14)	$\mathrm{St}_D \mathrm{Pr}^{2/3} = \dfrac{f}{2}$	Moody diagram	T_f (St_D at T_b)	[6]

Eq.	Correlation	Condition	Property	Ref.
(A.15)	$\mathrm{Nu}_D = C + 0.025 \mathrm{Pe}_D^{0.8}$	$\dfrac{L}{D} > 60$	T_b	[9]
	$C = 5.0$, constant T_w $C = 7.0$, constant q_w	$\mathrm{Pe}_D > 10^2$		
(A.16)	$\mathrm{Nu}_D = 0.625 \mathrm{Pe}_D^{0.4}$	$\dfrac{L}{D} > 60$ $10^2 < \mathrm{Pe}_D < 10^4$	T_b	[10]
(A.17)	$\mathrm{Nu}_D = 4.82 + 0.0185 \mathrm{Pe}_D^{0.827}$	$\mathrm{Pr} \cong 0.0153$ $\dfrac{L}{D} > 60$ $3.6 \times 10^3 < \mathrm{Re}_D < 9.05 \times 10^5$ $58 < \mathrm{Pe}_D < 1.31 \times 10^4$	T_b	[11]
(A.18)	$\mathrm{Nu}_D = C_1 + C_2 \mathrm{Re}_D^{0.85} \mathrm{Pr}^{0.93}$ $C_1 = 4.8$, $C_2 = 0.0156$, constant T_w $C_1 = 6.3$, $C_2 = 0.0167$, constant q_w	$\mathrm{Pr} < 0.1$	T_b (Pr at T_w) (Re_D at T_w)	[12]

Forced Convection of Fully Developed Laminar Flow in Ducts of Various Cross Sections [13]

Cross Section	b/a	D_H	$\mathrm{Nu}_D = hD_H/k$			$f\mathrm{Re}_{D_H}$
			Constant axial q_w Constant peripheral T_w	Constant q_w	Constant T_w	
Circle (diameter D)	—	D	4.364	4.364	3.657	16.000
Ellipse	1.11	~1.05a	5.099	4.35	3.66	18.700
Triangle	$2\sqrt{3}$	$a\sqrt{3}$	3.111	1.892	2.47	13.333
Square	1	a	3.608	3.091	2.976	14.227
Rectangle	2	$4a/3$	4.123	3.017	3.391	15.548
Rectangle	4	$8a/5$	5.331	2.930	4.439	18.233
Rectangle	8	$16a/9$	6.490	2.904	5.597	20.585
Parallel plates	∞	$2a$	8.235	8.235	7.541	24.000
Parallel plates (one insulated)	∞	$2a$	5.385	—	4.681	24.000
Hexagon	—	~$26a^2$	4.002	3.862	3.34	15.054

Forced Convection of Fully Developed Laminar Flow in Concentric Tube Annulus [14]

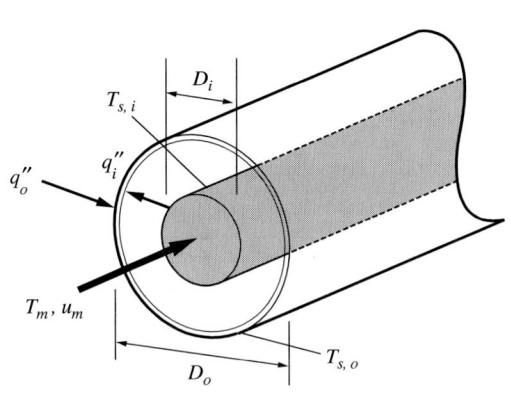

$$q_i'' = h_i(T_{s,i} - T_m)$$
$$q_o'' = h_o(T_{s,o} - T_m)$$

$$\text{Nu}_i \equiv \frac{h_i D_H}{k}$$

$$\text{Nu}_o \equiv \frac{h_o D_H}{k}$$

$$D_H = \frac{4(\pi/4)(D_o^2 - D_i^2)}{\pi D_o + \pi D_i} = D_o - D_i$$

$$\text{Nu}_i = \frac{\text{Nu}_{ii}}{1 - (q_o''/q_i'')\theta_i^\circ}$$

$$\text{Nu}_o = \frac{\text{Nu}_{oo}}{1 - (q_i''/q_o'')\theta_o^\circ}$$

Table of Nusselt numbers for fully developed laminar flow in a circular tube annulus with one surface insulated and the other at a constant temperature

D_i/D_o	Nu_i	Nu_o
0	—	3.66
0.05	17.46	4.06
0.10	11.56	4.11
0.25	7.37	4.23
0.50	5.74	4.43
1.00	4.86	4.86

Table of influence coefficients for fully developed laminar flow in a circular tube annulus with uniform heat flux maintained at both surfaces

D_i/D_o	Nu_{ii}	Nu_{oo}	θ_i°	θ_o°
0	—	4.364	∞	0
0.05	17.81	4.792	2.18	0.0294
0.10	11.91	4.834	1.383	0.0562
0.20	8.499	4.833	0.905	0.1041
0.40	6.583	4.979	0.603	0.1823
0.60	5.912	5.099	0.473	0.2455
0.80	5.58	5.24	0.401	0.299
1.00	5.385	5.385	0.346	0.346

Forced Convention of Flow Over a Flat Plate

Geometry			

Flow regime		Laminar	Turbulent
Boundary-layer thickness		(B.1)	(B.4)
Local friction factor		(B.2)	(B.5)
Average friction factor		(B.3)	(B.6)
Gases, water, viscous oils	Local Nusselt number	(B.8)	(B.11)
	Average Nusselt number	(B.9)	
Liquid metals	Local Nusselt number	(B.10)	
	Average Nusselt number		

Flow regime		Turbulent (laminar leading edge incuded)
Average friction factor		(B.7)
Gases, water, viscous oils	Average Nusselt number	(B.12)
Liquid metals	Average Nusselt number	(B.13)

Conventions used on the tables that follow are listed below.

$$f_x = \frac{\tau_{w,x}}{\frac{1}{2}\rho U_\infty^2}$$

$$f = \frac{(1/L)\int_0^L \tau_{w,x} dx}{(1/2)\rho U_\infty^2} = \frac{1}{L}\int_0^L f_x dx$$

$$f = \frac{1}{L}\left[\int_0^{x_c} f_{x,\text{lam}} dx + \int_{x_c}^L f_{x,\text{turb}} dx\right]$$

δ: Momentum boundary-layer thickness

$$\text{Re}_x = \frac{U_\infty x}{\nu}$$

$\text{Re}_{x,c}$: Transition (critical) Reynolds number

$$\text{Nu}_x = \frac{h_x x}{k}, \qquad \text{Pe}_x = \text{Re}_x \text{Pr} = \frac{U_\infty x}{\alpha}$$

$$\text{Nu}_L = \frac{h_L L}{k} = \frac{\left(\frac{1}{L}\int_0^L h_x dx\right)L}{k} = \int_0^L \left(\frac{\text{Nu}_x}{x}\right) dx$$

$$\text{Nu}_L = \int_0^{x_c} \left(\frac{\text{Nu}_{x,\text{Lam}}}{x}\right) dx + \int_{x_c}^L \left(\frac{\text{Nu}_{x,\text{Turb}}}{x}\right) dx$$

$$\text{Re}_D = \frac{U_\infty D}{\nu}, \qquad \text{Pe}_D = \text{Re}_D \text{Pr} = \frac{U_\infty D}{\alpha}, \qquad \text{Nu}_D = \frac{hD}{k}$$

Forced Convection of Flow Over a Flat Plate

Eq.	Correlation	Condition	Property	Ref.
(B.1)	$\dfrac{\delta}{x} = \dfrac{5}{\text{Re}_x^{1/2}}$	$0 < \text{Re}_x \leq 5 \times 10^5$	T_f	[15]
(B.2)	$f_x = \dfrac{0.664}{\text{Re}_x^{1/2}}$	$0 < \text{Re}_x \leq 5 \times 10^5$	T_f	[15]
(B.3)	$f = \dfrac{1.328}{\text{Re}_L^{1/2}}$	$0 < \text{Re}_x \leq 5 \times 10^5$	T_f	[15]

Eq.	Correlation	Condition	Property	Ref.
(B.4)	$\dfrac{\delta}{x} = \dfrac{0.37}{\text{Re}_x^{1/5}}$	$5 \times 10^5 \leq \text{Re}_x \leq 10^8$	T_f	[15]
(B.5)	$f_x = \dfrac{0.0592}{\text{Re}_x^{1/5}}$	$0 < \text{Re}_x \leq 5 \times 10^5$	T_f	[15]
(B.6)	$f = \dfrac{0.074}{\text{Re}_L^{1/5}}$	$0 < \text{Re}_x \leq 5 \times 10^5$	T_f	[15]
(B.7)	$f = \dfrac{0.074}{\text{Re}_L^{1/5}} - \dfrac{1742}{\text{Re}_L}$	$\text{Re}_L \leq 10^8$ $\text{Re}_{x,c} = 5 \times 10^5$	T_f	[15]
(B.8)	$\text{Nu}_x = 0.332 \text{Re}_x^{1/2} \text{Pr}^{1/3}$	$\text{Pr} \geq 0.6$	T_f	[6]
(B.9)	$\text{Nu}_L = 0.664 \text{Re}_L^{1/2} \text{Pr}^{1/3}$	$\text{Pr} \geq 0.6$	T_f	[6]
(B.10)	$\text{Nu}_x = 0.0565 Pe_x^{1/2}$	$\text{Pr} \leq 0.05$	T_f	[3]
(B.11)	$\text{Nu}_x = 0.0296 \text{Re}_x^{4/5} \text{Pr}^{1/3}$	$0.6 \leq \text{Pr} \leq 60$	T_f	[6]
(B.12)	$\text{Nu}_L = (0.037 \text{Re}_L^{4/5} - 871) \text{Pr}^{1/3}$	$0.6 < \text{Pr} < 60$ $\text{Re}_L \leq 10^8 \ \text{Re}_{x,c} = 5 \times 10^5$	T_f	
(B.13)	$\text{Nu}_L = 0.036(\text{Re}_L^{4/5} - 9200)\text{Pr}^{0.43}\left(\dfrac{\mu_\infty}{\mu_w}\right)^{1/4}$	$0.7 < \text{Pr} < 380$ $10^5 < \text{Re}_L < 5.5 \times 10^6$ $0.26 < \left(\dfrac{\mu_\infty}{\mu_w}\right) < 3.5$ $\text{Re}_{x,c} = 2 \times 10^5$	T_∞ (μ_w at T_w)	[16]

Forced Convention Over Other Geometries

Geometry	Correlation	
Circular cylinder $U_\infty, T_\infty \rightarrow$ T_w D	Gases, water, oil and Liquid metal	(C.1–7)
	Table C.1	
	Re_0 / C / m	
	0.4 – 4 / 0.989 / 0.330	
	4 – 40 / 0.911 / 0.385	
	40 – 4×10^3 / 0.683 / 0.466	
	$4 \times 10^3 – 4 \times 10^4$ / 0.193 / 0.618	
	$4 \times 10^4 – 4 \times 10^5$ / 0.0266 / 0.805	
	Table C.2	
	Re_0 / C / m	
	1 – 40 / 0.75 / 0.4	
	40 – 10^3 / 0.51 / 0.5	
	$10^3 – 2 \times 10^5$ / 0.26 / 0.6	
	$2 \times 10^5 – 10^6$ / 0.076 / 0.7	
Sphere $U_\infty, T_\infty \rightarrow$ T_w	Liquids	(C.8)
	Gases	(C.9)
	Water and oil	(C.10)
	Gases, water and oil	(C.11)
	Air	(C.12, 13)
	Liquid metal	(C.14)
Sphere $x \downarrow$ D $U \downarrow$	Free falling liquid sphere U: Terminal velocity	
	Idealized condition	(C.15)
	Droplet oscillations and distortions accounted for	(C.16)

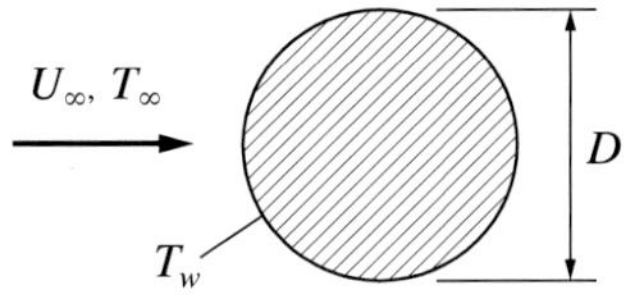

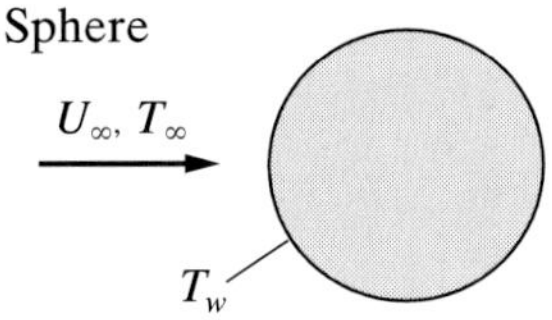

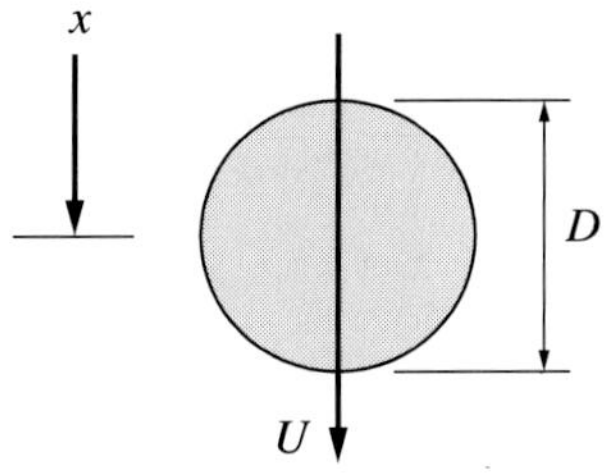

Note: Table C.1 and Table C.2 have three columns (Re_0, C, m) shown above separated by slashes for readability.

App. A Correlations

Forced Convention Over Other Geometries

Eq.	Correlation	Condition	Property	Ref.
(C.1)	$Nu_D = C Re_D^m Pr^{1/3}$ (Table C.1)	$Pr \le 0.7$ $0.4 < Re_D < 4 \times 10^5$	T_f	[17, 18]
(C.2)	$Nu_D = (0.4 Re_D^{1/2} + 0.06 Re_D^{2/3}) Pr^{0.4} \left(\dfrac{\mu_\infty}{\mu_w}\right)^{1/4}$	$0.67 < Pr < 300$ $10 < Re_D < 10^5$ $0.25 < \dfrac{\mu_\infty}{\mu_w} < 5.2$	T_f	[16]
(C.3)	$Nu_D = C Re_D^m Pr^n \left(\dfrac{Pr_\infty}{Pr_w}\right)^{1/4}$ $Pr < 10, \quad n = 0.37$ $Pr > 10, \quad n = 0.36$ (Table C.2)	$0.7 < Pr < 500$ $1 < Re_D < 10^6$	T_∞ Pr_w at T_w	[19]
(C.4)	$Nu_D = 0.3 + \dfrac{0.62 Re_D^{1/2} Pr^{1/3}}{[1 + (\dfrac{0.4}{Pr})^{2/3}]^{1/4}} \left[1 + (\dfrac{Re_D}{282000})^m\right]^n$ $10^2 < Re_D < 2 \times 10^4, \quad m = \dfrac{5}{8}, \quad n = \dfrac{4}{5}$ $4 \times 10^4 < Re_D < 4 \times 10^6, \quad m = \dfrac{1}{2}, \quad n = 1$	$Pe_D > 0.2$	T_f	[20]
(C.5)	$Nu_D = \dfrac{1}{0.8237 - \ln(Pe_D)^{1/2}}$	$Pe_D < 0.2$	T_f	[21]
(C.6)	$Nu_D = (0.43 + 0.50 Re_D^{0.5}) Pr^{0.38} \left(\dfrac{Pr_f}{Pr_w}\right)^{1/4}$	$1 < Re_D < 10^3$	T_f for gases T_∞ for liquids	[22]
(C.7)	$Nu_D = 0.25 Re_D^{0.6} Pr^{0.38} \left(\dfrac{Pr_f}{Pr_w}\right)^{1/4}$	$10^3 < Re_D < 2 \times 10^5$	T_f for gases T_∞ for liquids	[22]
(C.8)	$Nu_D = (0.97 + 0.68 Re_D^{1/2}) Pr^{0.3}$	Liquids $1 < Re_D < 2 \times 10^3$	T_f	[23]

Eq.	Correlation	Condition	Property	Ref.
(C.9)	$\mathrm{Nu}_D = 0.37 \mathrm{Re}_D^{0.6}$	Gases $17 < \mathrm{Re}_D < 7 \times 10^4$	T_f	[24]
(C.10)	$\mathrm{Nu}_D = (1.2 + 0.53 \mathrm{Re}_D^{0.54}) \mathrm{Pr}^{0.3} \left(\dfrac{\mu_\infty}{\mu_w}\right)^{1/4}$	Water and oils $1 < \mathrm{Re}_D < 2 \times 10^5$	T_∞ (μ_w at T_w)	[25]
(C.11)	$\mathrm{Nu}_D = 2 + (0.4 \mathrm{Re}_D^{1/2} + 0.06 \mathrm{Re}_D^{2/3}) \mathrm{Pr}^{0.4} \left(\dfrac{\mu_\infty}{\mu_w}\right)^{1/4}$	$0.71 < \mathrm{Pr} < 380$ $3.5 < \mathrm{Re}_D < 7.6 \times 10^4$ $1.0 < \left(\dfrac{\mu_\infty}{\mu_w}\right) < 3.2$	T_∞ (μ_w at T_w)	[16]
(C.12)	$\mathrm{Nu}_D = 2 + (0.25 + 3 \times 10^{-4} \mathrm{Re}_D^{1.6})^{1/2}$	$\mathrm{Pr} = 0.71$ $10^2 < \mathrm{Re}_D < 3 \times 10^5$	T_f	[26]
(C.13)	$\mathrm{Nu}_D = 4.30 + C_1 \mathrm{Re}_D + C_2 \mathrm{Re}_D^2 + C_3 \mathrm{Re}_D^3$ $C_1 = 5 \times 10^{-3}$, $C_2 = 0.25 \times 10^{-9}$ $C_3 = -3.1 \times 10^{-17}$	$\mathrm{Pr} = 0.71$ $3 \times 10^5 < \mathrm{Re}_D < 5 \times 10^6$	T_f	[26]
(C.14)	$\mathrm{Nu}_D = 2 + 0.386 \mathrm{Pe}_D^{1/2}$	Liquid metals T_f $35{,}600 < \mathrm{Re}_D < 152{,}500$		[28]
(C.15)	$\mathrm{Nu}_D = 2 + 0.6 \mathrm{Re}_D^{1/2} \mathrm{Pr}^{1/3}$		T_∞	[28]
(C.16)	$\mathrm{Nu}_D = 2 + 0.6 \mathrm{Re}_D^{1/2} \mathrm{Pr}^{1/3} \left[25 \left(\dfrac{x}{D}\right)^{-0.7}\right]$	Droplet oscillations and distortions accounted for	T_∞	[29]

Forced Convention: Tube Banks (30, 31, 32)

Figure shows tube arrangements in a bank.

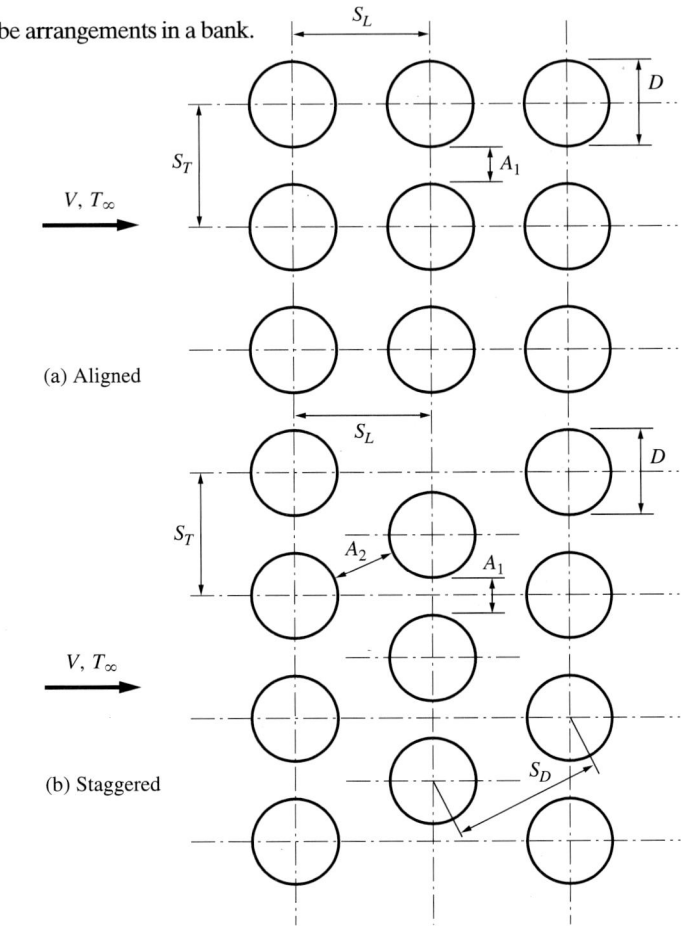

(a) Aligned

(b) Staggered

Forced Convention: Tube Banks [30, 31, 32]

Eq.	Correlation	Condition	Property	Ref.
(1)	$\mathrm{Nu}_D = C_1 \mathrm{Re}_{D,\max}^m \mathrm{Pr}^{1/3}$	$N \geq 10$ $2 \times 10^3 < \mathrm{Re}_{D,\max} < 4 \times 10^4$ $\mathrm{Pr} \geq 0.7$	T_f	[30]
(2)	$(\mathrm{Nu}_D)_{N<10} = C_2 (\mathrm{Nu}_D)_{N \geq 10}$	$N < 10$	T_f	[30]
(3)	$\mathrm{Nu}_D = 4.03 + 0.228(\mathrm{Pe}_{D,\max})^{0.67}$	Liquid metals $2 \times 10^4 < \mathrm{Re}_{D,\max} < 8 \times 10^4$	T_f	[31]

Constants of Eq. (1)

| | | \multicolumn{8}{c}{S_T/D} |
| | | 1.25 | | 1.5 | | 2.0 | | 3.0 | |
	S_L/D	C_1	m	C_1	m	C_1	m	C_1	m
In-line	1.25	0.348	0.592	0.275	0.608	0.100	0.704	0.0633	0.752
	1.50	0.367	0.586	0.250	0.620	0.101	0.702	0.0678	0.744
	2.00	0.418	0.570	0.299	0.602	0.229	0.632	0.198	0.608
	3.00	0.290	0.601	0.357	0.584	0.374	0.581	0.286	0.608
Staggered	0.600	—	—	—	—	—	—	0.213	0.636
	0.900	—	—	—	—	0.446	0.571	0.401	0.581
	1.000	—	—	0.497	0.558	—	—	—	—
	1.125	—	—	—	—	0.478	0.565	0.518	0.560
	1.250	0.518	0.556	0.505	0.554	0.519	0.556	0.522	0.567
	1.500	0.451	0.568	0.460	0.562	0.452	0.568	0.488	0.568
	2.000	0.404	0.572	0.416	0.568	0.482	0.556	0.449	0.570
	3.000	0.310	0.592	0.356	0.580	0.440	0.562	0.428	0.574

Constants of Eq. (2) [32]

N_L	1	2	3	4	5	6	7	8	9
In-line	0.64	0.80	0.87	0.90	0.92	0.94	0.96	0.98	0.99
Staggered	0.68	0.75	0.83	0.89	0.92	0.95	0.97	0.98	0.99

Natural Convection For Several Geometeries

Geometry	Condition		Constant T_w	Constant q_w
	Air and water	Laminar	(D.1)	(D.3)
		Turbulent	(D.2)	(D.4)
	Liquid metal, gases, water viscous oils	Laminar and Turbulent	(D.5)	(D.7)
		Laminar	(D.6)	(D.8)
	Air and water	Laminar		(D.12)
		Turbulent		(D.14)
	Liquid metal, gases, water viscous oils	Laminar and Turbulent	(D.15)	(D.17, 19, 20)
		Laminar	(D.16)	(D.18)
	Air and water	Laminar		(D.12)
		Turbulent		(D.13)
	Liquid metal, gases, water viscous oils	Laminar and Turbulent		(D.21)
		Laminar		

	Angle	Correlation
	$\theta = 180°$	(D.29)
	$\theta = 0°$	(D.30)
	$\theta = 90°$	(D.31–34)
	θ: arbitrary	(D.35–38)

Free Convections For Several Geometries

Geometry	Description	Geometry	Description
(heated plate, upward arrows)	Long horizontal plate of width L with heated surface facing upward. Constant T_w. Laminar (D.9). Turbulent (D.10).	(cooled plate, downward arrows)	Long horizontal plate of width L with cooled surface facing downward. Constant T_w. Laminar (D.9). Turbulent (D.10).
(cooled plate, downward arrows)	Long horizontal plate of width L with cooled surface facing upward. Constant T_w. Laminar and Turbulent (D.11).	(heated plate, upward arrows)	Long horizontal plate of width L with heated surface facing downward. Constant T_w. Laminar and Turbulent (D.11).
(horizontal cylinder cross-section)	Horizontal cylinders (D.22, 23, 24) (Table D.1). Spheres (D.26).	(inclined cylinder with angle θ)	Inclined cylinders (D.25).
(concentric cylinders)	Concentric cylinders (D.27).	(concentric spheres)	Concentric spheres (D.28).

Table D.1

Ra	10^{-10}	10^{-2}	10^{2}	10^{4}	10^{7}	10^{12}
C	0.675	1.02	0.850	0.480	0.125	
n	0.058	0.148	0.188	1/4	1/3	

$$\mathrm{Ra}_L = \mathrm{Gr}_L \mathrm{Pr} = \frac{g\beta(T_w - T_\infty)L^3}{\nu\alpha}, \quad \text{(for constant } T_w\text{)}$$

$$\mathrm{Ra}_x^* = \mathrm{Gr}_x^* \mathrm{Pr} = (\mathrm{Gr}_x \mathrm{Pr})\mathrm{Nu}_x = \mathrm{Ra}_x \mathrm{Nu}_x = \frac{g\beta\left(\frac{q_w'' x}{k}\right)x^3}{\nu^2}, \quad \text{(for constant } q_w''\text{)}$$

Table A.2 Correlation for free convection.

Free Convection For Several Geometeries

Eq.	Correlation	Condition	Property	Ref.
(D.1)	$\mathrm{Nu}_L = 0.59 \mathrm{Ra}_L^{1/4}$	$10^4 < \mathrm{Ra}_L < 10^9$	T_f	[33]
(D.2)	$\mathrm{Nu}_L = 0.10 \mathrm{Ra}_L^{1/3}$	$10^9 < \mathrm{Ra}_L < 10^{13}$	T_f	[34]
(D.3)	$\mathrm{Nu}_x = 0.60(\mathrm{Ra}_x^*)^{1/5}$	$10^5 < \mathrm{Ra}_x^* < 10^{11}$	T_f	[35]
(D.4)	$\mathrm{Nu}_x = 0.17(\mathrm{Ra}_x^*)^{1/4}$	$2 \times 10^{13} < \mathrm{Ra}_x^* < 10^{16}$	T_f	[35]
(D.5)	$\mathrm{Nu}_L^{1/2} = 0.825 + \dfrac{0.387 \mathrm{Ra}_L^{1/6}}{\left[1 + \left(\dfrac{0.492}{\mathrm{Pr}}\right)^{9/16}\right]^{8/27}}$	$10^{-1} < \mathrm{Ra}_L < 10^{12}$	T_f	[36]
(D.6)	$\mathrm{Nu}_L = 0.68 + \dfrac{0.670 \mathrm{Ra}_L^{1/4}}{\left[1 + \left(\dfrac{0.492}{\mathrm{Pr}}\right)^{9/16}\right]^{4/9}}$	$0 < \mathrm{Ra}_L < 10^9$	T_f	[36]
(D.7)	$\mathrm{Nu}_L^{1/6}(\mathrm{Nu}_L^{1/2} - 0.825) = \dfrac{0.387(\mathrm{Ra}_L^*)^{1/6}}{\left[1 + \left(\dfrac{0.492}{\mathrm{Pr}}\right)^{9/16}\right]^{8/27}}$	$10^{-1} < \mathrm{Ra}_L^* < 10^{12}$	T_f	[36]
(D.8)	$\mathrm{Nu}_L^{1/4}(\mathrm{Nu}_L - 0.68) = \dfrac{0.670(\mathrm{Ra}_L^*)^{1/4}}{\left[1 + \left(\dfrac{0.492}{\mathrm{Pr}}\right)^{9/16}\right]^{4/9}}$	$0 < \mathrm{Ra}_L^* < 10^9$	T_f	[36]

Eq.	Correlation	Condition	Property	Ref.
(D.9)	$\mathrm{Nu}_L = 0.54\mathrm{Ra}_L^{1/4}$	$10^5 \leq \mathrm{Ra}_L \leq 10^7$	T_f	[37]
(D.10)	$\mathrm{Nu}_L = 0.15\mathrm{Ra}_L^{1/3}$	$10^7 \leq \mathrm{Ra}_L \leq 10^{10}$	T_f	[37]
(D.11)	$\mathrm{Nu}_L = 0.27\mathrm{Ra}_L^{1/4}$	$10^5 \leq \mathrm{Ra}_L \leq 10^{10}$	T_f	[37]
(D.12)	$\mathrm{Nu}_x = 0.60(\mathrm{Ra}_x^* \cos\theta)^{1/5}$	$10^5 < \mathrm{Ra}_x^* \cos\theta < 10^{11}$	T_f	[38]
(D.13)	$\mathrm{Nu}_x = 0.17(\mathrm{Ra}_x^* \cos\theta)^{1/4}$	$2 \times 10^{13} < \mathrm{Ra}_x^* \cos\theta < 10^{16}$	T_f	[38]
(D.14)	$\mathrm{Nu}_x = 0.17(\mathrm{Ra}_x^* \cos\theta)^{1/4}$	$2 \times 10^{13} < \mathrm{Ra}_x^* \cos\theta < 10^{16}$	T_f	[38]
(D.15)	$\mathrm{Nu}_L^{1/2} = 0.825 + \dfrac{0.387(\mathrm{Ra}_L \cos\theta)^{1/6}}{\left[1 + \left(\dfrac{0.492}{\mathrm{Pr}}\right)^{9/16}\right]^{8/27}}$	$\mathrm{Ra}_L \leq 10^9, \quad \theta < 60°$	T_f	[36]
(D.16)	$\mathrm{Nu}_L = 0.68 + \dfrac{0.670(\mathrm{Ra}_L \cos\theta)^{1/4}}{\left[1 + \left(\dfrac{0.492}{\mathrm{Pr}}\right)^{9/16}\right]^{4/9}}$	$\mathrm{Ra}_L \leq 10^9, \quad \theta < 60°$	T_f	[36]
(D.17)	$\mathrm{Nu}_L^{1/6}(\mathrm{Nu}_L^{1/2} - 0.825) = \dfrac{0.387(\mathrm{Ra}_L^* \cos\theta)^{1/6}}{\left[1 + \left(\dfrac{0.492}{\mathrm{Pr}}\right)^{9/16}\right]^{8/27}}$	$\mathrm{Ra}_L^* \leq 10^9, \quad \theta < 60°$	T_f	[36]
(D.18)	$\mathrm{Nu}_L^{1/4}(\mathrm{Nu}_L - 0.68) = \dfrac{0.670(\mathrm{Ra}_L^* \cos\theta)^{1/4}}{\left[1 + \left(\dfrac{0.492}{\mathrm{Pr}}\right)^{9/16}\right]^{4/9}}$	$\mathrm{Ra}_L^* < 10^9, \quad \theta < 60°$	T_f	36]
(D.19)	$\mathrm{Nu} = 0.56(\mathrm{Ra}\cos\theta)^{1/4}$	$10^5 < \mathrm{Ra}\cos\theta < 10^{11}$	$\dfrac{T_w + T_f}{2}$	[34]
		$\theta < 88°$	(β at $\dfrac{T_\infty + T_f}{2}$)	
(D.20)	$\mathrm{Nu} = 0.58\mathrm{Ra}^{1/5}$	$10^6 < \mathrm{Ra} < 10^{11}$	$\dfrac{T_w + T_f}{2}$	[34]
		$87° \leq \theta < 90°$	(β at $\dfrac{T_\infty + T_f}{2}$)	

Eq.	Correlation	Condition	Property	Ref.
(D.21)	$Nu = 0.145 Pr^{1/3} \left[Gr^{1/3} - Gr_c^{1/3} \right] + 0.56(Ra \cos\theta)^{1/4}$	$10^5 < Ra \cos\theta < 10^{11}$	$\dfrac{T_w + T_f}{2}$	[39]
	θ: $-15°$ $-30°$ $-60°$ $-75°$	$-75° \leq \theta < -15°$	(β at $\dfrac{T_\infty + T_f}{2}$)	
	Gr_c: 5×10^9 2×10^9 10^8 10^6			
	For $Gr_c > Gr$, $Gr_c \equiv Gr$			
(D.22)	$Nu_D = C Ra_D^n$ (Table D.1)	Gases, liquids, and oils	T_f	[40]
(D.23)	$Nu_D^{1/2} = 0.6 + \dfrac{0.387 Ra_D^{1/6}}{\left[1 + \left(\dfrac{0.559}{Pr} \right)^{9/16} \right]^{8/27}}$	$10^{-5} < Ra_D < 10^{12}$	T_f	[36]
(D.24)	$Nu_D = 0.53 (Ra_D Pr)^{1/4}$	Liquid metal	T_f	[40]
(D.25)	$Nu_L = \left[0.60 - 0.488(\sin\theta)^{1.03} \right] Ra_L^{1/4} + \dfrac{(\sin\theta)^{1.75}}{2}$	$Ra_L < 2 \times 10^8$	T_f (β at T_∞)	[41]
(D.26)	$Nu_D = 2 + 0.56 \left(\dfrac{Pr}{0.846 + Pr} \right)^{1/4} Ra_D^{1/4}$	Laminar	T_f	[42]
(D.27)	$\dfrac{k_{\text{eff}}}{k} = 0.386 \left(\dfrac{Pr}{0.861 + Pr} \right)^{1/4} (Ra_c^*)^{1/4}$	$10^2 \leq Ra_c^* \leq 10^7$	$\dfrac{T_i + T_o}{2}$	[42]
(D.28)	$\dfrac{k_{\text{eff}}}{k} = 0.74 \left(\dfrac{Pr}{0.861 + Pr} \right)^{1/4} (Ra_s^*)^{1/4}$	$10^2 \leq Ra_s^* \leq 10^4$	$\dfrac{T_i + T_o}{2}$	[42]
(D.29)	$Nu_L = 1$ (pure conduction)	$\theta = 180°$ or $Ra_L < 1700$	$\dfrac{T_1 + T_2}{2}$	
(D.30)	$Nu_L = 0.069 Ra_L^{1/3} Pr^{0.074}$	$\theta = 0°$ ($\dfrac{L}{H}$ and $\dfrac{L}{W}$ small) $3 \times 10^5 < Ra_L < 7 \times 10^9$	$\dfrac{T_1 + T_2}{2}$	[43]
(D.31)	$Nu_L = 0.046 Ra_L^{1/3}$	$1 < \dfrac{H}{L} < 40$ $1 < Pr < 20$ $10^6 < Ra_L < 10^9$	$\dfrac{T_1 + T_2}{2}$	[46]

Eq.	Correlation	Condition	Property	Ref.
(D.32)	$\mathrm{Nu}_L = 0.42 \mathrm{Ra}_L^{1/4} \mathrm{Pr}^{0.012} \left(\dfrac{H}{L}\right)^{-0.3}$	$10 < \dfrac{H}{L} < 40$ $1 < \mathrm{Pr} < 2 \times 10^4$ $10^4 < \mathrm{Ra}_L < 10^7$	$\dfrac{T_1 + T_2}{2}$	[46]
(D.33)	$\mathrm{Nu}_L = 0.18 \left(\dfrac{\mathrm{Pr}}{0.2 + \mathrm{Pr}} \mathrm{Ra}_L\right)^{0.29}$	$1 < \dfrac{H}{L} < 2$ $10^{-3} < \mathrm{Pr} < 10^5$ $10^3 < \dfrac{\mathrm{Ra}_L \mathrm{Pr}}{0.2 + \mathrm{Pr}}$	$\dfrac{T_1 + T_2}{2}$	[44]
(D.34)	$\mathrm{Nu}_L = 0.22 \left(\dfrac{\mathrm{Pr}}{0.2 + \mathrm{Pr}} \mathrm{Ra}_L\right)^{0.28} \left(\dfrac{H}{L}\right)^{-1/4}$	$2 < \dfrac{H}{L} < 10$ $\mathrm{Pr} < 10^5$ $\mathrm{Ra}_L < 10^{10}$	$\dfrac{T_1 + T_2}{2}$	[44]
(D.35)	$\mathrm{Nu}_L = 1 + 1.44 \left[1 - \dfrac{1708}{\mathrm{Ra}_L \cos\theta}\right]^* \left[1 - \dfrac{1708(\sin 1.8\theta)^{1.6}}{\mathrm{Ra}_L \cos\theta}\right]$ $+ \left[\left(\dfrac{\mathrm{Ra}_L \cos\theta}{5830}\right)^{1/3} - 1\right]^*$ if $[\,]^* < 0$, set $[\,]^* \equiv 0$	$0 < \theta < \theta^*$, $\dfrac{H}{L} \geq 12$	$\dfrac{T_1 + T_2}{2}$	[45]
(D.36)	$\mathrm{Nu}_L = \mathrm{Nu}_L\Big|_{\theta=0^\circ} \left[\dfrac{\mathrm{Nu}_L\big|_{\theta=90^\circ}}{\mathrm{Nu}_L\big|_{\theta=0^\circ}}\right]^{\theta/\theta^*} (\sin\theta^*)^{\theta/4\theta^*}$ H/L: 1 3 6 12 > 12 θ^*: 25° 53° 60° 67° 70°	$\theta^* < \theta < 90^\circ$, $\dfrac{H}{L} \leq 12$	$\dfrac{T_1 + T_2}{2}$	[44]
(D.37)	$\mathrm{Nu}_L = \mathrm{Nu}_L\Big|_{\theta=90^\circ} (\sin\theta)^{1/4}$	$90^\circ < \theta < 180^\circ$	$\dfrac{T_1 + T_2}{2}$	[46]
(D.38)	$\mathrm{Nu}_L = 1 + \left[\mathrm{Nu}_L\Big|_{\theta=90^\circ} - 1\right] \sin\theta$	$90^\circ < \theta < 180^\circ$	$\dfrac{T_1 + T_2}{2}$	[47]

■ REFERENCES

A.1   [1] W. M. Rohsenow and H. Y. Choi. *Heat, Mass and Momentum Transfer*. Prentice Hall, Englewood Cliffs, NJ, 1961.

A.2   [2] V. S. Arpacı and P. S. Larsen. *Convection Heat Transfer*. Prentice Hall, Englewood Cliffs, NJ, 1984.

A.3   [3] W. M. Kays and M. E. Crawford. *Convective Heat and Mass Transfer*. McGraw Hill, New York, 2d ed., 1980.

A.4 [4] H. Hansen. Darstellung des warmeüberganges in rohren durch verallgemeinerte potez beziehungen. *VDIZ.*, 8, 1943.

A.5 [5] E. N. Sieder and C. E. Tate. Heat transfer and pressure drop of liquids in tubes. *Ind. Eng. Chem.*, 28, 1936.

A.6 [6] A. P. Colburn. A method of correlating forced convection heat transfer data and a comparison with fluid friction. *Trans. AIChE*, 29, 1933.

A.7 [7] F. W. Dittus and L. M. K. Boelter. vol. 2. University of California, Berkeley, 1930

A.8 [8] B. S. Petukhov. Heat transfer and friction in turbulent pipe flow with variable physical properties. In *Advances in Heat Transfer*. Academic Press, New York, 1970.

A.9 [9] R. A. Seban and T. T. Shimazaki. Heat transfer to fluid flowing turbulently in a smooth pipe with walls at constant temperature. *Trans. ASME*, 73, 1951.

A.10 [10] B. Lubarsky and S. J. Kaufman. *Review of Experimental Investigations of Liquid-Metal Heat Transfer*, 1956.

A.11 [11] E. S. Skupinski, J. Tortel, and L. Vautrey. Determination des coefficients de convection d'un alliage sodium-potassium dans un tube circulaire. *Int. J. Heat Mass Transfer*, 8, 1965.

A.12 [12] C. A. Sleicher and M. W. Rouse. A convenient correlation for heat transfer to constant and variable property fluids in turbulent pipe flow. *Int. J. Heat Mass Transfer*, 18, 1975.

A.13 [13] R. K. Shah and A. L. London. Thermal boundary conditions and some solutions for laminar duct flow forced convection. *Trans. ASME, J. Heat Transfer*, 96, 1974.

A.14 [14] W. M. Kays and H. C. Perkins. *Handbook of heat transfer*, volume 7. McGraw-Hill, New York, 1972.

A.15 [15] H. Schlichting. *Boundary Layer Theory*. McGraw-Hill, New York, 7th ed., 1979.

A.16 [16] S. Whitaker. Forced convection heat-transfer correlations for flow in pipes, past flat plates, single cylinders, single spheres, and flow in packed beds and tube bundles. *AIChE J.*, 18, 1972.

A.17 [17] R. Hilpert. Warmeabgabe vongeheizen drahten und rohren. *Forsch. Geb. Ingenieurwes*, 4, 1933.

A.18 [18] J. D. Knudsen and D. L. Katz. *Fluid Dynamics and Heat Transfer*. McGraw-Hill, New York, 1958.

A.19 [19] A. Zhukauskas. Heat transfer from tubes in cross flow. In *Advances in Heat Transfer*, vol. 8. Academic Press, New York, 1972.

A.20 [20] S. W. Churchill and M. Berstein. A correlating equation for forced convection from gases and liquids to a circular cylinder in crossflow. *J. Heat Transfer*, 99, 1977.

A.21 [21] S. Nakai and T. Okazaki. Heat transfer from a horizontal circular wire at small Re and Gr numbers—1 pure convection. *Int. J. Heat Mass Transfer*, 18, 1975.

A.22 [22] E. R. G. Eckert and R. M. Drake. *Analysis of Heat and Mass Transfer*. McGraw-Hill, New York, 1972.

A.23 [23] H. Kramers. Heat transfer from spheres to flowing media. *Physica*, 12, 1946.

A.24 [24] W. H. McAdams. *Heat Transmission*. McGraw-Hill, New York, 3rd ed., 1954.

A.25 [25] G. C. Vliet and G. Leppert. Forced convection heat transfer from an isotherrnal sphere to water. *J. Heat Transfer*, C83, 1961.

A.26 [26] E. Ackenbach. Heat transfer from spheres up to Re $= 6 \times 10^6$. In *Proc. Sucth Int. Heat Trans. Conf.*, vol. 5. Hemisphere, Washington, DC, 1978.

A.27 [27] L. C. Witte. An experimental study of forced convection heat transfer from a sphere to liquid sodium. *J. Heat Transfer*, 90, 1968.

A.28 [28] W. Ranz and W. Marshall. Evaporation from drops. *Chem. Eng. Progr.*, 48, 1952.

A.29 [29] S.-C. Yao and V. E. Shrock. *ASME Publication*, 1975.

A.30 [30] E. D. Grimson. Correlation and utilization of new data on flow resistance and heat transfer for cross flow of gases over tube banks. *Trans. ASME*, 59, 1937.

A.31 [31] S. Kalish and O. E. Dwyer. Heat transfer to NaK flowing through unbaffled rod bundles. *Int. J. Heat Mass Transfer*, 10, 1967.

A.32 [32] W. M. Kays and R. K. Lo. Basic heat transfer and friction data for gas flow normal to banks of staggered tubes: Use of a transient technique. *Stanford University, TR–15, Navy Contract N6-ONR251*, T.O. 6, 1952.

A.33 [33] C. Y. Warner and V. S. Arpacı. An experimental investigation of turbulent natural convection in air at low pressure for a vertical heated flat plate. *Int. J. Heat Mass Transfer*, 11, 1968.

A.34 [34] F. J. Bayley. An analysis of turbulent free convection heat transfer. *Proc. Inst. Mech. Eng.*, 169(20), 1955.

A.35 [35] G. C. Vliet. Natural convection local heat transfer on constant heat-flux inclined surfaces. *J. Heat Transfer*, 19, 1969. (Also see 517–531).

A.36 [36] S. W. Churchill and H. H. S. Chu. Correlating equations for laminar and turbulent free convection from a horizontal cylinder. *Int. J. Heat Mass Transfer*, 18, 1975.

A.37 [37] J . R. Lloyd and W. R. Moran. Natural convection adjacent to horizontal surface of various planforms. *ASME*, Paper 74-WA/HT 66, 1974.

A.38 [38] G. C. Vliet and D. C. Ross. Turbulent natural convection on upward and downward facing inclined constant heat flux surfaces. *ASME*, Paper 74-WA/HT–32, 1974.

A.39 [39] T. Fujii and H. Imura. Natural convection heat transfer from a plate with arbitrary inclination. *Int. J. Heat Mass Transfer*, 15, 1972.

A.40 [40] V. T. Morgan. The overall convective heat transfer from smooth circular cylinders. In *Advances in Heat Transfer*, vol. 11. Academic Press, New York, 1975.

A.41 [41] M. Al-Arabi and Y. K. Salman. Laminar natural convection from an inclined cylinder. *Int. J. Heat Mass Transfer*, 23, 1980.

A.42 [42] G. D. Raithby and K. G. T. Hollands. General method of obtaining approximate solutions to laminar and turbulent free convection problems. In *Advances in Heat Transfer*, vol. 11. Academic Press, New York, 1975.

A.43 [43] R. K. McGregor and A. P. Emery. Free convection through vertical plane layers: Moderate and high Pr number fluids. *J. Heat Transfer*, 91, 1969.

A.44 [44] I. Catton. Natural convection in enclosures. In *Proc. 6th Int. Heat Transfer Conf.*, volume 6, 13. Toronto, Canada, 1978.

A.45 [45] K. G. T. Hollands, S. E. Unny, G. D. Raithby, and L. Konicek. Free convective heat transfer across inclined air layers. *J. Heat Transfer*, 98, 1976.

A.46 [46] P. S. Ayyaswamy and I. Catton. The boundary layer regime for natural convection in a differentially heated, tilted rectangular cavity. *J. Heat Transfer*, 95, 1973.

A.47 [47] J. N. Arnold, I. Catton, and D. K. Edwards. Experimental investigation of natural convection in inclined rectangular regions of differing aspect ratios. *ASME*, Paper 75-HT–62, 1975.

Appendix B

THERMOPHYSICAL PROPERTIES

Table B.1 Solids

Metallic Solids

Composition	Melting point K	Properties at 300 K				Properties at various temperatures (K) k W/m·K / c_p J/kg·K					
		ρ kg/m^3	c_p J/kg·K	k W/m·K	$\alpha \cdot 10^6$ m^2/s	100	200	400	600	800	1200
Aluminum											
Pure	933	2702	903	237	97.1	302	237	240	231	218	
						482	798	949	1033	1146	
Alloy 2024-T6 (4.5% Cu, 1.5% Mg, 0.6% Mn)	775	2770	875	177	73.0	65	163	166	186		
						473	787	925	1042		
Alloy 195 Cast (4.5% Cu)		2790	883	168	68.2			174	135		
								—	—		
Beryllium	1550	1850	1825	200	59.2	990	301	161	125	106	78.7
						703	1114	2191	2504	2523	3227
Bismuth	545	9780	122	7.86	6.59	16.5	9.89	7.04			
						112	120	127			
Boron	2573	2500	1107	27.0	9.76	190	55.5	16.8	10.6	9.60	
						128	600	1463	1892	2160	
Cadmium	594	8650	231	96.8	48.4	203	99.3	94.7			
						196	222	242			
Chromium	2118	7160	449	93.7	29.1	150	111	90.9	80.7	71.3	61.5
						192	384	464	542	381	682
Cobalt	1760	6662	421	99.2	26.6	167	122	85.4	67.4	58.2	49.3
						236	379	450	503	550	733
Copper											
Pure	1358	8933	385	401	117	482	413	393	379	366	339
						252	356	397	417	423	480
Commercial bronze (90% Cu, 10% Al)	1293	8800	420	52	14		42	52	59		
						785	460	545			
Phosphor gear bronze (89% Cu, 11% Al)	1104	8780	355	54	17		41	65	74		
							—	—	—		
Cartridge brass (70% Cu, 30% Zn)	1188	8530	380	110	33.9	75	95	137	149		
						—	360	395	425		
Constantan (55% Cu, 45% Ni)	1493	8920	384	23	6.71	17	19				
						237	362				
Germanium	1211	5360	322	59.9	34.7	232	96.8	43.2	27.3	19.8	17.4
						190	290	337	348	357	395
Gold	1336	19300	129	317	127	327	323	311	298	284	255
						109	124	131	135	140	155
Iridium	2720	22500	130	147	50.3	172	153	144	138	132	120
						90	122	133	138	144	161
Iron											
Pure	1810	7870	447	80.2	23.1	134	94.0	69.5	54.7	43.3	28.2
						216	384	490	574	680	609
Armco (99.75% pure)		7870	447	72.7	20.7	95.6	80.6	65.7	53.1	42.2	28.7
						215	384	490	574	680	809

580 App. B Thermophysical Properties

Table B.1 Solids

Metallic Solids *(cont.)*

Composition	Melting point K	Properties at 300 K				Properties at various temperatures (K) k W/m·K / c_p J/kg·K					
		ρ kg/m³	c_p J/kg·K	k W/m·K	$\alpha \cdot 10^6$ m²/s	100	200	400	600	800	1200
Carbon steels											
Plain carbon (Min ≤ 0.1%, Si ≤ 0.1%)		7854	434	60.5	17.7			56.7 / 487	48.0 / 559	39.2 / 685	
AISI 1010		7832	434	63.9	18.8			58.7 / 487	48.7 / 559	39.2 / 685	
Carbon-silicon (Mn ≤ 1%, 0.1% < Si ≤ 0.6%		7817	446	51.9	14.9			49.8 / 50.1	44.0 / 582	37.4 / 699	
Carbon-manganese-silicon (1% < Mn ≤ 1.55%, 0.1% < Si ≤ 0.5%)		8131	434	41.0	11.6			42.2 / 487	39.7 / 559	35.0 / 685	
Chromium (low) steels											
Cr-Mo-Si (0.18% C, 0.65% Cr, 0.23% Mo, 0.5% Si)		7822	444	37.7	10.9			38.2 / 492	36.7 / 575	33.3 / 688	
Cr-Mo (0.15%C, 1% Cr, 0.54% Mo, 0.39% Si)		7858	442	42.3	12.2			42.0 / 492	39.1 / 575	34.5 / 688	
Cr-V (0.2% C, 1.02% Cr, 0.15% V)		7858	443	48.9	14.1			46.8 / 492	42.1 / 575	36.3 / 688	
Stainless steels											
AISI 302		8055	480	15.1	3.91		512	17.3 / 559	20.0 / 585	22.8	
AISI 304	1570	7900	477	14.9	3.95	9.2 / 272	12.6 / 402	16.6 / 515	19.8 / 557	22.5 / 582	29.0 / 640
AISI 316		8238	468	13.4	3.48			15.2 / 504	18.3 / 550	21.3 / 576	
AISI 347		7978	480	14.2	3.71			15.8 / 513	18.9 / 559	31.9 / 576	
Lead	631	11340	129	35.3	24.1	39.7 / 118	36.7 / 125	34.0 / 132	31.4 / 142		
Magnesium	923	1740	1024	156	87.6	169 / 649	159 / 934	153 / 1074	149 / 1170	146 / 1267	
Molybdenum	2594	10240	251	138	53.7	179 / 141	143 / 224	134 / 261	125 / 275	118 / 285	105 / 308
Nickel											
Pure	1728	8900	444	90.7	23.0	164 / 232	107 / 385	80.2 / 485	65.5 / 592	67.5 / 530	75.2 / 594
Nichrome (80% Ni, 20% Cr)	1572	8400	420	12	3.4			14 / 480	15 / 525	21 / 545	
Inconel X–750 (73% Ni, 15% Cr, 5.7% Fe 6.7% Fe)	1655	8510	439	11.7	3.1	8.7 / —	10.3 / 372	13.5 / 473	17.0 / 510	20.5 / 545	27.5 / —

Table B.1 Solids

Metallic Solids *(cont.)*

Composition	Melting point K	Properties at 300 K				Properties at various temperatures (K) k W/m·K / c_p J/kg·K					
		ρ kg/m³	c_p J/kg·K	k W/m·K	$\alpha \cdot 10^6$ m²/s	100	200	400	600	800	1200
Niobium	2741	8570	265	53.7	23.6	55.2	52.6	55.2	58.2	61.3	67.5
						188	249	274	283	292	310
Palladium	1827	12020	244	71.8	24.5	76.5	71.6	73.6	79.7	86.9	102
						168	227	251	261	271	291
Platinum											
Pure	2045	21450	133	71.6	25.1	77.5	72.6	71.8	73.2	75.6	82.6
						100	125	136	141	146	157
Alloy 60Pl–40Rh (60% Pl, 40% Rh)	1800	16830	162	47	17.4			52	59	65	73
								—	—	—	—
Rhenium	3453	21100	136	47.9	16.7	58.9	51.0	46.1	44.2	44.1	45.7
						97	127	139	145	151	162
Rhodium	2286	12450	243	150	49.6	186	154	146	136	127	116
						147	220	253	274	283	327
Silicon	1685	2330	712	148	89.2	864	264	96.9	61.9	42.2	25.7
						259	556	790	867	913	967
Tantalum	3269	16600	140	57.5	24.7	59.2	57.5	57.9	58.6	59.4	61.0
						110	133	144	145	149	155
Thorium	2923	11700	118	54.0	39.1	59.8	54.6	54.5	55.8	56.9	58.7
						99	112	124	134	145	167
Tin	505	7310	227	66.6	40.1	85.2	73.3	62.2			
						186	215	243			
Titanium	1953	4500	522	21.9	9.32	30.5	24.5	20.4	19.4	19.7	22.0
						300	465	551	591	633	620
Tungsten	3660	19300	132	174	68.3	206	166	159	137	125	113
						87	122	137	142	145	152
Uranium	1406	19070	116	27.6	12.5	21.7	25.1	29.6	34.9	38.6	49.0
						94	196	125	146	176	161
Vanadium	2192	6100	489	30.7	10.3	35.8	313	33.3	35.7	40.8	
						258	430	515	540	563	645
Zinc	693	7140	389	116	41.6	117	118	111	103		
						297	367	402	436		
Zirconium	2125	6570	276	22.7	12.5	33.2	25.2	21.6	20.7	21.6	28.0
						205	254	300	322	342	344

Table B.1 Solids

Nonmetallic Solids (cont.)

Composition	Melting point K	Properties at 300 K					Properties at various temperatures (K) k W/m·K / c_p J/kg·K					
		ρ kg/m^3	c_p J/kg·K	k W/m·K	$\alpha \cdot 10^6$ m^2/s	100	200	400	600	800	1200
Aluminum oxide, sapphire	2323	2970	765	46	15.1	450	82	32.4 / 940	18.9 / 1110	13.0 / 1180	
Aluminum oxide, polycrystalline	2323	2970	765	36.0	11.9	133	55	26.5 / 940	15.8 / 1110	10.4 / 1180	5.55 / —
Beryllium oxide	2725	3000	1030	272	88.0			196 / 1350	111 / 1690	70 / 1865	33 / 2055
Boron	2573	2500	1105	27.6	9.99	190 / —	52.5 / —	18.7 / 1490	11.3 / 1880	8. / 2135	5.2 / 2555
Boron fiber epoxy (30% vol) composite	590	2080									
k, ∥ to fibers				2.29		2.10	2.23	2.28			
k, ⊥ to fibers				0.59		0.37	0.49	0.60			
c_p			1122			364	757	1431			
Carbon											
Amorphous	1500	1950	—	1.60	—	0.67	1.18	1.89	2.19	2.37	2.94
Diamond, type IIe insulator	—	3500	509	2300		10000 / 21	4000 / 194	1540 / 853			
Graphite, pyrolytic	2273	2210									
k, ∥ to layers				1950		4970	3230	1390	892	667	448
k, ⊥ to layers				5.70		16.8	9.23	4.09	2.68	2.01	1.34
c_p			709			136	411	902	1406	1650	1890
Graphite fiber epoxy (25% vol) composite	450	1400									
k, heat flow ∥ to fibers				11.1		5.7	8.7	13.0			
k, heat flow ⊥ to fibers				0.87		0.46	0.68	1.1			
c_p			935			337	642	1216			
Pyroceram Corning 9606	1523	2500	808	2.98	1.89	5.25	4.78	3.64 / 906	3.28 / 1038	3.08 / 1122	2.87 / 1254
Silicon carbide	3100	3160	675	490	230			— / 680	— / 1050	— / 1135	58 / 1243
Silicon dioxide, crystalline (quartz)	1883	2550									
k, ∥ to c axis				10.4		39	16.4	7.6	5.0	4.2	
k, ⊥ to c axis				6.21		20.8	9.5	4.70	3.4	3.1	
k,			745			—	—	885	1075	1250	
silicon dioxide, polycrystalline (fused silica)	1883	2220	745	1.36	0.834	0.69 / —	1.14 / —	1.51 / 905	1.75 / 1040	2.17 / 1105	4.00 / 1195
Silicon nitride	2173	240	691	16.0	9.65	—	— / 578	13.9 / 778	11.3 / 937	9.88 / 1063	8.0 / 1226
Sulphur	392	2070	708	0.206	0.141	0.165 / 403	0.185 / 606				
Thorium dioxide	3573	9110	235	13	6.1			10.2 / 255	6.6 / 274	4.7 / 285	3.12 / 303
Titanium dioxide, polycrystalline	2133	4157	710	8.4	2.8			7.01 / 805	5.02 / 880	3.94 / 910	3.28 / 945

Table B.1 Solids

Insulation Materials

Description/composition	Max service temp K	Typical density kg/m³	\multicolumn{9}{c}{Typical thermal conductivity, k(W/m·K) at various temperatures (K)}
			200	240	270	300	310	365	420	530	645	750
Blankets												
Blanket, mineral fiber,	920	96 192					0.038	0.046	0.056	0.078		
reinforced metal	815	40 96					0.035	0.045	0.58	0.68		
Blanket, mineral fiber, glass	450	10		0.036	0.040	0.048	0.052	0.076				
fine fiber, organic bonded		12		0.035	0.039	0.046	0.049	0.069				
		16		0.033	0.936	0.042	0.046	0.062				
		24		0.030	0.033	0.039	0.040	0.053				
		32		0.029	0.032	0.036	0.038	0.048				
		48		0.027	0.030	0.033	0.035	0.045				
Blanket, alumina-silica fiber	1530	40								0.071	0.105	0.150
		64								0.059	0.067	0.125
		96								0.052	0.076	0.091
		128								0.049	0.068	0.091
Felt, semirigid, organic bonded	480	50 125				0.035	0.038	0.039	0.051	0.063		
	730	50	0.023	0.027	0.030	0.033	0.035	0.051	0.079			
Felt, laminated: no binder	920	120								0.051	0.065	0.087
Blocks, boards, and pipe insulations												
Asbestos paper, laminated and corrugated												
4-ply	420	190					0.078	0.082	0.096			
6-ply	420	255					0.071	0.074	0.085			
8-ply	420	300					0.068	0.071	0.082			
Magnesia, 85%	590	185					0.051	0.055	0.061			
Calcium silicate	920	190					0.055	0.059	0.063	0.075	0.049	0.104
Cellular glass	700	145		0.048	0.052	0.058	0.062	0.069	0.079			
Diatomaceous silica	1145	345								0.092	0.098	0.104
	1310	385								0.101	0.100	0.115
Polystyrene, rigid												
Extruded (R–12)	350	56	0.023	0.023	0.025	0.027	0.029					
Extruded (R–12)	350	35	0.023	0.023	0.036	0.029						
Molded beads	350	16	0.026	0.033	0.036	0.040						
Rubber, rigid foamed	340	70				0.029	0.032	0.033				
Insulating cement												
Mineral fiber (rock, slag or glass)												
with clay binder	1255	430					0.071	0.079	0.068	0.105	0.123	
with hydraulic setting binder	922	560					0.106	0.115	0.123	0.137		
Loose fill												
Cellulose, wood or paper pulp	—	45				0.039	0.042					
Perlite, expanded	—	105	0.038	0.043	0.049	0.053	0.056					
vermiculite, expanded	—	122		0.058	0.063	0.068	0.071					
				0.051	0.058	0.063	0.064					

Table B.1 Solids

Building Materials

Description/Composition	Typical Properties at 300 K		
	Density, ρ kg/m³	Thermal Conductivity, k W/m·K	Specific Heat, c_p J/kg·K
Building Boards			
Asbestos-cement board	1,920	0.58	—
Gypsum or plaster board	800	0.17	—
Plywood	545	0.12	1,215
Sheathing, regular density	290	0.055	1,300
Acoustic tile	290	0.058	1,340
Hardboard, siding	640	0.094	1,170
Hardboard, high density	1,010	0.15	1,380
Particle board, low density	590	0.078	1,300
Particle board, high density	1,000	0.170	1,300
Woods			
Hardwoods (oak, maple)	720	0.16	1,255
Softwoods (fir, pine)	510	0.12	1,380
Masonry Materials			
Cement mortar	1,860	0.72	780
Brick, common	1,920	0.72	835
Brick, face	2,083	1.3	—
Clay tile, hollow			
1 cell deep, 10 cm thick	—	0.52	—
3 cells deep, 30 cm thick	—	0.69	—
Concrete block, 3 oval cores			
sand/gravel, 20 cm thick	—	1.0	—
cinder aggregate, 20 cm thick	—	0.67	—
Concrete block, rectangular core			
2 core, 20 cm thick, 16 kg	—	1.1	—
same with filled cores	—	0.60	—
Plastering Materials			
Cement plaster, sand aggregate	1,860	0.72	—
Gypsum plaster, sand aggregate	1,680	0.22	1,085
Gypsum plaster, vermiculite aggregate	718420	0.25	—

Table B.1 Solids

	Insulation Materials		
	Typical Properties at 300 K		
Description/Composition	Density, ρ kg·m^{-3}	Thermal Conductivity, k W/m·K	Specific Heat, c_p J/kg·K
Blanket and Batt			
Glass fiber, paper faced	16	0.046	—
	28	0.038	—
	40	0.035	—
Glass fiber, coated: duct liner	32	0.038	835
Board and Slab			
Cellular glass	145	0.058	1,000
Glass fiber, organic bonded	105	0.036	795
Polystyrene, expanded			
extruded (R–12)	55	0.027	1,210
molded beads	16	0.040	1,210
Mineral fiberboard: roofing material	265	0.049	—
Wood, shredded/cemented	350	0.087	1,590
Cork, granulated	120	0.039	1,800
Loose Fill			
Cork, granulated	160	0.045	—
Diatomaceous silica, coarse powder	250	0.069	—
	400	0.091	—
Diatomaceous silica, fine powder	200	0.052	—
	275	0.061	—
Glass fiber, poured or blown	16	0.043	825
Vermiculite, flakes	80	0.068	835
	160	0.063	1,000
Formed/Foamed-in-Place			
Mineral wool granules with asbestos/inorganic binders, sprayed	190	0.046	—
Polyvinyl acetate cork mastic: sprayed or troweled	—	0.100	—
Urethane, two-part mixture: rigid foam	70	0.026	1,045
Reflective			
Aluminum foil separating fluffy glass mats: 10–12 layers: evacuated: for cryogenic application (150 K)	40	0.00016	—
Aluminum foil and glass paper laminate: 75–150 layers: evacuated: for cryogenic application (150 K)	120	0.000017	—
Typical silica powder/evacuated	160	0.0017	—

Table B.1 Solids

Miscellaneous Materials

Description/Composition	Temperature K	Density ρ kg/m^3	Thermal Conductivity, k W/m·K	Specific Heat, c_p J/kg·K
Asphalt	300	2,115	0.062	920
Bakelite	300	1,300	1.4	1,465
Brick, refractory				
Carborundum	872	—	18.5	—
	1,672	—	11.0	—
Chrome brick	473	3,010	2.3	835
	823		2.5	
	1,173		2.0	
Diatomaceous silica, fired	478	—	0.25	—
	1,145	—	0.30	
Fire clay, burnt 1600 K	773	2,050	1.0	960
	1,073	—	1.1	
	1,373	—	1.1	
Fire clay, burnt 1725 K	773	2,325	1.3	960
	1,073		1.4	
	1,373		1.4	
Fire clay brick	478	2,645	1.0	960
	922		1.5	
	1,478		1.8	
Magnesite	478	—	3.8	1,130
	922	—	2.8	
	1,478		1.9	
Clay	300	1,460	1.3	880
Coal, anthracite	300	1,350	0.26	1,260
Concrete (stone mix)	300	2,300	1.4	880
Cotton	300	80	0.06	1,300
Foodstuffs				
Banana (75.7% water content)	300	980	0.481	3,350
Apple, red (75% water content)	300	840	0.513	3,600
Cake, batter	300	720	0.223	—
Cake, fully done	300	280	0.121	—
Chicken meat, white	198	—	1.60	—
(74.4% water content)	233	—	1.49	
	253		1.35	
	263		1.20	
	273		0.476	
	283		0.480	
	293		0.489	

Table B.1 Solids

	Miscellaneous Materials *(cont.)*			
Description/Composition	Temperature K	Density ρ kg/m^3	Thermal Conductivity, k W/m·K	Specific Heat, c_p J/kg·K
Glass				
Plate (soda lime)	300	2,500	1.4	750
Pyrex	300	2,225	1.4	835
Ice	273	920	0.186	2,040
	253	—	0.203	1,945
Leather (sole)	300	998	0.013	—
Paper	300	930	0.011	1,340
Paraffin	300	900	0.020	2,890
Rock				
Granite, Barre	300	2,630	2.79	775
Limestone, Salem	300	2,320	2.15	810
Marble, Halston	300	2,680	2.80	830
Quartzite, Sioux	300	2,640	5.38	1,105
Sandstone, Berea	300	2,150	2.90	745
Rubber, vulcanized				
Soft	300	1,100	0.012	2,010
Hard	300	1,190	0.013	—
Sand	300	1,515	0.027	800
Soil	300	2,050	0.52	1,840
Snow	273	110	0.049	—
		500	0.190	—
Teflon	300	2,200	0.35	—
	400		0.45	—
Tissue, human				
Skin	300	—	0.37	—
Fat layer (adipose)	300	—	0.2	—
Muscle	300	—	0.41	—
Wood, cross grain				
Balsa	300	140	0.055	—
Cypress	300	465	0.097	—
Fir	300	415	0.11	2,720
Oak	300	545	0.17	2,385
Yellow pine	300	640	0.15	2,805
White pine	300	435	0.11	—
Wood, radial				
Oak	300	545	0.19	2,385
Fir	300	420	0.14	2,720

Table B.2 Gases

Air

T [K]	ρ [kg/m^3]	c_p [kJ/kg·K]	$\mu \cdot 10^7$ [N·s/m^2]	$k \cdot 10^3$ [W/m·K]	$\nu \cdot 10^6$ [m^2/s]	$\alpha \cdot 10^6$ [m^2/s]	Pr	$(g\beta/\nu\alpha) \cdot 10^{-6}$ [1/K·m^3]
100	3.5562	1.032	71.1	9.34	2.0	2.54	0.786	19,305
150	2.3364	1.012	103.4	13.8	4.426	5.84	0.758	2529
200	1.7458	1.007	132.5	18.1	7.590	10.3	0.737	627.2
250	1.3947	1.006	159.6	22.3	11.44	15.9	0.720	215.7
300	1.1614	1.007	184.6	26.3	15.89	22.5	0.707	91.43
350	0.9950	1.009	208.2	30.0	20.92	29.9	0.700	44.80
400	0.8711	1.014	230.1	33.8	26.41	38.3	0.690	24.24
450	0.7740	1.021	250.7	37.3	32.39	47.2	0.686	14.26
500	0.6964	1.030	270.1	40.7	38.79	56.7	0.684	8.918
550	0.6329	1.040	288.4	43.9	45.57	66.7	0.683	5.866
600	0.5804	1.051	305.8	46.9	52.69	76.9	0.685	4.034
650	0.5356	1.063	322.5	49.7	60.21	87.3	0.690	2.870
700	0.4975	1.075	338.8	52.4	68.10	98	0.695	2.099
750	0.4643	1.087	354.6	54.9	76.37	109	0.702	1.571
800	0.4354	1.099	369.8	57.3	84.93	120	0.709	1.203
850	0.4097	1.110	384.3	59.6	93.80	131	0.716	0.9390
900	0.3868	1.121	398.1	62.0	102.9	143	0.720	0.7405
950	0.3666	1.131	411.3	64.3	112.2	155	0.723	0.5936
1000	0.3482	1.141	424.4	66.7	121.9	168	0.726	0.4789
1100	0.3166	1.159	449	71.5	141.8	195	0.728	0.3224
1200	0.2902	1.175	473	76.3	162.9	224	0.728	0.2240
1300	0.2679	1.189	496	82	185.1	238	0.719	0.1712
1400	0.2488	1.207	530	91	213	303	0.703	0.1085
1500	0.2322	1.230	557	100	240	350	0.685	0.07783
1600	0.2177	1.248	584	106	268	390	0.688	0.05864
1700	0.2049	1.267	611	113	298	435	0.685	0.04450
1800	0.1935	1.286	637	120	329	482	0.683	0.03436
1900	0.1833	1.307	663	128	362	534	0.677	0.02670
2000	0.1741	1.337	689	137	396	589	0.672	0.02102
2100	0.1658	1.372	715	147	431	646	0.667	0.01677
2200	0.1582	1.417	740	160	468	714	0.655	0.01334
2300	0.1513	1.478	766	175	506	783	0.647	0.01076
2400	0.1448	1.538	792	196	547	869	0.630	0.008596
2500	0.1389	1.665	818	222	589	960	0.613	0.006938
3000	0.1135	2.726	955	486	841	1570	0.536	0.002476

Table B.2 Gases

Ammonia (NH_3)

T [K]	ρ [kg/m³]	c_p [kJ/kg·K]	$\mu \cdot 10^7$ [N·s/m²]	$k \cdot 10^3$ [W/m·K]	$\nu \cdot 10^6$ [m²/s]	$\alpha \cdot 10^6$ [m²/s]	Pr	$(g\beta/\nu\alpha) \cdot 10^{-6}$ [1/K·m³]
300	0.6894	2.158	101.5	24.7	14.7	16.6	0.887	134
320	0.6448	2.170	109	27.2	16.9	19.4	0.870	93.48
340	0.6059	2.192	116.5	29.3	19.2	22.1	0.872	67.98
360	0.5716	2.221	124	31.6	21.7	24.9	0.872	50.42
380	0.5410	2.254	131	34.0	24.2	27.9	0.869	38.22
400	0.5136	2.287	138	37.0	26.9	31.5	0.853	28.93
420	0.4888	2.322	145	40.4	29.7	35.6	0.833	22.08
440	0.4664	2.357	152.5	43.5	32.7	39.6	0.826	17.21
460	0.4460	2.393	159	46.3	35.7	43.4	0.822	13.76
480	0.4273	2.430	166.5	49.2	39.0	47.4	0.822	11.05
500	0.4101	2.467	173	52.5	42.2	51.9	0.813	8.955
520	0.3942	2.504	180	54.5	45.7	55.2	0.827	7.476
540	0.3795	2.540	186.5	57.5	49.1	59.7	0.824	6.196
560	0.3708	2.577	193	60.6	52.0	63.4	0.827	5.312
580	0.3533	2.613	199.5	63.8	56.5	69.1	0.817	4.331

Table B.2 Gases

Carbon Dioxide (CO_2)

T [K]	ρ [kg/m³]	c_p [kJ/kg·K]	$\mu \cdot 10^7$ [N·s/m²]	$k \cdot 10^3$ [W/m·K]	$\nu \cdot 10^6$ [m²/s]	$\alpha \cdot 10^6$ [m²/s]	Pr	$(g\beta/\nu\alpha) \cdot 10^{-6}$ [1/K·m³]
280	1.9022	0.830	140	15.20	7.36	9.63	0.765	494.2
300	1.7730	0.851	149	16.55	8.40	11.9	0.766	353.8
320	1.6609	0.872	156	18.05	9.39	12.5	0.754	261.1
340	1.5613	0.891	165	19.70	10.6	14.2	0.746	191.6
360	1.4743	0.908	173	21.2	11.7	15.8	0.741	147.4
380	1.3961	0.926	181	22.75	13.0	17.6	0.737	112.8
400	1.3257	0.942	190	24.3	14.3	19.5	0.737	87.92
450	1.1782	0.981	210	28.3	17.8	24.5	0.728	49.97
500	1.0594	1.02	231	32.5	21.8	30.1	0.725	29.89
550	0.9625	1.05	251	36.6	26.1	36.2	0.721	18.87
600	0.8826	1.08	270	40.7	30.6	42.7	0.717	12.51
650	0.8143	1.10	288	44.5	35.4	49.7	0.712	8.576
700	0.7564	1.13	305	48.1	40.3	56.3	0.717	6.175
750	0.7057	1.15	321	51.7	45.5	63.7	0.714	4.512
800	0.6614	1.17	337	55.1	51.0	71.2	0.716	3.376

Table B.2 Gases

Carbon Monoxide (CO)

T [K]	ρ [kg/m^3]	c_p [kJ/kg·K]	$\mu \cdot 10^7$ [N·s/m^2]	$k \cdot 10^3$ [W/m·K]	$\nu \cdot 10^6$ [m^2/s]	$\alpha \cdot 10^6$ [m^2/s]	Pr	$(g\beta/\nu\alpha) \cdot 10^{-6}$ [1/K·m^3]
200	1.6888	1.045	127	17.0	7.52	9.63	0.781	677.1
220	1.5341	1.044	137	19.0	8.93	11.9	0.753	419.5
240	1.4055	1.043	147	20.6	10.5	14.1	0.744	276.0
260	1.2967	1.043	157	22.1	12.1	16.3	0.741	191.2
280	1.2038	1.042	166	23.6	13.8	18.8	0.733	135.0
300	1.1233	1.043	175	25.0	15.6	21.3	0.730	98.38
320	1.0529	1.043	184	26.3	17.5	23.9	0.730	73.27
340	0.9909	1.044	193	27.8	19.5	26.9	0.725	54.99
360	0.9357	1.045	202	29.1	21.6	29.8	0.725	42.32
380	0.8864	1.047	210	30.5	23.7	32.9	0.729	33.10
400	0.8421	1.049	218	31.8	25.9	36.0	0.719	26.30
450	0.7483	1.055	237	35.0	31.7	44.3	0.714	15.52
500	0.67352	1.065	254	38.1	37.7	53.1	0.710	9.798
550	0.61226	1.076	271	41.1	44.3	62.4	0.710	6.450
600	0.56126	1.088	286	44.0	51.0	72.1	0.707	4.445
650	0.51806	1.101	301	47.0	58.1	82.4	0.705	3.152
700	0.48102	1.114	315	50.0	65.5	93.3	0.702	2.293
750	0.44899	1.127	329	52.8	73.3	104	0.702	1.715
800	0.42095	1.140	343	55.5	81.5	116	0.705	1.297

Table B.2 Gases

Helium (He)

T [K]	ρ [kg/m^3]	c_p [kJ/kg·K]	$\mu \cdot 10^7$ [N·s/m^2]	$k \cdot 10^3$ [W/m·K]	$\nu \cdot 10^6$ [m^2/s]	$\alpha \cdot 10^6$ [m^2/s]	Pr	$(g\beta/\nu\alpha) \cdot 10^{-6}$ [1/K·m^3]
100	0.4871	5.193	96.3	73.0	19.8	28.9	0.686	171.4
140	0.3481	5.193	118	90.7	33.9	50.2	0.676	41.6
180	0.2708	5.193	139	107.2	51.3	76.2	0.673	13.94
220	0.2216	5.193	160	123.1	72.2	107	0.675	5.770
260	0.1875	5.193	180	137	96.0	141	0.682	2.787
300	0.1625	5.193	199	152	122	180	0.680	1.489
400	0.1219	5.193	243	187	199	295	0.675	0.4176
500	0.09754	5.193	283	220	290	434	0.668	0.1558
600	—	5.193	320	252	—	—	—	—
700	0.06969	5.193	350	278	502	768	0.654	0.03634
800	—	5.193	382	304	—	—	—	—
1000	0.04879	5.193	446	354	914	1400	0.654	0.007664

Table B.2 Gases

Hydrogen (H$_2$)

T [K]	ρ [kg/m^3]	c_p [kJ/kg·K]	$\mu \cdot 10^7$ [N·s/m^2]	$k \cdot 10^3$ [W/m·K]	$\nu \cdot 10^6$ [m^2/s]	$\alpha \cdot 10^6$ [m^2/s]	Pr	$(g\beta/\nu\alpha) \cdot 10^{-6}$ [1/K·m^3]
100	0.24255	11.23	42.1	67.0	17.4	24.6	0.707	229.1
150	0.16156	12.60	56.0	101	34.7	49.6	0.699	37.99
200	0.12115	13.54	68.1	131	56.2	79.9	0.704	10.92
250	0.09693	14.06	78.9	157	81.4	115	0.707	4.191
300	0.08078	14.31	89.6	183	111	158	0.701	1.864
350	0.06924	14.43	98.8	204	143	204	0.700	0.9605
400	0.06059	14.48	108.2	226	179	258	0.695	0.5309
450	0.05386	14.50	117.2	247	218	316	0.689	0.3164
500	0.04848	14.52	126.4	266	261	378	0.691	0.1988
550	0.04407	14.53	134.3	285	305	445	0.685	0.1314
600	0.04040	14.55	142.4	305	352	519	0.678	0.08947
700	0.03463	14.61	157.8	342	456	676	0.675	0.04545
800	0.03030	14.70	172.4	378	569	849	0.670	0.02538
900	0.02694	14.83	186.5	412	692	1030	0.671	0.01529
1000	0.02424	14.99	201.3	448	830	1230	0.673	0.009606
1100	0.02204	15.17	213.0	488	966	1460	0.662	0.006321
1200	0.02020	15.37	226.2	528	1120	1700	0.659	0.004292
1300	0.01865	15.59	238.5	568	1279	1955	0.655	0.003017
1400	0.01732	15.81	250.7	610	1447	2230	0.650	0.002171
1500	0.01616	16.02	262.7	655	1626	2530	0.643	0.001589
1600	0.0152	16.28	273.7	697	1801	2815	0.639	0.001209
1700	0.0143	16.58	284.9	742	1992	3130	0.637	0.0009252
1800	0.0135	16.96	296.1	786	2193	3435	0.639	0.0007233
1900	0.0128	17.49	307.2	835	2400	3730	0.643	0.0005766
2000	0.0121	18.25	318.2	878	2630	3975	0.661	0.0004690

Table B.2 Gases

Nitrogen (N$_2$)

T [K]	ρ [kg/m^3]	c_p [kJ/kg·K]	$\mu \cdot 10^7$ [N·s/m^2]	$k \cdot 10^3$ [W/m·K]	$\nu \cdot 10^6$ [m^2/s]	$\alpha \cdot 10^6$ [m^2/s]	Pr	$(g\beta/\nu\alpha) \cdot 10^{-6}$ [1/K·m^3]
100	3.4388	1.070	68.8	9.58	2.00	2.60	0.768	18,860
150	2.2594	1.050	100.6	13.9	4.45	5.86	0.759	2507
200	1.6883	1.043	129.2	18.3	7.65	10.4	0.736	616.3
250	1.3488	1.042	154.9	22.2	11.48	15.8	0.727	216.2
300	1.1233	1.041	178.2	25.9	15.86	22.1	0.716	93.27
350	0.9625	1.042	200.0	29.3	20.78	29.2	0.711	46.18
400	0.8425	1.045	220.4	32.7	26.16	37.1	0.704	25.26
450	0.7485	1.050	239.6	35.8	32.01	45.6	0.703	14.93
500	0.6739	1.056	257.7	38.9	38.24	54.7	0.700	9.377
550	0.6124	1.065	274.7	41.7	44.86	63.9	0.702	6.220
600	0.5615	1.075	290.8	44.6	51.79	73.9	0.701	4.271
700	0.4812	1.098	321.0	49.9	66.71	94.4	0.706	2.225
800	0.4211	1.122	349.1	54.8	82.90	116	0.715	1.275
900	0.3743	1.146	375.3	59.7	100.3	139	0.721	0.7816
1000	0.3368	1.167	399.9	64.7	118.7	165	0.721	0.5007
1100	0.3062	1.187	423.2	70.0	138.2	193	0.718	0.3343
1200	0.2807	1.204	445.3	75.8	158.6	224	0.707	0.2300
1300	0.2591	1.219	466.2	81.0	179.9	256	0.701	0.1638

Table B.2 Gases

Oxygen (O$_2$)

T [K]	ρ [kg/m^3]	c_p [kJ/kg·K]	$\mu \cdot 10^7$ [N·s/m^2]	$k \cdot 10^3$ [W/m·K]	$\nu \cdot 10^6$ [m^2/s]	$\alpha \cdot 10^6$ [m^2/s]	Pr	$(g\beta/\nu\alpha) \cdot 10^{-6}$ [1/K·m^3]
100	3.945	0.962	76.4	9.25	1.94	2.44	0.796	20,718
150	2.585	0.921	114.8	13.8	4.44	5.80	0.766	2539
200	1.930	0.915	147.5	15.3	7.64	10.4	0.737	617.1
250	1.542	0.915	178.6	22.6	11.58	16.0	0.723	211.7
300	1.284	0.920	207.2	26.8	16.14	22.7	0.711	89.22
350	1.100	0.929	233.5	29.6	21.23	29.0	0.733	45.51
400	0.9620	0.942	258.2	33.0	26.84	36.4	0.737	25.09
450	0.8554	0.936	281.4	36.3	32.90	44.4	0.741	14.92
500	0.7698	0.972	303.3	41.2	39.40	55.1	0.716	9.035
550	0.6998	0.988	324.0	44.1	46.30	63.8	0.726	6.036
600	0.6414	1.003	343.7	47.3	53.59	73.5	0.729	4.150
700	0.5498	1.031	380.8	52.8	69.26	93.1	0.744	2.173
800	0.4810	1.054	415.2	58.9	86.32	116	0.743	1.224
900	0.4275	1.074	447.2	64.9	104.6	141	0.740	0.7388
1000	0.3848	1.090	477.0	71.0	124.0	169	0.733	0.4680
1100	0.3498	1.103	505.5	75.8	144.5	196	0.736	0.3148
1200	0.3206	1.115	532.5	81.9	166.1	229	0.725	0.2149
1300	0.2960	1.125	588.4	87.1	188.6	262	0.721	0.1527

Table B.2 Gases

Water Vapor (Superheated)

T [K]	ρ [kg/m³]	c_p [kJ/kg·K]	$\mu \cdot 10^2$ [N·s/m²]	$k \cdot 10^3$ [W/m·K]	$\nu \cdot 10^6$ [m²/s]	$\alpha \cdot 10^7$ [m²/s]	Pr	$(g\beta/\nu\alpha) \cdot 10^{-6}$ [1/K·m³]
380	0.5863	2.060	127.1	24.6	21.68	20.4	1.06	58.35
400	0.5542	2.014	134.4	26.1	24.25	23.4	1.04	43.21
450	0.4902	1.980	152.4	29.9	31.11	30.8	1.01	22.74
500	0.4405	1.985	170.4	33.9	38.68	38.8	0.998	13.07
550	0.4005	1.997	188.4	37.9	47.04	47.4	0.993	7.997
600	0.3852	2.026	206.7	42.2	56.60	57.0	0.993	5.066
650	0.3380	2.056	224.7	46.4	66.48	66.8	0.996	3.397
700	0.3140	2.085	242.6	50.5	77.26	77.1	1.00	2.352
750	0.2931	2.119	260.4	54.9	88.84	88.4	1.00	1.665
800	0.2739	2.152	278.6	59.2	101.7	100	1.01	1.205
850	0.2579	2.186	296.9	63.7	115.1	113	1.02	0.8871

Table B.3 Liquids

Engine Oil (New)

T [K]	ρ [kg/m³]	c_p [kJ/kg·K]	$\mu \cdot 10^2$ [N·s/m²]	$k \cdot 10^3$ [W/m·K]	$\nu \cdot 10^6$ [m²/s]	$\alpha \cdot 10^7$ [m²/s]	Pr	$(g\beta/\nu\alpha) \cdot 10^{-9}$ [1/K·m³]
273	899.1	1.796	385	147	4280	0.910	47,000	0.01763
280	895.3	1.827	217	144	2430	0.880	27,500	0.03210
290	890.0	1.868	99.9	145	1120	0.872	12,900	0.07029
300	884.1	1.909	48.6	145	550	0.859	6400	0.1453
310	877.9	1.951	25.3	145	288	0.847	3400	0.2814
320	871.8	1.993	14.1	143	161	0.823	1965	0.5181
330	865.8	2.035	8.36	141	96.6	0.800	1205	0.8883
340	859.9	2.076	5.31	139	61.7	0.779	793	1.428
350	853.9	2.118	3.56	137	41.7	0.763	546	2.158
360	847.8	2.161	2.52	138	29.7	0.753	295	3.070
370	841.8	2.206	1.86	137	22.0	0.738	300	4.228
380	836.0	2.250	1.41	136	16.9	0.723	233	5.618
390	830.6	2.294	1.10	135	13.3	0.709	187	7.280
400	825.1	2.337	0.874	134	10.6	0.695	152	9.318
410	818.9	2.361	0.698	133	8.52	0.682	125	11.814
420	812.1	2.426	0.564	133	6.94	0.675	103	14.654
430	806.5	2.471	0.470	132	5.83	0.662	88	17.787

Table B.3 Liquids

Ethylene Glycol ($C_2H_4(OH)_2$)

T [K]	ρ [kg/m^3]	c_p [kJ/kg·K]	$\mu \cdot 10^2$ [N·s/m^2]	$k \cdot 10^3$ [W/m·K]	$\nu \cdot 10^6$ [m^2/s]	$\alpha \cdot 10^7$ [m^2/s]	Pr	$(g\beta/\nu\alpha) \cdot 10^{-9}$ [1/K·m^3]
273	1130.8	2.294	6.51	242	57.6	0.933	617	1.186
280	1125.8	2.323	4.20	244	37.3	0.933	400	1.832
290	1118.8	2.368	2.47	248	22.1	0.936	236	3.082
300	1111.4	2.415	1.57	252	14.1	0.939	151	4.815
310	1103.7	2.460	1.07	255	9.65	0.939	103	7.035
320	1096.2	2.505	0.757	258	6.91	0.940	73.5	9.814
330	1089.5	2.549	0.561	260	5.15	0.936	55.0	13.224
340	1083.8	2.592	0.431	261	3.98	0.929	42.8	17.241
350	1079.0	2.637	0.342	261	3.17	0.917	34.6	21.93
360	1074.0	2.682	0.278	261	2.59	0.906	28.6	27.17
370	1066.7	2.728	0.228	262	2.14	0.900	23.7	33.10
373	1058.5	2.742	0.215	263	2.03	0.906	22.4	34.66

Table B.3 Liquids

Glycerin ($C_3H_5(OH)_3$)

T [K]	ρ [kg/m^3]	c_p [kJ/kg·K]	$\mu \cdot 10^2$ [N·s/m^2]	$k \cdot 10^3$ [W/m·K]	$\nu \cdot 10^6$ [m^2/s]	$\alpha \cdot 10^7$ [m^2/s]	Pr	$(g\beta/\nu\alpha) \cdot 10^{-9}$ [1/K·m^3]
273	1276.0	2.261	1060	282	8310	0.977	85,000	0.00568
280	1271.9	2.298	534	284	4200	0.972	43,200	0.0113
290	1265.8	2.367	185	286	1460	0.955	15,300	0.0338
300	1259.9	2.427	79.9	286	634	0.925	6780	0.0794
310	1253.9	2.490	35.2	286	281	0.916	3060	0.1867
320	1247.2	2.564	21.0	287	168	0.897	1870	0.3254

Table B.3 Liquids

Freon–12 (R–12) (CCl_2F_2)

T [K]	ρ [kg/m³]	c_p [kJ/kg·K]	$\mu \cdot 10^2$ [N·s/m²]	$k \cdot 10^3$ [W/m·K]	$\nu \cdot 10^6$ [m²/s]	$\alpha \cdot 10^7$ [m²/s]	Pr	$(g\beta/\nu\alpha) \cdot 10^{-9}$ [1/K·m³]
230	1525.4	0.8816	0.0457	68	0.299	0.505	5.9	1201.6
240	1498.0	0.8923	0.0385	69	0.237	0.516	5.0	1405.1
250	1469.5	0.9037	0.0354	70	0.241	0.527	4.6	1544.3
260	1439.0	0.9163	0.0322	73	0.224	0.554	4.0	1659.6
270	1407.2	0.9301	0.0304	73	0.216	0.555	3.9	1830.8
280	1374.4	0.9450	0.0253	73	0.206	0.562	3.7	1990.7
290	1340.5	0.9609	0.0265	73	0.198	0.567	3.5	2227.6
300	1305.8	0.9781	0.0254	72	0.195	0.564	3.5	2452.2
310	1268.9	0.9963	0.0244	69	0.192	0.546	3.4	2853.3
320	1225.6	1.0155	0.0233	68	0.190	0.545	3.5	3314.8

Table B.3 Liquids

Water (Saturated Liquid)

T [K]	$P \cdot 10^{-5}$ [Pa]	ρ [kg/m^3]	c_p [kJ/kg·K]	$\mu \cdot 10^6$ [N·s/m^2]	$k \cdot 10^3$ [W/m·K]	$\nu \cdot 10^6$ [m^2/s]	$\alpha \cdot 10^7$ [m^2/s]	Pr	$(g\beta/\nu\alpha) \cdot 10^{-9}$ [1/K·m^3]
273.15	0.00611	1000	4.217	1750	659	1.750	1.56	12.99	2.440
275	0.00697	1000	4.211	1652	574	1.652	1.36	12.22	1.426
280	0.00990	1000	4.198	1422	582	1.422	1.39	10.26	2.290
285	0.91387	1000	4.189	1225	590	1.225	1.41	8.81	6.485
290	0.91917	999	4.184	1080	598	1.081	1.43	7.56	11.033
295	0.02617	998	4.181	959	606	0.961	1.45	6.62	15.987
300	0.03531	997	4.179	855	613	0.858	1.47	5.83	21.46
305	0.04712	995	4.178	769	620	0.773	1.49	5.20	27.28
310	0.06221	993	4.178	695	628	0.700	1.51	4.62	33.50
315	0.98132	991	4.179	631	634	0.637	1.53	4.16	40.29
320	0.1053	989	4.180	577	640	0.583	1.55	3.77	47.43
325	0.1351	987	4.182	528	645	0.535	1.56	3.42	55.30
330	0.1719	984	4.184	489	650	0.497	1.58	3.15	63.03
335	0.2167	982	4.186	453	656	0.461	1.60	2.88	51.38
340	0.2713	979	4.188	420	660	0.429	1.61	2.66	80.45
345	0.3372	976	4.191	389	668	0.398	1.63	2.45	89.81
350	0.4163	974	4.195	365	668	0.375	1.64	2.29	99.86
355	0.5100	971	4.199	343	671	0.353	1.65	2.14	110.0
360	0.6109	967	4.203	324	674	0.335	1.66	2.02	123.2
365	0.7514	963	4.209	306	677	0.317	1.67	1.91	130.8
370	0.9040	961	4.214	289	679	0.301	1.68	1.80	141.6
373.15	1.0133	958	4.217	279	680	0.291	1.68	1.76	150.0
375	1.0815	957	4.220	274	681	0.286	1.69	1.70	154.6
380	1.2869	953	4.226	260	683	0.273	1.70	1.61	167.1
385	1.5233	950	4.232	248	685	0.261	1.70	1.53	179.4
390	1.794	945	4.239	237	686	0.251	1.71	1.47	192.1
400	2.455	937	4.256	217	688	0.232	1.72	1.34	220.0
410	3.302	929	4.278	200	688	0.215	1.73	1.24	250.2
420	4.370	919	4.302	185	688	0.201	1.74	1.16	282.8
430	5.699	910	4.331	173	685	0.190	1.74	1.09	
440	7.333	901	4.36	162	682	0.180	1.74	1.04	
450	9.319	890	4.40	152	678	0.171	1.73	0.99	
460	11.71	880	4.44	143	673	0.163	1.72	0.95	
470	14.55	868	4.48	136	667	0.157	1.72	0.92	
480	17.90	857	4.53	129	660	0.151	1.70	0.89	
490	21.83	845	4.59	125	651	0.147	1.68	0.87	
500	26.40	831	4.66	118	642	0.142	1.66	0.86	
510	31.66	818	4.74	113	631	0.138	1.63	0.85	
520	37.70	804	4.84	108	621	0.134	1.60	0.84	
530	44.58	789	4.95	104	608	0.132	1.56	0.85	
540	52.38	773	5.08	101	594	0.131	1.51	0.86	
550	61.19	736	5.24	97	580	0.128	1.46	0.87	
560	71.08	738	5.43	94	563	0.127	1.41	0.90	
570	82.16	718	3.68	91	548	0.127	1.34	0.94	
580	94.51	698	6.00	88	528	0.126	1.26	0.99	
590	108.3	675	6.41	84	513	0.125	1.19	1.05	
600	123.5	649	7.00	81	497	0.125	1.09	1.14	
610	137.3	620	7.85	77	467	0.124	0.96	1.30	
620	159.1	587	9.35	72	444	0.123	0.81	1.52	
625	169.1	562	10.6	70	430	0.125	0.72	1.65	
630	179.7	539	12.6	67	412	0.124	0.61	2.0	
635	190.9	517	16.4	74	392	0.124	0.46	2.7	
640	202.7	482	26	59	367	0.122	0.29	4.2	
645	215.2	425	90	54	331	0.127	0.14	12	
647.3	221.2	315	∞	45	238	0.143	0.00	∞	

Table B.4 Liquid Metals

Bismuth (Bi)

T [K]	ρ [kg/m³]	c_p [kJ/kg · K]	$\mu \cdot 10^2$ [N · s/m²]	$k \cdot 10^3$ [W/m · K]	$\nu \cdot 10^6$ [m²/s]	$\alpha \cdot 10^7$ [m²/s]	Pr	$(g\beta/\nu\alpha) \cdot 10^{-9}$ [1/K · m³]
589	10,011	0.1445	0.1622	16440	0.157	114	0.014	0.6411
700	9867	0.1495	0.1339	15380	0.135	106	0.013	0.8361
811	9739	0.1545	0.1101	15580	0.108	103	0.011	1.1108
922	9611	0.1595	0.0923	15580	0.0903	101	0.009	
1033	9467	0.1645	0.0789	15580	0.0813	101	0.008	

Table B.4 Liquid Metals

Mercury (Hg)

T [K]	ρ [kg/m³]	c_p [kJ/kg · K]	$\mu \cdot 10^2$ [N · s/m²]	$k \cdot 10^3$ [W/m · K]	$\nu \cdot 10^6$ [m²/s]	$\alpha \cdot 10^7$ [m²/s]	Pr	$(g\beta/\nu\alpha) \cdot 10^{-9}$ [1/K · m³]
273	13,595	0.1404	0.1688	8180	0.1240	42.85	0.0290	3.341
300	13,529	0.1393	0.1523	8540	0.1125	45.30	0.0248	3.483
350	13,407	0.1377	0.1309	9180	0.0976	49.75	0.0196	3.656
400	13,287	0.1365	0.1171	9800	0.0882	54.05	0.0163	3.723
450	13,167	0.1357	0.1075	10,400	0.0816	58.10	0.0140	3.744
500	13,048	0.1353	0.1007	10,950	0.0771	61.90	0.0125	3.740
550	12,929	0.1352	0.0953	11,450	0.0737	65.55	0.0112	3.735
600	12,509	0.1355	0.0911	11,950	0.0711	68.80	0.0103	3.749

Table B.4 Liquid Metals

Sodium (Na)

T [K]	ρ [kg/m³]	c_p [kJ/kg · K]	$\mu \cdot 10^2$ [N · s/m²]	$k \cdot 10^3$ [W/m · K]	$\nu \cdot 10^6$ [m²/s]	$\alpha \cdot 10^7$ [m²/s]	Pr	$(g\beta/\nu\alpha) \cdot 10^{-9}$ [1/K · m³]
367	929	1.382	0.0699	86,200	0.731	671	0.0110	0.0540
478	902	1.340	0.0432	80,300	0.460	671	0.0072	0.1144
644	860	1.298	0.0283	72,400	0.316	645	0.0051	
811	820	1.256	0.0208	65,400	0.244	619	0.0040	
978	778	1.256	0.0179	59,700	0.226	619	0.0038	

APPENDIX C

SI UNITS

Table C.1 Fundamental Units

Quantity	Name of Unit	Symbol
Length	meter	m
Mass	kilogram	kg
Time	second	s
Electrical current	ampere	A
Thermodynamic temperature	kelvin	K
Luminous intensity	candela	cd
Amount of a substance	mole	mol

Table C.2 Derived Units

Quantity	Name of Unit	Symbol
Acceleration	meters per second squared	m/s^2
Area	square meters	m^2
Capacitance	farad	F
Density	kilogram per cubic meter	kg/m^3
Dynamic viscosity	newton-second per square meter	$N \cdot s/m^2$
Electrical resistance	ohm	Ω
Force	newton	N
Frequency	hertz	Hz
Kinematic viscosity	square meter per second	m^2/s
Plane angle	radian	rad
Potential difference	volt	V
Power	watt	W
Pressure	pascal	Pa
Radiant intensity	watts per steradian	W/sr
Solid angle	steradian	sr
Specific heat	joules per kilogram $\cdot$ Kelvin	$J/kg \cdot K$
Thermal conductivity	watts per meter $\cdot$ Kelvin	$W/m \cdot K$
Velocity	meters per second	m/s
Volume	cubic meter	m^3
Work, energy, heat	joule	J

Table C.3 Defined Units

Quantity	Unit	Defining Equation
Capacitance	farad, F	$1 F = 1 A \cdot s/V$
Electrical resistance	ohm, Ω	$1 \Omega = 1 V/A$
Force	newton, N	$1 N = 1 kg \cdot m/s^2$
Potential difference	volt, V	$1 V = 1 W/A$
Power	watt, W	$1 W = 1 J/s$
Pressure	pascal, Pa	$1 Pa = 1 N/m^2$
Temperature	kelvin, K	$K = °C + 273.15$
Work, heat, energy	joule, J	$1 J = 1 N \cdot m$

Table C.4 Prefixes

Multiplier	Symbol	Prefix
10^{12}	T	tera
10^{9}	G	giga
10^{6}	M	mega
10^{3}	k	kilo
10^{2}	h	hecto
10^{1}	da	deka
10^{-1}	d	deci
10^{-2}	c	centi
10^{-3}	m	milli
10^{-6}	μ	micro
10^{-9}	n	nano
10^{-12}	p	pico
10^{-15}	f	femto
10^{-18}	a	atto

Appendix D

Heisler Charts

App. D Heisler Charts

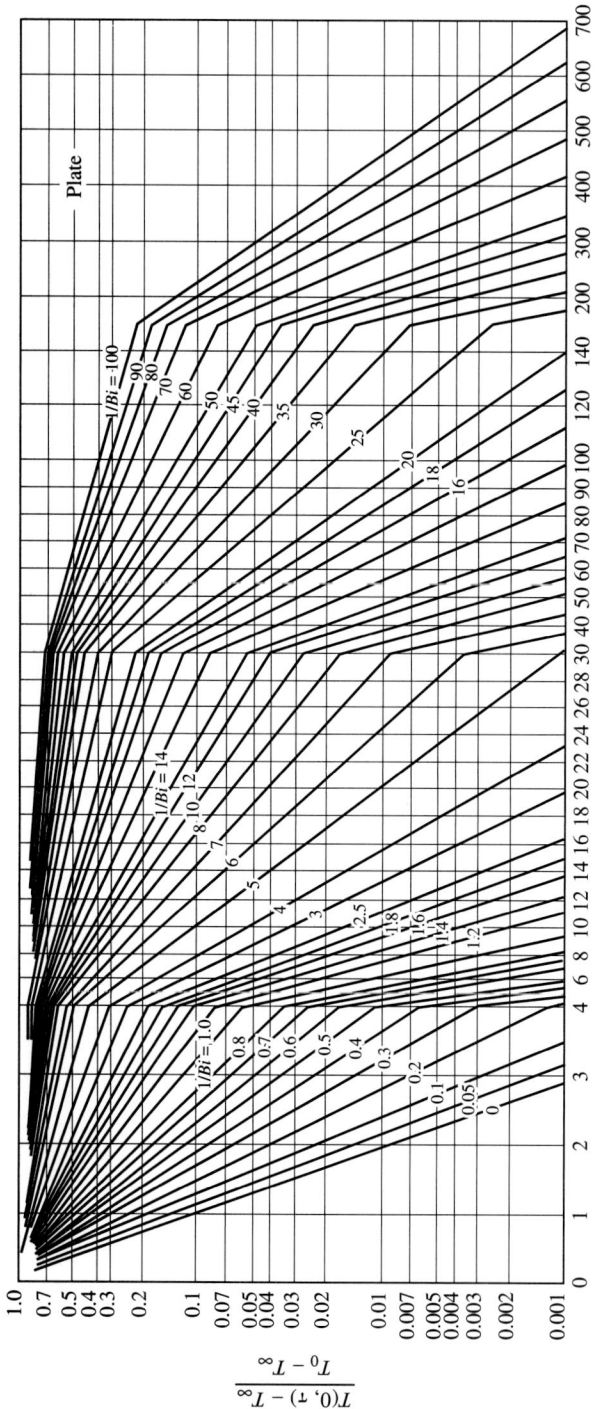

Figure D.1 Instantaneous middle plane temperature °C

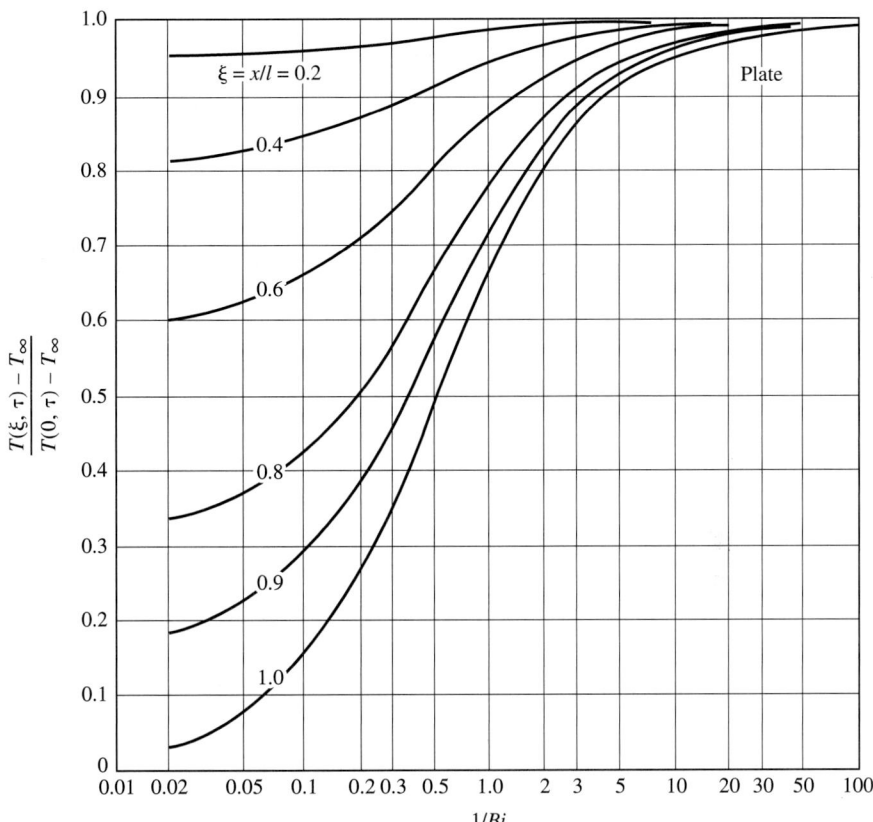

Figure D.2 Instantaneous temperature at various locations of $\xi = x/\ell$

604 App. D Heisler Charts

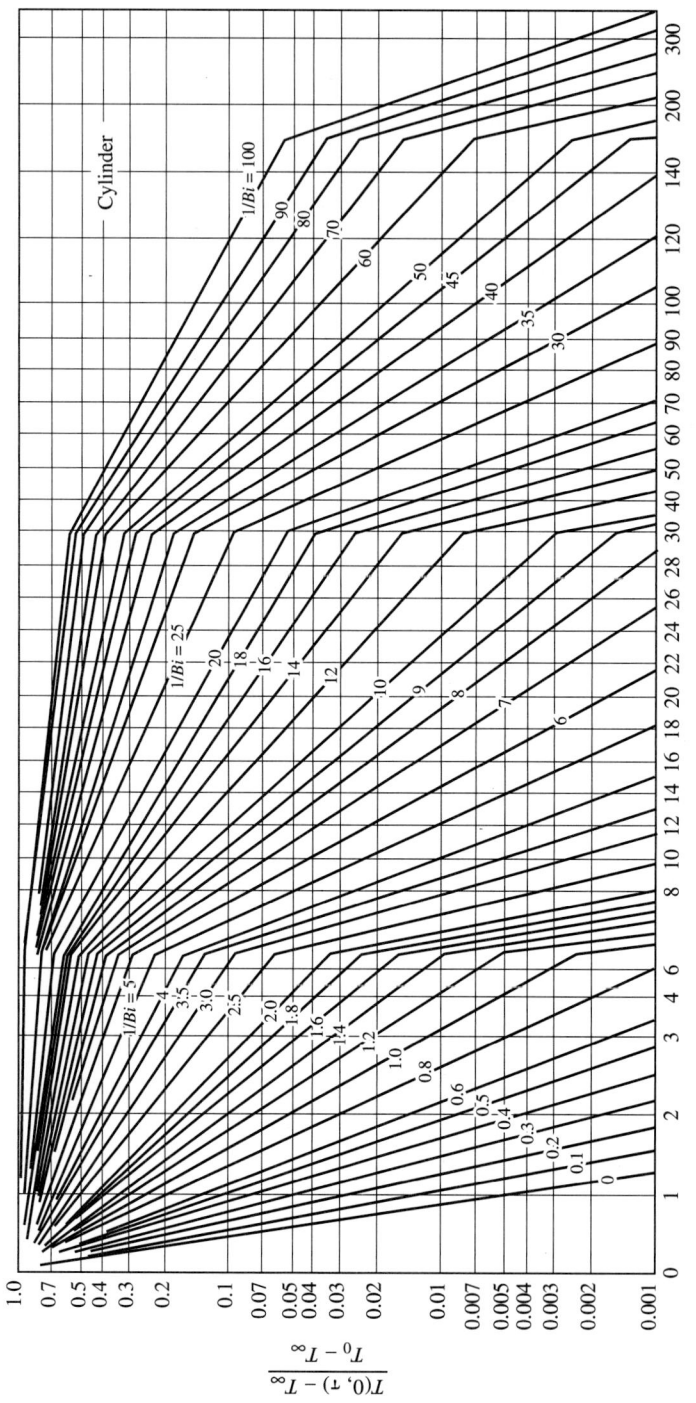

Figure D.3 Instantaneous centerline temperature of cylinder

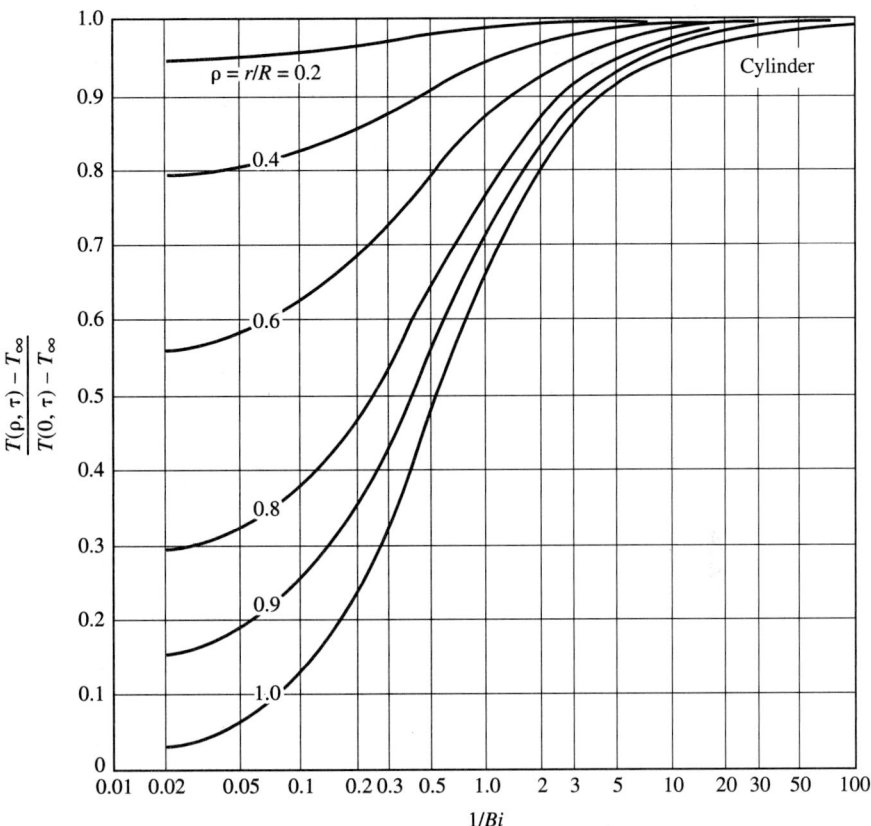

Figure D.4 Instantaneous temperature at various locations of $\rho = r/R$

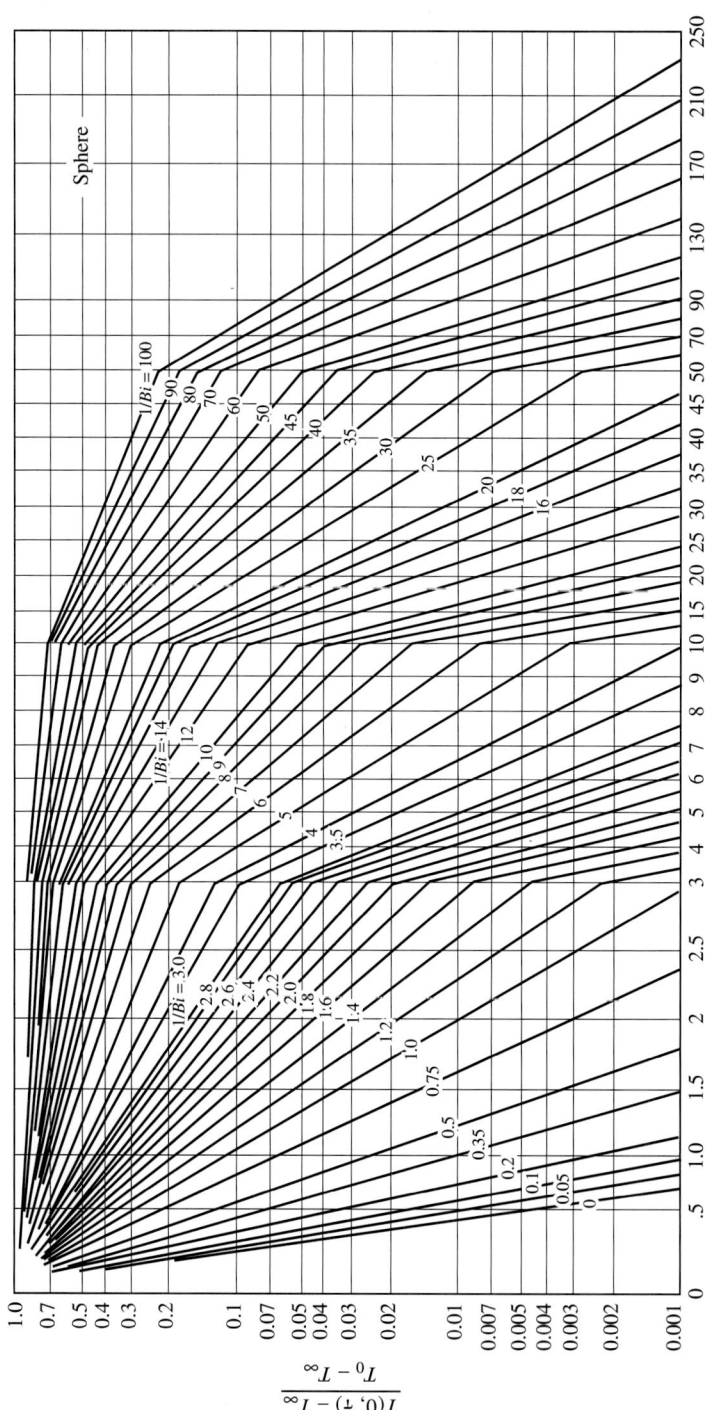

Figure D.5 Instantaneous center temperature of sphere

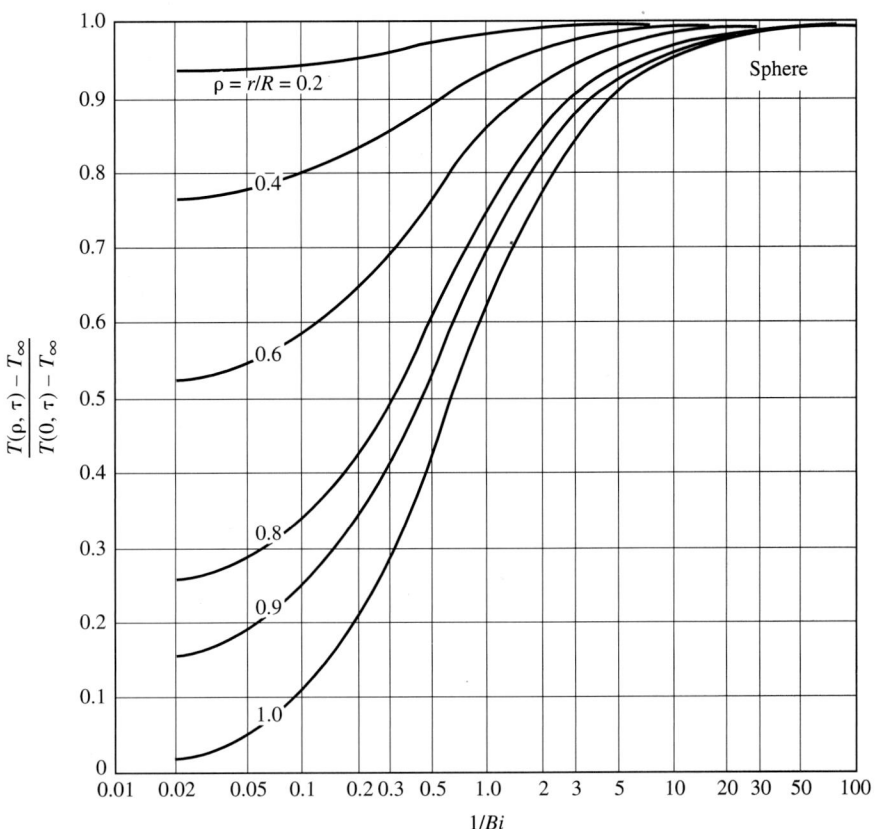

Figure D.6 Instantaneous temperature at various locations of $\rho = r/R$

INDEX

A

Absorption coefficient, 515
   monochromatic, 508
Absorptivity, 422
Averaged temperature,
   logarithmic-mean, 44
   geometric-mean, 44
   arithmetic-mean, 45

B

Black surface, 24

C

Composite structures, 45
   cylinder, 49
   slab, 45
   sphere, 57
Conduction,
   computational, 184
      discrete formulation, 184
         exact, 185

      finite difference, 186
      finite volume, 186
   Euler's method, 224
   multi-dimensional, 194
      enthalpy flow, 207
      non-uniform grid, 202
   truncation error, 209
   unsteady, 212
      Crank-Nicolson method, 222
      explicit finite difference, 212
      implicit scheme, 220
      stability, 213
conductivity, 17
   variable, 42
Fourier's law, 15
multi-dimensional, 125
one-dimensional, 15
original problem, 14
steady, 40
unsteady, 125
   analog solution, 168
   approximate solution, 152
   charted solution, 156

Conduction (*continued*)
  unsteady (*continued*)
    flat plate (key problem), 156
    semi-infinite plate, 165
    solid cylinder (key problem), 161
    solid sphere (key problem), 163
  distributed, 144
  integral formulation, 152
  lumped, 124
  periodic, 141
  steady periodic, 149
Control surface, 12
Control volume, 4
Convection,
  correlation, 288
    drag coefficient, 288
    forced, 295
      external flow, 301
      internal flow, 295
    friction factor, 289
  foundations, 240
    boundary layer, 244
      laminar forced, 244
      laminar natural, 258
  two key problems, 90
    first problem, 90
    second problem, 95
Critical radius,
  cylinder, 52
  sphere, 58

D

Dimensional analysis, 266
  forced convection, 275
  forced flow, 270
  free fall, 273
  natural convection, 278
  Π-theorem, 266
  physical similitude, 266
Dimensionless numbers,
  Biot, 62, 126, 156
  Froude, 286
  Grashof, 274
  Jacob, 537
  Nusselt, 21, 249, 255
  Peclet, 93
  Π_N, 286
  Rayleigh, 261, 280
  Reynolds, 271
  Stanton, 255

E

Emissive power, 402
  black body, 403
  monochromatic, 405
Energy generation, 58
  cylinder, 70
  flat plate, 58
  sphere, 70
Extended surface, 74
  performance, 89
  thermal length, 80

F

First law, 44
First law for a,
  control volume, 6
  rate of, 6
  system, 7
Five steps of,
  computation of heat transfer coefficient,
    forced,
      external flow, 305
      internal flow, 297
    natural, 314, 320
  formulation, 32
  formulation of enclosure radiation, 445

G

Gray gas, 515, 517
Gray surface, 422

H

Heat exchangers, 346
  condenser, 365
  correction factor, 359
  evaporator (boiler), 365
  fouling factor, 385
  LMTD method, 349
  NTU method, 370
  performance, 370
  thermal design, 349
  variable coefficient, 385

Heat transfer,
  combined modes, 27
  formulation of, 3
  foundations of, 1
  inductive formulation, 32
  origin of, 13

I

Integral formulation,
  steady, 80, 249, 253, 256, 262–264
  unsteady, 129, 152
Intensity, 401

K

Key problems,
  conduction,
    energy generation, 58
      flat plate, 58
      solid cylinder, 70
      solid sphere, 70
  convection,
    first key, 90
    second key, 95
Kirchhoff's law, 417

M

Mass conservation for a,
  control volume, 7
  rate of, 7

N

Newton's law, 19
Nuclear reactor, 102

O

Optical thickness, 517

P

Phase change, 535
  boiling, 548
    regimes of, 548
  dimensionless number for, 546
  laminar, 537

R

Radiation,
  approximation, 400
  electromagnetic waves, 397
  enclosure, 430
    effect of conduction/convection, 475
    electrical analogy, 443
    net radiation, 472
    view factor, 434
  foundations, 396
  gas, 506
    distributed, 517
      boundary effect, 521
      thick gas, 519
      thin gas, 518
    energy balance, 517
    properties, 509
  monochromatic, 405
  optical rays, 400
  origin, 396
  properties of, 413
  quantum mechanics, 406
Reflectivity, 422

S

Solar collector, 99
Stefan-Boltzmann's law, 24, 403

T

Thick gas, 519–520
Thin gas, 518–519
Time constant, 128
Transmissivity, 422
  monochromatic, 413

V

View factor, 434

W

Wien's law, 406

Physical Constants

Universal gas constant:
$$R = 8.205 \times 10^{-2} \, m^3 \cdot atm/kmol \cdot K$$
$$= 8.314 \times 10^{-2} \, m^3 \cdot bar/kmol \cdot K$$
$$= 8.315 \, kJ/kmol \cdot K$$
$$= 1545 \, ft \cdot lb_f/lbmole \cdot °R$$
$$= 1.986 \, Btu/lbmole \cdot °R$$

Avogadro's number:
$$N = 6.024 \times 10^{23} \, molecules/mol$$

Planck's constant:
$$h = 6.623 \times 10^{-34} \, J \cdot s/molecule$$

Boltzmann's constant:
$$k = 1.380 \times 10^{-23} \, J/K \cdot molecule$$

Speed of light in vacuum:
$$c_o = 2.998 \times 10^8 \, m/s$$

Stefan-Boltzmann constant:
$$\sigma = 5.670 \times 10^{-8} \, W/m^2 \cdot K^4$$
$$= 0.1714 \times 10^{-8} \, Btu/h \cdot ft^2 \cdot °R^4$$

Blackbody radiation constants:
$$C_1 = 3.7420 \times 10^8 \, W \cdot \mu m^4/m^2$$
$$= 1.187 \times 10^8 \, Btu \cdot \mu m^4/h \cdot ft^2$$
$$C_2 = 1.4388 \times 10^4 \, \mu m \cdot K$$
$$= 2.5897 \times 10^4 \, \mu m \cdot °R$$
$$C_3 = 2897.7 \, \mu m \cdot K$$
$$= 5215.6 \, \mu m \cdot °R$$

Gravitational acceleration (sea level):
$$g = 9.807 \, m/s^2$$

Normal atmospheric pressure:
$$p = 101,325 \, N/m^2$$

Conversion Factors

Acceleration	$1\ \text{m/s}^2$	$= 4.2520 \times 10^7\ \text{ft/h}^2$
Area	$1\ \text{m}^2$	$= 1550.0\ \text{in}^2$
		$= 10.764\ \text{ft}^2$
Energy	$1\ \text{J}$	$= 9.4787 \times 10^{-4}\ \text{Btu}$
Force	$1\ \text{N}$	$= 0.22481\ \text{lb}_f$
Heat transfer rate	$1\ \text{W}$	$= 3.4123\ \text{Btu/h}$
Heat flux	$1\ \text{W/m}^2$	$= 0.3171\ \text{Btu/h} \cdot \text{ft}^2$
Heat generation rate	$1\ \text{W/m}^3$	$= 0.09665\ \text{Btu/h} \cdot \text{ft}^3$
Heat transfer coefficient	$1\ \text{W/m}^2 \cdot \text{K}$	$= 0.17612\ \text{Btu/h} \cdot \text{ft}^2 \cdot {}^\circ\text{F}$
Kinematic viscosity and diffusivities	$1\ \text{m}^2/\text{s}$	$= 3.875 \times 10^4\ \text{ft}^2/\text{h}$
Latent heat	$1\ \text{J/kg}$	$= 4.2995 \times 10^{-4}\ \text{Btu/lb}_m$
Length	$1\ \text{m}$	$= 39.370\ \text{in.}$
		$= 3.2808\ \text{ft}$
	$1\ \text{km}$	$= 0.62137\ \text{mile}$
Mass	$1\ \text{kg}$	$= 2.2046\ \text{lb}_m$
Mass density	$1\ \text{kg/m}^3$	$= 0.062428\ \text{lb}_m\text{ft}^3$
Mass flow rate	$1\ \text{kg/s}$	$= 7936.6\ \text{lb}_m/\text{h}$
Mass transfer coefficient	$1\ \text{m/s}$	$= 1.1811 \times 10^4\ \text{ft/h}$
Pressure and stress	$1\ \text{N/m}^2$	$= 0.020886\ \text{lb}_f/\text{ft}^2$
		$= 1.4504 \times 10^{-4}\ \text{lb}_f/\text{in.}^2$
		$= 4.015 \times 10^{-3}\ \text{in. water}$
		$= 2.953 \times 10^{-4}\ \text{in. Hg}$
	$1.0133 \times 10^5\ \text{N/m}^2$	$= 1\ \text{standard atmosphere}$
	$1 \times 10^5\ \text{N/m}^2$	$= 1\ \text{bar}$
Specific heat	$1\ \text{J/kg} \cdot \text{K}$	$= 2.3886 \times 10^{-4}\ \text{Btu/lb}_m \cdot {}^\circ\text{F}$
Temperature	K	$= (5/9){}^\circ\text{R}$
		$= (5/9)({}^\circ\text{F} + 459.67)$
		$= {}^\circ\text{C} + 273.15$
Temperature difference	$1\ \text{K}$	$= 1\ {}^\circ\text{C}$
		$= (9/5){}^\circ\text{R} = (9/5){}^\circ\text{F}$
Thermal conductivity	$1\ \text{W/m} \cdot \text{K}$	$= 0.57782\ \text{Btu/h} \cdot \text{ft} \cdot {}^\circ\text{F}$
Thermal resistance	$1\ \text{K/W}$	$= 0.52750\ {}^\circ\text{F/h} \cdot \text{Btu}$
Viscosity (dynamic)	$1\ \text{N} \cdot \text{s/m}^2$	$= 2419.1\ \text{lb ft} \cdot \text{h}$
		$= 5.8016 \times 10^{-6}\ \text{lb}_f \cdot \text{h/ft}^2$
Volume	$1\ \text{m}^3$	$= 6.1023 \times 10^4\ \text{in.}^3$
		$= 35.314\ \text{ft}^3$
		$= 264.17\ \text{gal}$
Volume flow rate	$1\ \text{m}^3/\text{s}$	$= 1.2713 \times 10^5\ \text{ft}^3/\text{h}$
		$= 2.1189 \times 10^3\ \text{ft}^3/\text{min}$
		$= 1.5850 \times 10^4\ \text{gal/min}$

Typical Values of Heat Transfer Coefficient

Condition		$h, \dfrac{W}{m^2 \cdot {}^\circ K}$
Natural Convection	Gases	5–12
	Oils	10–120
	Water	100–1,200
	Liquid metals	1,000–7,000
Forced Convection	Gases	10–300
	Oils	50–1,200
	Water	300–12,000
	Liquid metals	5,000–120,000
Phase Change	Boiling	3,000–50,000
	Condensation	5,000–120,000

Typical Values of Heat Transfer Coefficient

Condition		$h, \dfrac{W}{m^2 \cdot °K}$
Natural Convection	Gases	5–12
	Oils	10–120
	Water	100–1,200
	Liquid metals	1,000–7,000
Forced Convection	Gases	10–300
	Oils	50–1,200
	Water	300–12,000
	Liquid metals	5,000–120,000
Phase Change	Boiling	3,000–50,000
	Condensation	5,000–120,000